Topics in Advanced Quantum Mechanics

BARRY R. HOLSTEIN
University of Massachusetts at Amherst

DOVER PUBLICATIONS, INC.
Mineola, New York

Copyright

Copyright © 1992 by Barry R. Holstein
All rights reserved.

Bibliographical Note

This Dover edition, first published in 2014, is an unabridged and corrected republication of the work originally published by Addison-Wesley, Reading, Massachusetts, in 1992.

Library of Congress Cataloging-in-Publication Data

Holstein, Barry R., 1943– author.
   Topics in advanced quantum mechanics / by Barry Holstein.
     pages cm
   "This Dover edition, first published in 2014, is an unabridged and corrected republication of the work originally published by Addison-Wesley, Reading, Massachusetts, in 1992."
   Includes bibliographical references and index.
   ISBN-13: 978-0-486-49985-7 (alk. paper)
   ISBN-10: 0-486-49985-5 (alk. paper)
     1. Quantum theory. I. Title.
QC174.12.H64 2014
530.12—dc23

                                                                       2013045282

Manufactured in the United States by Courier Corporation
49985502   2015
www.doverpublications.com

To the memory of Julius Ashkin

Acknowledgments

It is a pleasure to acknowledge the contributions of those who have helped to make this book a reality—the late Julius Ashkin who first instructed me in these matters and on whose unpublished lecture notes a good deal of chapters I, II, IV, and VII are based, Mrs. Claire Ashkin for granting permission to use this material, a generation of students who participated in the development of this course, Roger "T_EX" Gilson who transformed my handwritten notes into a typeset manuscript, and finally Margaret "M_AC" Donald, who skillfully generated the many figures.

Prologue

About fifteen years ago I was asked by the chairman to teach the graduate course in advanced quantum mechanics. This was the third semester of the quantum sequence and was to include relativistic quantum mechanics as well as an introduction to quantum field theory. I had intended to begin my discussion with the Dirac equation and to develop covariant perturbation theory via Green's function methods. However, I soon found that the students needed a better grounding in the related nonrelativistic techniques before launching into a full relativistic treatment. Thus I developed material on nonrelativistic Feynman diagrams and their application to electromagnetic processes. I searched at that time for a suitable text, but was unable to find any that really fit my needs—Bjorken and Drell's *Relativistic Quantum Mechanics* was excellent for the relativistic aspects but contained no corresponding nonrelativistic material; Sakurai's *Advanced Quantum Mechanics* was much better in this regard but used the "wrong" metric; *etc.* Since I did not wish to see the students buried in their notebooks trying to keep up with my backboard work, I developed a set of lecture notes and problems covering this material, some being substantially based upon a similar course which I had taken from Julius Ashkin at Carnegie-Mellon University a decade earlier. My notes were duplicated and handed out to the students.

Since that time, I have taught the course on four or five additional occasions. Each time I included some new material with additional problems and related notes. It is these notes and problems which are collected below into what I hope is a coherent volume. Problems are inserted not at the end of each chapter but rather in the flow of discussion. Since this was a graduate course, I have taken the liberty of assuming a thorough grounding in basic quantum mechanics at the level of, say, Merzbacher's *Quantum Mechanics* and have used $\hbar = 1$ and $c = 1$ throughout, except where they are required for clarity. The charge on the electron is denoted by e and is taken to be negative: $e = -|e|$. Contraction of four vectors is accomplished via the metric tensor

$$\eta_{\mu\nu} = \begin{pmatrix} 1 & 0 & 0 & 0 \\ 0 & -1 & 0 & 0 \\ 0 & 0 & -1 & 0 \\ 0 & 0 & 0 & -1 \end{pmatrix}$$

i.e. $A \cdot B = A^\mu B_\mu = A^\mu \eta_{\mu\nu} B^\nu = A^0 B^0 - \vec{A} \cdot \vec{B}$ and the conventions of Bjorken and Drell with respect to Dirac matrices are followed, except for γ_5 for which the negative of their definition is chosen. Unit vectors are denoted by a hat symbol, as are abstract operators, but it should be clear from the context which is which.

There is clearly too much discussed here to be included in a single semester course, and there is a good deal of material, usually included at the end of each chapter and designated by an asterisk, which is supplementary to the major content—intended rather to whet the appetite of adventuresome readers. It is my hope, nevertheless, that other instructors will find useful material herein to supplement their own presentations of advancd quantum mechanics and that in turn a new generation of students will be exposed to the excitement which I first felt twenty-five years ago.

Amherst, Massachusetts
December, 1991

Contents

Acknowledgments   v
Prologue   vii

I. Propagator Methods   1
1 Basic Quantum Mechanics   1
2 The Propagator   6
3 Harmonic Oscillator Propagator   14
4 Time-dependent Perturbation Theory   27
5 Propagator as the Green's Function   37
6 Functional Techniques*   39

II. Scattering Theory   51
1 Basic Formalism   51
2 The Optical Theorem   64
3 Path Integral Approach   70
4 Two-Body Scattering   73
5 Two-Particle Scattering Cross Section   76
7 Scattering Matrix   85
8 Partial Wave Expansion   92

III. Charged Particle Interactions   99
1 Charged Particle Lagrangian   99
2 Review of Maxwell Equations and Gauge Invariance   103
3 The Bohm–Aharonov Effect   111
4 The Maxwell Field Lagrangian   116
5 Quantization of the Radiation Field   123
6 The Vacuum Energy*   126

IV. Charged Particle Interactions: Applications   133
1 Radiative Decay: Formal   133
2 Radiative Decay: Intuitive   135
3 Angular Distribution of Radiative Decay   140
4 Line Shape Problem: Wigner–Weisskopf Approach   149
5 Compton Scattering via Feynman Diagrams   158
6 Resonant Scattering   164
7 Line Shape via Feynman Diagrams   174

8 The Lamb Shift	178
9 Dispersion Relations	186
10 Effective Lagrangians*	195
11 Complex Energy and Effective Lagrangians*	203

V. Alternate Approximate Methods   211
1 WKP Approximation   211
2 Semiclassical Propagator   215
3 The Adiabatic Approximation   232
4 Berry's Phase   247

VI. The Klein-Gordon Equation   259
1 Derivation and Covariance   259
2 Klein's Paradox and Zitterbewegung   269
3 The Coulomb Solution: Mesonic Atoms   274

VII. The Dirac Equation   281
1 Derivation and Covariance   281
2 Bilinear Forms   291
3 Nonrelativistic Reduction   294
4 Coulomb Solution   305
5 Plane Wave Solutions   308
6 Negative Energy Solutions and Antiparticles   321
8 Dirac Propagator   332
9 Covariant Perturbation Theory   338
10 Electromagnetic Interactions   346

VIII. Advanced Topics   369
1 Radiative Corrections   369
2 Spinless Particles: Electromagnetic Interactions   395
3 Path Integrals and Quantum Field Theory*   402
4 Pion Exchange and Strong Interactions*   411

Epilogue	427
Notation	428
References	432
Index	435

CHAPTER I

PROPAGATOR METHODS

I.1 BASIC QUANTUM MECHANICS

The fundamental problem of quantum mechanics is to determine the time development of quantum states. That is, given a state vector $|\psi(0)\rangle$ at time $t=0$, what is the state at a later time $t - |\psi(t)\rangle$? The answer is provided by the Schrödinger equation

$$i\frac{\partial}{\partial t}|\psi(t)\rangle = \hat{H}|\psi(t)\rangle \qquad (1.1)$$

where $\hat{H}$ is the Hamiltonian operator.[†] Usually one sees this equation expressed in terms of the coordinate space projection of the state vector — i.e. the wavefunction $\psi(x,t)$ where[††]

$$\psi(x,t) \equiv \langle x|\psi(t)\rangle \; . \qquad (1.2)$$

The time-evolution of the wavefunction is then given by

$$i\frac{\partial}{\partial t}\langle x|\psi(t)\rangle = \left\langle x\left|\hat{H}\right|\psi(t)\right\rangle \; . \qquad (1.3)$$

In order to evaluate the matrix element on the right we can insert a complete set of co-ordinate states

$$1 = \int_{-\infty}^{\infty} dx' \, |x'\rangle\langle x'| \qquad (1.4)$$

yielding

$$i\frac{\partial}{\partial t}\langle x|\psi(t)\rangle = \int_{-\infty}^{\infty} dx' \left\langle x\left|\hat{H}\right|x'\right\rangle \langle x'|\psi(t)\rangle \; . \qquad (1.5)$$

Finally we need to interpret the operator matrix element $\left\langle x|\hat{H}|x'\right\rangle$. In general, the Hamiltonian $\hat{H}$ can be written in terms of kinetic and potential energy components as

$$\hat{H} = \frac{\hat{p}^2}{2m} + V(\hat{x}) \; . \qquad (1.6)$$

Here $\hat{x}|x\rangle = x|x\rangle$ with $\langle x|x'\rangle = \delta(x-x')$ so

$$\langle x\,|V(\hat{x})|\,x'\rangle = V(x)\,\langle x|x'\rangle = V(x)\delta(x-x') \; . \qquad (1.7)$$

[†] Note that thoughout this book, we set $\hbar=1$.
[††] For simplicity of notation, we shall work here in one dimension. However, generalization to three dimensions is obvious.

In order to represent the kinetic energy piece we can insert a complete set of momentum states such that $\hat{p}|p\rangle = p|p\rangle$ with $\langle p|p'\rangle = 2\pi\delta(p-p')$. Then

$$1 = \int_{-\infty}^{\infty} \frac{dp}{2\pi} |p\rangle\langle p| \tag{1.8}$$

yielding

$$\left\langle x \left| \frac{\hat{p}^2}{2m} \right| x' \right\rangle = \int_{-\infty}^{\infty} \frac{dp}{2\pi} \langle x|p\rangle \frac{p^2}{2m} \langle p|x'\rangle \ . \tag{1.9}$$

Since $\langle x|p\rangle$ is simply a plane wave

$$\langle x|p\rangle = e^{ipx} \tag{1.10}$$

we have

$$\begin{aligned}
\left\langle x \left| \frac{\hat{p}^2}{2m} \right| x' \right\rangle &= \int_{-\infty}^{\infty} \frac{dp}{2\pi} \frac{p^2}{2m} e^{ip(x-x')} \\
&= -\frac{1}{2m} \frac{\partial^2}{\partial x^2} \int_{-\infty}^{\infty} \frac{dp}{2\pi} e^{ip(x-x')} \\
&= -\frac{1}{2m} \frac{\partial^2}{\partial x^2} \delta(x-x') \ .
\end{aligned} \tag{1.11}$$

Substitution back into Eq. 1.3 yields

$$\begin{aligned}
i\frac{\partial}{\partial t}\langle x|\psi(t)\rangle &= i\frac{\partial}{\partial t}\psi(x,t) \\
&= \left(-\frac{1}{2m}\frac{\partial^2}{\partial x^2} + V(x)\right) \int_{-\infty}^{\infty} dx' \, \delta(x-x') \langle x'|\psi(t)\rangle \\
&= H(x)\psi(x,t)
\end{aligned} \tag{1.12}$$

which is the usual version of the Schrödinger equation, where

$$H(x) = -\frac{1}{2m}\frac{\partial^2}{\partial x^2} + V(x) \tag{1.13}$$

provides the representation of the operator $\hat{H}$ in coordinate space. For a free particle this reduces to the simple form

$$H_0(x) = -\frac{1}{2m}\frac{\partial^2}{\partial x^2} \ . \tag{1.14}$$

Time Development Operator

An alternative formulation of this problem is in terms of the time development operator $\hat{U}(t,t')$ defined via

$$\hat{U}(t,t')|\psi(t')\rangle = \begin{cases} |\psi(t)\rangle & t \geq t' \\ 0 & t < t' \end{cases} \tag{1.15}$$

with the boundary condition

$$\lim_{t \to t'^+} \hat{U}(t,t') = 1 \ . \tag{1.16}$$

For the case of a free particle, obeying

$$i\frac{\partial}{\partial t}|\psi(t)\rangle = \hat{H}_0|\psi(t)\rangle \tag{1.17}$$

the solution for $\hat{U}^{(0)}(t,0)$ is

$$\hat{U}^{(0)}(t,0) = \theta(t)\exp\left(-i\hat{H}_0 t\right), \tag{1.18}$$

where

$$\theta(t) = \begin{cases} 1 & t > 0 \\ 0 & t < 0 \end{cases} \tag{1.19}$$

is the usual theta function. For example, if

$$\psi(x,0) = \frac{1}{(2\pi\sigma^2)^{1/4}}\exp\left(-\frac{(x-a)^2}{4\sigma^2}\right) \tag{1.20}$$

we find

$$\psi(x,t) = \langle x|\hat{U}^{(0)}(t,0)|\psi(0)\rangle = \int_{-\infty}^{\infty} dx'\,\langle x|\hat{U}^{(0)}(t,0)|x'\rangle\langle x'|\psi(0)\rangle$$

$$= e^{-iH_0(x)t}\psi(x,0) = e^{-iH_0(x)t}\frac{1}{(2\pi\sigma^2)^{1/4}}\exp\left(-\frac{(x-a)^2}{4\sigma^2}\right) \tag{1.21}$$

$$= \sum_{n=0}^{\infty}\frac{1}{n!}\left(\frac{it}{2m}\right)^n \frac{\partial^{2n}}{\partial x^{2n}}\exp\left(-\frac{(x-a)^2}{4\sigma^2}\right)\frac{1}{(2\pi\sigma^2)^{1/4}}.$$

Although one could straightforwardly evaluate this power series, it is easier to note the identity [Bl 68]

$$\frac{\partial^2}{\partial x^2}\frac{1}{\sqrt{\rho}}\exp\left(-\frac{(x-a)^2}{4\rho}\right) = \frac{\partial}{\partial \rho}\frac{1}{\sqrt{\rho}}\exp\left(-\frac{(x-a)^2}{4\rho}\right). \tag{1.22}$$

Then using

$$\exp\left(\alpha\frac{\partial}{\partial z}\right)f(z) = \sum_{n=0}^{\infty}\frac{\alpha^n}{n!}\frac{\partial^n}{\partial z^n}f(z) = f(z+\alpha) \tag{1.23}$$

we find

$$\psi(x,t) = \left(\frac{\sigma^2}{2\pi}\right)^{1/4}\frac{1}{\left(\sigma^2+i\frac{t}{2m}\right)^{1/2}}\exp\left(-\frac{(x-a)^2}{4\left(\sigma^2+i\frac{t}{2m}\right)}\right). \tag{1.24}$$

We note that

$$|\psi(x,t)|^2 = \frac{1}{\sqrt{2\pi}\,\sigma(t)}\exp\left(-\frac{(x-a)^2}{2\sigma^2(t)}\right) \quad \text{with} \quad \sigma(t) = \left(\sigma^2 + \frac{t^2}{4m^2\sigma^2}\right)^{1/2} \tag{1.25}$$

which obviously exhibits the canonical spreading experienced by such a wavepacket.

We can equivalently perform the above calculation in momentum space, where the time development operator has the simple form

$$\left\langle p \left| \hat{U}^{(0)}(t,t') \right| p' \right\rangle = \langle p | \exp(-i\hat{p}^2(t-t')/2m)|p'\rangle \theta(t-t')$$

$$= \exp-i\frac{p^2}{2m}(t-t')\langle p|p'\rangle \theta(t-t') = \exp-i\frac{p^2}{2m}(t-t')2\pi\delta(p-p')\theta(t-t') \ . \quad (1.26)$$

If

$$\langle x|\psi(0)\rangle = \frac{1}{(2\pi\sigma^2)^{1/4}} \exp\left(-\frac{(x-a)^2}{4\sigma^2}\right) \quad (1.27)$$

we have

$$\langle p|\psi(0)\rangle = \int_{-\infty}^{\infty} dx \ \langle p|x\rangle \langle x|\psi(0)\rangle$$

$$= \int_{-\infty}^{\infty} dx \ e^{-ipx} \frac{1}{(2\pi\sigma^2)^{1/4}} \exp\left(-\frac{(x-a)^2}{4\sigma^2}\right) \quad (1.28)$$

$$= \left(8\pi\sigma^2\right)^{1/4} \exp\left(-\sigma^2 p^2\right) \exp\left(-ipa\right) \ .$$

Then

$$\langle p|\psi(t)\rangle = \int_{-\infty}^{\infty} \frac{dp'}{2\pi} \langle p|\hat{U}^{(0)}(t,0)|p'\rangle \langle p'|\psi(0)\rangle$$

$$= \left(8\pi\sigma^2\right)^{1/4} \exp\left(-\sigma^2 p^2 - ipa - i\frac{p^2}{2m}t\right) \theta(t) \ . \quad (1.29)$$

We can return to coordinate space via

$$\langle x|\psi(t)\rangle = \int_{-\infty}^{\infty} \frac{dp}{2\pi} \langle x|p\rangle \langle p|\psi(t)\rangle$$

$$= \int_{-\infty}^{\infty} \frac{dp}{2\pi} e^{ipx} \left(8\pi\sigma^2\right)^{1/4} \exp\left(-\sigma^2 p^2 - ipa - i\frac{p^2}{2m}t\right) \theta(t) \quad (1.30)$$

$$= \left(\frac{\sigma^2}{2\pi}\right)^{1/4} \frac{1}{\left(\sigma^2 + i\frac{t}{2m}\right)^{1/2}} \exp-\frac{(x-a)^2}{4\left(\sigma^2 + i\frac{t}{2m}\right)} \theta(t)$$

which agrees precisely with Eq. 1.24 found via coordinate space methods.

PROBLEM I.1.1

Wave Packet Spreading: A Paradox

It was demonstrated above using the identity

$$\left(\frac{\partial^2}{\partial x^2} - \frac{\partial}{\partial z}\right) z^{-1/2} \exp\left(-\frac{x^2}{4z}\right) = 0$$

that a Gaussian wavepacket

$$\psi(x, t=0) = \left(2\pi\sigma^2\right)^{-1/4} \exp\left(-\frac{x^2}{4\sigma^2}\right)$$

evolves in time via

$$\psi(x,t) = \xi^{-1/2} (2\pi\sigma^2)^{-1/4} \exp\left(-\frac{x^2}{4\sigma^2\xi}\right)$$

where

$$\xi = 1 + i\frac{t}{2m\sigma^2}.$$

Then

$$|\psi(x,t)|^2 = (2\pi\sigma^2(t))^{-1/2} \exp\left(-\frac{x^2}{2\sigma^2(t)}\right)$$

where

$$\sigma^2(t) = \sigma^2 + \frac{t^2}{4m^2\sigma^2}.$$

i) Show that

$$\int_{-\infty}^{\infty} dx \, |\psi(x,t)|^2 = 1$$

$$\int_{-\infty}^{\infty} dx \, x^2 \, |\psi(x,t)|^2 = \sigma^2(t)$$

so that the wavepacket remains normalized to unity but has a width

$$\sigma(t) = \sqrt{\sigma^2 + \frac{t^2}{4m^2\sigma^2}}$$

which evolves with time. This is simply the usual "spreading" of a quantum mechanical wave packet.

ii) Derive the time evolution of the Gaussian wavepacket without exploiting the identity by using a power series expansion

$$\psi(x,t) = e^{-iH_0(x)t}\psi(x,0)$$

$$= (2\pi\sigma^2)^{-1/4} \sum_{\ell=0}^{\infty} \frac{1}{\ell!} \left(i\frac{t}{2m}\right)^\ell \frac{\partial^{2\ell}}{\partial x^{2\ell}} \sum_{n=0}^{\infty} \frac{1}{n!} \left(-\frac{x^2}{4\sigma^2}\right)^n.$$

iii) Now suppose that

$$\psi(x,0) = N \begin{cases} \exp\left(-\frac{a^2}{a^2-x^2}\right) & |x| < a \\ 0 & |x| > a \end{cases}$$

where N is a normalization constant. Although this functional form may look a bit strange, a little thought should convince one that the wavefunction and all its derivatives are continuous at any point on the real line. However, it is easy to see that

$$e^{-iH_0(x)t}\psi(x,0) = \sum_{n=0}^{\infty} \frac{1}{n!} \left(i\frac{t}{2m}\right)^n \frac{\partial^{2n}}{\partial x^{2n}}\psi(x,0)$$

vanishes for all time if $|x| \geq a$ since

$$\frac{\partial^{2n}}{\partial x^{2n}} 0 = 0 \ .$$

Hence, this type of wavepacket apparently does not undergo spreading. Is this assertion correct? If not, where have we made an error in our analysis and what does the actual time evolved wavefunction look like [HoS 72]?

I.2 THE PROPAGATOR

One can evaluate the co-ordinate space matrix element of the time development operator by transforming to momentum space and back again.

$$\begin{aligned}
D_F^{(0)}(x',t;x,0) &\equiv \langle x'|\hat{U}^{(0)}(t,0)|x\rangle = \left\langle x'\left|e^{-i\hat{H}_0 t}\right|x\right\rangle \theta(t) \\
&= \theta(t) \int_{-\infty}^{\infty} \frac{dp}{2\pi} \langle x'|p\rangle e^{-i\frac{p^2}{2m}t} \langle p|x\rangle = \theta(t) \int_{-\infty}^{\infty} \frac{dp}{2\pi} e^{ip(x'-x) - i\frac{p^2}{2m}t} \\
&= \theta(t) \sqrt{\frac{m}{2\pi i t}} \exp i\frac{m(x'-x)^2}{2t} \ .
\end{aligned} \tag{2.1}$$

D_F is usually called the "propagator," as it gives the amplitude for a particle produced at position x at time 0 to "propagate" to position x' at time t.

Just as a check we can verify that this form of the propagator does indeed generate the time development of the freely moving Gaussian wavefunction

$$\begin{aligned}
\psi(x',t) &= \int_{-\infty}^{\infty} dx\, D_F^{(0)}(x',t;x,0)\psi(x,0) \\
&= \int_{-\infty}^{\infty} dx \sqrt{\frac{m}{2\pi i t}} \exp \frac{im(x'-x)^2}{2t} \frac{1}{(2\pi\sigma^2)^{1/4}} \exp\left(-\frac{(x-a)^2}{4\sigma^2}\right) \\
&= \left(\frac{\sigma^2}{2\pi}\right)^{1/4} \exp\left(-\frac{(x'-a)^2}{4\left(\sigma^2 + i\frac{t}{2m}\right)}\right) \frac{1}{\left(\sigma^2 + i\frac{t}{2m}\right)^{1/2}}
\end{aligned} \tag{2.2}$$

in complete agreement with expression derived in sect. I.1.

Path Integrals and the Propagator

Before going further, it is useful to note an alternative way by which the propagator can be calculated—the Feynman path integral [FeH 65]

$$D_F(x',t;x,0) = \int \mathcal{D}[x(t)] \exp\frac{iS[x(t)]}{\hbar} \tag{2.3}$$

where the notation is that the integral represents a sum over *all* paths $x(t)$ connecting the initial and final spacetime points — $x, 0$ and x', t respectively. For each path there is a weighting factor given by $\exp\frac{iS}{\hbar}$ where $S = \int dt L[x(t)]$ is the classical action associated with that path. The path integration can be carried out by

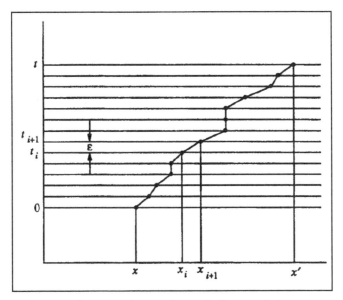

Fig. I.1: *A particular time slice used in calculation of the propagator.*

dividing the time interval $0 - t$ into n slices of width ϵ. This provides a set of times t_i spaced a distance ϵ apart between the values 0 and t. At each time t_i we select a point x_i. A path is constructed by connecting all possible x_i points so selected by straight lines as shown in Figure I.1 and the path integral is written (setting $\hbar = 1$) as

$$D_F(x', t; x, 0) = \lim_{n \to \infty} \frac{1}{A^n} \prod_{i=1}^{n-1} \left(\int_{-\infty}^{\infty} dx_i \right) \exp i S^{(0)} \qquad (2.4)$$

where A is a normalization constant which defines the measure— note that there is one factor of A for each straight line segment. In the limit as $\epsilon \to 0$ we can evaluate the action for each line segment in the infinitesimal approximation. For the free particle we have

$$\begin{aligned} S^{(0)} &= \int_0^t dt' L\left(x(t'), \dot{x}(t'), t'\right) = \int_0^t dt' \frac{1}{2} m \dot{x}^2(t') \\ &= \sum_{i=1}^{n} \frac{1}{2} m \frac{(x_i - x_{i-1})^2}{\epsilon} \quad \text{with} \quad x_0 = x \quad \text{and} \quad x_n = x' \end{aligned}$$

The integrations may be performed sequentially

$$\int_{-\infty}^{\infty} dx_1 \exp i \frac{m}{2\epsilon} \left((x_1 - x_0)^2 + (x_2 - x_1)^2\right) = \sqrt{\frac{2\pi i \epsilon}{2m}} \exp i \frac{m}{2 \cdot 2\epsilon} (x_2 - x_0)^2$$

$$\int_{-\infty}^{\infty} dx_2 \exp i \frac{m}{2\epsilon} \left(\frac{1}{2}(x_2 - x_0)^2 + (x_3 - x_2)^2\right) = \sqrt{\frac{2\pi i \epsilon \cdot 2}{3m}} \exp i \frac{m}{3 \cdot 2\epsilon} (x_3 - x_0)^2$$

$$\vdots$$

$$\int_{-\infty}^{\infty} dx_{n-1} \exp i \frac{m}{2\epsilon} \left(\frac{1}{n-1}(x_{n-1} - x_0)^2 + (x_n - x_{n-1})^2\right)$$

$$= \sqrt{\frac{2\pi i \epsilon (n-1)}{nm}} \exp i \frac{m}{n \cdot 2\epsilon} (x_n - x_0)^2 \tag{2.5}$$

yielding

$$D_F^{(0)}(x', t; x, 0) = \left(\frac{2\pi i \epsilon}{m}\right)^{(n-1)/2} \frac{1}{A^n} \frac{1}{\sqrt{n}} \exp i \frac{m}{2n\epsilon} (x' - x)^2 \tag{2.6}$$

The constant A may be determined by use of the completeness condition

$$\psi(x', t) = \int_{-\infty}^{\infty} dx \langle x' | \hat{U}^{(0)}(t, 0) | x \rangle \langle x | \psi(0) \rangle$$
$$= \int_{-\infty}^{\infty} dx D_F^{(0)}(x', t; x, 0) \psi(x, 0) \; . \tag{2.7}$$

If we pick $t = \epsilon \lll 1$ then

$$\psi(x', \epsilon) \cong \int_{-\infty}^{\infty} dx \frac{1}{A} \exp i \frac{m(x' - x)^2}{2\epsilon} \psi(x, 0) + \ldots \; . \tag{2.8}$$

Since ϵ is small, the exponential will rapidly oscillate and thereby wash out the integral unless $x \cong x'$. Thus, we can write

$$\psi(x', \epsilon) \cong \psi(x', 0) \frac{1}{A} \int_{-\infty}^{\infty} dx \exp i \frac{m(x' - x)^2}{2\epsilon} + \ldots$$
$$= \psi(x', 0) \frac{1}{A} \sqrt{\frac{2\pi i \epsilon}{m}} + \ldots \; . \tag{2.9}$$

Hence in order to have the correct behavior as $\epsilon \to 0$ we must pick

$$A = \sqrt{\frac{2\pi i \epsilon}{m}} \tag{2.10}$$

so that the free propagator becomes, using $t = n\epsilon$

$$D_F^{(0)}(x', t; x, 0) = \sqrt{\frac{m}{2\pi i n \epsilon}} \exp \frac{im}{2n\epsilon} (x' - x)^2 = \sqrt{\frac{m}{2\pi i t}} \exp \frac{im(x' - x)^2}{2t} \tag{2.11}$$

in complete agreement with the expression derived via more conventional means (*cf.* Eq. 2.1).

The reason that the propagator can be written as a path integral can be understood by using the completeness relation

$$\mathbf{1} = \int_{-\infty}^{\infty} dx_i \, |x_i\rangle\langle x_i| \ . \tag{2.12}$$

For later use, we shall give the derivation here for the general case involving interaction with a potential $V(\hat{x})$. Starting with the definition

$$D_F(x_f, t_f; x_i, t_i) = \left\langle x_f \left| \exp -i\hat{H}(t_f - t_i) \right| x_i \right\rangle \theta(t_f - t_i) \tag{2.13}$$

and breaking the time interval $t_f - t_i$ (assumed to be positive) into n discrete steps of size

$$\epsilon = \frac{t_f - t_i}{n} \tag{2.14}$$

we can write

$$D_F(x_f, t_f; x_i, t_i) = \int_{-\infty}^{\infty} dx_1 \ldots dx_{n-1} \left\langle x_n \left| e^{-i\epsilon\hat{H}} \right| x_{n-1} \right\rangle$$
$$\cdot \left\langle x_{n-1} \left| e^{-i\epsilon\hat{H}} \right| x_{n-2} \right\rangle \ldots \left\langle x_1 \left| e^{-i\epsilon\hat{H}} \right| x_0 \right\rangle \ . \tag{2.15}$$

In the limit of large n the time slices become infinitesimal and

$$\left\langle x_\ell \left| e^{-i\epsilon\hat{H}} \right| x_{\ell-1} \right\rangle = \left\langle x_\ell \left| \exp -i\epsilon \left(\frac{\hat{p}^2}{2m} + V(\hat{x}) \right) \right| x_{\ell-1} \right\rangle$$
$$\approx \exp -i\epsilon V(x_\ell) \left\langle x_\ell \left| e^{-i\epsilon \frac{\hat{p}^2}{2m}} \right| x_{\ell-1} \right\rangle + \mathcal{O}(\epsilon^2) \ . \tag{2.16}$$

Introducing a complete set of momentum states, we have

$$\left\langle x_\ell \left| e^{-i\epsilon \frac{\hat{p}^2}{2m}} \right| x_{\ell-1} \right\rangle = \int_{-\infty}^{\infty} \frac{dp}{2\pi} e^{ip(x_\ell - x_{\ell-1})} e^{-i\epsilon \frac{p^2}{2m}}$$
$$= \sqrt{\frac{m}{2\pi i \epsilon}} \exp i \frac{m}{2\epsilon} (x_\ell - x_{\ell-1})^2 \tag{2.17}$$

and, taking the continuum limit, we find the path integral prescription

$$D_F(x_f, t_f; x_i, t_i) = \lim_{n \to \infty} \left(\frac{m}{2\pi i \epsilon} \right)^{\frac{n}{2}} \int_{-\infty}^{\infty} dx_{n-1} \ldots dx_1$$
$$\times \exp i \sum_{\ell=1}^{n} \left(m \frac{(x_\ell - x_{\ell-1})^2}{2\epsilon} - \epsilon V(x_\ell) \right) \tag{2.18}$$
$$= \int \mathcal{D}[x(t)] \exp iS[x(t)]$$

where

$$S[(t)] = \int_{t_1}^{t_2} dt \left(\frac{1}{2} m \dot{x}^2(t) - V(x(t)) \right) \tag{2.19}$$

is the classical action.

Classical Connection

Perhaps the most peculiar and fascinating aspect of this prescription is that *all* paths connecting the spacetime endpoints must be included in the summation. This appears to be in total contradiction with the classical mechanics result that a particle traverses a well-defined trajectory. The resolution of this apparent paradox may be found by explicitly restoring the dependence on $\hbar$ and noting that the path integral prescription is given by

$$\sum_{x(t)} \exp iS[x(t)]/\hbar \qquad (2.20)$$

Classical physics results as $\hbar \to 0$, and in this limit a slight change in the path $x(t)$ produces a huge change in phase and hence little or no contribution to the path summation except for trajectories $\bar{x}(t)$ for which the action is stationary—*i.e.*, Hamilton's principle

$$\left. \frac{\delta S[x(t)]}{\delta x} \right|_{x(t)=\bar{x}(t)} = 0 \ . \qquad (2.21)$$

In order to find such a path we take

$$0 = S\left[\bar{x}(t) + \delta x(t)\right] - S\left[\bar{x}(t)\right]$$

$$= \int_0^t dt' \left[\frac{m}{2} (\dot{\bar{x}}(t') + \delta \dot{x}(t'))^2 - \frac{m}{2}(\dot{\bar{x}}(t'))^2 - V(\bar{x}(t') + \delta x(t')) + V(\bar{x}(t')) \right]$$

$$= \int_0^t dt' \left(m\dot{\bar{x}}(t')\delta \dot{x}(t') - V'(\bar{x}(t'))\delta x(t') \right) + \mathcal{O}((\delta x)^2) \ , \qquad (2.22)$$

integrate by parts and use the feature that the endpoints of the path are fixed, *i.e.*, $\delta x(0) = \delta x(t) = 0$. Then

$$0 = -\int_0^t dt' (m\ddot{\bar{x}}(t') + V'(\bar{x}(t')))\delta x(t') \ . \qquad (2.23)$$

so that the trajectory which satisfies the stationary phase condition for arbitrary $\delta x(t')$ must obey

$$m\ddot{\bar{x}} + V'(\bar{x}) = 0 \qquad (2.24)$$

which is just the classical mechanics prescription for the motion of a freely moving particle, *i.e.*, $\bar{x}(t) = x_{\text{cl}}(t)$. In the limit $\hbar \to 0$ the classical trajectory represents the *only* path contributing to the path integral and the paradox is resolved.

One can also get a feel for the meaning of the propagator by noting that since

$$\langle x|\psi(t)\rangle = \left\langle x \left| e^{-i\hat{H}_o t} \right| \psi(0) \right\rangle = \int_{-\infty}^{\infty} dx' \left\langle x \left| e^{-i\hat{H}_o t} \right| x' \right\rangle \langle x'|\psi(0)\rangle$$

$$= \int_{-\infty}^{\infty} dx' \, D_F^{(0)}(x,t;x',0) \langle x'|\psi(0)\rangle \qquad (2.25)$$

if we take
$$\langle x'|\psi(0)\rangle = \delta(x') \tag{2.26}$$
so that at $t=0$ the particle is located precisely at the origin, then
$$D_F^{(0)}(x,t;0,0) = \sqrt{\frac{m}{2\pi i t}} \exp\frac{imx^2}{2t} = \langle x|\psi(t)\rangle \ . \tag{2.27}$$

That is to say, $D_F^{(0)}(x,t;0,0)$ is just the Schrödinger wavefunction of a freely moving particle which started at the origin at time zero. If we look at a specific location x_0, t_0 we would say classically that if a particle is observed at this point then it must have momentum
$$p_{\rm cl} = mv_0 = m\frac{x_0}{t_0} \tag{2.28}$$
and energy
$$E_{\rm cl} = \frac{1}{2}mv_0^2 = \frac{1}{2}m\frac{x_0^2}{t_0^2} \ . \tag{2.29}$$
Examining the variation of the phase of the wavefunction in the vicinity of x_0, t_0, we find
$$\begin{aligned}\langle x|\psi(t)\rangle &= \sqrt{\frac{m}{2\pi i t}}\exp i\frac{mx^2}{2t}\\ &= \sqrt{\frac{m}{2\pi i t}}\exp i\frac{m}{2}\left(\frac{x_0^2}{t_0} + (x-x_0)\frac{\partial}{\partial x}\frac{x^2}{t}\bigg|_{x=x_0} + (t-t_0)\frac{\partial}{\partial t}\frac{x^2}{t}\bigg|_{t=t_0} + \cdots\right)\\ &= \sqrt{\frac{m}{2\pi i t}}\exp i\frac{m}{2}\left(\frac{x_0^2}{t_0} + 2\frac{x_0}{t_0}(x-x_0) - \frac{x_0^2}{t_0^2}(t-t_0) + \cdots\right) \ .\end{aligned} \tag{2.30}$$
Thus in the vicinity of this point we can write
$$\langle x|\psi(t)\rangle \approx \sqrt{\frac{m}{2\pi i t}}\exp\left(ip_{\rm cl}x - iE_{\rm cl}t\right) \tag{2.31}$$
so that both the wavelength associated with the particle
$$\lambda = \frac{2\pi}{p_{\rm cl}} \tag{2.32}$$
and the corresponding frequency
$$\nu = \frac{E_{\rm cl}}{2\pi} \tag{2.33}$$
are given by the usual quantum mechanical relations.

Finally, the probability that the particle is located between x and $x+dx$ at time t is
$$P(x, x+dx) = |\langle x|\psi(t)\rangle|^2 dx = \frac{m\,dx}{2\pi t} \tag{2.34}$$
and is independent of x. All momenta then are equally likely at $t=0$, as would be expected from the momentum space representation of the co-ordinate space wavefunction
$$\delta(x) = \int\frac{dp}{2\pi}e^{ipx} \tag{2.35}$$
and the momentum density is $\frac{dp}{2\pi}$ with $dp = \frac{m\,dx}{t}$. We conclude that all our intuitive notions are satisfied by the propagator, Eq. 2.27.

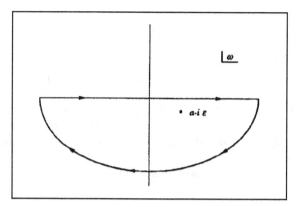

Fig. I.2: *When $t > 0$ the contour is closed by means of a large semicircle in the lower half plane.*

Frequency Space Representation

Before moving on to the more interesting case of motion in the presence of a potential, it is important to note that the time development operator is often used in Fourier transform or frequency space form rather than in its time representation. Before examining this result, however, it is useful to prove a simple mathematical identity. Consider the integral

$$I(a) = \int_{-\infty}^{\infty} \frac{d\omega}{2\pi} e^{-i\omega t} \frac{i}{\omega - a + i\epsilon} \ . \tag{2.36}$$

If $t > 0$ the contour can be closed in the lower half plane by means of a large semicircle (*cf.* Figure I.2) which contributes nothing to the integral because of the exponential damping introduced by the factor $e^{-(\mathrm{Im}\omega)t}$. The integral can then be evaluated by means of the residue theorem [MaW 64]. There exists a single pole at $\omega = a - i\epsilon$ and the integral is found to be

$$I(a) = -2\pi i \times \frac{i}{2\pi} e^{-iat} = e^{-iat} \qquad t > 0 \ . \tag{2.37}$$

On the other hand, if $t < 0$ exponential damping of the semicircular contribution demands that we close the contour in the upper half plane. In this case there is no singularity so

$$I(a) = 0 \qquad t < 0 \ . \tag{2.38}$$

We have in general then

$$I(a) = \int_{-\infty}^{\infty} \frac{d\omega}{2\pi} e^{-i\omega t} \frac{i}{\omega - a + i\epsilon} = \theta(t) e^{-iat} \ , \tag{2.39}$$

so that, replacing a by the operator $\hat{H}_0$, an alternative way to represent the free time development operator is

$$\hat{U}^{(0)}(t, 0) = e^{-i\hat{H}_0 t} \theta(t) = \int_{-\infty}^{\infty} \frac{d\omega}{2\pi} e^{-i\omega t} \frac{i}{\omega - \hat{H}_0 + i\epsilon} \tag{2.40}$$

1.2 THE PROPAGATOR

i.e., $\hat{U}^{(0)}(t,0)$ can be written as a Fourier transform with

$$\hat{U}^{(0)}(\omega) = \frac{i}{\omega - \hat{H}_0 + i\epsilon} . \qquad (2.41)$$

Other Representations

It is often useful to represent $\hat{U}^{(0)}(\omega)$ in terms either of its momentum space

$$\left\langle p \left| \hat{U}^{(0)}(\omega) \right| p' \right\rangle = 2\pi\delta(p-p') \frac{i}{\omega - \frac{p^2}{2m} + i\epsilon} \qquad (2.42)$$

or coordinate space matrix elements

$$\left\langle x \left| \hat{U}^{(0)}(\omega) \right| x' \right\rangle = \int_{-\infty}^{\infty} \frac{dp}{2\pi} e^{ip(x-x')} \frac{i}{\omega - \frac{p^2}{2m} + i\epsilon} . \qquad (2.43)$$

Defining

$$\omega = \frac{p_0^2}{2m} \qquad (2.44)$$

we can explicitly evaluate the latter

$$\begin{aligned}
\left\langle x \left| \hat{U}^{(0)}(\omega) \right| x' \right\rangle &= \int_{-\infty}^{\infty} \frac{dp}{2\pi} e^{ip(x-x')} \frac{i}{\omega - \frac{p^2}{2m} + i\epsilon} \\
&= -\int_{-\infty}^{\infty} \frac{dp}{2\pi} e^{ip(x-x')} \frac{2mi}{p^2 - p_0^2 - i\epsilon} \\
&= -2m \int_{-\infty}^{\infty} \frac{dp}{2\pi} e^{ip(x-x')} \frac{i}{(p-p_0-i\epsilon)(p+p_0+i\epsilon)} .
\end{aligned} \qquad (2.45)$$

If $x - x' > 0$ we close the contour in the upper half plane and pick up the pole at $p_0 + i\epsilon$, yielding

$$\left\langle x \left| \hat{U}^{(0)}(\omega) \right| x' \right\rangle = 2\pi i \times \frac{-2mi}{2\pi} \times \frac{1}{2p_0} e^{ip_0(x-x')} = \frac{m}{p_0} e^{ip_0(x-x')} , \qquad (2.46)$$

while if $x - x' < 0$ we must close the contour in the lower half plane and pick up the pole at $-p_0 - i\epsilon$, yielding

$$\left\langle x \left| \hat{U}^{(0)}(\omega) \right| x' \right\rangle = -2\pi i \times \frac{-2mi}{2\pi} \times \frac{1}{-2p_0} e^{-ip_o(x-x')} = \frac{m}{p_0} e^{-ip_0(x-x')} . \qquad (2.47)$$

The general result can be written as

$$\left\langle x \left| \hat{U}^{(0)}(\omega) \right| x' \right\rangle = \frac{m}{p_0} e^{ip_0|x-x'|} \qquad (2.48)$$

and will be useful later.

PROBLEM I.2.1

The Hamiltonian Path Integral

An alternative—Hamiltonian—version of the path integral is often useful when one is dealing with non-Cartesian variables or with constrained systems.

i) Show that

$$\int_{-\infty}^{\infty} \frac{dp}{2\pi} \exp\left(ipx - i\frac{p^2}{2m}\epsilon\right) = \sqrt{\frac{m}{2\pi i\epsilon}} \exp\left(i\frac{mx^2}{2\epsilon}\right)$$

ii) Using the result from i) demonstrate that the propagator may be written as

$$D_F(x_f, t_f; x_i, t_i) = \lim_{n\to\infty} \int_{-\infty}^{\infty} \frac{dp_0}{2\pi} dx_1 \frac{dp_1}{2\pi} dx_2 \ldots dx_{n-1} \frac{dp_{n-1}}{2\pi}$$
$$\times \exp i \sum_{\ell=0}^{n-1} \left(p_\ell(x_{\ell+1} - x_\ell) - \left(\frac{p_\ell^2}{2m} + V(x_\ell)\right)\epsilon\right)$$
$$\equiv \int \mathcal{D}[x(t)]\mathcal{D}[p(t)] \exp i \int dt \, (p\dot{x} - H(p,x))$$

which is the form we were seeking. Note that here p, x are considered as independent variables.

I.3 HARMONIC OSCILLATOR PROPAGATOR

Having examined the form of the free propagator in the previous section, we now consider motion under the influence of a potential $V(\hat{x})$. In this case the time development operator becomes (hereafter assuming $t > 0$)

$$\hat{U}(t,0) = e^{-i\hat{H}t} = e^{-i(\hat{H}_0 + V(\hat{x}))t} \tag{3.1}$$

which has the coordinate space representation

$$D_F(x', t; x, 0) = \left\langle x' \left| e^{-i(\hat{H}_0 + V(\hat{x}))t} \right| x \right\rangle . \tag{3.2}$$

Provided that the Hamiltonian can be solved exactly to yield eigenvalues and eigenfunctions

$$\hat{H}|\phi_n\rangle = E_n|\phi_n\rangle \quad n = 0, 1, 2, \ldots \tag{3.3}$$

we can find an exact representation of the propagator

$$D_F(x', t; x, 0) = \sum_n \langle x'|\phi_n\rangle e^{-iE_n t} \langle \phi_n|x\rangle$$
$$= \sum_n \phi_n(x')\phi_n^*(x) e^{-iE_n t} . \tag{3.4}$$

(Note that the free particle propagator is of this form since

$$D_F(x',t;x,0) = \sum_p \langle x'|p\rangle e^{-i\frac{p^2}{2m}t}\langle p|x\rangle$$
$$= \int_{-\infty}^{\infty}\frac{dp}{2\pi}e^{ip(x'-x)-i\frac{p^2}{2m}t} \qquad (3.5)$$

as before.) That Eq. 3.4 generates the time development of an arbitrary wavefunction is clear since, assuming $t > 0$

$$\psi(x,t) = \int_{-\infty}^{\infty}dx\,\langle x'|e^{-i\hat{H}t}|x\rangle\langle x|\psi(0)\rangle$$
$$= \int_{-\infty}^{\infty}dx\,D_F(x',t;x,0)\psi(x,0) = \sum_n \phi_n(x')e^{-iE_n t}c_n \qquad (3.6)$$

where

$$c_n = \int_{-\infty}^{\infty}dx\,\phi_n^*(x)\psi(x,0)\;. \qquad (3.7)$$

is the projection of the initial state wavefunction $\psi(x,0)$ onto eigenstate $\phi_n(x)$. Eq. 3.6 then is simply the usual expansion of the wavefunction at later time t in terms of eigenstates of the Hamiltonian.

For soluble problems the propagator can generally be given simply and in closed form, and below we shall show how this is done for the case of the harmonic oscillator

$$\hat{H} = \frac{\hat{p}^2}{2m} + \frac{1}{2}m\omega^2\hat{x}^2\;. \qquad (3.8)$$

However, before deriving the explicit form it is useful to review the solution of the harmonic oscillator problem using the technique due to Dirac [Di 58].

Harmonic Oscillator Review

We begin by defining the so called "creation" and "annihilation" operators

$$\hat{a} = \sqrt{\frac{m\omega}{2}}\hat{x} + \frac{i\hat{p}}{\sqrt{2m\omega}} \qquad \hat{a}^\dagger = \sqrt{\frac{m\omega}{2}}\hat{x} - \frac{i\hat{p}}{\sqrt{2m\omega}}\;. \qquad (3.9)$$

Then defining the "number operator" $\hat{N} \equiv \hat{a}^\dagger\hat{a}$ we find

$$\hat{N} = \frac{\hat{p}^2}{2m\omega} + \frac{1}{2}m\omega\hat{x}^2 + \frac{i}{2}[\hat{x},\hat{p}]$$
$$= \frac{\hat{H}}{\omega} - \frac{1}{2} \qquad (3.10)$$

or

$$\hat{H} = \omega\left(\hat{N} + \frac{1}{2}\right)\;. \qquad (3.11)$$

Now look for eigenstates $|n\rangle$ such that

$$\hat{N}|n\rangle = n|n\rangle \ . \tag{3.12}$$

We can determine the properties of the eigenvalues n as follows. Since $\hat{N}$ is hermitian—

$$\hat{N}^\dagger = \left(\hat{a}^\dagger \hat{a}\right)^\dagger = \hat{a}^\dagger \hat{a} = \hat{N} \tag{3.13}$$

—we have

$$n = \left\langle n|\hat{N}|n\right\rangle = \left\langle n|\hat{N}^\dagger|n\right\rangle^* = \left\langle n|\hat{N}|n\right\rangle^* = n^* \ . \tag{3.14}$$

i.e., n is real. Also n is non-negative since it can be written as the inner product of a state with itself.

$$n = \left\langle n|\hat{N}|n\right\rangle = \langle n|\hat{a}^\dagger \hat{a}|n\rangle = \left(\langle n|\hat{a}^\dagger\right)\hat{a}|n\rangle \geq 0 \ . \tag{3.15}$$

Commutation relations are easily found

$$[\hat{a}, \hat{a}^\dagger] = \frac{1}{2}\left(i\,[\hat{p}, \hat{x}] - i\,[\hat{x}, \hat{p}]\right) = 1 \ . \tag{3.16}$$

Also

$$\begin{aligned}[] [\hat{N}, \hat{a}] &= \hat{a}^\dagger \hat{a}\hat{a} - \hat{a}\hat{a}^\dagger \hat{a} = [\hat{a}^\dagger, \hat{a}]\hat{a} = -\hat{a} \\ [\hat{N}, \hat{a}^\dagger] &= \hat{a}^\dagger \hat{a}\hat{a}^\dagger - \hat{a}^\dagger \hat{a}^\dagger \hat{a} = \hat{a}^\dagger [\hat{a}, \hat{a}^\dagger] = \hat{a}^\dagger \ . \end{aligned} \tag{3.17}$$

The state $\hat{a}|n\rangle$ is an eigenstate of $\hat{N}$ with eigenvalue $n - 1$, since

$$\hat{N}\hat{a}|n\rangle = (\hat{a}\hat{N} - \hat{a})|n\rangle = (n-1)\hat{a}|n\rangle \ . \tag{3.18}$$

The normalization of this state $\hat{a}|n\rangle$ is given by

$$\left(\langle n|\hat{a}^\dagger\right)\hat{a}|n\rangle = \left\langle n|\hat{N}|n\right\rangle = n \tag{3.19}$$

so that

$$\hat{a}|n\rangle = \sqrt{n}\,|n-1\rangle \ . \tag{3.20}$$

Similarly, operating repeatedly with $\hat{a}$ we can lower the eigenvalue even further

$$\hat{a}^m|n\rangle = \sqrt{n(n-1)\ldots(n-m+1)}\,|n-m\rangle \ . \tag{3.21}$$

From Eq. 3.15, however, negative eigenvalues are not permitted, so we must eventually reach a state $|1\rangle$ such that

$$\hat{N}|1\rangle = |1\rangle \tag{3.22}$$

and

$$\hat{a}|1\rangle = |0\rangle \quad \text{with} \quad \hat{N}|0\rangle = \hat{a}|0\rangle = 0 \ . \tag{3.23}$$

We conclude that the eigenvalue n must be an integer—$n = 0, 1, 2, \ldots$. Likewise, we can increase the eigenvalues by utilizing the adjoint operator $\hat{a}^\dagger$

$$\hat{N}\hat{a}^\dagger |n\rangle = \left(\hat{a}^\dagger \hat{N} + \hat{a}^\dagger\right) |n\rangle = (n+1)\hat{a}^\dagger |n\rangle \tag{3.24}$$

with the normalization condition

$$\begin{aligned}(\langle n|\hat{a}) \hat{a}^\dagger |n\rangle &= \langle n|\hat{a}\hat{a}^\dagger |n\rangle \\ &= \left\langle n \left| \hat{N} + [\hat{a}, \hat{a}^\dagger] \right| n \right\rangle = n + 1 \ .\end{aligned} \tag{3.25}$$

Application of $\hat{a}^\dagger$ then yields

$$\hat{a}^\dagger |n\rangle = \sqrt{n+1}\, |n+1\rangle \ . \tag{3.26}$$

Starting with the lowest energy (ground) state $|0\rangle$ we find

$$\left(\hat{a}^\dagger\right)^n |0\rangle = \sqrt{n(n-1)\ldots 1}\, |n\rangle \ . \tag{3.27}$$

so that an arbitrary state $|n\rangle$ can be written as

$$|n\rangle = \frac{1}{\sqrt{n!}} \left(\hat{a}^\dagger\right)^n |0\rangle \ . \tag{3.28}$$

The energy of the eigenstate $|n\rangle$ is given by

$$\hat{H}|n\rangle = \omega \left(\hat{N} + \frac{1}{2}\right)|n\rangle = \omega\left(n + \frac{1}{2}\right)|n\rangle \ . \tag{3.29}$$

as expected.

Thus far we have been dealing with the eigenstates in an abstract—ket—representation. In order to make connection with the usual Schrödinger wavefunction, we take the coordinate space projection of the equation $\hat{a}|0\rangle = 0$, namely

$$\begin{aligned}0 = \langle x|\hat{a}|0\rangle &= \left\langle x \left| \sqrt{\frac{m\omega}{2}}\hat{x} + i\frac{\hat{p}}{\sqrt{2m\omega}} \right| 0 \right\rangle \\ &= \left(\sqrt{\frac{m\omega}{2}}x + \frac{1}{\sqrt{2m\omega}}\frac{d}{dx}\right)\langle x|0\rangle = \frac{1}{\sqrt{2m\omega}}\left(\frac{d}{dx} + m\omega x\right)\langle x|0\rangle \ .\end{aligned} \tag{3.30}$$

The solution to this differential equation yields the familiar ground state wavefunction

$$\langle x|0\rangle = \left(\frac{m\omega}{\pi}\right)^{1/4} \exp\left(-\frac{1}{2}m\omega x^2\right) \ . \tag{3.31}$$

Excited state wavefunctions may be generated by repeated applications of $\hat{a}^\dagger$. For the first excited state, we find

$$\begin{aligned}\langle x|1\rangle = \langle x|\hat{a}^\dagger|0\rangle &= \left\langle x \left| \sqrt{\frac{m\omega}{2}}\hat{x} - i\frac{\hat{p}}{\sqrt{2m\omega}} \right| 0 \right\rangle \\ &= -\frac{1}{\sqrt{2m\omega}}\left(\frac{d}{dx} - m\omega x\right)\langle x|0\rangle = \sqrt{2m\omega}\, x\, \langle x|0\rangle \ ,\end{aligned} \tag{3.32}$$

and one can generate the wavefunction of an arbitrary excited state via

$$\langle x|n\rangle = \frac{1}{\sqrt{n!}}\langle x|(\hat{a}^\dagger)^n|0\rangle = \frac{(-)^n}{\sqrt{n!}(2m\omega)^{\frac{n}{2}}}(\frac{d}{dx} - m\omega x)^n \langle x|0\rangle \\ = \frac{1}{\sqrt{2^n n!}} H_n\left(\sqrt{m\omega}\,x\right)\langle x|0\rangle \ . \tag{3.33}$$

where $H_n(x)$ represents the Hermite polynomial of order n.

Harmonic Oscillator Propagator

We now return to the problem of the harmonic oscillator propagator. There exist a number of techniques by which this result may be obtained. For example, Itzykson and Zuber [ItZ 80] use a traditional time slice procedure in order to yield the closed form

$$D_F(x',t;x,0) = \sqrt{\frac{m\omega}{2\pi i \sin\omega t}} \exp i\frac{m\omega}{2}\left[\left(x'^2+x^2\right)\operatorname{ctn}\omega t - 2x'x\csc\omega t\right] \ . \tag{3.34}$$

(Note that Eq. 3.34 reduces to the free propagator result in the limit $\omega \to 0$.) However this procedure, while straightforward, is also lengthy and cumbersome. An alternate way to obtain the same result is by use of the Feynman path integral

$$D_F(x',t;x,0) = \int \mathcal{D}[x(t)] \exp iS[x(t)] \ , \tag{3.35}$$

but with the arbitrary trajectory $x(t)$ characterized in terms of its deviation $\delta x(t)$ from the classical path $x_{cl}(t)$, which satisfies

$$\ddot{x}_{cl}(t) = -\omega^2 x_{cl}(t) \\ x_{cl}(0) = x \\ x_{cl}(t) = x' \tag{3.36}$$

Then

$$D_F(x',t;x,0) = \int \mathcal{D}[\delta x(t)] \exp iS[x_{cl}(t)+\delta x(t)] \ . \tag{3.37}$$

where

$$S[x_{cl}(t)+\delta x(t)] = S[x_{cl}(t)] + 2\int_0^t dt'\frac{1}{2}m\left(\dot{x}_{cl}(t')\delta\dot{x}(t') - \omega^2 x_{cl}(t')\delta x(t')\right) \\ + \int_0^t dt'\frac{1}{2}m\left((\delta\dot{x}(t'))^2 - \omega^2(\delta x(t'))^2\right) \ . \tag{3.38}$$

Integrating the term linear in δx by parts, we find

$$\int_0^t dt'\,m\left(\dot{x}_{cl}(t')\delta\dot{x}(t') - \omega^2 x_{cl}(t')\delta x(t')\right) = -\int_0^t dt'\,m\delta x(t')\left(\ddot{x}_{cl}(t') + \omega^2 x_{cl}(t')\right) = 0 \tag{3.39}$$

1.3 HARMONIC OSCILLATOR PROPAGATOR

where we have used the classical equation of motion and the fixed endpoint constraint— $\delta x(t) = \delta x(0) = 0$. We have then

$$S[x(t)] = S[x_{\text{cl}}(t)] + S[\delta x(t)] \tag{3.40}$$

and

$$D_F(x', t; x, 0) = \int [\mathcal{D}\delta x(t)] \exp iS[x(t)] = \exp iS[x_{\text{cl}}] \times D_F(0, t; 0, 0) \tag{3.41}$$

i.e., the phase of the exponential is simply the action for the classical path! Writing

$$x_{\text{cl}}(t) = A\sin(\omega t + \phi) \tag{3.42}$$

we require the boundary conditions

$$\begin{aligned} x_{\text{cl}}(0) &= A\sin\phi = x \\ x_{\text{cl}}(t) &= A\sin(\omega t + \phi) = x' \end{aligned}, \tag{3.43}$$

and the action is found to be

$$\begin{aligned} S[x_{\text{cl}}(t)] &= \frac{1}{2}m \int_0^t dt' \left(\dot{x}_{\text{cl}}^2(t') - \omega^2 x_{\text{cl}}^2(t') \right) \\ &= \frac{1}{2}mA^2\omega^2 \int_0^t dt' \left(\cos^2(\omega t' + \phi) - \sin^2(\omega t' + \phi) \right) \\ &= \frac{1}{2}mA^2\omega^2 \int_0^t dt' \cos(2\omega t' + 2\phi) = \frac{m\omega}{4} A^2 \left(\sin(2\omega t + 2\phi) - \sin 2\phi \right) \end{aligned}. \tag{3.44}$$

We must now eliminate A, ϕ in favor of x, x'. Noting that

$$\frac{x}{x'} = \frac{\sin\phi}{\sin(\omega t + \phi)} = \frac{\sin\phi}{\sin\omega t \cos\phi + \cos\omega t \sin\phi} \tag{3.45}$$

and solving for $\cos\phi$ we have

$$A\cos\phi = \frac{x'}{\sin\omega t} - \frac{x\cos\omega t}{\sin\omega t}. \tag{3.46}$$

Thus

$$\begin{aligned} A^2 \sin 2\phi &= 2A\cos\phi \cdot A\sin\phi = 2x\left(\frac{x'}{\sin\omega t} - \frac{x\cos\omega t}{\sin\omega t}\right) \\ A^2\sin(2\omega t + 2\phi) &= 2A^2\sin(\omega t + \phi)\cos(\omega t + \phi) = 2x'A(\cos\omega t\cos\phi - \sin\omega t\sin\phi) \\ &= 2x'\left(\cos\omega t\left(\frac{x'}{\sin\omega t} - \frac{x\cos\omega t}{\sin\omega t}\right) - x\sin\omega t\right) = 2x'\left(\frac{x'\cos\omega t}{\sin\omega t} - \frac{x}{\sin\omega t}\right) \end{aligned} \tag{3.47}$$

and
$$S[x_{\text{cl}}(t)] = \frac{m\omega}{2}\left[\left(x^2 + x'^2\right)\operatorname{ctn}\omega t - 2xx'\csc\omega t\right] . \quad (3.48)$$

The time dependent prefactor $D_F(0,t;0,0)$ can be evaluated either via standard path integral techniques or via a shortcut. We first demonstrate the latter. Using the completeness property

$$\left\langle x' \left| e^{-i\hat{H}t} \right| x \right\rangle = \left\langle x' \left| e^{-i\hat{H}(t-t_1)} e^{-i\hat{H}t_1} \right| x \right\rangle$$
$$= \int_{-\infty}^{\infty} dx'' \left\langle x' \left| e^{-i\hat{H}(t-t_1)} \right| x'' \right\rangle \left\langle x'' \left| e^{-i\hat{H}t_1} \right| x \right\rangle . \quad (3.49)$$

Defining
$$D_F(0,t;0,0) \equiv J(t) \quad (3.50)$$

we have, combining Eqs. 3.34 and 3.49

$$J(t)\exp i\frac{m\omega}{2}\left[\left(x^2+x'^2\right)\operatorname{ctn}\omega t - 2xx'\csc\omega t\right]$$
$$= J(t-t_1)J(t_1)\int_{-\infty}^{\infty} dx'' \exp\frac{im\omega}{2}\left[\left(x'^2+x''^2\right)\operatorname{ctn}\omega(t-t_1) \quad (3.51)$$
$$-2x'x''\csc\omega(t-t_1)\right]\exp\frac{im\omega}{2}\left[\left(x^2+x''^2\right)\operatorname{ctn}\omega t_1 - 2xx''\csc\omega t_1\right] .$$

To simplify things pick $x = x' = 0$. Then

$$\frac{J(t)}{J(t-t_1)J(t_1)} = \int_{-\infty}^{\infty} dx'' \exp i\frac{m\omega}{2}\left(\operatorname{ctn}\omega(t-t_1) + \operatorname{ctn}\omega t_1\right)x''^2$$
$$= \sqrt{\frac{2\pi i}{m\omega}\frac{1}{\operatorname{ctn}\omega(t-t_1)+\operatorname{ctn}\omega t_1}} = \sqrt{\frac{2\pi i}{m\omega}\frac{\sin\omega(t-t_1)\sin\omega t_1}{\sin\omega t}} \quad (3.52)$$

whose solution is
$$J(t) = \sqrt{\frac{m\omega}{2\pi i \sin\omega t}} . \quad (3.53)$$

in agreement with Eq. 3.34.

Determinant Methods and the Prefactor

A technique by which to evaluate the prefactor which is more generally useful is given below. Writing the path integral as

$$D_F(0,t_f;0,t_i) = \int \mathcal{D}[\delta x(t)] \exp iS[\delta x(t)] \quad (3.54)$$

where $\delta x(t_i) = \delta x(t_f) = 0$ and integrating by parts we find

$$S[\delta x(t)] = \int_{t_i}^{t_f} dt\left(\frac{1}{2}m(\delta\dot{x}(t))^2 - \frac{1}{2}m\omega^2(\delta x(t))^2\right)$$
$$= \int_{t_i}^{t_f} dt \frac{m}{2}\delta x(t)\left(-\frac{d^2}{dt^2} - \omega^2\right)\delta x(t) \quad (3.55)$$
$$\equiv \int_{t_i}^{t_f} dt\,\delta x(t)\mathcal{O}\delta x(t) .$$

1.3 HARMONIC OSCILLATOR PROPAGATOR

Now expand $\delta x(t)$ in terms of eigenfunctions of the operator $\mathcal{O} = -\frac{d^2}{dt^2} - \omega^2$

$$\delta x(t) = \sum_n a_n x_n(t) \tag{3.56}$$

where $x_n(t)$ satisfies

$$\mathcal{O} x_n(t) = \lambda_n x_n(t) \tag{3.57}$$

with $x_n(t_i) = x_f(t_f) = 0$ and is subject to the orthogonality condition

$$\int_{t_i}^{t_f} dt\, x_n(t) x_m(t) = \delta_{nm} \ . \tag{3.58}$$

The sum over all possible trajectories can then be performed by summation over *all* expansion coefficients a_n

$$\begin{aligned} D_F(0, t_f; 0, t_i) &= N \prod_{j=1}^{\infty} \left(\int_{-\infty}^{\infty} da_j \right) \exp i \int_{t_i}^{t_f} dt\, \delta x(t) \mathcal{O} \delta x(t) \\ &= N \prod_{j=1}^{\infty} \left(\int_{-\infty}^{\infty} da_j \right) \exp i \int_{t_i}^{t_f} \sum_{k=1}^{\infty} a_k x_k(t) \sum_{\ell=1}^{\infty} a_\ell x_\ell(t) \lambda_\ell dt \\ &= N \prod_{j=1}^{\infty} \left(\int_{-\infty}^{\infty} da_j \exp i \lambda_j a_j^2 \right) \\ &\equiv N' (\det \mathcal{O})^{-1/2} \end{aligned} \tag{3.59}$$

where N, N' are normalization coefficients and

$$\det \mathcal{O} \equiv \prod_{j=1}^{\infty} \lambda_j \tag{3.60}$$

is the product of operator eigenvalues. For the harmonic oscillator

$$x_n(t) = \sqrt{\frac{2}{t_f - t_i}} \sin \omega_n (t - t_i) \ ,$$

$$\text{with} \quad \omega_n = \left(\frac{n\pi}{t_f - t_i} \right) \quad n = 1, 2, \ldots \tag{3.61}$$

Then $\lambda_n = \omega_n^2 - \omega^2$ and the determinant can be evaluated using the identity [GrR 65]

$$\prod_{n=1}^{\infty} \left(1 - \frac{x^2}{n^2 \pi^2} \right) = \frac{\sin x}{x} \ , \tag{3.62}$$

i.e.,

$$\det \mathcal{O} = \prod_{n=1}^{\infty}\left(\left(\frac{n\pi}{t_f - t_i}\right)^2 - \omega^2\right)$$

$$= \text{Const.} \times \prod_{n=1}^{\infty}\left(1 - \frac{\omega^2(t_f - t_i)^2}{n^2\pi^2}\right) \quad (3.63)$$

$$= \text{Const.} \times \frac{\sin\omega(t_f - t_i)}{\omega(t_f - t_i)}$$

where the constant is independent of ω, and may be absorbed into N'. The normalization constant N' can in turn be determined by demanding that

$$\lim_{\omega \to 0} N'(\det \mathcal{O})^{-1/2} = \left(\frac{m}{2\pi i(t_f - t_i)}\right)^{1/2} \quad (3.64)$$

i.e. that the free particle result be obtained in the limit $\omega \to 0$. Hence

$$N'(\det \mathcal{O})^{-1/2} = D_F(0, t_f; 0, t_i) = \left(\frac{m\omega}{2\pi i \sin\omega(t_f - t_i)}\right)^{1/2} \quad (3.65)$$

which agrees with Eq. 3.53 found via the completeness property.

Wavefunction Connection

We can explicitly verify that Eq. 3.34 represents the correct form of the harmonic oscillator propagator by comparing with the sum over eigenstates, Eq. 3.4. We first examine the spectrum by taking the coordinate space trace

$$\int_{-\infty}^{\infty} dx\, D_F(x, t; x, 0) = \sum_n e^{-iE_n t} \int_{-\infty}^{\infty} dx\, \phi_n^*(x)\phi_n(x)$$
$$= \sum_n e^{-iE_n t} \quad . \quad (3.66)$$

Using Eq. 3.34 we have

$$\int_{-\infty}^{\infty} dx\, D_F(x, t; x, 0) = \int_{-\infty}^{\infty} dx\, \sqrt{\frac{m\omega}{2\pi i \sin\omega t}} \exp(-im\omega x^2 \tan\frac{\omega t}{2})$$
$$= \sqrt{\frac{m\omega}{2\pi i \sin\omega t}} \times \sqrt{\frac{\pi}{im\omega \tan\frac{\omega t}{2}}} = \frac{1}{2i\sin\frac{\omega t}{2}} \quad (3.67)$$
$$= \sum_{n=0}^{\infty} e^{-i\frac{\omega}{2}(2n+1)t} \quad .$$

Thus

$$E_n = \omega(n + \frac{1}{2}) \qquad n = 0, 1, 2, \ldots.$$

as expected. The wavefunctions may be obtained by expansion of the propagator itself:

$$\begin{aligned}D_F(x,t;x,0) &= \sqrt{\frac{m\omega}{\pi}} e^{-i\frac{\omega t}{2}} \left(1 + e^{-2i\omega t} + e^{-4i\omega t} + \ldots\right)^{1/2} \\ &\quad \times \exp\left[-m\omega x^2 \left(1 - e^{-i\omega t}\right)\left(1 - e^{-i\omega t} + e^{-2i\omega t} - \ldots\right)\right] \\ &= \left(\frac{m\omega}{\pi}\right)^{1/2} \exp\left(-m\omega x^2\right) \left\{e^{-i\frac{\omega t}{2}} + 2m\omega x^2 e^{-i\frac{3\omega t}{2}} + \ldots\right\}.\end{aligned} \quad (3.68)$$

Comparing with Eq. 3.4 we identify

$$\begin{aligned}\phi_0(x) &= \left(\frac{m\omega}{\pi}\right)^{1/4} \exp\left(-\frac{m\omega x^2}{2}\right) \\ \phi_1(x) &= \sqrt{2m\omega}\, x \phi_0(x)\end{aligned} \quad (3.69)$$

etc., as the familiar harmonic oscillator wavefunctions.

PROBLEM I.3.1

Operator Solution of the Harmonic Oscillator

i) Demonstrate, using the commutation relations between $\hat{a}$, $\hat{a}^\dagger$ that

$$|n\rangle = \frac{1}{\sqrt{n!}} \left(\hat{a}^\dagger\right)^n |0\rangle$$

is a normalized eigenfunction of the harmonic oscillator.
Also, we know from elementary quantum mechanics that

$$\langle x|n\rangle = \left(\frac{m\omega}{\pi}\right)^{1/4} \frac{1}{\sqrt{2^n n!}} H_n\left(\sqrt{m\omega}\, x\right) \exp\left(-\frac{1}{2}m\omega x^2\right)$$

where $H_n(x)$ is a Hermite polynomial.

ii) Use these results to prove the recursion relation

$$\left(\frac{d}{dz} - 2z\right) H_n(z) = -H_{n+1}(z).$$

iii) Use the recursion relation and $H_0(z) = 1$ to generate the first four Hermite polynomials.

PROBLEM I.3.2

Propagator for the Linear Potential

In order to gain experience dealing with the Feynman path integral, consider a particle of mass m moving in one dimension under the influence of a constant gravitational acceleration g. The corresponding potential energy is

$$V(x) = mgx$$

and the Lagrangian becomes

$$L = \frac{1}{2}m\dot{x}^2 - mgx \ .$$

The propagator for such a situation is exactly calculable and can be evaluated in at least two different ways:

i) Solve for the classical trajectory $x_{cl}(t)$ which obeys

$$\frac{d}{dt}\frac{\partial L}{\partial \dot{x}_{cl}} = \frac{\partial L}{\partial x_{cl}}$$

and calculate the classical action

$$S[x_{cl}(t)] = \int_{t_1}^{t_2} dt' \, L(x_{cl}(t'), \dot{x}_{cl}(t'))$$

for a path satisfying the boundary conditions

$$x_{cl}(t_1) = x_1 \qquad x_{cl}(t_2) = x_2 \ .$$

Show that

$$S[x_{cl}(t)] = \frac{m}{2}\frac{(x_2-x_1)^2}{t_2-t_1} - \frac{mg}{2}(t_2-t_1)(x_2+x_1) - \frac{mg^2}{24}(t_2-t_1)^3 \ .$$

Now imagine performing the path integral by expanding about this trajectory

$$x(t) = x_{cl}(t) + \delta x(t) \ .$$

Demonstrate that

$$D_F(x_2, t_2; x_1, t_1) = J(t_2-t_1)\exp(iS[x_{cl}(t)]) \ .$$

Here

$$J(t_2-t_1) = \int \mathcal{D}[\delta x(t)] \exp\left(i\int_{t_1}^{t_2} dt' \frac{1}{2}m(\delta\dot{x}(t'))^2\right)$$
$$= D_F^{(0)}(0, t_2; 0, t_1)$$

where $D_F^{(0)}$ is the free propagator. However, we already know the form of the free propagator, yielding

$$J(t_2-t_1) = \sqrt{\frac{m}{2\pi i(t_2-t_1)}} \ .$$

Thus the propagator for the linear potential takes the form

$$D_F(x_2, t_2; x_1, t_1) = \sqrt{\frac{m}{2\pi i(t_2-t_2)}} \exp(iS[x_{cl}(t)]) \ .$$

ii) Evaluate the path integral by breaking the paths into infinitesimal time slices and performing successive integrations, *i.e.*

$$D_F(x_2, t_2; x_1, t_1) = \lim_{n \to \infty} \left(\frac{m}{2\pi i \epsilon}\right)^{n/2} \prod_{i=1}^{n-1} \left(\int_{-\infty}^{\infty} dx_i\right)$$

$$\times \exp i \left[\frac{m}{2} \sum_{j=1}^{n-1} \frac{(x_j - x_{j-1})^2}{\epsilon} - \sum_{j=1}^{n-1} \epsilon m g x_j\right].$$

Demonstrate that the form after k integrations is

$$\left(\frac{m}{2\pi i(k+1)\epsilon}\right)^{1/2} \exp\left[\frac{im}{2\epsilon(k+1)}(x_{k+1} - x_0)^2 - i\epsilon m g \frac{k}{2}(x_{k+1} + x_0)\right.$$

$$\left. - i\frac{\epsilon^3}{24} k(k+1)(k+2) m g^2\right]$$

and that this expression reduces to the previously derived result in the limit that $k \to \infty$.

PROBLEM I.3.3

The Forced Harmonic Oscillator

An example of a problem which one can solve exactly to obtain the propagator is that of a harmonic oscillator which is acted upon by an external time varying force $j(t)$. The corresponding Lagrangian is

$$L(x, \dot{x}, t) = \frac{1}{2} m \dot{x}^2 - \frac{1}{2} m \omega^2 x^2 + x j(t) \ .$$

i) Show that the classical solution to this problem which satisfies the boundary conditions

$$x(t_1) = x_1 \qquad x(t_2) = x_2$$

is given by

$$x_{\mathrm{cl}}(t) = x_1 \cos\omega(t - t_1)$$

$$+ \frac{x_2 - x_1 \cos\omega(t_2 - t_1) - \int_{t_1}^{t_2} dt' \frac{1}{m\omega} j(t') \sin\omega(t_2 - t')}{\sin\omega(t_2 - t_1)} \sin\omega(t - t_1)$$

$$+ \int_{t_1}^{t} dt' \frac{1}{m\omega} j(t') \sin\omega(t - t') \ .$$

ii) Demonstrate that the corresponding action associated with this path is given by

$$S[x_{\mathrm{cl}}(t)] = \frac{m\omega}{2 \sin\omega(t_2 - t_1)} \left[(x_2^2 + x_1^2) \cos\omega(t_2 - t_1) - 2x_2 x_1\right.$$

$$+ 2\frac{x_2}{m\omega} \int_{t_1}^{t_2} dt\, j(t) \sin\omega(t - t_1) + 2\frac{x_1}{m\omega} \int_{t_1}^{t_2} dt\, j(t) \sin\omega(t_2 - t)$$

$$\left. - \frac{2}{m^2\omega^2} \int_{t_1}^{t_2} dt \int_{t_1}^{t} ds\, j(t) j(s) \sin\omega(t_2 - t) \sin\omega(s - t_1)\right] \ .$$

iii) Show by expanding an arbitrary path $x(t)$ as

$$x(t) = x_{cl}(t) + \delta x(t)$$

that

$$D_F^j(x_2, t_2; x_1, t_1) = J(t_2 - t_1) \exp iS[x_{cl}(t)]$$

where

$$J(t_2 - t_1) = \sqrt{\frac{m\omega}{2\pi i \sin\omega(t_2 - t_1)}}$$

is the prefactor for the free harmonic oscillator.

PROBLEM I.3.4

Wavefunction of a Particle Acted Upon by a Constant Force

Consider a particle which obeys the Schrödinger equation

$$\left(-\frac{1}{2m}\frac{\partial^2}{\partial x^2} + xF\right)\psi(x,t) = i\frac{\partial}{\partial t}\psi(x,t)$$

corresponding to a constant force $-F$.

i) Show that classically, if at time $t = 0$ the particle is at the origin with momentum p and energy $p^2/2m$, then at later time t the momentum becomes $p - Ft$ while the energy remains unchanged.

ii) Construct the quantum mechanical solution to this problem. That is, if at time $t = 0$ the wavefunction is

$$\psi(x, 0) = e^{ipx}$$

show, using the constant force propagator, that at time t the wavefunction assumes the form

$$\psi(x, t) = e^{i(p-Ft)x - i\frac{p^2}{2m}t} e^{i\phi}$$

where

$$\phi = \frac{p}{2m}Ft^2 - \frac{F^2}{6m}t^3$$

is a (time dependent) phase factor.

A formal prescription for the time development is, of course,

$$\psi(x, t) = \exp(-iH(x)t)\psi(x, 0) \ .$$

However, since

$$\left[\frac{\partial^2}{\partial x^2}, xF\right] \neq 0$$

we have

$$\exp(-iH(x)t) \neq \exp\left(i\frac{t}{2m}\frac{\partial^2}{\partial x^2}\right)\exp(-iFtx) \ .$$

Nevertheless, we can still utilize this formalism. One can prove that if $\hat{A}, \hat{B}$ are operators such that
$$[\hat{A}, \hat{B}] = \hat{C} \neq 0$$
but
$$\left[\hat{A}, [\hat{C}, \hat{A}]\right] = \left[\hat{A}, [\hat{C}, \hat{B}]\right] = \left[\hat{B}, [\hat{C}, \hat{B}]\right] = 0$$
then
$$\exp(\hat{A} + \hat{B}) = e^{\hat{A}} e^{\hat{B}} e^{-\frac{\hat{C}}{2}} e^{\frac{[\hat{B},\hat{C}]}{3}} e^{\frac{[\hat{A},\hat{C}]}{6}} .$$

iii) Prove this by expanding both sides and comparing.

iv) Using this result demonstrate that the time development of the wavefunction
$$\psi(x, 0) = e^{ipx}$$
agrees with that found via use of the propagator in part ii).

I.4 TIME-DEPENDENT PERTURBATION THEORY

The number of exactly soluble problems in quantum mechanics (and hence the number of exactly soluble propagators) is limited to the familiar cases

i) Free particle

ii) Harmonic oscillator

iii) Hydrogen atom

iv) $\cosh^{-2} ax$ potential, *etc.*

known from elementary quantum texts. For application to realistic situations we require a technique which is applicable to a more general class of problems. A common case is that wherein the full Hamiltonian $\hat{H}$ can be separated into two components — $\hat{H}_0$ for which exact solutions are available and $\hat{V}$ which is in some sense small – *e.g.*

$$\frac{\left\langle \phi | \hat{V} | \psi \right\rangle}{\left\langle \phi | \hat{H}_0 | \psi \right\rangle} \ll 1 . \tag{4.1}$$

In this case one can use the methods of time dependent perturbation theory, which we now develop.

TDPT and the Interaction Representation

An exact form for the time development operator is, of course, known (*cf.* sect. I.2)

$$\hat{U}(t, 0) = e^{-i(\hat{H}_0 + \hat{V})t} \theta(t) . \tag{4.2}$$

If $\hat{V}$ is small, it is natural to attempt some sort of Taylor series expansion of the exponential. However, this is not directly feasible since in general $\hat{H}_0$ and $\hat{V}$ do not commute. The solution is provided by going to the "interaction representation."

That is, if $|\psi_S(t)\rangle$ defines the usual state in the Schrödinger representation, the corresponding interaction representation state is defined to be

$$|\psi_I(t)\rangle \equiv e^{i\hat{H}_0 t}|\psi_S(t)\rangle \quad . \tag{4.3}$$

We introduce a time development operator in the interaction representation as

$$|\psi_I(t)\rangle = \hat{U}_I(t,0)|\psi_I(0)\rangle \tag{4.4}$$

in analogy to

$$|\psi_S(t)\rangle = \hat{U}_S(t,0)|\psi_S(0)\rangle \quad . \tag{4.5}$$

Since

$$i\frac{\partial}{\partial t}|\psi_S(t)\rangle = \hat{H}|\psi_S(t)\rangle \tag{4.6}$$

we find that $\hat{U}_S(t,0)$ obeys the differential equation

$$i\frac{\partial}{\partial t}\hat{U}_S(t,0) = \hat{H}\hat{U}_S(t,0) \quad . \tag{4.7}$$

Similarly we have

$$\begin{aligned} i\frac{\partial}{\partial t}|\psi_I(t)\rangle &= i\frac{\partial}{\partial t}e^{i\hat{H}_0 t}|\psi_S(t)\rangle = e^{i\hat{H}_0 t}\left(-\hat{H}_0 + i\frac{\partial}{\partial t}\right)|\psi_S(t)\rangle \\ &= e^{i\hat{H}_0 t}\hat{V}e^{-i\hat{H}_0 t}|\psi_I(t)\rangle \end{aligned} \tag{4.8}$$

so that

$$i\frac{\partial}{\partial t}\hat{U}_I(t,0) = \hat{V}_I(t)\hat{U}_I(t,0) \quad . \tag{4.9}$$

where we have defined

$$\hat{V}_I(t) \equiv e^{i\hat{H}_0 t}\hat{V}e^{-i\hat{H}_0 t} \tag{4.10}$$

We can convert Eq. 4.9 to an integral equation via

$$\hat{U}_I(t,0) = \mathbf{1} - i\int_0^t dt'\, \hat{V}_I(t')\hat{U}_I(t',0) \quad . \tag{4.11}$$

with the unit operator being required by the boundary condition $\hat{U}_I(t,0) \xrightarrow[t\to 0^+]{} \mathbf{1}$. Eq. 4.11 can then be solved by successive iteration, yielding an expansion in "powers" of the interaction potential $\hat{V}_I$.

$$\begin{aligned} \hat{U}_I(t,0) &= \mathbf{1} - i\int_0^t dt_1 \hat{V}_I(t_1) + (-i)^2 \int_0^t dt_2 \int_0^{t_2} dt_1 \hat{V}_I(t_2)\hat{V}_I(t_1) + \ldots \\ &= \sum_{n=0}^{\infty} \hat{U}_I^{(n)}(t,0) \end{aligned} \tag{4.12}$$

where

$$\hat{U}_I^{(0)}(t,0) = 1$$
$$\hat{U}_I^{(1)}(t,0) = -i \int_0^t dt_1 \hat{V}_I(t_1) \qquad (4.13)$$
$$\vdots$$
$$\hat{U}_I^{(n)}(t,0) = (-i)^n \int_0^t dt_n \int_0^{t_n} dt_{n-1} \ldots \int_0^{t_2} dt_1 \hat{V}_I(t_n)\hat{V}_I(t_{n-1})\ldots \hat{V}_I(t_1) \ .$$

Note the time ordering here

$$t \geq t_n \geq t_{n-1} \ldots \geq t_1 \geq 0 \ . \qquad (4.14)$$

This ordering is very important, as the $\hat{V}_I(t)$ do *not* commute.

The relationship between $\hat{U}_I(t,0)$ and its Schrödinger counterpart $\hat{U}_S(t,0)$ is given via

$$\hat{U}_I(t,t')|\psi_I(t')\rangle = |\psi_I(t)\rangle = e^{i\hat{H}_0 t}|\psi_S(t)\rangle$$
$$= e^{i\hat{H}_0 t}\hat{U}_S(t,t')|\psi_S(t')\rangle = e^{i\hat{H}_0 t}\hat{U}_S(t,t')e^{-i\hat{H}_0 t'}|\psi_I(t')\rangle \ . \qquad (4.15)$$

Since $|\psi(t')\rangle$ is arbitrary, we require

$$\hat{U}_I(t,t') = e^{i\hat{H}_0 t}\hat{U}_S(t,t')e^{-i\hat{H}_0 t'} \qquad (4.16a)$$

or

$$\hat{U}_S(t,t') = e^{-i\hat{H}_0 t}\hat{U}_I(t,t')e^{i\hat{H}_0 t'} \qquad (4.16b)$$

so that we can write a corresponding perturbation series for $\hat{U}_S(t,t')$

$$\hat{U}_S(t,t') = \sum_{n=0}^{\infty} \hat{U}_S^{(n)}(t,t') = \sum_{n=0}^{\infty} e^{-iH_0 t}\hat{U}_I^{(n)}(t,t')e^{i\hat{H}_0 t'} \qquad (4.17)$$

with

$$\hat{U}_S^{(0)}(t,t') = e^{-i\hat{H}_0(t-t')}$$
$$\hat{U}_S^{(1)}(t,t') = -i \int_{t'}^t dt_1 \, e^{-i\hat{H}_0(t-t_1)} \hat{V}(t_1) \, e^{-i\hat{H}_0(t_1-t')} \qquad (4.18)$$
$$\ldots$$

Transition Amplitude and Feynman Diagrams

We can also generate a series representation for the *amplitude* to make a transition from state $|i\rangle$ at $t = 0$ to state $|f\rangle$ at time t. We have (hereafter we shall remain in the Schrödinger representation but drop the subscript S)

$$\mathrm{Amp}_{fi}(t) = \left\langle f|\hat{U}(t,0)|i\right\rangle = \sum_{n=0}^{\infty} \mathrm{Amp}_{fi}^{(n)}(t) = \sum_{n=0}^{\infty} \left\langle f\left|\hat{U}^{(n)}(t,0)\right|i\right\rangle \ , \qquad (4.19)$$

It is extremely easy and useful to employ diagrams in order to give a pictorial history of the time evolution. We associate a line

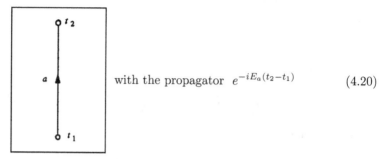

with the propagator $e^{-iE_a(t_2-t_1)}$ (4.20)

and a "vertex"

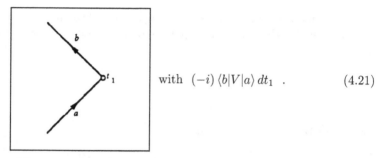

with $(-i)\langle b|V|a\rangle\, dt_1$. (4.21)

When an internal line a joins two vertices we speak of intermediate state $|a\rangle$, while i and f designate "external" lines. Each diagram represents an independent route by which the system can evolve from state $|i\rangle$ at time 0 to state $|f\rangle$ at time t. Each route has its own corresponding amplitude and the total amplitude is given by the sum over all possible routes

$$\text{Amp}_{\text{tot}} = \sum_{\text{route}} \text{Amp (route)} . \tag{4.22}$$

(Note the similarity here to the Feynman path integral in the sum over *all* trajectories.) We have then, assuming $\hat{V}$ to be independent of time,

$$\begin{aligned}
\text{Amp}_{fi}^{(0)}(t) &= \langle f|i\rangle\, e^{-iE_i t} \\
\text{Amp}_{fi}^{(1)}(t) &= -i \int_0^t dt_1\, e^{-iE_f(t-t_1)} \left\langle f|\hat{V}|i\right\rangle e^{-iE_i t_1} \\
&= -\frac{i}{i(E_f - E_i)} e^{-iE_f t} \left(e^{i(E_f - E_i)t} - 1 \right) \left\langle f|\hat{V}|i\right\rangle \\
&= \frac{1}{E_f - E_i} \left(e^{-iE_f t} - e^{-iE_i t} \right) \left\langle f|\hat{V}|i\right\rangle \\
\text{Amp}_{fi}^{(2)}(t) &= (-i)^2 \sum_a \int_0^t dt_2 \int_0^{t_2} dt_1\, e^{-iE_f(t-t_2)} \left\langle f|\hat{V}|a\right\rangle \\
&\quad \times e^{-iE_a(t_2-t_1)} \left\langle a|\hat{V}|i\right\rangle e^{-iE_i t_1}
\end{aligned} \tag{4.23}$$

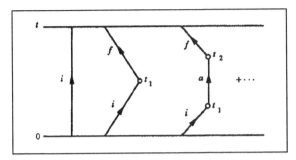

Fig. 1.3: Lowest order diagrams for the transition amplitude $\text{Amp}_{fi}(t)$.

etc., and may associate a "Feynman diagram" with each order of perturbation theory, as shown in Figure I.3.

TDPT and Frequency Space

It is also possible (and useful) to derive the time-dependent perturbation expansion using frequency space methods. We write the time development operator (assuming $\hat{V}$ is time-independent) as

$$\hat{U}(t,0)\theta(t) = e^{-i(\hat{H}_0+\hat{V})t}\theta(t)$$
$$= \int_{-infty}^{\infty} \frac{d\omega}{2\pi} e^{-i\omega t} \frac{i}{\omega - \hat{H}_0 - \hat{V} + i\epsilon} \equiv int_{-\infty}^{\infty} \frac{d\omega}{2\pi} e^{-i\omega t} \hat{\mathcal{K}}(\omega) \quad (4.24)$$

where we have defined the "full propagator" in frequency space as

$$\hat{\mathcal{K}}(\omega) = \frac{i}{\omega - \hat{H}_0 - \hat{V} + i\epsilon} \quad . \quad (4.25)$$

We may generate a perturbative expansion by use of the operator identity

$$\frac{1}{\hat{A}-\hat{B}} = \frac{1}{\hat{A}} + \frac{1}{\hat{A}}\hat{B}\frac{1}{\hat{A}} + \frac{1}{\hat{A}}\hat{B}\frac{1}{\hat{A}}\hat{B}\frac{1}{\hat{A}} + \ldots \quad (4.26)$$

[Note: The proof of Eq. 4.26 is provided by multiplication by $\hat{A} - \hat{B}$

$$(\hat{A}-\hat{B})\left[\frac{1}{\hat{A}} + \frac{1}{\hat{A}}\hat{B}\frac{1}{\hat{A}} + \frac{1}{\hat{A}}\hat{B}\frac{1}{\hat{A}}\hat{B}\frac{1}{\hat{A}} + \ldots\right]$$
$$= 1 - \hat{B}\frac{1}{\hat{A}} + \hat{B}\frac{1}{\hat{A}} - \hat{B}\frac{1}{\hat{A}}\hat{B}\frac{1}{\hat{A}} + \hat{B}\frac{1}{\hat{A}}\hat{B}\frac{1}{\hat{A}} - \ldots = 1]$$

The full propagator can then be written as

$$-i\hat{\mathcal{K}}(\omega) = \frac{1}{\omega - \hat{H}_0 - \hat{V} + i\epsilon} = \frac{1}{\omega - \hat{H}_0 + i\epsilon} + \frac{1}{\omega - \hat{H}_0 + i\epsilon}\hat{V}\frac{1}{\omega - \hat{H}_0 + i\epsilon}$$
$$+ \frac{1}{\omega - \hat{H}_0 + i\epsilon}\hat{V}\frac{1}{\omega - \hat{H}_0 + i\epsilon}\hat{V}\frac{1}{\omega - \hat{H}_0 + i\epsilon} + \ldots \quad (4.27)$$
$$= -i\left(\hat{K}^{(0)}(\omega) + \hat{K}^{(0)}(\omega)(-i\hat{V})\hat{K}^{(0)}(\omega)\right.$$
$$\left. + \hat{K}^{(0)}(\omega)(-i\hat{V})\hat{K}^{(0)}(\omega)(-i\hat{V})\hat{K}^{(0)}(\omega) + \ldots\right)$$

where
$$\hat{K}^{(0)}(\omega) = \frac{i}{\omega - \hat{H}_0 + i\epsilon} \tag{4.28}$$

is the free propagator. Returning to a real time rather than frequency representation we have

$$\hat{U}(t,0) = \int_{-\infty}^{\infty} \frac{d\omega}{2\pi} e^{-i\omega t} \hat{K}(\omega)$$
$$= \hat{U}^{(0)}(t,0) - i \int_0^t dt_1 \hat{U}^{(0)}(t,t_1) \hat{V} \hat{U}^{(0)}(t_1,0) \tag{4.29}$$
$$+ (-i)^2 \int_0^t dt_2 \int_0^{t_2} dt_1 \hat{U}^{(0)}(t,t_2) \hat{V} \hat{U}^{(0)}(t_2,t_1) \hat{V} \hat{U}^{(0)}(t_1,0) + \ldots$$

where
$$\hat{U}^{(0)}(t,0) = \int_{-\infty}^{\infty} \frac{d\omega}{2\pi} e^{-i\omega t} \frac{i}{\omega - \hat{H}_0 + i\epsilon} = \theta(t) e^{-i\hat{H}_0 t} \tag{4.30}$$

is the free propagator. [Eq. 4.29 follows via, e.g.

$$\int_{-\infty}^{\infty} \frac{d\omega}{2\pi} e^{-i\omega t} \frac{1}{\omega - \hat{H}_0 + i\epsilon} \hat{V} \frac{1}{\omega - \hat{H}_0 + i\epsilon} = \int_{-\infty}^{\infty} \frac{d\omega}{2\pi} \int_{-\infty}^{\infty} \frac{d\omega'}{2\pi} \int_{-\infty}^{\infty} dt_1 e^{-i\omega(t-t_1)}$$
$$\times e^{-i\omega' t_1} \frac{1}{\omega - \hat{H}_0 + i\epsilon} \hat{V} \frac{1}{\omega' - \hat{H}_0 + i\epsilon} = \int_{-\infty}^{\infty} dt_1 \hat{U}^{(0)}(t,t_1) \hat{V} \hat{U}^{(0)}(t_1,0)$$
$$= \int_0^t dt_1 e^{-i\hat{H}_0(t-t_1)} \hat{V} e^{-i\hat{H}_0 t_1}$$
$$\tag{4.31}$$

and similarly for the other terms.] Then for a transition from state $|i\rangle$ at time $t=0$ to state $|f\rangle$ at (later) time t we have

$$\text{Amp}_{fi}(t) = \left\langle f \left| \hat{U}(t,0) \right| i \right\rangle = i \int_{-\infty}^{\infty} \frac{d\omega}{2\pi} e^{-i\omega t} \left\{ \delta_{fi} \frac{1}{\omega - E_i + i\epsilon} \right.$$
$$+ \frac{1}{\omega - E_f + i\epsilon} \langle f|\hat{V}|i\rangle \frac{1}{\omega - E_i + i\epsilon} + \sum_a \frac{1}{\omega - E_f + i\epsilon} \langle f|\hat{V}|a\rangle$$
$$\times \frac{1}{\omega - E_a + i\epsilon} \langle a|\hat{V}|i\rangle \frac{1}{\omega - E_i + i\epsilon} + \ldots \right\}$$
$$= \delta_{fi} e^{-iE_i t} + \frac{\langle f|\hat{V}|i\rangle}{E_f - E_i + i\epsilon} \left(e^{-iE_f t} - e^{-iE_i t} \right) + \ldots$$
$$\tag{4.32}$$

as found previously (cf. Eq. 4.23).

TDPT and Path Integrals

It is particularly instructive to derive the time-dependent perturbation expansion within the path integral formalism, where we have

$$\left\langle x' \left| \hat{U}(t,0) \right| x \right\rangle = \int \mathcal{D}[x(t)] \exp i \int_0^t dt' \left(\frac{1}{2} m\dot{x}^2(t') - V(x(t')) \right)$$

$$= \int \mathcal{D}[x(t)] \exp i \int_0^t dt' \frac{1}{2} m\dot{x}^2(t')$$

$$\times \left(1 - i \int_0^t dt_1 V(x(t_1)) + \frac{1}{2!} \left(-i \int_0^t dt_1 V(x(t_1)) \right)^2 + \ldots \right) .$$
(4.33)

Note that here since we are dealing only with *classical* quantities — no operators — we do not have to worry about lack of commutativity and can expand the exponential involving the potential straightforwardly. The first term in the expansion is recognized to be the free particle propagator

$$\left\langle x' \left| \hat{U}^{(0)}(t,0) \right| x \right\rangle = \int \mathcal{D}[x(t)] \exp i \int_0^t dt' \frac{1}{2} m\dot{x}^2(t') .$$
(4.34)

For the term linear in the potential, we interchange orders of integration over time t_1 and paths $x(t)$ yielding

$$-i \int_0^t dt_1 \int \mathcal{D}[x(t)] V(x(t_1)) \exp i \int_0^t \frac{1}{2} m\dot{x}^2(t') dt' ,$$
(4.35)

and in the path integration we separate the paths into two pieces:

i) a path which begins at x at time $t=0$ and connects in all possible ways with point $x'' = x(t_1)$ at time t_1; and

ii) a path which begins at $x'' = x(t_1)$ at time t_1 and connects in all possible ways with point x' at time t.

Of course, x'' is not fixed but must take on all possible values, so that the linear term becomes

$$-i \int_0^t dt_1 \int_{-\infty}^{\infty} dx'' \left\langle x' \left| \hat{U}^{(0)}(t,t_1) \right| x'' \right\rangle V(x'') \left\langle x'' \left| \hat{U}^{(0)}(t_1,0) \right| x \right\rangle$$
(4.36)
$$= \text{(by completeness)} - i \int_0^t dt_1 \left\langle x' \left| \hat{U}^{(0)}(t,t_1) \hat{V} \hat{U}^{(0)}(t_1,0) \right| x \right\rangle .$$

Likewise we can analyze the quadratic term. In this case, however, we must divide the time integration into two regions depending on whether $t_1 > t_2$ or $t_1 < t_2$:

$$\frac{(-i)^2}{2!} \int_0^t dt_2 \int_0^{t_2} dt_1 \int_{-\infty}^{\infty} dx'' \int_{-\infty}^{\infty} dx''' \left\langle x' \left| \hat{U}^{(0)}(t,t_2) \right| x''' \right\rangle V(x''')$$

$$\times \left\langle x''' \left| \hat{U}^{(0)}(t_2,t_1) \right| x'' \right\rangle V(x'') \left\langle x'' \left| \hat{U}^{(0)}(t_1,0) \right| x \right\rangle$$

$$+ \frac{(-i)^2}{2!} \int_0^t dt_2 \int_{t_2}^t dt_1 \int_{-\infty}^{\infty} dx'' \int_{-\infty}^{\infty} dx''' \left\langle x' \left| \hat{U}^{(0)}(t,t_1) \right| x'' \right\rangle V(x'')$$
(4.37)

$$\times \left\langle x'' \left| \hat{U}^{(0)}(t_1,t_2) \right| x''' \right\rangle V(x''') \left\langle x''' \left| \hat{U}^{(0)}(t_2,0) \right| x \right\rangle .$$

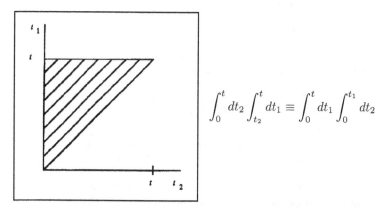

Fig. I.4: The t_1, t_2 integration region: Either integration can be performed first, provided the limits indicated above are used.

In the second term we can change the order of integration (*cf.* Figure I.4), and if we now interchange the identities of the variables $t_2 \leftrightarrow t_1$ the second term in Eq. 4.37 is seen to be identical to the first, canceling the 2!. The quadratic piece of the expansion becomes then

$$(-i)^2 \int_0^t dt_2 \int_0^{t_2} dt_1 \left\langle x' \left| \hat{U}^{(0)}(t,t_2) \hat{V}(t_2) \hat{U}^{(0)}(t_2,t_1) \hat{V}(t_1) \hat{U}^{(0)}(t_1,0) \right| x \right\rangle \quad (4.38)$$

and we recognize the general form of the propagator to be

$$\left\langle x' \left| \hat{U}(t,0) \right| x \right\rangle = \left\langle x' \left| \hat{U}^{(0)}(t,0) - i \int_0^t dt_1 U^{(0)}(t,t_1) \hat{V}(t_1) \hat{U}^{(0)}(t_1,0) \right. \right.$$
$$\left. \left. + (-i)^2 \int_0^t dt_2 \int_0^{t_2} dt_1 \hat{U}^{(0)}(t,t_2) \hat{V}(t_2) \hat{U}^{(0)}(t_2,t_1) \hat{V}(t_1) \hat{U}^{(0)}(t_1,0) + \ldots \right| x \right\rangle \quad (4.39)$$

which is identical in form to the perturbative expansion derived by the conventional technique—Eq. 4.18.

PROBLEM I.4.1

Time-Dependent Perturbation Theory and the Two State System [Sa 67]

Consider a two state system with unperturbed eigenfunctions

$$|\phi_1\rangle = \begin{pmatrix} 1 \\ 0 \end{pmatrix} \qquad |\phi_2\rangle = \begin{pmatrix} 0 \\ 1 \end{pmatrix}$$

such that the zeroeth order Hamiltonian is

$$H_0 = \begin{pmatrix} E_1 & 0 \\ 0 & E_2 \end{pmatrix}$$

with $E_1 < E_2$. Now add a time-dependent potential of the form

$$V = \begin{pmatrix} 0 & \gamma e^{i\omega t} \\ \gamma e^{-i\omega t} & 0 \end{pmatrix}$$

with γ real. Suppose that at $t = 0$ only the lower level is populated — i.e. expanding a general wavefunction as

$$|\psi(t)\rangle = c_1(t)e^{-iE_1 t}|\phi_1\rangle + c_2(t)e^{-iE_2 t}|\phi_2\rangle$$

take the initial conditions to be

$$c_1(0) = 1 \qquad c_2(t) = 0 .$$

i) Show that an exact solution of the Schrödinger equation requires

$$i\dot{c}_k = \sum_{n=1}^{2} V_{kn}(t) e^{i\omega_{kn} t} c_n \qquad (k = 1, 2)$$

with $\omega_{kn} = E_k - E_n$.

ii) Find $|c_1(t)|^2$ and $|c_2(t)|^2$ by exactly solving the coupled differential equation found in part i). Show that Rabi's formula

$$|c_2(t)|^2 = \frac{\gamma^2}{\gamma^2 + \frac{1}{4}(\omega - \omega_{21})^2} \sin^2\left(\sqrt{\gamma^2 + \frac{1}{4}(\omega - \omega_{21})^2}\, t\right)$$

$$|c_1(t)|^2 = 1 - |c_2(t)|^2$$

results.

iii) Now do the same problem using time-dependent perturbation theory to lowest non-vanishing order. Compare the two approaches for small values of γ. Treat the following two cases separately: a) ω very different from ω_{21} and b) ω close to ω_{21}.

PROBLEM I.4.2

Hydrogen Atom in a Time-Dependent Electric Field [DaM 86]

Suppose that a hydrogen atom in its ground state — $n = 1, m = 0$ — is placed between the plates of a capacitor and a time-dependent (but spatially uniform) electric field

$$\vec{E} = \begin{cases} 0 & t < 0 \\ \vec{E}_0 \exp(-t/\tau) & t > 0 \end{cases}$$

is applied.

i) Using first order time-dependent perturbation theory calculate the probability for the atom to be found at $t \gg \tau$ in each of the $2p$ states — $n, \ell, m = 2, 1, \pm 1 \atop 0$ — and in the $2s$ state — $n, \ell, m = 2, 0, 0$.

ii) Express your result in terms of the total energy of the perturbing field and discuss the relation between the time duration τ of the electric field "pulse" and the transition probability for fixed total power.

Suggestion: Use the electric dipole approximation to write the interaction Hamiltonian as
$$H = e\vec{E}(t) \cdot \vec{x}$$

PROBLEM I.4.3

The Forced Harmonic Oscillator: Time-Dependent Perturbation Approach

Consider a one-dimensional harmonic oscillator with mass m and frequency ω which is acted upon by an external force $F(t)$. The Hamiltonian is then of the form
$$H(x) = -\frac{1}{2m}\frac{\partial^2}{\partial x^2} + \frac{1}{2}m\omega^2 x^2 - xF(t) \ .$$

Suppose that $F(t)$ is non-vanishing only for $t_1 < t < t_2$ and consider evaluation of the probability to make a transition from the free oscillator state $|m\rangle$ at time $t' < t_1$ to the free oscillator state $|n\rangle$ at time $t > t_2$. This calculation has been performed by a number of authors (see, e.g. [FeH 65] Ch. 8.9), yielding

$$P_{m,n} = \begin{cases} \frac{m!}{n!} z^{n-m} e^{-z} \left(L_m^{n-m}(x)\right)^2 & n \geq m \\ P_{n,m} & n < m \end{cases}$$

where
$$L_m^{n-m}(z) = \sum_{k=0}^{m} \binom{n}{m-k} \frac{(-z)^k}{k!}$$

is a Laguerre polynomial and
$$z = \frac{|\gamma|^2}{2m\omega}$$

with
$$\gamma = \int_{t_1}^{t_2} dt\, F(t) e^{-i\omega t}$$

being the Fourier transform of the force. We expect that lowest order perturbation theory with
$$H_0(x) = -\frac{1}{2m}\frac{\partial^2}{\partial x^2} + \frac{1}{2}m\omega^2 x^2$$

and
$$V = -xF(t)$$

should give a reasonable representation of this transition probability provided that the force is weak — i.e. $z \ll 1$.

i) Show that in this limit the exact result reduces to

$$P_{m,n} = \begin{cases} z^{n-m} \frac{n!}{m!((n-m)!)^2} & n > m \\ 1 - (2m+1)z & n = m \\ P_{n,m} & n < m \end{cases}.$$

ii) Calculate the transition probability using first order time-dependent perturbation theory and show that

$$P^{(1)}_{m,n} = [(m+1)\delta_{n,m+1} + m\delta_{n,m-1}]z + \delta_{n,m}$$

Compare your answer with part i).

iii) Calculate the transition probability using second order time-dependent perturbation theory and show that

$$P^{(2)}_{m,n} = P^{(1)}_{m,n} + [(m+2)(m+1)\delta_{n,m+2} + m(m-1)\delta_{n,m-2}]\frac{z^2}{4} - (2m+1)z\delta_{n,m}.$$

Compare your answer with part i).

Hint for part iii): Prove the identity

$$g(\omega) = \int_{t_1}^{t_2} dt'' \int_{t_1}^{t''} dt' F(t'')F(t')e^{i\omega(t''-t')}$$
$$= \int_{t_1}^{t_2} dt'' \int_{t''}^{t_2} dt' F(t'')F(t')e^{-i\omega(t''-t')}.$$

Hence, show that

$$\text{Re}\, g(\omega) = \frac{1}{2}|\gamma|^2.$$

I.5 PROPAGATOR AS THE GREEN'S FUNCTION

Most readers have no doubt already have recognized the propagator as the Green's function for the Schrödinger equation, satisfying

$$\left(-\frac{1}{2m}\vec{\nabla}_2^2 + V(\vec{x}_2) - i\frac{\partial}{\partial t_2}\right) D_F(\vec{x}_2, t_2; \vec{x}_1, t_1) = -i\delta^3(\vec{x}_2 - \vec{x}_1)\delta(t_2 - t_1) \quad (5.1)$$

with the condition that

$$D_F(\vec{x}_2, t_2; \vec{x}_1, t_1) = 0 \quad \text{for} \quad t_2 < t_1. \quad (5.2)$$

This assertion is easily checked using the representation in terms of a sum over a complete set of states:

$$D_F(\vec{x}_2, t_2; \vec{x}_1, t_1) = \begin{cases} \sum_n \psi_n(\vec{x}_2)\psi_n^*(\vec{x}_1)e^{-iE_n(t_2-t_1)} & t_2 > t_1 \\ 0 & t_2 < t_1 \end{cases} \quad (5.3)$$

Here $\psi_n(\vec{x})$ is a solution to the time-independent Schrödinger equation

$$\left(-\frac{\vec{\nabla}^2}{2m} + V(\vec{x})\right)\psi_n(\vec{x}) = E_n\psi_n(\vec{x}) \tag{5.4}$$

so that

$$\left(-\frac{1}{2m}\vec{\nabla}_2^2 + V(\vec{x}_2) - i\frac{\partial}{\partial t_2}\right)\sum_n \psi_n(\vec{x}_2)\psi_n^*(\vec{x}_1)e^{-iE_n(t_2-t_1)} = 0 \quad \text{if} \quad t_2 > t_1 \; . \tag{5.5}$$

Also for $t_2 < t_1$ Eq. 5.1 is obviously satisfied, so that we need only verify the normalization associated with the step function. If we perform a time integration of Eq. 5.1 over the interval

$$t_1 - \epsilon < t_2 < t_1 + \epsilon \tag{5.6}$$

and subsequently take the limit as $\epsilon \to 0$, then on the right-hand side we have

$$\lim_{\epsilon \to 0} -i \int_{t_1-\epsilon}^{t_1+\epsilon} dt_2 \delta^3(\vec{x}_2 - \vec{x}_1)\delta(t_2 - t_1) = -i\delta^3(\vec{x}_2 - \vec{x}_1) \tag{5.7}$$

while on the left-hand side

$$\lim_{\epsilon \to 0} \int_{t_1-\epsilon}^{t_1+\epsilon} dt_2 \left(-\frac{\vec{\nabla}_2^2}{2m} + V(\vec{x}_2) - i\frac{\partial}{\partial t_2}\right) D_F(\vec{x}_2, t_2; \vec{x}_1, t_1)$$

$$= \lim_{\epsilon \to 0} \left[2\epsilon \left(-\frac{\vec{\nabla}_2^2}{2m} + V(\vec{x}_2)\right) D_F(\vec{x}_2, \bar{t}; \vec{x}_1, t_1) \right. \tag{5.8}$$

$$\left. - iD_F(\vec{x}_2, t_1 + \epsilon; \vec{x}_1, t_1) + iD_F(\vec{x}_2, t_1 - \epsilon; \vec{x}_1, t_1)\right]$$

$$= -iD_F(\vec{x}_2, t_1; \vec{x}_1, t_1)$$

where $t_1 - \epsilon < \bar{t} < t_1 + \epsilon$ and we have used the mean value theorem. Finally, because of the completeness of the set of states n we have

$$-iD_F(\vec{x}_2, t_1; \vec{x}_1, t_1) = -i\sum_n \psi_n(\vec{x}_2)\psi_n^*(\vec{x}_1) = -i\delta^3(\vec{x}_2 - \vec{x}_1) \tag{5.9}$$

so that our assertion—Eq. 5.1—is verified. This identification of the propagator with the Green's function will prove to be useful in the relativistic problem also, as we shall see later.

With this result a perturbation expansion is easily derived. If we define the free propagator $D_F^{(0)}(\vec{x}_2, t_2; \vec{x}_1, t_1)$ via

$$\left(-\frac{\vec{\nabla}_2^2}{2m} - i\frac{\partial}{\partial t_2}\right) D_F^{(0)}(\vec{x}_2, t_2; \vec{x}_1, t_1) = -i\delta^3(\vec{x}_2 - \vec{x}_1)\delta(t_2 - t_1) \tag{5.10}$$

then it is easy to see that an expression which satisfies the strictures required of the full propagator is[†]

$$D_F(\vec{x}_2,t_2;\vec{x}_1,t_1) = D_F^{(0)}(\vec{x}_2,t_2;\vec{x}_1,t_1)$$
$$- i \int d^4x_3\, D_F^{(0)}(\vec{x}_2,t_2;\vec{x}_3,t_3)V(\vec{x}_3)D_F(\vec{x}_3,t_3;\vec{x}_1,t_1) \quad (5.11)$$

which may be solved by iteration to yield the usual perturbative expansion

$$D_F(\vec{x}_2,t_2;\vec{x}_1,t_1) = D_F^{(0)}(\vec{x}_2,t_2;\vec{x}_1,t_1)$$
$$- i \int d^4x_3\, D_F^{(0)}(\vec{x}_2,t_2;\vec{x}_3,t_3)V(\vec{x}_3)D_F^{(0)}(\vec{x}_3,t_3;\vec{x}_1,t_1) \quad (5.12)$$
$$+ \ldots$$

I.6 FUNCTIONAL TECHNIQUES*

An alternative formulation of perturbation theory is provided by so-called functional techniques, which are based upon the Feynman path integral. Such methods are commonplace in modern quantum field theory and have a simple non-relativistic analog, as shown here.

The concept of a function is a familiar one in physics. Thus the function $f(t)$ involves a mapping from one set of numbers t to another set described by the function $f(t)$. Given a *number* t, application of $f(t)$ produces another *number*. The idea of a functional is not so familiar and involves a mapping from a function to a number. An example is provided by Feynman path integral for the forced oscillator as calculated in problem I.3.3. For fixed space-time points, x_2, t_2 and x_1, t_1 the value of the propagator is a *number* whose value depends on the functional form of the driving force $j(t)$. Thus $D_F^j(x_1,t_2;x_1,t_1)$ is said to be a *functional* of $j(t)$. The use of such a functional in perturbation theory applications will now be developed.

Large Time Limit

The importance of the large time limit of the propagator has already been explored in Section I.3 where we showed that

$$D_F(x_2,t_2;x_1,t_1) \underset{t_2-t_1\to\infty}{\sim} \psi_0(x_2)\psi_0^*(x_1)e^{-iE_0(t_2-t_1)} + \mathcal{O}\left(e^{-iE_1(t_2-t_1)}\right) \quad (6.1)$$

[†] Note that

$$\left(-\frac{\vec{\nabla}_2^2}{2m} - i\frac{\partial}{\partial t_2}\right)D_F(\vec{x}_2,t_2;\vec{x}_1,t_1) = -i\delta^3(\vec{x}_2-\vec{x}_1)\delta(t_2-t_1)$$

$$- \int d^4x_3\delta^3(\vec{x}_2-\vec{x}_3)\delta(t_2-t_3)V(\vec{x}_3)D_F(\vec{x}_3,t_3;\vec{x}_1,t_1)$$

$$= -i\delta^3(\vec{x}_2-\vec{x}_1)\delta(t_2-t_1) - V(\vec{x}_2)D_F(\vec{x}_2,t_2;\vec{x}_1,t_1)$$

Here $\psi_0(x), E_0$ are the ground state wavefunction, energy, respectively, and E_1 represents the energy of the first excited state. This procedure is often carried out by taking the "imaginary time" limit

$$t_2 \to -i\frac{T}{2} \quad , \quad t_1 \to i\frac{T}{2} \tag{6.2}$$

and expanding in the small quantity e^{-ET} for T large. Eq. 6.1 then assumes the form

$$D_F\left(x_2, -i\frac{T}{2}; x_1, i\frac{T}{2}\right) = \psi_0(x_2)\psi_0^*(x_1)e^{-E_0 T} + \mathcal{O}\left(e^{-E_1 T}\right) \tag{6.3}$$

which is of the familiar form of a perturbation expansion and is usually easier to employ than the real time technique. For example, for the harmonic oscillator

$$D_F\left(x_2, -i\frac{T}{2}; x_1, i\frac{T}{2}\right) = \left(\frac{m\omega}{\pi\left(e^{\omega T} - e^{-\omega T}\right)}\right)^{1/2} \exp\left[-\frac{m\omega}{e^{\omega T} - e^{-\omega T}}\right.$$
$$\left. \times \left(\frac{1}{2}\left(x_2^2 + x_1^2\right)\left(e^{\omega T} + e^{-\omega T}\right) - 2x_1 x_2\right)\right] \tag{6.4}$$

$$= \left(\frac{m\omega}{\pi}\right)^{1/2} \exp\left(-\frac{1}{2}m\omega\left(x_1^2 + x_2^2\right)\right) e^{-\frac{1}{2}\omega T}\left(1 + 2m\omega x_1 x_2 e^{-\omega T} + \ldots\right)$$

whereby we identify

$$\psi_0(x) = \left(\frac{m\omega}{\pi}\right)^{1/4} \exp\left(-\frac{1}{2}m\omega x^2\right) \quad , \quad E_0 = \frac{1}{2}\omega$$
$$\psi_1(x) = \sqrt{2m\omega}\, x\psi_0(x) \quad , \quad E_1 = \frac{3}{2}\omega \tag{6.5}$$

etc., as before. Matrix elements can also be evaluated. For example, we note

$$\int_{-\infty}^{\infty} dx\, x^2 D_F\left(x, -i\frac{T}{2}; x, i\frac{T}{2}\right) = \left(\frac{m\omega}{2\pi \sinh \omega T}\right)^{\frac{1}{2}} \left(\frac{\pi}{4m^3\omega^3}\operatorname{ctnh}^3 \frac{1}{2}\omega T\right)^{\frac{1}{2}}$$
$$= \frac{1}{2m\omega} e^{-\frac{1}{2}\omega T}\left(1 + 3e^{-\omega T} + 5e^{-2\omega T} + \ldots\right) \quad . \tag{6.6}$$

Comparison with the form

$$\int_{-\infty}^{\infty} dx\, x^2 D_F\left(x, i\frac{T}{2}; x, i\frac{T}{2}\right) = \sum_{n=0}^{\infty} e^{-E_n T} \int_{-\infty}^{\infty} dx\, x^2 |\psi_n(x)|^2 \tag{6.7}$$

yields

$$\langle n|\hat{x}^2|n\rangle = \int_{-\infty}^{\infty} dx\, x^2 |\psi_n(x)|^2 = \frac{2n+1}{2m\omega} \tag{6.8}$$

in agreement with the result obtained by more conventional techniques

$$\langle n|\hat{x}^2|n\rangle = \frac{1}{2m\omega}\langle n|(\hat{a}+\hat{a}^\dagger)(\hat{a}+\hat{a}^\dagger)|n\rangle = \frac{2n+1}{2m\omega} \quad . \tag{6.9}$$

In a similar fashion, non-diagonal matrix elements can be determined by use of the coordinate space creation and annihilation operators. We find, for example,

$$\int_{-\infty}^{\infty} dx\, x \sqrt{\frac{m\omega}{2}} \left(x + \frac{1}{m\omega}\frac{\partial}{\partial x}\right) D_F\left(x, -i\frac{T}{2}; x', i\frac{T}{2}\right)\bigg|_{x=x'}$$

$$= \left(\frac{m\omega}{2\pi \sinh \omega T}\right)^{\frac{1}{2}} \times \left(\frac{\pi}{8m^2\omega^2}\operatorname{ctnh}^3 \omega T\right)^{\frac{1}{2}} \tag{6.10}$$
$$\times [1 - \operatorname{ctnh}\omega T + \operatorname{csch}\omega T]$$

$$= \frac{1}{\sqrt{2m\omega}} e^{-\frac{3}{2}\omega T}\left(1 + 2e^{-\omega T} + 3e^{-2\omega T} + \ldots\right)\ .$$

Comparison with the form

$$\int_{-\infty}^{\infty} dx\, x \sqrt{\frac{m\omega}{2}} \left(x + \frac{1}{m\omega}\frac{\partial}{\partial x}\right) D_F\left(x, -i\frac{T}{2}; x', i\frac{T}{2}\right)\bigg|_{x=x'}$$

$$= \sum_{n=1}^{\infty} e^{-E_n T} \sqrt{n} \int_{-\infty}^{\infty} dx\, \psi_{n-1}^*(x) x \psi_n(x) \tag{6.11}$$

then yields

$$\langle n-1|\hat{x}|n\rangle = \int_{-\infty}^{\infty} dx\, \psi_{n-1}^*(x) x \psi_n(x) = \sqrt{\frac{n}{2m\omega}} \tag{6.12}$$

in agreement with the result

$$\langle n-1|\hat{x}|n\rangle = \frac{1}{\sqrt{2m\omega}}\langle n-1|\hat{a}+\hat{a}^\dagger|n\rangle = \sqrt{\frac{n}{2\omega m}}\ . \tag{6.13}$$

The Generating Functional

Thus far we have done nothing new, having merely derived, in a different and somewhat more cumbersome fashion, results which are well-known from alternative procedures. This warm-up will serve us well, however, in our discussion of functional methods, which we now undertake. We begin by inserting a "driving term" $xj(t)$ into the Lagrangian. The equation of motion for the corresponding classical system becomes

$$m\frac{d^2x}{dt^2} + m\omega^2 x = j(t) \tag{6.14}$$

so that $j(t)$ represents an external force (or "current," hence the symbol j) that acts upon the oscillator. The modified path integral becomes

$$D_F^j(x_f, t_f; x_i, t_i) = \int \mathcal{D}[x(t)] \exp i \int_{t_i}^{t_f} dt \tag{6.15}$$
$$\times \left(\frac{1}{2}m\dot{x}^2(t) - \frac{1}{2}m\omega x^2(t) + x(t)j(t)\right)$$

where we assume that $j(t)$ vanishes at both endpoints. We recognize this as the propagator of the forced oscillator, which was evaluated in Problem I.3.3. It is sufficient for our purposes to determine its value in the limit as $t_2 \to \infty$ and $t_1 \to -\infty$, which isolates the ground state contribution. The resulting quantity, which we represent by $W[j(t)]$

$$W[j(t)] \underset{T \to \infty}{\sim} D_F^j(x_f, \frac{T}{2}; x_i, -\frac{T}{2}) \ . \tag{6.16}$$

is called the generating functional. As mentioned above, the term functional is used in that W depends not just upon the value of a single variable, but rather upon the shape of the function $j(t)$ over the entire real axis. We shall see that the generating functional contains all the information about the harmonic oscillator through its response to an arbitrary external probe.

The key to this procedure is the identity

$$(-i)^n \frac{\delta^n}{\delta j(t_1) \ldots \delta j(t_n)} W[j(t)] \bigg|_{j=0}$$
$$= \lim_{T \to \infty} \int \mathcal{D}[x(t)] \, x(t_n) \ldots x(t_1) \exp i \int_{-\frac{T}{2}}^{\frac{T}{2}} dt \left(\frac{1}{2} m\dot{x}^2(t) - \frac{1}{2} m\omega x^2(t) \right)$$
$$= \lim_{T \to \infty} \left\langle x_f, \frac{T}{2} \left| T\left(x(t_n) \ldots x(t_1)\right) \right| x_i, -\frac{T}{2} \right\rangle \tag{6.17}$$

where the symbol T denotes the time ordered product

$$T(A(t_2)B(t_1)) = \theta(t_2 - t_1)A(t_2)B(t_1) + \theta(t_1 - t_2)B(t_1)A(t_2) \tag{6.18}$$

and $\delta/\delta j(t)$ describes functional differentiation, as defined via

$$\frac{\delta j(t)}{\delta j(t')} = \frac{\delta}{\delta j(t')} \int dt' \delta(t-t')j(t') = \delta(t-t') \tag{6.19}$$

[One can prove Eq. 6.17 by choosing a particular ordering, say

$$t_i < t_1 < t_2 < \ldots < t_f$$

and noting that

$$\langle x_f, t_f | T(x(t_1) \ldots x(t_n)) | x_i, t_i \rangle = \langle x_f, t_f | x(t_n)x(t_{n-1}) \ldots x(t_1) | x_i, t_i \rangle$$
$$= \prod_{k=1}^{N} \int_{-\infty}^{\infty} dx_k \, \langle x_f, t_f | x_n, t_n \rangle x_n \, \langle x_n, t_n | x_{n-1}, t_{n-1} \rangle x_{n-1} \tag{6.20}$$
$$\times \ldots x_1 \, \langle x_1, t_1 | x_i, t_i \rangle$$

where we have used completeness and have defined $x_k = x(t_k)$ $(k = 1, 2, \ldots, n)$. The amplitudes $\langle x_k, t_k | x_{k-1}, t_{k-1} \rangle$ are simply free propagators and can be evaluated

1.6 FUNCTIONAL TECHNIQUES*

by means of time slice methods. Thus the above expression is identical to Eq. 6.17. In the case of a different time ordering the same result goes through provided one places the times such that the later time always appears to the left of an earlier counterpart. However, this is simply the definition of the time ordered product and hence the proof holds in general.]

Before seeing how this identity can be utilized, we need the form of $W[j(t)]$, which can be easily obtained by means of Fourier transform techniques— (equivalently cf. Problem I.6.1). We define

$$x(t) = \int_{-\infty}^{\infty} \frac{dE}{2\pi} e^{-iEt} \tilde{x}(E)$$
$$j(t) = \int_{-\infty}^{\infty} \frac{dE}{2\pi} e^{-iEt} \tilde{j}(E)$$
(6.21)

so that

$$S[x(t)] = \int_{-\infty}^{\infty} dt \left(\frac{1}{2} m \dot{x}^2(t) - \frac{1}{2} m \omega^2 x^2(t) + x(t) j(t) \right)$$
$$= \int_{-\infty}^{\infty} \frac{dE}{2\pi} \left(\frac{m}{2} \left(E^2 - \omega^2 \right) \tilde{x}(E) \tilde{x}(-E) + \frac{1}{2} \tilde{j}(E) \tilde{x}(-E) + \frac{1}{2} \tilde{j}(-E) \tilde{x}(E) \right).$$
(6.22)

Completing the square, we find

$$S[x(t)] = \int_{-\infty}^{\infty} \frac{dE}{2\pi} \left[\frac{m}{2} \left(E^2 - \omega^2 \right) \left(\tilde{x}(E) + \frac{1}{m(E^2 - \omega^2 + i\epsilon)} \tilde{j}(E) \right) \right.$$
$$\left. \times \left(\tilde{x}(-E) + \frac{1}{m(E^2 - \omega^2 + i\epsilon)} \tilde{j}(-E) \right) - \frac{1}{2m} \tilde{j}(E) \frac{1}{E^2 - \omega^2 + i\epsilon} \tilde{j}(-E) \right]$$
(6.23)

where the $+i\epsilon$ has been inserted to assure convergence. Defining the Green's function

$$D(t - t') = \int_{-\infty}^{\infty} \frac{dE}{2\pi} e^{-iE(t-t')} \frac{1}{E^2 - \omega^2 + i\epsilon}$$
(6.24)

which satisfies

$$\left(\frac{d^2}{dt^2} + \omega^2 \right) D(t - t') = -\delta(t' - t)$$
(6.25)

we have

$$S[x(t)] = \int_{-\infty}^{\infty} dt \left(\frac{1}{2} m \dot{x}'^2(t) - \frac{1}{2} m \omega^2 x'^2(t) \right)$$
$$- \frac{1}{2m} \int_{-\infty}^{\infty} dt \int_{-\infty}^{\infty} dt' j(t) D(t - t') j(t')$$
(6.26)

where

$$x'(t) \equiv x(t) + \frac{1}{m} \int_{-\infty}^{\infty} dt' D(t - t') j(t') .$$
(6.27)

Now change the path integration "variable" from $x(t)$ to $x'(t)$. Since the path integral involves a sum over *all* $x(t)$ the Jacobian of the change of variables is unity, i.e.

$$\int \mathcal{D}[x(t)] = \int \mathcal{D}[x'(t)] \tag{6.28}$$

and

$$W[j(t)] = \exp\left(-\frac{i}{2m}\int_{-\infty}^{\infty}dt\int_{-\infty}^{\infty}dt'\, j(t)D(t-t')j(t')\right)$$
$$\times \int \mathcal{D}[x'(t)]\exp i\int_{-\infty}^{\infty} dt\left(\frac{1}{2}m\dot{x}'^2 - \frac{1}{2}m\omega'^2 x'^2(t)\right) . \tag{6.29}$$

The path integration above is simply that for a harmonic oscillator in the absence of a driving force $j(t)$ so that our final result for the generating functional is

$$W[j(t)] = W[0]\exp -\frac{i}{2m}\int_{-\infty}^{\infty}dt\int_{-\infty}^{\infty}dt'\, j(t)D(t-t')j(t') . \tag{6.30}$$

The meaning of this quantity is that

$$W[j(t)] = \lim_{T\to\infty}\left\langle x_f, \frac{T}{2}\Big|x_i, -\frac{T}{2}\right\rangle_{j(t)}$$
$$= \lim_{T\to\infty}\left\langle x_f, \frac{T}{2}\Big|x_i, -\frac{T}{2}\right\rangle_{j=0}\exp -\frac{i}{2m}\int_{-\frac{T}{2}}^{\frac{T}{2}}dt\int_{-\frac{T}{2}}^{\frac{T}{2}}dt'\, j(t)D(t-t')j(t') . \tag{6.31}$$

That is, the transition amplitude in the presence of a driving current $j(t)$ is equal to the corresponding amplitude for the free oscillator times a simple phase factor.

The Green's function $D(t-t')$ can be evaluated by the usual contour integration methods. If $t > t'$ we must close the contour in the lower half plane in which case there exists a contribution from the pole at $E = \omega - i\epsilon$. On the other hand, if $t < t'$ the contour must be closed in the upper half plane and the pole at $E = -\omega + i\epsilon$ is the one which contributes. In general we have

$$D(t-t') = \int_{-\infty}^{\infty}\frac{dE}{2\pi}e^{-iE(t-t')}\frac{1}{E^2 - \omega^2 + i\epsilon}$$
$$= -\frac{i}{2\omega}\theta(t-t')e^{-i\omega(t-t')} - \frac{i}{2\omega}\theta(t'-t)e^{-i\omega(t'-t)} \tag{6.32}$$
$$= -\frac{i}{2\omega}e^{-i\omega|t-t'|} .$$

Oscillator Matrix Elements

Armed with an explicit form for the generating functional, let's see how it can be employed. One simple application involves the calculation of free oscillator matrix elements. For example, since taking the limit $t_f \to \infty$, $t_i \to -\infty$ projects out the ground state amplitudes we obtain

$$(-i)^2\frac{1}{W[0]}\frac{\delta^2 W[j]}{\delta j(t_2)\delta j(t_1)}\bigg|_{j=0} = \frac{\langle\psi_0|T(x(t_2)x(t_1))|\psi_0\rangle}{\langle\psi_0|\psi_0\rangle}$$
$$= \frac{i}{m}D(t_2 - t_1) = \frac{1}{2m\omega}e^{-i\omega|t_2-t_1|} . \tag{6.33}$$

Taking the limit $t_2 \to t_1$ we reproduce the familiar result (cf. Eq. 6.8)

$$\langle 0|\hat{x}^2|0\rangle = \frac{1}{2m\omega} . \qquad (6.34)$$

In similar fashion arbitrary ground state matrix elements can be found.

Despite the apparent restriction to the ground state in $W[j]$ excited state matrix elements can also be obtained by reducing them to ground state results using functional analogs of creation and annihilation operators. For example, the matrix element

$$\langle 0|\hat{x}|1\rangle = \left\langle 0 \left| \frac{1}{\sqrt{2m\omega}} (\hat{a} + \hat{a}^\dagger) \right| 1 \right\rangle = \frac{1}{\sqrt{2m\omega}} \qquad (6.35)$$

can be written as

$$\langle 0|\hat{x}a^\dagger|0\rangle = \left\langle 0 \left| \hat{x} \sqrt{\frac{m\omega}{2}} \left(\hat{x} - i\frac{1}{m\omega}\hat{p} \right) \right| 0 \right\rangle . \qquad (6.36)$$

Using the classical analog $p = m\dot{x}$ we find

$$\langle 0|\hat{x}|1\rangle = \lim_{t_2 \to t_1^+} (-i)^2 \sqrt{\frac{m\omega}{2}} \left(1 - \frac{i}{\omega}\frac{\partial}{\partial t_1}\right) \frac{\delta^2}{\delta j(t_1)\delta j(t_2)} \frac{W[j]}{W[0]}\bigg|_{j=0} \qquad (6.37)$$

$$= \lim_{t_2 \to t_1^+} \sqrt{\frac{m\omega}{2}} \left(1 - \frac{i}{\omega}\frac{\partial}{\partial t_1}\right) \frac{i}{m} D(t_2 - t_1) = \frac{1}{\sqrt{2m\omega}} .$$

Similarly

$$\langle 0|\hat{p}|1\rangle = \lim_{t_2 \to t_1^+} m\frac{\partial}{\partial t_2} \sqrt{\frac{m\omega}{2}} \left(1 - \frac{i}{\omega}\frac{\partial}{\partial t_1}\right) \frac{i}{m} D(t_2 - t_1) = -i\sqrt{\frac{m\omega}{2}} \qquad (6.38)$$

and even more complex results may be found, e.g.

$$\langle 1|\hat{x}^2|1\rangle = \frac{m\omega}{2} \lim_{\substack{t_1 \to t^+ \\ t_2 \to t^-}} \left(1 + \frac{i}{\omega}\frac{\partial}{\partial t_1}\right)\left(1 - \frac{i}{\omega}\frac{\partial}{\partial t_2}\right) \langle 0|x(t_1)x^2(t)x(t_2)|0\rangle$$

$$= (-i)^4 \lim_{\substack{t_1 \to t^+ \\ t_2 \to t^-}} \left(\frac{m\omega}{2}\right) \left(1 + \frac{i}{\omega}\frac{\partial}{\partial t_1}\right)\left(1 - \frac{i}{\omega}\frac{\partial}{\partial t_2}\right)$$

$$\times \frac{\delta^4}{\delta j(t_1)\ldots\delta j(t)} \frac{W[j]}{W[0]}\bigg|_{j=0} \qquad (6.39)$$

$$= \frac{m\omega}{2} \lim_{\substack{t_1 \to t^+ \\ t_2 \to t^-}} \left(1 + \frac{i}{\omega}\frac{\partial}{\partial t_1}\right)\left(1 - \frac{i}{\omega}\frac{\partial}{\partial t_2}\right) \left(\frac{i}{m}\right)^2$$

$$\times [D(t_1 - t_2)D(0) + 2D(t_1 - t)D(t - t_2)] = \frac{3}{2m\omega} .$$

In like fashion arbitrary harmonic oscillator matrix elements can be reduced to ground state expectation values, which in turn can be evaluated using the generating functional $W[j]$. The ground state amplitude in the presence of an arbitrary source $j(t)$ thus contains *all* the information about the harmonic oscillator.

Functional Perturbation Theory

If calculation of matrix elements were all that functional techniques were good for, they would be of little interest since more efficient methods are available. However, functional methods can also be used in order to determine the propagator for more general interactions. Consider a particle moving under the influence of the potential

$$V(x) = \frac{1}{2}m\omega^2 x^2 - f(x) \ . \tag{6.40}$$

The propagator for such a particle is given, at least formally, by the expression

$$\begin{aligned}D_F(x_f,t_f;x_i,t_i;f(x)) &= \int \mathcal{D}\left[x(t)\right] \exp i \int_{t_i}^{t_f} dt \left[\frac{1}{2}m\dot{x}^2(t) - \frac{1}{2}m\omega^2 x^2(t)\right.\\ &\left.+f\left(x(t)\right)\right] = \exp\left(i\int_{t_i}^{t_f} dt \, f\left(-i\frac{\delta}{\delta j(t)}\right)\right) D_F^j\left(x_f,t_f;x_i,t_i\right)\bigg|_{j=0}\end{aligned} \tag{6.41}$$

where D_F^j denotes the propagator in the presence of a driving term $j(t)$ and has been given in Problem I.3.3. In general, this exact result is only of formal interest, as it is no more subject to evaluation than is the full propagator which it represents. However, in the circumstance that $f(x)$ is in some sense small the exponential in Eq. 6.41 can be expanded, yielding a perturbative expansion of the propagator

$$D_F\left(x_f,t_f;x_i,t_i;f(x)\right) = \sum_{n=0}^{\infty} \frac{i^n}{n!} \left(\int_{t_i}^{t_f} dt \, f\left(-i\frac{\delta}{\delta j(t)}\right)\right)^n D_F^j\left(x_f,t_f;x_i,t_i\right)\bigg|_{j=0} . \tag{6.42}$$

As a simple example of the use of such methods, we examine a problem which is exactly soluble — $f(x) = \lambda x$. The exact solution is found by noting that the potential can be rewritten by completing the square

$$V(x) = \frac{1}{2}m\omega^2 x^2 - \lambda x = \frac{1}{2}m\omega^2 x'^2 - \frac{\lambda^2}{2m\omega^2} \tag{6.43}$$

where

$$x' = x - \frac{\lambda}{m\omega^2} \ . \tag{6.44}$$

Changing variables to x', the Hamiltonian becomes

$$H = -\frac{1}{2m}\frac{d^2}{dx'^2} + \frac{1}{2}m\omega^2 x'^2 - \frac{\lambda^2}{2m\omega^2} \tag{6.45}$$

whose ground state wavefunction is

$$\psi_0(x,t) = \left(\frac{m\omega}{\pi}\right)^{1/4} \exp -\frac{1}{2}m\omega\left(x - \frac{\lambda}{m\omega^2}\right)^2 \exp -i\left(\frac{\omega}{2} - \frac{\lambda^2}{2m\omega^2}\right)t \ . \tag{6.46}$$

1.6 FUNCTIONAL TECHNIQUES*

Now look at the same problem using the functional formalism. Noting that

$$D_F^j(x_f, t_f; x_i, t_i) = D_F^{(0)}(x_f, t_f; x_i, t_i)$$
$$\times \exp\left(i \int_{t_i}^{t_f} dt\, j(t) x_{\text{cl}}(t) + \frac{i}{2} \int_{t_i}^{t_f} dt \int_{t_i}^{t_f} ds\, j(t) g(t,s) j(s)\right)$$
(6.47)

where

$$x_{\text{cl}}(t) = x_i \frac{\sin \omega(t_f - t)}{\sin \omega(t_f - t_i)} + x_f \frac{\sin \omega(t - t_i)}{\sin \omega(t_f - t_i)} \tag{6.48}$$

is the classical solution for the free oscillator and

$$g(t,s) = \begin{cases} -\frac{\sin \omega(t_f - t) \sin \omega(s - t_i)}{m\omega \sin \omega(t_f - t_i)} & t > s \\ -\frac{\sin \omega(t_f - s) \sin \omega(t - t_i)}{m\omega \sin \omega(t_f - t_i)} & t < s \end{cases} \tag{6.49}$$

is the Green's function, we find

$$-i\frac{\delta}{\delta j(t)} D^j(x_f, t_f; x_i, t_i)\bigg|_{j=0} = x_{\text{cl}}(t) D^{(0)}(x_f, t_f; x_i, t_i)$$
$$\frac{(-i)^2 \delta^2}{\delta j(t) \delta j(s)} D^j(x_f, t_f; x_i, t_i)\bigg|_{j=0} = (x_{\text{cl}}(t) x_{\text{cl}}(s) - i g(t,s)) D_F^{(0)}(x_f, t_f; x_i, t_i)$$
(6.50)

We require the integrals

$$\int_{t_i}^{t_f} dt\, x_{\text{cl}}(t) = -\frac{x_f}{\omega \sin \omega(t_f - t_i)} \cos \omega(t - t_i) + \frac{x_i}{\omega \sin \omega(t_f - t_i)} \cos \omega(t_f - t)\bigg|_{t_i}^{t_f}$$
$$= \frac{x_i + x_f}{\omega \sin \omega(t_f - t_i)}(1 - \cos \omega(t_f - t_i))$$

$$\int_{t_i}^{t_f} dt \int_{t_i}^{t} ds\, g(t,s) = \frac{1}{m\omega^2 \sin \omega(t_f - t_i)} \int_{t_i}^{t_f} dt\, \sin \omega(t_f - t) \cos \omega(s - t_i)\bigg|_{t_i}^{t}$$
$$= \frac{1}{m\omega^2 \sin \omega(t_f - t_i)}\left(-\frac{1}{\omega} \cos \omega(t_f - t)\bigg|_{t_i}^{t_f} + \int_{t_i}^{t_f} dt\, \sin \omega(t_f - t) \cos \omega(t - t_i)\right)$$
$$= \frac{-1}{m\omega^3 \sin \omega(t_f - t_i)}(1 - \cos \omega(t_f - t_i)) + \frac{t_f - t_i}{2m\omega^2}$$
(6.51)

which yield the result

$$D_F(x_f, t_f; x_i, t_i; -\lambda x) = D_F^{(0)}(x_f, t_f; x_i, t_i)\left(1 + i\lambda \frac{x_f + x_i}{\omega} \tan \frac{1}{2}\omega(t_f - t_i)\right.$$
$$-\frac{\lambda^2}{2\omega^2}(x_f + x_i)^2 \tan^2 \frac{1}{2}\omega(t_f - t_i)$$
$$\left.+ i\lambda^2 \frac{t_f - t_i}{2m\omega^2} - \frac{i}{m\omega^3}\lambda^2 \tan \frac{1}{2}\omega(t_f - t_i) + \mathcal{O}(\lambda^3)\right)$$
(6.52)

We can check Eq. 6.52 by taking the limit $t_f \to -iT/2$, $t_i \to iT/2$ so that

$$D_F\left(x_f, -i\frac{T}{2}; x_i, i\frac{T}{2}; -\lambda x\right) = D_F^{(0)}\left(x_f, -i\frac{T}{2}, x_i, i\frac{T}{2}\right)$$
$$\times \left(1 + \frac{\lambda}{\omega}(x_f + x_i) + \frac{\lambda^2}{2\omega^2}(x_f + x_i)^2 + \lambda^2 \frac{T}{2m\omega^2} - \frac{\lambda^2}{m\omega^3} + \mathcal{O}(\lambda^3)\right) \quad (6.53)$$

Using Eq. 6.3 this becomes

$$\phi_0(x_f)\phi_0(x_i)\exp(-E_0 T) = \phi_0^{(0)}(x_f)\phi_0^{(0)}(x_i)\exp\left(-\frac{1}{2}\omega T\right)$$
$$\times \left(1 + \frac{\lambda}{\omega}(x_f + x_i) + \frac{\lambda^2}{2\omega^2}(x_f + x_i)^2 + \frac{\lambda^2}{2m\omega^2}T - \frac{\lambda^2}{m\omega^3} + \mathcal{O}(\lambda^3)\right)$$
$$= \phi^{(0)}(x_f)\exp\left(\frac{\lambda}{\omega}x_f - \frac{\lambda^2}{2m\omega^3}\right)\phi^{(0)}(x_i)\exp\left(\frac{\lambda}{\omega}x_i - \frac{\lambda^2}{2m\omega^3}\right) \quad (6.54)$$
$$\times \exp\left(-\left(\frac{1}{2}\omega - \frac{\lambda^2}{2m\omega^2}\right)T\right) + \mathcal{O}(\lambda^3)$$

which agrees to this order with the exact ground state wavefunction and energy given in Eq. 6.46.

PROBLEM I.6.1

The Generating Functional and the Forced Harmonic Oscillator

In an earlier problem the propagator for a forced harmonic oscillator was calculated and found to be

$$D_F^j(x_f, t_f; x_i, t_i) = \left(\frac{m\omega}{2\pi i \sin\omega(t_f - t_i)}\right)^{1/2} \exp\frac{im\omega}{2\sin\omega(t_f - t_i)}$$
$$\times \left[\cos\omega(t_f - t_i)\left(x_f^2 + x_i^2\right) - 2x_f x_i + \frac{2x_f}{m\omega}\int_{t_i}^{t_f} dt\, j(t)\sin\omega(t - t_i)\right. \quad (6.55)$$
$$+ \frac{2x_i}{m\omega}\int_{t_i}^{t_f} dt\, j(t)\sin\omega(t_f - t)$$
$$\left. - \frac{2}{m^2\omega^2}\int_{t_i}^{t_f} dt \int_{t_i}^{t} ds\, j(t)j(s)\sin\omega(t_f - t)\sin\omega(s - t_i)\right] \, .$$

Note that the form of the propagator is

$$D_F^j(x_f, t_f; x_i, t_i) = D_F^{HO}(x_f, t_f; x_i, t_i)\exp i\Phi^j(x_f, t_f; x_i, t_i) \quad (6.56)$$

where Φ is a simple phase factor and D_F^{HO} is the propagator for the free harmonic oscillator.

i) Take the limit $t_f \to \infty$, $t_i \to -\infty$ and show that

$$\lim_{\substack{t_f \to \infty \\ t_i \to -\infty}} \Phi = \frac{i}{4m\omega}\int_{-\infty}^{\infty} dt \int_{-\infty}^{\infty} ds\, j(t)j(s)e^{-i\omega|t-s|}$$

in agreement with Eq. 6.29. Note: You may find it easier to use imaginary time methods. Also, remember that $j(t)$ vanishes as $t \to \pm\infty$.

ii) Show in the large (but not infinite) time limit that

$$\Phi^j\left(x_f, \frac{T}{2}; x_i, -\frac{T}{2}\right) = \frac{i}{4m\omega} \int_{-T/2}^{T/2} dt \int_{-T/2}^{T/2} ds\, j(t)j(s) e^{-i\omega|t-s|}$$
$$+ \int_{-T/2}^{T/2} dt\, j(t) x_{\text{cl}}(t)$$

where

$$x_{\text{cl}}(t) = x_i \frac{\sin\omega(T/2 - t)}{\sin\omega T} + x_f \frac{\sin\omega(t + T/2)}{\sin\omega T}$$

is the classical solution to the free (non-driven) oscillator.

PROBLEM I.6.2

Perturbation Theory and the Anharmonic Oscillator

Consider the potential

$$V(x) = \frac{1}{2} m\omega^2 x^2 + \alpha x^3 + \beta x^4$$

i) Using time independent perturbation theory show that the energy of the nth eigenstate is given by

$$E_n = \omega\left(n + \frac{1}{2}\right) + \frac{3}{2} \frac{\beta}{m^2\omega^2}\left(n^2 + n + \frac{1}{2}\right) - \frac{15}{4} \frac{\alpha^2}{m^3\omega^4}\left(n^2 + n + \frac{11}{30}\right) + \ldots$$

so that

$$E_0 = \frac{1}{2}\omega + \frac{3}{4} \frac{\beta}{m^2\omega^2} - \frac{11}{8} \frac{\alpha^2}{m^3\omega^4} + \ldots$$

ii) Set $\alpha = 0$ and use functional perturbation theory to evaluate the first order shift arising from the quartic potential on the the ground state energy.

iii) Set $\beta = 0$ and use functional perturbative methods to evaluate the energy shift of the ground state due to the cubic potential. (Note that this effect arises only in second order—why?)

iv) Compare your answers with those found from i).

CHAPTER II

SCATTERING THEORY

II.1 BASIC FORMALISM

An application wherein the frequency space representation of perturbation theory proves particularly useful is that of quantum mechanical scattering. Since the scattering process will play an important role in the remainder of this volume, we shall treat it in some detail.

The Scattering Amplitude

We imagine a transition from some plane wave state $|i\rangle$ in the remote past to a plane wave state $|f\rangle$ in the remote future. Defining

$$|i\rangle = e^{-iE_i t}|\phi_i\rangle$$
$$|f\rangle = e^{-iE_f t'}|\phi_f\rangle \qquad (1.1)$$

the relevant transition amplitude is

$$\text{Amp}_{fi}(t',t) = e^{iE_f t'} \left\langle \phi_f \left| \hat{U}(t',t) \right| \phi_i \right\rangle e^{-iE_i t}$$
$$= \int_{-\infty}^{\infty} \frac{d\omega}{2\pi} e^{iE_f t' - iE_i t} e^{-i\omega(t'-t)} \left\langle \phi_f \left| \hat{\mathcal{K}}(\omega) \right| \phi_i \right\rangle \qquad (1.2)$$

where

$$-i\left\langle \phi_f \left| \hat{\mathcal{K}}(\omega) \right| \phi_i \right\rangle = \langle \phi_f | \phi_i \rangle \frac{1}{\omega - E_i + i\epsilon} + \frac{1}{\omega - E_f + i\epsilon} \mathcal{F}(\omega) \frac{1}{\omega - E_i + i\epsilon} \qquad (1.3)$$

with

$$\mathcal{F}(\omega) = \left\langle \phi_f | \hat{V} | \phi_i \right\rangle + \left\langle \phi_f \left| \hat{V} \frac{1}{\omega - \hat{H}_0 + i\epsilon} \hat{V} \right| \phi_i \right\rangle + \ldots \quad . \qquad (1.4)$$

We have then

$$\text{Amp}_{fi}(t',t) = \langle \phi_f | \phi_i \rangle \times i \int_{-\infty}^{\infty} \frac{d\omega}{2\pi} \frac{e^{i(E_f-\omega)t'-i(E_i-\omega)t}}{\omega - E_i + i\epsilon}$$
$$+ i \int_{-\infty}^{\infty} \frac{d\omega}{2\pi} \frac{e^{i(E_f-\omega)t'}}{\omega - E_f + i\epsilon} \mathcal{F}(\omega) \frac{e^{-i(E_i-\omega)t}}{\omega - E_i + i\epsilon} \quad . \qquad (1.5)$$

Now in a scattering problem generally one is interested in the situation that the initial time t is long before the interaction with the potential takes place and the final time t' is long after. In order to proceed we consider the integral

$$J = \lim_{t \to -\infty} \int_{-\infty}^{\infty} d\omega\, f(\omega) \frac{e^{i\omega t}}{\omega + i\epsilon} \quad . \qquad (1.6)$$

In the large $|t|$ limit, the exponential oscillates so rapidly that there is virtually no contribution to the integral except from the region $|\omega| \lesssim \frac{1}{|t|}$. Then we can write

$$J \simeq f(0) \times \lim_{t \to -\infty} \int_{-\infty}^{\infty} d\omega \, \frac{e^{i\omega t}}{\omega + i\epsilon} \tag{1.7}$$

i.e., under an integral the quantity

$$\lim_{t \to -\infty} \frac{e^{i\omega t}}{\omega + i\epsilon} \tag{1.8}$$

behaves effectively like a Dirac delta function $\delta(\omega)$. In order to get the proper normalization, we note

$$\lim_{\substack{t \to -\infty \\ \epsilon \to 0}} \int_{-\infty}^{\infty} d\omega \, \frac{e^{i\omega t}}{\omega + i\epsilon} = \lim_{t \to -\infty} \lim_{\epsilon \to 0} -2\pi i e^{\epsilon t} = -2\pi i \tag{1.9}$$

where we have closed the contour in the lower half plane and taken the limit $\epsilon \to 0$ *before* the limit $t \to -\infty$. Thus we deduce that

$$\lim_{t \to -\infty} \frac{e^{i\omega t}}{\omega + i\epsilon} \simeq -2\pi i \delta(\omega) \tag{1.10}$$

so

$$\lim_{\substack{t' \to \infty \\ t \to -\infty}} \mathrm{Amp}_{fi}(t',t) = \lim_{t' \to \infty} i \int_{-\infty}^{\infty} \frac{d\omega}{2\pi} e^{i(E_f - \omega)t'} \times -2\pi i \delta(\omega - E_i) \langle \phi_f | \phi_i \rangle$$

$$+ \lim_{t' \to \infty} i \int \frac{d\omega}{2\pi} \frac{e^{i(E_f - \omega)t'}}{\omega - E_f + i\epsilon} \times -2\pi i \delta(\omega - E_i) \mathcal{F}(\omega) \tag{1.11}$$

$$= \lim_{t' \to \infty} \left[e^{i(E_f - E_i)t'} \langle \phi_f | \phi_i \rangle + \frac{e^{i(E_f - E_i)t'}}{E_i - E_f + i\epsilon} \mathcal{F}(E_i) \right] .$$

Since by orthogonality

$$\langle \phi_f | \phi_i \rangle = \delta_{fi} \tag{1.12}$$

and by our previous result

$$\lim_{t' \to \infty} \frac{e^{-i(E_i - E_f)t'}}{E_i - E_f + i\epsilon} \simeq -2\pi i \delta(E_i - E_f) \tag{1.13}$$

we can write

$$\lim_{\substack{t' \to \infty \\ t \to -\infty}} \mathrm{Amp}_{fi}(t',t) \simeq \delta_{fi} - 2\pi i \delta(E_f - E_i) \mathcal{F}(E_i) . \tag{1.14}$$

The transition amplitude between initial and final states at time $t = -\infty$ and $t' = +\infty$, respectively, is called the *S*- or scattering-matrix, *i.e.*,

$$\lim_{\substack{t' \to \infty \\ t \to -\infty}} \left\langle f \left| \hat{U}(t',t) \right| i \right\rangle \equiv \langle f | S | i \rangle \tag{1.15}$$

and from above we have

$$S = 1 - 2\pi i \delta(E_f - E_i) T \tag{1.16}$$

where T is the *T*- or transition-matrix

$$\langle f | T | i \rangle \equiv \mathcal{F}(E_i) \tag{1.17}$$

The quantity

$$\mathcal{F}(E_i) = \langle f|\hat{V}|i\rangle + \left\langle f\left|\hat{V}\frac{1}{E_i - \hat{H}_0 + i\epsilon}\hat{V}\right|i\right\rangle + \ldots \quad (1.18)$$

is the transition amplitude and can be written as

$$\mathcal{F}(E_i) = \left\langle \phi_f|\hat{V}|\psi_i^{(+)}\right\rangle \quad (1.19)$$

where

$$|\psi_i^{(+)}\rangle = |\phi_i\rangle + \frac{1}{E_i - \hat{H}_0 + i\epsilon}\hat{V}|\phi_i\rangle + \frac{1}{E_i - \hat{H}_0 + i\epsilon}\hat{V}\frac{1}{E_i - \hat{H}_0 + i\epsilon}\hat{V}|\phi_i\rangle$$
$$+ \ldots \quad (1.20)$$
$$= |\phi_i\rangle + \frac{1}{E_i - \hat{H}_0 + i\epsilon}\hat{V}|\psi_i^{(+)}\rangle \ .$$

If $|\phi_i\rangle$ is a solution of the "free" Schrödinger equation with $\hat{H}_0$ and energy E_i

$$(\hat{H}_0 - E_i)|\phi_i\rangle = 0 \quad (1.21)$$

then $|\psi_i^{(+)}\rangle$ is a solution of the *full* Schrödinger equation with the same energy E_i, since

$$(E_i - \hat{H}_0 - \hat{V})|\psi_i^{(+)}\rangle = (E_i - \hat{H}_0)\left[|\phi_i\rangle + \frac{1}{E_i - \hat{H}_0 + i\epsilon}\hat{V}|\psi_i^{(+)}\rangle\right] - \hat{V}|\psi_i^{(+)}\rangle$$
$$= 0 + \hat{V}|\psi_i^{(+)}\rangle - \hat{V}|\psi_i^{(+)}\rangle = 0 \ .$$
$$(1.22)$$

The coordinate space projection of $|\psi_i^{(+)}\rangle$

$$\left\langle \vec{r}|\psi_i^{(+)}\right\rangle \quad (1.23)$$

is the scattering wavefunction. The above definition then yields

$$\left\langle \vec{r}|\psi_i^{(+)}\right\rangle = \langle \vec{r}|\phi_i\rangle + \int d^3r' \left\langle \vec{r}\left|\frac{1}{E_i - \hat{H}_0 + i\epsilon}\right|\vec{r}'\right\rangle V(\vec{r}')\left\langle \vec{r}'|\psi_i^{(+)}\right\rangle \ . \quad (1.24)$$

The coordinate space representation of $|\phi_i\rangle$ is a simple plane wave—$\langle \vec{r}|\phi_i\rangle = e^{i\vec{p}_i\cdot\vec{r}}$—so the key to the problem is the evaluation of the amplitude

$$\left\langle \vec{r}\left|\frac{1}{E_i - \hat{H}_0 + i\epsilon}\right|\vec{r}'\right\rangle \quad (1.25)$$

In section I.2 we evaluated a similar quantity in a one dimensional situation. We now require its three dimensional analog. As before we can write

$$K^{(0)}(\vec{r},\vec{r}') \equiv \left\langle \vec{r}\left|\frac{1}{E_i - \hat{H}_0 + i\epsilon}\right|\vec{r}'\right\rangle = \int \frac{d^3p}{(2\pi)^3} \langle \vec{r}|\vec{p}\rangle \frac{1}{E_i - \frac{\vec{p}^2}{2m} + i\epsilon} \langle \vec{p}|\vec{r}'\rangle$$
$$= \int \frac{d^3p}{(2\pi)^3} e^{i\vec{p}\cdot(\vec{r}-\vec{r}')} \frac{1}{E_i - \frac{\vec{p}^2}{2m} + i\epsilon} \ . \quad (1.26)$$

Defining $E_i \equiv \frac{p_i^2}{2m}$ and $R \equiv |\vec{r} - \vec{r}'|$ we can perform the angular integration, yielding

$$K^{(0)}(\vec{r}, \vec{r}') = \frac{2m}{(2\pi)^2 iR} \int_0^\infty dp\, p \frac{e^{ipR} - e^{-ipR}}{p_i^2 - p^2 + i\epsilon} \;. \tag{1.27}$$

Since the integrand is an even function of p we may extend the integration interval to $(-\infty, \infty)$ and perform the integral by contour methods. The poles are located at $p = p_i + i\epsilon$ and $p = -p_i - i\epsilon$. For the e^{ipR} term we must close the contour in the upper half plane, picking up the pole at $p_i + i\epsilon$, while for the e^{-ipR} component the contour must be closed in the lower half plane so that the appropriate pole is that at $-p_i - i\epsilon$. We find

$$K^{(0)}(\vec{r}, \vec{r}') = \frac{m}{(2\pi)^2 iR} \times -2\pi i e^{ip_i R} = -\frac{m}{2\pi R} e^{ip_i R} \;. \tag{1.28}$$

The scattering wavefunction then satisfies the integral equation

$$\langle \vec{r} | \psi_i^{(+)} \rangle = e^{i\vec{p}_i \cdot \vec{r}} - \frac{m}{2\pi} \int d^3 r' \frac{e^{ip_i R}}{R} V(\vec{r}') \langle \vec{r}' | \psi_i^{(+)} \rangle \;. \tag{1.29}$$

Now imagine $\vec{r}$ to become very large with respect to the range of the potential, so that we may approximate

$$R = \sqrt{r^2 - 2\vec{r} \cdot \vec{r}' + r'^2} \simeq r\sqrt{1 - 2\vec{r} \cdot \vec{r}'/r^2} \simeq r - \hat{r} \cdot \vec{r}' \;. \tag{1.30}$$

Then

$$e^{ip_i R} \approx e^{ip_i r} \times e^{-i\vec{p}_f \cdot \vec{r}'} \tag{1.31}$$

where we have defined $\vec{p}_f = p_i \hat{r}$, and hence

$$\begin{aligned}
\langle \vec{r} | \psi_i^{(+)} \rangle &= e^{i\vec{p}_i \cdot \vec{r}} + \frac{1}{r} e^{ip_i r} \left(-\frac{m}{2\pi} \int d^3 r' e^{-i\vec{p}_f \cdot \vec{r}'} V(\vec{r}') \langle \vec{r}' | \psi_i^{(+)} \rangle \right) \\
&= \langle \vec{r} | \phi_i \rangle + \frac{1}{r} e^{ip_i r} \left(-\frac{m}{2\pi} \langle \phi_f | \hat{V} | \psi_i^{(+)} \rangle \right) \\
&= \langle \vec{r} | \phi_i \rangle + \frac{1}{r} e^{ip_i r} f_{p_i}(\theta, \phi)
\end{aligned} \tag{1.32}$$

with

$$f_{p_i}(\theta, \phi) \equiv -\frac{m}{2\pi} \langle \phi_f | \hat{V} | \psi_i^{(+)} \rangle = -\frac{m}{2\pi} \mathcal{F}(E_i) \tag{1.33}$$

being the "scattering amplitude."

Differential Scattering Cross Section: Wave Packet Approach

The connection with the differential scattering cross section can now be made by constructing a wave packet incident on a scattering potential localized say about $\vec{r} = 0$ [Me 70]. Suppose that at time t=0 the wavepacket is located at $\vec{r} = \vec{r}_o$ and

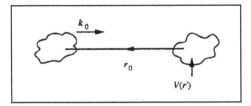

Fig. II.1: At time t=0 the incident wavepacket is located at position $\vec{r}_0$ and is moving toward the target with velocity $\vec{k}_0/m$.

is moving toward the target with velocity $\vec{v}_i = \vec{k}_o/m$, as shown in Figure II.1 This can be described by the wavefunction

$$\psi^{(+)}(\vec{r}, t=0) = \int \frac{d^3k}{(2\pi)^3} \phi(\vec{k}) e^{-i\vec{k}\cdot\vec{r}_o} \left\langle \vec{r} | \psi_{\vec{k}}^{(+)} \right\rangle \tag{1.34}$$

where $\phi(\vec{k})$ is localized about $\vec{k} = \vec{k}_o$. We find for the wavefunction at later time t

$$\psi^{(+)}(\vec{r}, t) = \int \frac{d^3k}{(2\pi)^3} \phi(\vec{k}) e^{-i\vec{k}\cdot\vec{r}_o - iE_k t} \left\langle \vec{r} | \psi_{\vec{k}}^{(+)} \right\rangle$$

$$\underset{r\to\infty}{\sim} \int \frac{d^3k}{(2\pi)^3} \phi(\vec{k}) \left(e^{i\vec{k}\cdot(\vec{r}-\vec{r}_o) - iE_k t} + \frac{e^{ikr}}{r} e^{-i\vec{k}\cdot\vec{r}_o - iE_k t} f_k(\theta, \phi) \right) \tag{1.35}$$

with $E_k = k^2/2m$. We find then that the plane wave component of the wavefunction is located where the phase $\vec{k}\cdot(\vec{r}-\vec{r}_o) - \frac{k^2}{2m}t$ is independent of $\vec{k}$[†]

$$\frac{\partial}{\partial \vec{k}}(\vec{k}\cdot(\vec{r}-\vec{r}_o) - \frac{k^2}{2m}t) = \vec{r} - \vec{r}_o - \frac{\vec{k}}{m}t = 0 \tag{1.36}$$

since in other regions of phase space the argument is rapidly varying and contributes little to the integral. As $\phi(\vec{k})$ is localized near $\vec{k} = \vec{k}_0$ we see that the wavepacket is centered at

$$\vec{r} = \vec{r}_0 + \frac{\vec{k}_0}{m}t \ . \tag{1.37}$$

At $t = 0$ then the packet is located at $\vec{r} = \vec{r}_0$ as shown and is moving toward the origin with velocity $\frac{\vec{k}_0}{m}$. Writing

$$\vec{k} = \vec{k}_0 + \Delta\vec{k} \tag{1.38}$$

and neglecting terms in $(\Delta k)^2$ we have for the plane wave piece of Eq. 1.35

$$\psi_{\text{plane-wave}}(\vec{r}, t) = e^{i\vec{k}_0\cdot(\vec{r}-\vec{r}_0) - i\frac{k_0^2}{2m}t} \int \frac{d^3\Delta k}{(2\pi)^3} \phi(\vec{k}_0 + \Delta\vec{k}) e^{i\Delta\vec{k}\cdot\left(\vec{r}-\vec{r}_0 - \frac{\vec{k}_0}{m}t\right)} \tag{1.39}$$

[†] This procedure is called the stationary phase approximation and will be used again in chapter V.

so that in this approximation the wavepacket moves without even changing shape (spreading comes from the small terms in $(\Delta k)^2$). Now look at the second component of the scattering wavefunction—Eq. 1.35. The phase of the exponential in this case is given by

$$kr - \vec{k} \cdot \vec{r}_0 - \frac{k^2}{2m}t \quad . \tag{1.40}$$

However,

$$\vec{k} \cdot \vec{r}_0 = kr_0 \cos\theta = -kr_0\left(1 + \mathcal{O}\left(\frac{\Delta k}{k}\right)^2 + \ldots\right) \tag{1.41}$$

so that, again dropping terms in $\left(\frac{\Delta k}{k}\right)^2$, the phase can be written as

$$kr + kr_0 - \frac{k^2}{2m}t \quad . \tag{1.42}$$

Writing

$$k = k_0 + \Delta k \tag{1.43}$$

we have

$$kr + kr_0 - \frac{k^2}{2m}t = k_0(r + r_0) - \frac{k_0^2}{2m}t + \Delta k\left(r + r_0 - \frac{k_0}{m}t\right) + \mathcal{O}(\Delta k^2) \tag{1.44}$$

and the second component of Eq. 1.35 becomes, in this approximation,

$$\psi^{(+)}_{\text{scattered}} \approx \frac{1}{r}e^{ik_o(r+r_0) - i\frac{k_0^2}{2m}t} f_{k_0}(\theta, \phi) \int \frac{d^3 \Delta k}{(2\pi)^3} \phi(\vec{k}_0 + \Delta \vec{k}) e^{i\Delta k\left(r + r_0 - \frac{k_0}{m}t\right)} \tag{1.45}$$

which is a spherical wavepacket centered at the origin and located at

$$r = -r_0 + \frac{k_0}{m}t \quad . \tag{1.46}$$

No spherical wave exists until $r > 0$ — i.e., until

$$t > \frac{r_0}{k_0/m} \equiv t_c \tag{1.47}$$

which is the time at which the planewave piece strikes the scattering center, consistent with the causality requirement that the scattered wave cannot exist until after this point in time. After t_c the plane wave and spherical wave both move away from the origin with speed $\frac{k_0}{m}$. [Note that this result depends upon the $+i\epsilon$ prescription — had we used $-i\epsilon$ the scattered wave—$|\psi^{(-)}\rangle$—would had been associated with an exponential $\frac{1}{r}e^{-ikr}$ and would exist even *before* the plane wave hits $\vec{r} = 0$. The use of $+i\epsilon$ is essential and is required by causality.]

Actually, it appears that we have an excess of probability since the plane wavepacket normalization is preserved as it moves, while additional probability associated with the outgoing spherical wave is generated for $t > t_c$. However, interference terms in the region of space where the plane and spherical waves overlap precisely cancel the "extra" probability associated with the spherical wave. We shall discuss this point in section II.2 but before doing so we evaluate the differential scattering cross section, which is defined as

$$\frac{d\sigma}{d\Omega} = \frac{\text{Probability scattered into solid angle } d\Omega}{d\Omega} \bigg/ \frac{\text{Probability}}{\text{Area in } \psi_{\text{planewave}}} . \quad (1.48)$$

Suppose the packet is incident along the $+z$ direction and from the negative z side. We can write then

$$\text{Probability} = |\psi_{\text{planewave}}(z, x=0, y=0, t_0)|^2 \, dz \, \Delta x \, \Delta y \equiv I_0 \Delta x \, \Delta y \quad (1.49)$$

where I_0 is the probability per unit area and we have assumed that $\psi_{\text{planewave}}$ does not vary much over the area of the target $\Delta x \, \Delta y$. For the probability scattered into solid angle $d\Omega$ we have (away from the forward direction)

$$\text{Probability} = \int |\psi_{\text{planewave}}(z, x=0, y=0, t=0)|^2 \frac{|f_{k_0}(\theta, \phi)|^2}{r^2} r^2 \, dr \, d\Omega$$
$$= I_0 |f_{k_0}(\theta, \phi)|^2 \, d\Omega \; . \quad (1.50)$$

Then

$$\frac{d\sigma}{d\Omega} = \frac{I_0 |f_{k_0}(\theta, \phi)|^2}{I_0} = |f_{k_0}(\theta, \phi)|^2 \quad (1.51)$$

which is the conventional result.

Differential Scattering Cross Section: Intuitive Approach

The proper way of dealing with quantum mechanical transition probabilities is, of course, using wavepackets as above and, as shown definitively in Goldberger and Watson's *Collision Theory*, for example, the results of such a careful treatment yield finite values for such things as transition rates and scattering cross sections [GoW 64]. In quantum mechanical calculations, however, particularly in perturbation theory, one often chooses to do the calculation with plane waves, yielding for a typical transition amplitude

$$\text{Amp}_{fi} = -2\pi i \delta(E_f - E_i) T_{fi} \quad (1.52)$$

which is formally infinite. Nevertheless, one can generate finite and correct expressions for the physically relevant quantities by a simple intuitive (but hardly rigorous) argument, as we now demonstrate.

The probability of making a transition from initial state $|i\rangle$ to final state $|f\rangle$ is given by the square of the transition amplitude. Thus the total decay probability is

$$\sum_f P_{fi} = \sum_f |\text{Amp}_{fi}|^2 = \sum_f (2\pi\delta(E_f - E_i))^2 |T_{fi}|^2 \quad (1.53)$$

and appears to be infinite. Now, however, rewrite this expression as

$$\sum_f P_{fi} \cong \sum_f \int_{-T/2}^{T/2} dt\, e^{i(E_f - E_i)t} 2\pi\delta(E_f - E_i)|T_{fi}|^2$$
$$= \sum_f \int_{-\frac{T}{2}}^{\frac{T}{2}} dt\, 2\pi\delta(E_f - E_i)|T_{fi}|^2 = \sum_f T\, 2\pi\delta(E_f - E_i)|T_{fi}|^2 \quad (1.54)$$

where T is the total time of interaction. The transition *rate*—(*i.e.*, probability per unit time)—is

$$\sum_f R_{fi} = \sum_f \frac{1}{T} P_{fi} = \sum_f 2\pi\delta(E_f - E_i)|T_{fi}|^2 \quad (1.55)$$

which is finite and is recognized as Fermi's golden rule. A rigorous derivation in terms of wavepackets yields an identical result.

Application of this mnemonic to a cross section calculation is equally straightforward. Consider scattering from a potential V. Let the initial (final) states be represented in terms of plane waves $|\vec{p}_i\rangle$ ($|\vec{p}_f\rangle$) and sum over a group of final states with $\vec{p}_f$ in some range of solid angle. Then

$$\sum_f R_{fi} = \int \frac{d^3 p_f}{(2\pi)^3} 2\pi\delta(E_f - E_i)|T_{fi}|^2$$
$$= \int \frac{d\Omega_f p_f^2 dp_f}{(2\pi)^3} 2\pi\delta(E_f - E_i)|T_{fi}|^2 \quad (1.56)$$
$$= \frac{d\Omega_f p_f^2}{(2\pi)^2} \frac{1}{dE_f/dp_f} |T_{fi}|^2 \quad ,$$

with

$$\frac{dE_f}{dp_f} = \frac{d}{dp_f} \frac{p_f^2}{2m} = \frac{p_f}{m} \quad . \quad (1.57)$$

The scattering cross section $d\sigma$ is defined as

$$d\sigma = \frac{1}{\text{Flux}} \sum_f R_{fi} \quad (1.58)$$

with the incident flux being

$$\text{Flux} = v_i = p_i/m \quad . \quad (1.59)$$

By energy conservation $p_i = p_f$ and

$$d\sigma = \frac{d\Omega_f p_f^2}{(2\pi)^2} \frac{1}{\frac{p_f}{m} \cdot \frac{p_i}{m}} |T_{fi}|^2$$
$$= d\Omega_f \left| -\frac{m}{2\pi} T_{fi} \right|^2 \quad . \quad (1.60)$$

Thus the scattering amplitude $f_{p_i}(\theta)$, defined by

$$\frac{d\sigma}{d\Omega} \equiv |f_{p_i}(\theta)|^2 \quad (1.61)$$

is given by

$$f_{p_i}(\theta) = -\frac{m}{2\pi}T_{fi} = -\frac{m}{2\pi}\left\langle\phi_f|\hat{V}|\psi_i^{(+)}\right\rangle \quad , \tag{1.62}$$

as found above by more rigorous means.

Perturbative Expansion of the Transition Amplitude

A perturbative expansion for the transition (scattering) amplitude is easily generated. To lowest order in $\hat{V}$, we have

$$T_{fi}^{(1)} = \left\langle f|\hat{V}|i\right\rangle \equiv \left\langle\phi_f|\hat{V}|\phi_i\right\rangle \tag{1.63}$$

which is the usual Born approximation amplitude. However, higher order contributions can also be easily represented by means of (frequency space) Feynman diagrams as shown below: †

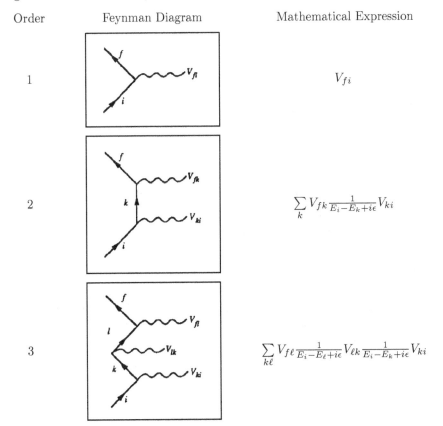

Order	Feynman Diagram	Mathematical Expression
1		V_{fi}
2		$\sum_k V_{fk}\frac{1}{E_i-E_k+i\epsilon}V_{ki}$
3		$\sum_{k\ell} V_{f\ell}\frac{1}{E_i-E_\ell+i\epsilon}V_{\ell k}\frac{1}{E_i-E_k+i\epsilon}V_{ki}$

etc. The general term in the expansion can be found from the result

---

† Henceforth all Feynman diagrams will be of the frequency space form, wherein the arrow indicates the direction from initial to final states and a straight line (propagator) for state k stands for $\frac{1}{E_i-E_k+i\epsilon}$.

$$T_{fi} = \left\langle f|\hat{V}|\psi_i^{(+)}\right\rangle = \left\langle f|\hat{V}|i\right\rangle + \left\langle f\left|\hat{V}\frac{1}{E_i - \hat{H}_0 + i\epsilon}\hat{V}\right|i\right\rangle$$
$$+ \left\langle f\left|\hat{V}\frac{1}{E_i - \hat{H}_0 + i\epsilon}\hat{V}\frac{1}{E_i - \hat{H}_0 + i\epsilon}\hat{V}\right|i\right\rangle + \ldots \quad (1.64)$$

which can be considered to be a manifestation of the Feynman prescription

$$\text{Amp}_{\text{tot}} = \sum_{\text{route}} \text{Amp (route)} \quad (1.65)$$

Here each route is specified to begin in state i and end in state f, passing through a succession of intermediate states $k, \ell, m, \ldots$ For each change of state the interaction enters via the matrix element $V_{fm}, V_{m\ell}, \ldots, V_{ki}$ and while in intermediate that ℓ the system "propagates" via the amplitude

$$\frac{1}{E_i - E_\ell + i\epsilon} \quad (1.66)$$

where the $+i\epsilon$ signifies the *outgoing* waves generated by the scattering. Actually the $+i\epsilon$ has a deeper significance. Consider the second order amplitude in Eq. 1.64, for which we can write

$$T_{fi}^{(2)} = \sum_b \frac{V_{fb}V_{bi}}{E_i - E_b + i\epsilon} \quad (1.67)$$

Although the $+i\epsilon$ appears innocuous here, its presence is essential. Originally introduced as a simple convergence factor in the integration, it plays a critical role also in expressions such as that above—if we imagine $\sum_b$ to be over states in the continuum, then $+i\epsilon$ *defines the meaning* of the sum. The summation in this case implies an integration

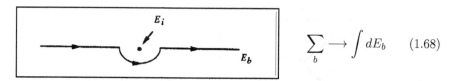

$$\sum_b \longrightarrow \int dE_b \quad (1.68)$$

and the $+i\epsilon$ requires the integration contour to deflect about the pole at E_i in the manner shown above. The integral

$$\int_{E_b^{\min}}^{\infty} dE_b \frac{f(E_b)}{E_b - E_i - i\epsilon}$$

becomes

$$\int_{E_b^{\min}}^{E_i-\epsilon} + \int_{E_i+\epsilon}^{\infty} dE_b \frac{f(E_b)}{E_b - E_i} + \int_{\mathcal{C}} dE_b \frac{f(E_b)}{E_b - E_i} \quad (1.69)$$

where $\mathcal{C}$ indicates the semicircular contour shown above. In the limit $\epsilon \to 0$ the first piece of Eq. 1.69 becomes the principal value integral

$$\int_{E_b^{\min}}^{\infty} dE_b P\left(\frac{f(E_b)}{E_b - E_i}\right) \qquad (1.70)$$

while for the semicircular contour, we define

$$\begin{aligned} E_b - E_i &= \eta\, e^{i\theta} \\ dE_b &= i\eta\, e^{i\theta} d\theta \end{aligned} \qquad (1.71)$$

Then

$$\lim_{\eta \to 0} \int_{\mathcal{C}} dE_b \frac{f(E_b)}{E_b - E_i} = \lim_{\eta \to 0} f(E_i) \int_{\pi}^{2\pi} \frac{i\eta\, e^{i\theta} d\theta}{\eta\, e^{i\theta}} = i\pi f(E_i)$$
$$= i\pi \int_{E_b^{\min}}^{\infty} dE_b F(E_b) \delta(E_b - E_i) \ . \qquad (1.72)$$

so we can write

$$\sum_b f(E_b) \frac{1}{E_b - E_i - i\epsilon} \equiv \sum_b f(E_b) \left(P\frac{1}{E_b - E_i} + i\pi \delta(E_b - E_i)\right) \ . \qquad (1.73)$$

The presence of the $+i\epsilon$ thus produces an imaginary piece of the transition amplitude whose presence is required by unitarity considerations, as we shall see in the next section.

PROBLEM II.1.1

Elastic Scattering of Fast Electrons by an Atom

Apply the Born approximation to the scattering of a fast electron by an atom, using as a perturbation the interaction

$$V(\vec{r}) = -\frac{Z\alpha}{r} + \sum_{i=1}^{Z} \frac{\alpha}{|\vec{r} - \vec{r}_i|}$$

which represents the Coulomb potential between the incident electron and the nucleus of charge $Z|e|$ and the individual electrons which make up the atom.

For the initial state wavefunction take

$$\langle \vec{r}, \vec{r}_1, \ldots \vec{r}_Z | i \rangle = e^{i\vec{k}_i \cdot \vec{r}} \psi_0(\vec{r}_1, \ldots, \vec{r}_Z)$$

and for the final state

$$\langle \vec{r}, \vec{r}_1, \ldots \vec{r}_Z | f \rangle = e^{i\vec{k}_f \cdot \vec{r}} \psi_0(\vec{r}_1, \ldots, \vec{r}_Z)$$

where $\psi_0(\vec{r}_1, \ldots, \vec{r}_Z)$ is the wavefunction of the ground state of the atom, assumed to be the same before and after the collision. [Note: In principle, the electron wavefunction should be antisymmetrized. However, at high energies this can be shown to be relatively unimportant.]

i) Demonstrate that the differential cross section takes the form

$$\frac{d\sigma}{d\Omega} = \left.\frac{d\sigma}{d\Omega}\right|_{\text{point}} \times |1 - F(\vec{q})|^2$$

where

$$\left.\frac{d\sigma}{d\Omega}\right|_{\text{point}} = \frac{Z^2 \alpha^2}{4m^2 v^4 \sin^4 \frac{1}{2}\theta}$$

is the scattering cross section corresponding to a point particle target of charge $Z|e|$ and $F(\vec{q})$ is the "form factor" for the atomic ground state $|\psi_0\rangle$.

$$F(\vec{q}) = \frac{1}{Z}\left\langle \psi_0 \left| \sum_{i=1}^{Z} e^{i\vec{q}\cdot\vec{r}_i} \right| \psi_0 \right\rangle$$

$$= \frac{1}{Z}\prod_{i=1}^{Z}\left(\int d^3 r_i\right) \sum_{i=1}^{Z} e^{i\vec{q}\cdot\vec{r}_i} |\psi_0(\vec{r}_1, \ldots, \vec{r}_Z)|^2$$

ii) For the specific case of electron scattering from the ground state of a hydrogen atom show that the differential scattering cross section has the form

$$\frac{d\sigma}{d\Omega} = \frac{\alpha^2}{4m^2 v^4 \sin^4 \frac{1}{2}\theta}\left(1 - \frac{1}{\left(1+\frac{q^2 a_0^2}{4}\right)^2}\right)^2$$

where $a_0 = 1/m\alpha$ is the Bohr radius.

In general, a form factor arises in problems wherein scattering is due to a composite system made up of identical scatterers (electrons in our case) which are distributed in space. The form factor expresses the feature that the scattering amplitude generated by different scattering centers differ from one another because of purely *geometric* considerations — variation of the phase of the incident wave at the scattering center and of the scattered wave at the detector. Measurement of the form factor thus yields information about the distribution of scattering centers within the composite system, *i.e.*, the Fourier transform of the spatial distribution. For a given momentum transfer $\vec{q}$ the form factor is sensitive to variations in the geometric distribution over distances of order d such that

$$qd \sim 1 \quad , \quad i.e., \quad d \sim \frac{1}{q} \quad .$$

This is the motivation behind the use of high energy scattering experiments to map the distribution of constituent scatterers.

iii) More quantitatively consider the situation diagrammed below: a source S and detector D with scattering centers at O,O'

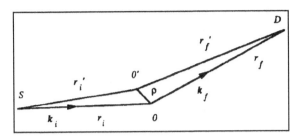

The full scattering amplitude is evidently

$$\text{Amp} = \frac{1}{r_i} e^{ikr_i} A\left(\vec{k}_i \to \vec{k}_f\right) \frac{1}{r_f} e^{ikr_f} + \frac{1}{r'_i} e^{ikr'_i} A\left(\vec{k}'_i \to \vec{k}'_f\right) \frac{1}{r'_f} e^{ikr'_f}.$$

Show that if $\vec{r}_i, \vec{r}'_i \gg |\rho|$ then it is reasonable to take the intrinsic scattering amplitude $A(\vec{k}_i \to \vec{k}_f)$ as being the *same* at both centers and that

$$\text{Amp} \approx \frac{1}{r_i} e^{ikr_i} A(\vec{k}_i \to \vec{k}_f) \left(1 + e^{i\vec{q}\cdot\vec{\rho}}\right) \frac{1}{r_f} e^{ikr_f}$$

where

$$\vec{q} = \vec{k}_i - \vec{k}_f$$

is the momentum transfer. Explain your result for atomic scattering in part i) in light of this form.

iv) Large momentum transfer: Show that if

$$q \gg \frac{1}{\text{Atomic Radius}}$$

then $|F(\vec{q})| \ll 1$ so that $\frac{d\sigma}{d\Omega}$ is essentially the Rutherford cross section for scattering of the electron by the Coulomb field of the nucleus.

v) Small momentum transfer: Assume for simplicity that atomic ground state wavefunction ψ_0 has total angular momentum zero. Expand $F(\vec{q})$ in powers of $\vec{q}$ and show that

$$ZF(\vec{q}) = Z - \frac{1}{6}q^2 \left\langle \psi_0 \left| \sum_{i=1}^{Z} r_i^2 \right| \psi_0 \right\rangle + \ldots$$

If the differential scattering cross section finite or infinite at $\theta = 0$? Explain this result.

vi) Analyze the hydrogen atom cross section calculated in part ii) in terms of the large and small q limits discussed in part iv).

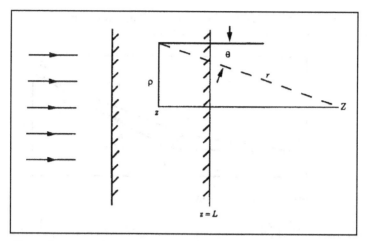

Fig. II.2 : Geometry for plane wave scattering from a slab of thickness L.

II.2 THE OPTICAL THEOREM

We commented above on the cancellation which must obtain between the incident plane wave and the outgoing spherical wave in the forward direction in order to conserve probability. This requirement leads to a result called the optical theorem. We shall derive this relation here in a slightly different context.

Intuitive Derivation

Consider a plane wave incident perpendicularly on a thin slab of matter extending from $z = 0$ to $z = L$. Assume that the slab is thin compared to the mean free path for interaction, and that the target atoms are distributed homogeneously at random (*i.e.*, no crystalline structure). Suppose that collisions result only in elastic scattering (this is not a necessary assumption and will be removed momentarily) described by a scattering amplitude $f_k(\theta)$ which is azimuthally symmetric (this is *also* not a necessary assumption).

The essential idea of the proof is that scattering out of the incident beam reduces the intensity of the wave transmitted in the forward direction. The transmitted amplitude at some point $Z > L$, say, must be regarded as the sum of the incident amplitude at Z plus the resultant of all elastically scattered waves arriving at Z

$$\psi(Z) = e^{ikZ} + \psi_{\text{scatt}}(Z) \ . \tag{2.1}$$

Then

$$|\psi(Z)|^2 = 1 + 2\,\text{Re}\,e^{-ikZ}\psi_{\text{scatt}}(Z) + |\psi_{\text{scatt}}(Z)|^2 \ . \tag{2.2}$$

But this expression must correspond to the usual reduction in intensity as a result of the scattering, namely[†]

$$|\psi(Z)|^2 = e^{-N\sigma L} \qquad Z \geq L \tag{2.3}$$

[†] Recall that the reduction in probability as a result of scattering out of the

where

$$N = \text{number density of target atoms} \qquad (2.4)$$
$$\sigma = \text{total scattering cross section} \,.$$

This loss in intensity is accomplished by *destructive* interference between $\psi_{\text{scatt}}(Z)$ and $\psi_{\text{inc}}(Z) = e^{ikZ}$, implying that a relation must exist between the scattering amplitude and the total cross section. This is the optical theorem.

In order to be more quantitative we must calculate the scattered wave $\psi_{\text{scatt}}(Z)$. The contribution to $\psi_{\text{scatt}}(Z)$ due to the target atoms of the slab between z and $z + dz$ is given by

$$d\psi_{\text{scatt}} = e^{ikz} N \, dz \int_0^\infty \frac{1}{r} e^{ikr} f_k(\theta) 2\pi \rho \, d\rho \qquad (2.5)$$

where the notation is as given in Fig. II.2. [Note that we neglect attenuation of the incident amplitude e^{ikz} falling on each atom in slice $z, z + dz$ since we assume the slab to be sufficiently thin that at most a single scattering occurs.]

From Figure II.2 we have

$$r^2 = \rho^2 + (Z - z)^2 \quad \text{so} \quad \rho \, d\rho = r \, dr \,, \qquad (2.6)$$

and

$$\int_0^\infty d\rho \, \rho \frac{1}{r} e^{ikr} f_k(\theta) = \int_{Z-z}^\infty dr \, e^{ikr} f_k(\theta) \,. \qquad (2.7)$$

This integral diverges as $r \to \infty$ so we introduce a convergence factor — $e^{-\epsilon r}$, representing absorption — and integrate by parts

$$\int_{Z-z}^\infty dr \, e^{ikr} f_k(\theta) = f_k(\theta) \frac{e^{ikr-\epsilon r}}{ik} \bigg|_{Z-z}^\infty - \frac{1}{ik} \int_{Z-z}^\infty dr \, e^{ikr-\epsilon r} \frac{df_k(\theta)}{dr} \,. \qquad (2.8)$$

Since $f_k(\theta)$ changes very slowly when r changes by one wavelength

$$\left| \frac{1}{k} \frac{d}{dr} f_k(\theta) \right| \lll |f_k(\theta)| \,, \qquad (2.9)$$

the second piece of Eq. 2.8 can be neglected yielding

$$\int_{Z-z}^\infty dr \, e^{ikr} f_k(\theta) \approx \frac{i}{k} f_k(0) e^{-ik(Z-z)} \qquad (2.10)$$

---

incident beam is given by

$$\frac{dP}{dx} = -N\sigma_{\text{tot}} P$$

where N is the number of scattering centers per unit volume and σ_{tot} denotes the total scattering cross section. Thus after passing through a thickness L of material the resulting probability is reduced by the factor $\exp(-N\sigma_{\text{tot}} L)$.

and

$$\int d\psi_{\text{scatt}} = \int_0^L 2\pi N \, dz \frac{i}{k} e^{ikz} f_k(0) \, e^{ik(Z-z)}$$
$$= \int_0^L 2\pi i N \frac{1}{k} \, dz \, f_k(0) \, e^{ikZ} = \frac{2\pi i N L}{k} f_k(0) \, e^{ikZ} = \psi_{\text{scatt}}(z) \ . \quad (2.11)$$

The total amplitude at Z becomes

$$\psi(Z) = e^{ikZ} \left(1 + \frac{2\pi i N L}{k} f_k(0) \right) \approx e^{ikZ} \, e^{i\frac{2\pi N L}{k} f_k(0)} \ . \quad (2.12)$$

Then

$$|\psi(Z)|^2 = e^{-\frac{4\pi N L}{k} \text{Im} f_k(0)} \equiv e^{-N\sigma L} \quad (2.13)$$

which requires

$$\sigma = \frac{4\pi}{k} \text{Im} f_k(0) \quad (2.14)$$

This result is the optical theorem and is the relationship which we were seeking.

Note that in the case that not all scattering is elastic, inelastic processes will also contribute to the attenuation via

$$\sigma_{\text{tot}} = \sigma_{\text{elastic}} + \sigma_{\text{absorption}} \quad (2.15)$$

where $\sigma_{\text{absorption}}$ is the total cross section for all processes other than elastic scattering. On the other hand, cancellation between the incident and scattered wave is relevant only for that component of the scattered wave which can interfere with the incident wave. Thus the optical theorem becomes

$$\sigma_{\text{tot}} = \frac{4\pi}{k} \text{Im} f_{k,\text{el}}(0) \quad (2.16)$$

where $f_{k,\text{el}}(0)$ is the amplitude for coherent forward elastic scattering — scattering with no change in the energy or spin state of the incident and target particles. Eq. 2.16 implies that if there is a large inelastic cross section for some particular process then there must also exist a correspondingly large elastic scattering cross section as well (at least in the forward direction).

Formal Derivation

Although we have derived the optical theorem above in a physically intuitive fashion, it is also possible and desirable to get the same result via formal techniques. This procedure is quicker but less revealing. The idea again is that probability must be conserved. That is, if one begins in a state $|a\rangle$ which is normalized to unity so that

$$\langle a|a\rangle = 1 \ , \quad (2.17)$$

then as time evolves into the remote future the state $|a\rangle$ evolves into other states $|b\rangle$. However, the total probability of being in *some* state must remain unity — i.e.,

$$\lim_{\substack{t'\to\infty \\ t\to-\infty}} \sum_b \left|\langle b|\hat{U}(t',t)|a\rangle\right|^2 = \sum_b |\langle b|S|a\rangle|^2$$

$$= \sum_b \langle a|S^\dagger|b\rangle \langle b|S|a\rangle \quad (2.18)$$

$$= \langle a|S^\dagger S|a\rangle = 1$$

where we have used the completeness property of the states $|b\rangle$

$$1 = \sum_b |b\rangle\langle b| \quad . \quad (2.19)$$

Since Eq. 2.18 must hold for an arbitrary initial state $|a\rangle$ we must have

$$S^\dagger S = 1 \quad . \quad (2.20)$$

[Note: If the final state $|b\rangle$ is kept fixed and we sum over initial states $|a\rangle$ which lead to this state then

$$\sum_a |\langle b|S|a\rangle|^2 = \sum_a \langle b|S|a\rangle \langle a|S^\dagger|b\rangle$$

$$= \langle b|SS^\dagger|b\rangle = 1 \quad (2.21)$$

so that we also have

$$SS^\dagger = 1 \quad (2.22)$$

i.e., the S-matrix is unitary. For a finite matrix either one of these relations implies the other. For an infinite-dimensional matrix this is not always the case.]

We have then a quick but rigorous derivation of the optical theorem via

$$\langle \vec{p}_c|S^\dagger S|\vec{p}_a\rangle = \int \frac{d^3 p_b}{(2\pi)^3} \langle \vec{p}_c|S^\dagger|\vec{p}_b\rangle \langle \vec{p}_b|S|\vec{p}_a\rangle$$

$$= \langle \vec{p}_c|1|\vec{p}_a\rangle = (2\pi)^3 \delta^3(\vec{p}_c - \vec{p}_a) \quad . \quad (2.23)$$

Since

$$\langle \vec{p}_b|S|\vec{p}_a\rangle = (2\pi)^3 \delta^3(\vec{p}_b - \vec{p}_a) - 2\pi i \delta(E_b - E_a)\langle \vec{p}_b|\hat{V}|\psi_a^{(+)}\rangle$$

$$\langle \vec{p}_c|S^\dagger|\vec{p}_b\rangle = (2\pi)^3 \delta^3(\vec{p}_c - \vec{p}_b) + 2\pi i \delta(E_c - E_b)\langle \vec{p}_b|\hat{V}|\psi_c^{(+)}\rangle^* \quad (2.24)$$

we have

$$0 = -2\pi i \delta(E_c - E_a)\langle \vec{p}_c|\hat{V}|\psi_a^{(+)}\rangle + 2\pi i \delta(E_c - E_a)\langle \vec{p}_a|\hat{V}|\psi_c^{(+)}\rangle^*$$

$$+ \int \frac{d^3 p_b}{(2\pi)^3} 2\pi \delta(E_a - E_b) 2\pi \delta(E_b - E_c)\langle \vec{p}_b|\hat{V}|\psi_c^{(+)}\rangle^* \langle \vec{p}_b|\hat{V}|\psi_a^{(+)}\rangle \quad (2.25)$$

or, integrating over one of the product of delta functions,

$$0 = i\left(\left\langle\vec{p}_a|\hat{V}|\psi_c^{(+)}\right\rangle^* - \left\langle\vec{p}_c|\hat{V}|\psi_a^{(+)}\right\rangle\right)$$
$$+ 2\pi p^2 \frac{dp}{dE}\frac{1}{(2\pi)^3}\int \left\langle\vec{p}_b|\hat{V}|\psi_c^{(+)}\right\rangle^* \left\langle\vec{p}_b|\hat{V}|\psi_a^{(+)}\right\rangle d\Omega_b \ . \quad (2.26)$$

Now specialize to the forward direction — $\vec{p}_a = \vec{p}_c$. Then since

$$\frac{dE}{dp} = \frac{p}{m} \quad (2.27)$$

we find

$$\frac{4\pi}{m}\operatorname{Im} f_k(0) = \frac{p}{m}\int d\Omega\,|f_k(\theta)|^2 \quad (2.28)$$

or

$$\operatorname{Im} f_k(0) = \frac{p}{4\pi}\sigma \quad (2.29)$$

as given previously.

PROBLEM II.2.1

The Born Approximation and the Optical Theorem

i) In the situation that the scattering potential is radial—$V(\vec{r}) = V(r)$—show that
$$f_k^{\text{Born}}(\theta,\phi) = -\frac{m}{2\pi}\left\langle\phi_f\left|\hat{V}\right|\phi_i\right\rangle = -\frac{m}{2\pi}\int d^3r\, V(r) e^{i(\vec{p}_i-\vec{p}_f)\cdot\vec{r}}$$
is real.

Thus $\operatorname{Im} f_k(0) = 0$ so that the optical theorem seems not to be satisfied. However, this is to be expected, since the cross section to which $\operatorname{Im} f_k(0)$ is proportional is second order in the potential. Thus we must go to one higher level of approximation.

ii) Show that, correct to second order in the potential

$$f_k(\theta,\phi) = -\frac{m}{2\pi}\left(\left\langle\phi_f\left|\hat{V}\right|\phi_i\right\rangle + \sum_j \frac{\left\langle\phi_f\left|\hat{V}\right|\phi_j\right\rangle\left\langle\phi_j\left|\hat{V}\right|\phi_i\right\rangle}{E_i - E_j + i\epsilon} + \cdots\right)$$

iii) Evaluate $\operatorname{Im} f_k(0)$ and verify that the optical theorem is satisfied, provided that the Born cross section is used.

PROBLEM II.2.2

The Optical Theorem: Another Approach

We have shown (*cf.* Eq. 1.73) that under an integral $\frac{1}{E-E_0+i\epsilon} = P\frac{1}{E-E_0} - i\pi\delta(E - E_0)$. An interesting application of this result is to prove the golden rule *and* the optical theorem.

II.2 THE OPTICAL THEOREM

The counter in a scattering experiment can be considered as a device for measuring the probability that a particle described by wavefunction $\psi_a^{(+)}(\vec{r})$ will be in state $|\phi_b\rangle$ which takes it into the counter. Here $|\phi_b\rangle$ is an eigenstate of the free Hamiltonian

$$\hat{H}_0|\phi_b\rangle = E_b|\phi_b\rangle$$

while $|\psi_a^{(+)}\rangle$ is an eigenstate of the full Hamiltonian $\hat{H} = \hat{H}_0 + \hat{V}$

$$\hat{H}|\psi_a^{(+)}\rangle = E_a|\psi_a^{(+)}\rangle \ .$$

i) Show that the state

$$|\psi_a^{(+)}\rangle = |\phi_a\rangle + \frac{1}{E_a - \hat{H}_0 + i\epsilon}\hat{V}|\psi_a^{(+)}\rangle$$

has this property. The overlap of state $|\psi_a^{(+)}\rangle$ with $|\phi_b\rangle$ is then found to be

$$\left\langle\phi_b|\psi_a^{(+)}\right\rangle = \langle\phi_b|\left(|\phi_a\rangle + \frac{1}{E_a - E_b + i\epsilon}\hat{V}\left|\psi_a^{(+)}\right\rangle\right)$$

$$= \delta_{ab} + \frac{1}{E_a - E_b + i\epsilon}\mathcal{F}(E_a)$$

where $\mathcal{F}(E_A) = \left\langle\phi_b|\hat{V}|\psi_a^{(+)}\right\rangle$ is the transition matrix element.

ii) If $w_{ba} = \left|\left\langle\phi_b|\psi_a^{(+)}\right\rangle\right|^2$ then the rate is

$$\frac{dw_{ba}}{dt} = \left(\left\langle\dot\phi_b|\psi_a^{(+)}\right\rangle + \left\langle\phi_b|\dot\psi_a^{(+)}\right\rangle\right)\left\langle\phi_b|\psi_a^{(+)}\right\rangle^* + \text{c.c.} \ .$$

Use the equations satisfied by $|\psi_a^{(+)}\rangle$ and $|\phi_b\rangle$ to show

$$\frac{dw_{ba}}{dt} = -i\left[\left\langle\phi_b|\hat{V}|\psi_a^{(+)}\right\rangle\left\langle\phi_b|\psi_a^{(+)}\right\rangle^* - \text{c.c.}\right] \ .$$

iii) Show that if $b \neq a$

$$\frac{dw_{ba}}{dt} = 2\pi\delta(E_a - E_b)\,|\mathcal{F}(E_a)|^2$$

which is Fermi's Golden Rule.

iv) Use the fact that $\sum_b \frac{dw_{ba}}{dt} = 0$ to derive

$$\text{Im}\left\langle\phi_a|\hat{V}|\psi_a^{(+)}\right\rangle \cong -\frac{p_i}{2m}\sigma_a$$

where σ_a is the total scattering cross section. Since

$$f_k(\theta) = -\frac{m}{2\pi}\left\langle\phi_b|\hat{V}|\psi_a^{(+)}\right\rangle$$

is the scattering amplitude, we have

$$\text{Im}\,f_k(0) = \frac{p_i}{4\pi}\sigma_a$$

which is the standard optical theorem.

II.3 PATH INTEGRAL APPROACH

It is interesting to compare the traditional — wave function — approach to the derivation of the scattering cross section with that given by the path integral or propagator approach [FeH 65].

Born Approximation

Consider the first order (Born approximation) term for which the propagator is

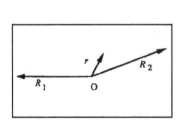

$$
\begin{aligned}
K^{(1)}(2,1) &= -i \int_0^t dt' \int d^3 r \, K^{(0)}(\vec{R}_2, t; \vec{r}, t') \\
&\quad \times V(\vec{r}) K^{(0)}(\vec{r}, t'; \vec{R}_1, 0) \\
&= -i \int_0^t dt' \int d^3 r \left(\frac{m}{2\pi i(t-t')}\right)^{3/2} \left(\frac{m}{2\pi i t'}\right)^{3/2} \\
&\quad \times \exp\left(\frac{im(\vec{R}_2 - \vec{r})^2}{2(t-t')} + \frac{im(\vec{r} - \vec{R}_1)^2}{2t'}\right) V(\vec{r}) \ .
\end{aligned}
\tag{3.1}
$$

Although we could imagine performing the time integration via stationary phase methods, it is possible to perform the quadrature exactly, yielding [†]

$$
\begin{aligned}
K^{(1)}(2,1) &= -i \left(\frac{m}{2\pi i t}\right)^{5/2} t \int d^3 r \, V(\vec{r}) \left(\frac{1}{|\vec{R}_2 - \vec{r}|} + \frac{1}{|\vec{r} - \vec{R}_1|}\right) \\
&\quad \times \exp \frac{im}{2t}\left(|\vec{R}_2 - \vec{r}| + |\vec{r} - \vec{R}_1|\right)^2 \ .
\end{aligned}
\tag{3.2}
$$

Assuming that the initial and final spatial locations are far outside the (localized) region of the potential, we write

$$
\begin{aligned}
|\vec{R}_2 - \vec{r}| &= \left(R_2^2 - 2\vec{R}_2 \cdot \vec{r} + r^2\right)^{1/2} \cong R_2 - \hat{p}_f \cdot \vec{r} \\
|\vec{r} - \vec{R}_1| &= \left(R_1^2 - 2\vec{R}_1 \cdot \vec{r} + r^2\right)^{1/2} \cong R_1 + \hat{p}_i \cdot \vec{r}
\end{aligned}
\tag{3.3}
$$

where $\hat{p}_f = \hat{R}_2$, $\hat{p}_i = -\hat{R}_1$ are unit vectors in the direction of final, initial momenta respectively. We have then

$$
\begin{aligned}
K^{(1)}(2,1) &\cong -i \left(\frac{m}{2\pi i t}\right)^{5/2} t \left(\frac{1}{R_1} + \frac{1}{R_2}\right) \exp \frac{im}{2t}(R_1 + R_2)^2 \\
&\quad \times \int d^3 r \, V(\vec{r}) \exp \frac{im}{t}(R_1 + R_2)\vec{r} \cdot (\hat{p}_i - \hat{p}_f) \ .
\end{aligned}
\tag{3.4}
$$

[†] The result

$$
\int_0^T \exp\left(-\frac{a}{T-t} - \frac{b}{t}\right) \frac{dt}{(T-t)^{\frac{3}{2}} t^{\frac{3}{2}}} = \sqrt{\frac{\pi}{T^3}} \frac{\sqrt{a} + \sqrt{b}}{\sqrt{ab}} \exp\left[-\frac{1}{T}(\sqrt{a} + \sqrt{b})^2\right]
$$

is given by Feynman and Hibbs [FeH 65].

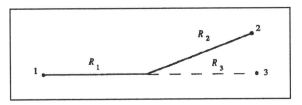

Fig. II.3: Geometry for the path integral discussion of scattering.

If the interaction region is small compared to R_1, R_2 the particle velocity is

$$v \approx \frac{R_1 + R_2}{t} \tag{3.5}$$

so

$$K^{(1)}(2,1) \cong -i\left(\frac{m}{2\pi i}\right)^{5/2} \frac{v}{t^{1/2} R_1 R_2} \exp(iEt)\tilde{V}(\vec{q}) \tag{3.6}$$

where

$$E = \frac{1}{2}mv^2 \tag{3.7}$$

is the asymptotic particle energy and

$$\tilde{V}(\vec{q}) = \int d^3r\, V(\vec{r}) e^{i\vec{q}\cdot\vec{r}} \tag{3.8}$$

is the Fourier transform of the potential energy, with $\vec{q} = \vec{p}_i - \vec{p}_f$ being the momentum transfer. The probability per unit volume of the particle being scattering into the region about point 2 is given by

$$\frac{P(2)}{\text{Volume}} = \left(\frac{m}{2\pi}\right)^5 \frac{v^2}{tR_1^2 R_2^2} \left|\tilde{V}(\vec{q})\right|^2 \tag{3.9}$$

On the other hand the propagator to travel directly to point 3, as shown in Figure II.3, without scattering is given by

$$K^{(0)}(3,1) = \left(\frac{m}{2\pi it}\right)^{3/2} \exp\frac{im(R_1+R_2)^2}{2t} \tag{3.10}$$

so

$$\frac{P(3)}{\text{Volume}} = \left(\frac{m}{2\pi t}\right)^3 = \left(\frac{m}{2\pi}\right)^3 \frac{v^2}{t(R_1+R_2)^2} \cdot \tag{3.11}$$

Then

$$\frac{P(2)}{P(3)} = \left|-\frac{m}{2\pi}\tilde{V}(\vec{q})\right|^2 \frac{(R_1+R_2)^2}{R_1^2 R_2^2} \cdot \tag{3.12}$$

We may interpret this ratio in terms of the differential scattering cross section as follows: If particles starting at position #1 were to strike a small target of area $d\sigma$ at the origin, they would be removed from an area $d\sigma \times (R_1+R_2)^2/R_1^2$ at position #3. Instead, they are redirected into solid angle $d\Omega$ toward point #2 where they cover an area $R_2^2 d\Omega$. Thus the ratio of probabilities for finding them at 2 vs. finding them at 3 is

$$\frac{P(2)}{P(3)} = \frac{d\sigma(R_1+R_2)^2/R_1^2}{R_2^2 d\Omega} = \left|-\frac{m}{2\pi}\tilde{V}(\vec{q})\right|^2 \frac{(R_1+R_2)^2}{R_1^2 R_2^2} \quad , \qquad (3.13)$$

yielding

$$\frac{d\sigma}{d\Omega} = \left|-\frac{m}{2\pi}\tilde{V}(\vec{q})\right|^2 \qquad (3.14)$$

in exact agreement with the Born approximation cross section derived via wave function methods.

Higher Order Contributions

Generalization to higher orders is straightforward. The second order term is written as

$$K^{(2)}(2,1) = (-i)^2 \int_0^t dt_2 \int_0^{t_2} dt_1 \int d^3r_2 \int d^3r_1 \, K^{(0)}(\vec{R}_2,t;\vec{r}_2,t_2)V(\vec{r}_2)$$
$$\times K^{(0)}(\vec{r}_2,t_2;\vec{r}_1,t_1)V(\vec{r}_1)K^{(0)}(\vec{r}_1,t_1;\vec{R}_1,0)$$
$$= (-i)^2 \int_0^t dt_2 \int_0^{t_2} dt_1 \int d^3r_2 \int d^3r_1 \left(\frac{m}{2\pi i(t-t_2)}\right)^{3/2} \exp\frac{im(\vec{R}_2-\vec{r}_2)^2}{2(t-t_2)}$$
$$\times V(\vec{r}_2)\left(\frac{m}{2\pi i(t_2-t_1)}\right)^{3/2} \exp\frac{im(\vec{r}_2-\vec{r}_1)^2}{2(t_2-t_1)}V(\vec{r}_1)\left(\frac{m}{2\pi i t_1}\right)^{3/2} \exp\frac{im(\vec{r}_1-\vec{R}_1)^2}{2t_1}$$
$$(3.15)$$

The time integrations may be performed sequentially, yielding

$$K^{(2)}(2,1) = (-i)^2 \left(\frac{m}{2\pi i t}\right)^{7/2} t^2 \int d^3r_1 \int d^3r_2 \frac{|\vec{R}_2-\vec{r}_2|+|\vec{r}_2-\vec{r}_1|+|\vec{r}_1-\vec{R}_1|}{|\vec{R}_2-\vec{r}_2|\times|\vec{r}_2-\vec{r}_1|\times|\vec{r}_1-\vec{R}_1|}$$
$$\times V(\vec{r}_1)V(\vec{r}_2)\exp\frac{im}{2t}\left(|\vec{R}_2-\vec{r}_2|+|\vec{r}_2-\vec{r}_1|+|\vec{r}_1-\vec{R}_1|\right)^2 \quad . \qquad (3.16)$$

Under the assumption that $R_1, R_2 \ggg r_1, r_2$ as before, we have

$$K^{(2)}(2,1) = \left(\frac{m}{2\pi i}\right)^{3/2} \frac{v}{t^{1/2}R_1 R_2}\left(\frac{m}{2\pi}\right)^2 \exp(iEt)\int d^3r_1 d^3r_2 \, V(\vec{r}_1)V(\vec{r}_2)$$
$$\times \exp\left(i\vec{p}_i\cdot\vec{r}_1 - i\vec{p}_f\cdot\vec{r}_2\right) \times \frac{\exp ip_i|\vec{r}_1-\vec{r}_2|}{|\vec{r}_1-\vec{r}_2|} \quad . \qquad (3.17)$$

Hence, to this order

$$\frac{d\sigma}{d\Omega} = |\mathcal{F}(E_i)|^2 \qquad (3.18)$$

with

$$\mathcal{F}(E_i) = -\frac{m}{2\pi}\int d^3r \, e^{i(\vec{p}_i-\vec{p}_f)\cdot\vec{r}}V(\vec{r})$$
$$+ \left(\frac{m}{2\pi}\right)^2 \int d^3r_2 d^3r_1 \, e^{-i\vec{p}_f\cdot\vec{r}_2}V(\vec{r}_2)\frac{\exp ip_i|\vec{r}_1-\vec{r}_2|}{|\vec{r}_1-\vec{r}_2|}V(\vec{r}_1)e^{i\vec{p}_i\cdot\vec{r}_1} \quad . \qquad (3.19)$$

Noting that (*cf.* Eq. 1.28)

$$\left\langle \vec{r}_2 \left| \frac{1}{E_i - \hat{H}_0 + i\epsilon} \right| \vec{r}_1 \right\rangle = -\frac{m}{2\pi} \frac{1}{|\vec{r}_2 - \vec{r}_1|} \exp i p_i |\vec{r}_2 - \vec{r}_1| \qquad (3.20)$$

where $E_i \equiv \frac{p_i^2}{2m}$ we have

$$\mathcal{F}(E_i) = -\frac{m}{2\pi} \left\langle \phi_f \left| \hat{V} + \hat{V} \frac{1}{E_i - \hat{H}_0 + i\epsilon} \hat{V} + \ldots \right| \phi_i \right\rangle \qquad (3.21)$$

and the generalization to higher orders is obvious.

II.4 TWO-BODY SCATTERING

Thus far we have been proceeding as if the scattering process occurs off some immovable potential source. However, realistically one deals with the scattering of two particles under the influence of their mutual interaction.

Two-Body Matrix Element

Consider then a pair of particles of mass m_1 and m_2, respectively. Provided that they interact through a potential V which depends only upon the relative separation of the particles — $\vec{r}_1 - \vec{r}_2$ — we expect that, based upon the corresponding classical mechanics problem, overall momentum will be conserved. Mathematically this must be manifested in the requirement that the momentum space matrix element

$$\left\langle \vec{p}_1^f \vec{p}_2^f | \hat{V} | \vec{p}_1^i \vec{p}_2^i \right\rangle = 0 \quad \text{unless} \quad \vec{p}_1^i + \vec{p}_2^i = \vec{p}_1^f + \vec{p}_2^f \ . \qquad (4.1)$$

Representing in coordinate space

$$\left\langle \vec{r}_1' \vec{r}_2' | \hat{V} | \vec{r}_1 \vec{r}_2 \right\rangle = \delta^3(\vec{r}_1' - \vec{r}_1) \delta^3(\vec{r}_2' - \vec{r}_2) V(\vec{r}_1 - \vec{r}_2) \qquad (4.2)$$

we calculate

$$\begin{aligned}\left\langle \vec{p}_1^f \vec{p}_2^f | V | \vec{p}_1^i \vec{p}_2^i \right\rangle &= \int d^3 r_1 \, d^3 r_2 \left\langle \vec{p}_1^f \vec{p}_2^f | \vec{r}_1 \vec{r}_2 \right\rangle V(\vec{r}_1 - \vec{r}_2) \left\langle \vec{r}_1 \vec{r}_2 | \vec{p}_1^i \vec{p}_2^i \right\rangle \\ &= \int d^3 r_1 \, d^2 r_2 \, e^{-i \vec{p}_1^f \cdot \vec{r}_1 - i \vec{p}_2^f \cdot \vec{r}_2 + i \vec{p}_1^i \cdot \vec{r}_1 + i \vec{p}_2^i \cdot \vec{r}_2} V(\vec{r}_1 - \vec{r}_2) \ .\end{aligned} \qquad (4.3)$$

In order to evaluate the integral we change variables to center of mass coordinates

$$\vec{r} = \vec{r}_1 - \vec{r}_2 \quad \text{and} \quad \vec{R} = \frac{m_1 \vec{r}_1 + m_2 \vec{r}_2}{m_1 + m_2} \ . \qquad (4.4)$$

Since the Jacobian for this transformation is unity — $d^3 r_1 \, d^3 r_2 = d^3 r \, d^3 R$ — we find

$$\begin{aligned}\left\langle \vec{p}_1^f \vec{p}_2^f | \hat{V} | \vec{p}_1^i \vec{p}_2^i \right\rangle &= \int d^3 R \, \exp i \vec{R} \cdot (\vec{p}_1^i + \vec{p}_2^i - \vec{p}_1^f - \vec{p}_2^f) \\ &\quad \times \int d^3 r \, V(\vec{r}) \exp i \vec{r} \cdot \left(\frac{m_2 \vec{p}_1^i - m_1 \vec{p}_2^i}{m_1 + m_2} - \frac{m_2 \vec{p}_1^f - m_1 \vec{p}_2^f}{m_1 + m_2} \right) \\ &= (2\pi)^3 \delta^2(\vec{p}_1^i + \vec{p}_2^i - \vec{p}_1^f - \vec{p}_2^f) \int d^3 r \, V(\vec{r}) \\ &\quad \times \exp i \vec{r} \cdot \left(\frac{m_2 \vec{p}_1^i - m_1 \vec{p}_2^i}{m_1 + m_2} - \frac{m_2 \vec{p}_1^f - m_1 \vec{p}_2^f}{m_1 + m_2} \right)\end{aligned} \qquad (4.5)$$

so that overall momentum conservation is apparent. This result is most easily expressed in terms of center of mass momenta

$$\vec{P} = \vec{p}_1 + \vec{p}_2$$
$$\left(\frac{1}{m_1} + \frac{1}{m_2}\right)\vec{p} = \frac{1}{m_1}\vec{p}_1 - \frac{1}{m_2}\vec{p}_2 , \qquad (4.6)$$

in terms of which the matrix element can be written

$$\left\langle \vec{p}_1^f \vec{p}_2^f | \hat{V} | \vec{p}_1^i \vec{p}_2^i \right\rangle = (2\pi)^3 \delta^3(\vec{P}_i - \vec{P}_f) V_{\vec{p}_f, \vec{p}_i} \qquad (4.7)$$

where

$$V_{\vec{p}_f, \vec{p}_i} \equiv \int d^3 r \, e^{-i\vec{p}_f \cdot \vec{r}} V(\vec{r}) e^{i\vec{p}_i \cdot \vec{r}} \qquad (4.8)$$

is the Fourier transform of the potential energy.

Similarly one can express the energy in terms of these momenta

$$E = \frac{\vec{p}_1^2}{2m_1} + \frac{\vec{p}_2^2}{2m_2} = \frac{\vec{P}^2}{2(m_1 + m_2)} + \frac{\vec{p}^2}{2\mu} \qquad (4.9)$$

where

$$\mu = \frac{1}{\frac{1}{m_1} + \frac{1}{m_2}} \qquad (4.10)$$

is the reduced mass.

Two-Body Scattering Amplitude

We can analyze the two particle scattering process in terms of the formalism already developed. That is, we take the initial and final states to be eigenstates of $\hat{H}_0$ with

$$\begin{aligned}|\phi_i\rangle = |\vec{p}_1^i \vec{p}_2^i\rangle && E_i = \frac{\vec{p}_1^{i2}}{2m_1} + \frac{\vec{p}_2^{i2}}{2m_2} \\ |\phi_f\rangle = |\vec{p}_1^f \vec{p}_2^f\rangle && E_f = \frac{\vec{p}_1^{f2}}{2m_1} + \frac{\vec{p}_2^{f2}}{2m_2} .\end{aligned} \qquad (4.11)$$

The transition amplitude can be represented diagrammatically as indicated below

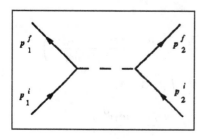

$$\text{Amp}^{(1)} = -2\pi i \delta(E_f - E_i)$$
$$\times (2\pi)^3 \delta^3(\vec{P}_f - \vec{P}_i) \left\langle \vec{p}_f | \hat{V} | \vec{p}_i \right\rangle$$

$$(4.12)$$

II.4 TWO-BODY SCATTERING

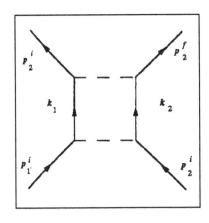

$$\text{Amp}^{(2)} = -2\pi i \delta(E_f - E_i) \int \frac{d^3k_1}{(2\pi)^3} \frac{d^3k_2}{(2\pi)^3}$$
$$\times \frac{\langle \vec{p}_1^f \vec{p}_2^f | \hat{V} | \vec{k}_1 \vec{k}_2 \rangle \langle \vec{k}_1 \vec{k}_2 | \hat{V} | \vec{p}_1^i \vec{p}_2^i \rangle}{E_i - \frac{1}{2m_1}\vec{k}_1^2 - \frac{1}{2m_2}\vec{k}_2^2 + i\epsilon} \ .$$

(4.13)

Switching to center of mass momenta $\vec{k}, \vec{K}$, and noting that the momentum transformation Jacobian is unity — $d^3k_1\, d^3k_2 = d^3k\, d^3K$ — we find

$$\text{Amp}^{(2)} = -2\pi i \delta(E_f - E_i) \int \frac{d^3k\, d^3K}{(2\pi)^6} \frac{(2\pi)^3 \delta^3(\vec{P}_f - \vec{K}) V_{\vec{p}_f, \vec{k}} (2\pi)^3 \delta^3(\vec{K} - \vec{P}_i) V_{\vec{k}, \vec{p}_i}}{\frac{\vec{P}_i^2}{2(m_1+m_2)} + \frac{\vec{p}_i^2}{2\mu} - \frac{\vec{K}^2}{2(m_1+m_2)} - \frac{\vec{k}^2}{2\mu} + i\epsilon}$$
$$= -2\pi i \delta(E_f - E_i)(2\pi)^3 \delta^3(\vec{P}_f - \vec{P}_i) \int \frac{d^3k}{(2\pi)^3} \frac{V_{\vec{p}_f, \vec{k}} V_{\vec{k}, \vec{p}_i}}{\frac{\vec{p}_i^2}{2\mu} - \frac{\vec{k}^2}{2\mu} + i\epsilon} \ .$$

(4.14)

Likewise, the third order term is easily determined

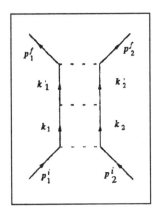

$$\text{Amp}^{(3)} = -i 2\pi \delta(E_f - E_i)(2\pi)^3 \delta^3(\vec{P}_f - \vec{P}_i)$$
$$\times \int \frac{d^3k}{(2\pi)^3} \frac{d^3k'}{(2\pi)^3} \frac{V_{\vec{p}_f, \vec{k}'} V_{\vec{k}', \vec{k}} V_{\vec{k}, \vec{p}_i}}{\left(\frac{\vec{p}_i^2}{2\mu} - \frac{\vec{k}^2}{2\mu} + i\epsilon\right)\left(\frac{\vec{p}_i^2}{2\mu} - \frac{\vec{k}'^2}{2\mu} + i\epsilon\right)}$$

(4.15)

and the generalization to arbitrary order is clear:

$$\text{Amp}_{fi} = \langle \vec{p}_1^f \vec{p}_2^f | \vec{p}_1^i \vec{p}_2^i \rangle - i 2\pi \delta(E_f - E_i)(2\pi)^3 \delta^3(\vec{P}_f - \vec{P}_i) T_{fi}$$
$$= (2\pi)^6 \delta^3(\vec{p}_1^f - \vec{p}_1^i) \delta^3(\vec{p}_2^f - \vec{p}_2^i) - i(2\pi)^4 \delta^4(p_f - p_i) T_{fi} \ .$$

(4.16)

Here the first term, giving the inner product of the initial and final scattering states, vanishes except in the very forward direction and is often omitted. The second piece involves the T-matrix element multiplied by an energy and momentum conserving delta function and describes the scattering in terms of a "fictitious" particle of mass μ interacting with a central potential $V(\vec{r})$

$$T_{fi} = V_{\vec{p}_f,\vec{p}_i} + \int \frac{d^3k}{(2\pi)^3} \frac{V_{\vec{p}_f,\vec{k}} V_{\vec{k},\vec{p}_i}}{\frac{\vec{p}_i^2}{2\mu} - \frac{\vec{k}^2}{2\mu} + i\epsilon} + \cdots \qquad (4.17)$$

The two-body problem has been reduced to an effective one-body problem, just as in classical mechanics.

II.5 TWO-PARTICLE SCATTERING CROSS SECTION

Calculation of the two-body differential scattering cross section takes place parallel to that given above for the corresponding one-body problem. The presence of the four-dimensional delta function suggests that some caution is in order, however. In reality the calculation *should* be performed by the use of wavepackets. However, a generalization of the simplified one-body "derivation" gives the same result and will suffice.

Scattering Cross Section

Suppose that we are not in the forward direction. Then the zeroeth order term $\langle f|i\rangle$ vanishes and the transition probability is given by

$$\text{Prob}(i \to f) = \left[(2\pi)^4 \delta^4(P_f - P_i)\right]^2 |T_{fi}|^2 \quad . \qquad (5.1)$$

Now write one of the four-dimensional delta functions in terms of its Fourier transform and notice that the presence of the remaining delta function requires $P_i = P_f$

$$\begin{aligned}
\text{Prob}(i \to f) &= (2\pi)^4 \delta^4(P_i - P_f) \int d^4x \, \exp i(P_f - P_i) \cdot x \times |T_{fi}|^2 \\
&= (2\pi)^4 \delta^4(P_i - P_f) \int d^4x \, |T_{fi}|^2 \\
&= (2\pi)^4 \delta^4(P_i - P_f) VT \, |T_{fi}|^2
\end{aligned} \qquad (5.2)$$

where $V = 1$ is the quantization volume and T is the total interaction time. The cross section involves the *transition rate* to a group of final states, *i.e.*,

$$\sum_f \begin{array}{c}\text{Transition probability}\\ \text{per unit time}\end{array} = \int \frac{d^3 p_1^f}{(2\pi)^3} \frac{d^3 p_2^f}{(2\pi)^3} (2\pi)^4 \delta^4(P_i - P_f) |T_{fi}|^2 \qquad (5.3)$$

divided by the incident flux

$$\text{Flux} = \rho_1 \rho_2 \left|\vec{v}_1^i - \vec{v}_2^i\right| \quad . \qquad (5.4)$$

In our case the wavefunctions are normalized to one particle per unit volume so $\rho_1 = \rho_2 = 1$ and so the cross section is given by

$$d\sigma = \frac{1}{\left|\vec{v}_1^i - \vec{v}_2^i\right|} \int \frac{d^3 p_1^f}{(2\pi)^3} \frac{d^3 p_2^f}{(2\pi)^3} (2\pi)^4 \delta^4(P_i - P_f) |T_{fi}|^2 \ . \tag{5.5}$$

Galilean Invariance

It is interesting to note that this expression is invariant under a shift of coordinates to a moving frame. That is, suppose we view the scattering process as seen by an observer $\mathcal{O}'$ who moves with relative velocity $\vec{u}$ with respect to the original observer $\mathcal{O}$. Then

$$\vec{p}_1' = \vec{p}_1 - m_1 \vec{u} \qquad \vec{p}_2' = \vec{p}_2 - m_2 \vec{u} \ . \tag{5.6}$$

In terms of center of mass variables the relative momentum is unaffected

$$\vec{p}' = \frac{m_2 \vec{p}_1' - m_1 \vec{p}_2'}{m_1 + m_2} = \frac{m_2 \vec{p}_1 - m_1 \vec{p}_2}{m_1 + m_2} = \vec{p} \tag{5.7}$$

as would be expected, but

$$\vec{P}' = \vec{p}_1' + \vec{p}_2' = \vec{P} - (m_1 + m_2)\vec{u} \ . \tag{5.8}$$

For the energy we find

$$E' = \frac{\vec{p}_1'^2}{2m_1} + \frac{\vec{p}_2'^2}{2m_2} = \frac{\vec{p}_1^2}{2m_1} + \frac{\vec{p}_2^2}{2m_2} + \frac{1}{2}(m_1 + m_2)\vec{u}^2 - (\vec{p}_1 + \vec{p}_2) \cdot \vec{u}$$
$$= E + \frac{1}{2}(m_1 + m_2)\vec{u}^2 - \vec{P} \cdot \vec{u} \ . \tag{5.9}$$

Since $\vec{u}$ is fixed

$$d^3 p_1' = d^3 p_1 \qquad d^3 p_2' = d^3 p_2$$
$$\vec{P}_f' - \vec{P}_i' = \vec{P}_f - \vec{P}_i = 0$$
$$E_f' - E_i' = E_f - E_i - \vec{u} \cdot (\vec{P}_f - \vec{P}_i) = E_f - E_i = 0 \tag{5.10}$$
$$\vec{v}_1' - \vec{v}_2' = \frac{\vec{p}_1'}{m_1} - \frac{\vec{p}_2'}{m_2} = \frac{\vec{p}_1}{m_1} - \frac{\vec{p}_2}{m_2} = \vec{v}_1 - \vec{v}_2 \ .$$

Noting that $T_{fi}' = T_{fi}$ and that only *relative* momenta are involved, we have

$$d\sigma' = \frac{1}{\left|\vec{v}_1^{i'} - \vec{v}_2^{i'}\right|} \int \frac{d^3 p_1^{f'}}{(2\pi)^3} \frac{d^3 p_2^{f'}}{(2\pi)^3} (2\pi)^4 \delta^4(P_i' - P_f') |T_{fi}'|^2 = d\sigma \tag{5.11}$$

as claimed. (Interestingly this same invariance holds relativistically, as might be expected since $d\sigma$ represents an area transverse to the beam direction.)

Returning now to Eq. 5.5 we may perform the integration over $d^3p_2^f$ by use of the momentum conserving delta function—

$$\int \frac{d^3p_1^f}{(2\pi)^3} \frac{d^3p_2^f}{(2\pi)^3} (2\pi)^4 \delta^4(P_i - P_f) \ldots = \int \frac{d^3p_1^f}{(2\pi)^3} 2\pi \delta(E_i - E_1^f - E_2^f) \ldots$$

$$= \frac{1}{(2\pi)^2} p_1^{f^2} d\Omega_{1_f} \frac{1}{\left| d\left(\sqrt{m_1^2 + \vec{p}_1^{f^2}} + \sqrt{m_2^2 + (\vec{p}_1^f - \vec{P}_i)^2}\right) / dp_1^f \right|} \ldots \quad (5.12)$$

$$= \frac{1}{(2\pi)^2} p_1^{f^2} d\Omega_{1_f} \frac{1}{\frac{p_1^f}{m_1} + \frac{p_1^f - P_i \cos\theta}{m_2}} \ldots$$

where θ is the angle between $\hat{P}_i$ and $\hat{p}_1^f$, and we can write the general result as

$$\frac{d\sigma}{d\Omega_{1_f}} = \frac{1}{|\vec{v}_1^i - \vec{v}_2^i|} \frac{p_1^{f^2}}{(2\pi)^2} \frac{|T_{fi}|^2}{\frac{p_1^f}{m_1} + \frac{p_1^f - P_i \cos\theta}{m_2}} . \quad (5.13)$$

This form, valid in an arbitrary frame, is somewhat cumbersome to use. Instead one generally expresses the cross section as measured in the center of momentum frame wherein $\vec{P}_i = \vec{P}_f = 0$. Then

$$|\vec{v}_1^i - \vec{v}_2^i| = \left| \frac{\vec{p}_1^i}{m_1} - \frac{\vec{p}_2^i}{m_2} \right|$$

$$= \left| \frac{\vec{p}_1^i}{m_1} + \frac{\vec{p}_1^i}{m_2} \right| = \frac{p_1^i}{m_1 m_2}(m_1 + m_2) = \frac{p_1^i}{\mu} \quad (5.14)$$

and $p_1^i = p_1^f$ so that

$$\frac{d\sigma}{d\Omega_{1_f}} = \frac{\mu^2}{(2\pi)^2} |T_{fi}|^2 \quad (5.15)$$

which can alternatively be expressed in terms of the scattering amplitude

$$f_k(\theta) = -\frac{\mu}{2\pi} T_{fi} . \quad (5.16)$$

If one requires the form of the differential scattering cross section in a different frame one need only effect a coordinate transformation. Since $d\sigma$ is invariant we find

$$\frac{d\sigma}{d\Omega_{1_f}^F} = \frac{d\sigma}{d\Omega_{1_f}^{CM}} \cdot \frac{d\Omega_{1_f}^{CM}}{d\Omega_{1_f}^F} \quad (5.17)$$

II.6 SCATTERING OF IDENTICAL PARTICLES

Now consider what happens when the two particles which are scattering from one another are identical. First imagine, however, a different but related situation wherein two particles of *identical* mass, charge, and other internal quantum numbers but *different*, say, "color" scatter under their mutual interaction. (Here "color" signifies merely some property which distinguishes one particle from the other in this toy problem.) Consider the following scattering reactions in the center of mass system

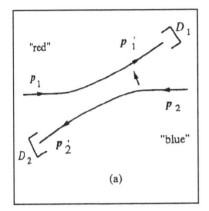 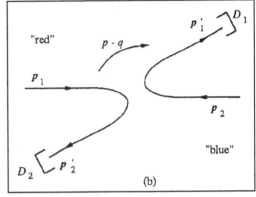

Fig. II.4: Idealized scattering of "colored" (non-identical) particles.

Suppose that the particle whose initial momentum is $\vec{p}_1$ is say "red" while that with original momentum $\vec{p}_2$ is "blue." Suppose also that detectors D_1 and D_2 are equipped to distinguish color. Then the processes shown above are clearly distinct from one another in that

$$\frac{d\sigma}{d\Omega}(\text{"red" in } D_1, \text{"blue" in } D_2) = |f_k(\theta)|^2$$
$$\frac{d\sigma}{d\Omega}(\text{"red" in } D_2, \text{"blue" in } D_1) = |f_k(\pi - \theta)|^2 \ . \quad (6.1)$$

On the other hand, if the detectors are colorblind then process a is not distinguishable from process b and the cross section becomes

$$\frac{d\sigma}{d\Omega}(\text{count in } D_1, \text{count in } D_2) = |f_k(\theta)|^2 + |f_k(\pi - \theta)|^2 \ . \quad (6.2)$$

(There is no interference here because even though the detectors do *not* distinguish color, there exist in principle color sensitive detectors which *could* separate reactions a and b. Thus we add probabilities—*not* amplitudes.)

Identical Boson Scattering

Now, however, consider a similar scattering process but for identical particles. As the particles are completely indistinguishable there is no way *even in principle* that a detector could tell them apart. In this case there exist two possible "routes" for the scattering to occur. One, "route θ" is given by a sum of amplitudes of the form

$f_k(\theta)$:
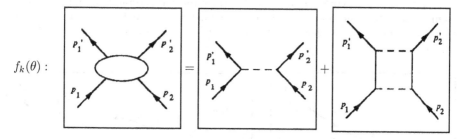

Fig. II.5: *Identical particle scattering via (direct) route "θ".*

as would be used in a non-identical particle situation to calculate the conventional scattering amplitude $f_k(\theta)$.

Clearly for *each* diagram above there exists a corresponding diagram for "route $\pi - \theta$" wherein the final state particles are simply exchanged with one another

$f_k(\pi - \theta)$:
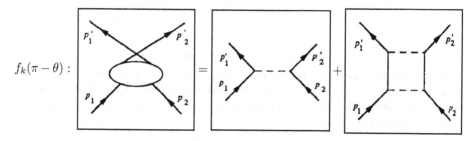

Fig. II.6: *Identical particle scattering via (exchange) route "$\pi - \theta$".*

as would be used in a non-identical particle situation to calculate the conventional scattering amplitude $f_k(\pi - \theta)$. Since there exists no way, even in principle, to distinguish these routes from one another and since quantum mechanics requires *all* possible routes to be summed, we must add these two transition *amplitudes* before taking the square to yield the probability and interference between the two will occur. The two terms $f_k(\theta)$ and $f_k(\pi - \theta)$ are called the "direct-" and "exchange-" amplitudes and it is found experimentally (and demanded by quantum field theory) that for particles of integral spin — $S = 0, 1, 2, \ldots$ — the direct and exchange pieces should be added together, producing a resulting differential scattering cross section

$$\frac{d\sigma}{d\Omega} = |\text{Direct} + \text{Exchange}|^2 \qquad S = 0, 1, \ldots$$
$$= |f_k(\theta) + f_k(\pi - \theta)|^2 \qquad (6.3)$$
$$= |f_k(\theta)|^2 + |f_k(\pi - \theta)|^2 + 2\,\text{Re}\, f_k^*(\theta) f_k(\pi - \theta) \ .$$

This interference is very clear in such scattering experiments. Consider, for example, the scattering of two alpha particles ($S = 0$) at very low energies so that the only interaction between them is due to the Coulomb repulsion. (At higher energies, *i.e.*, such that $E_\alpha > \frac{4\alpha}{R_\alpha}$ the particles can actually make physical contact and the strong nuclear force can also come into play. However, the strong force is short-ranged and vanishes exponentially when the α-particle separation R is greater than the particle radius R_α). The Coulomb scattering amplitude is easily calculated and can be compared to experiment. At $\theta = \frac{\pi}{2}$, for example, the cross section with interference included is a factor of two larger than the corresponding non-identical particle result. This "enhancement" is observed experimentally.

Readers may wonder about particle exchange in intermediate states, as shown in Figure II.7.

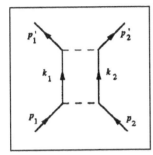 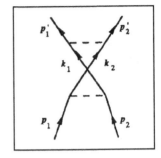

Fig. II.7: Diagrams related by intermediate state particle exchange.

Since these two routes are also indistinguishable, shouldn't these amplitudes also be added before squaring to find the cross section? The answer is yes, but the second diagram is clearly taken into account by our procedure of adding direct and exchange terms since it can be redrawn as shown in Figure II.8

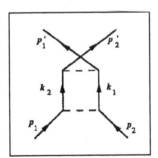

Fig. II.8: Final state particle exchange is topologically equivalent to Figure II.7b.

and this contribution is already included as part of the exchange diagram when all $\vec{k}_1, \vec{k}_2$ are summed over.

An alternative way in which to understand the need for both direct and exchange contributions is in terms of the properties of the configuration space wavefunction. Since the particles are identical, Bose symmetry requires the wavefunction to be invariant under interchange of particle identity. *i.e.*,

$$\Psi \sim \sqrt{\frac{1}{2}} \left(\psi(\vec{p}_1, \vec{p}_2) + \psi(\vec{p}_2, \vec{p}_1) \right) \quad . \tag{6.4}$$

In order to see how this requirement imposes restrictions upon the scattering amplitude consider first the scattering of our two distinguishable ("colored") particles. In terms of center of mass coordinates we can write the solution of the Schrödinger equation in the form

$$\Psi(\vec{r}_1, \vec{r}_2) = \exp\left(i\vec{P} \cdot \vec{R}\right) \psi^{(+)}(\vec{r}) \qquad \begin{aligned} \vec{R} &= \frac{1}{2}(\vec{r}_1 + \vec{r}_2) \\ \vec{r} &= \vec{r}_1 - \vec{r}_2 \end{aligned} \qquad (6.5)$$

where the exponential expresses the plane wave motion of the center of mass, with $\vec{P} = \vec{p}_1 + \vec{p}_2$ being the total momentum, and $\psi^{(+)}(\vec{r})$ describes the relative motion. In the large r limit we have

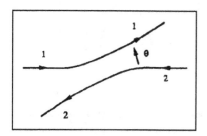

$$\Psi^{(+)}(\vec{r}) \underset{r \to \infty}{\sim} e^{ikz} + \frac{1}{r} e^{ikr} f_k(\theta) \qquad (6.6)$$

Obviously $e^{ikz} = e^{ik(z_1 - z_2)}$ which indicates

$$\begin{aligned} \vec{p}_1 &\quad \text{along} \quad +z \\ \vec{p}_2 &\quad \text{along} \quad -z \\ \vec{p}_1' &\quad \text{along} \quad \theta \\ \vec{p}_2' &\quad \text{along} \quad \pi - \theta \;. \end{aligned}$$

On the other hand, if the particles are indistinguishable the wavefunction must be *symmetric* in the coordinates $\vec{r}_1, \vec{r}_2$ so that

$$\Psi \sim \psi(\vec{r}_1, \vec{r}_2) + \psi(\vec{r}_2, \vec{r}_1) = \exp i\vec{P} \cdot \vec{R} \left(\psi^{(+)}(\vec{r}) + \psi^{(+)}(-\vec{r}) \right) \;. \qquad (6.7)$$

For $\psi^{(+)}(-\vec{r})$ the plane wave component of the wavefunction

$$e^{-ikz} = e^{-ik(z_1 - z_2)} \qquad (6.8)$$

indicates that particle #1 is initially moving in the $-z$ direction while particle #2 is moving toward $+z$. We can look for the asymptotic amplitude as before with particle #1 moving in direction θ and particle #2 moving toward $\pi - \theta$. It is clear

from the figure that

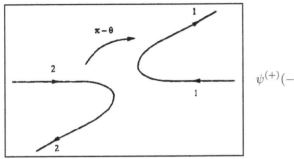

$$\psi^{(+)}(-\vec{r}) = e^{-ikz} + \frac{1}{r} e^{ikr} f_k(\pi - \theta) \tag{6.9}$$

We find then for the full wavefunction the result

$$\psi^{(+)}(\vec{r}) + \psi^{(+)}(-\vec{r}) \underset{r \to \infty}{\sim} e^{ikz} + e^{-ikz} + \frac{1}{r} e^{ikr} (f_k(\theta) + f_k(\pi - \theta)) \tag{6.10}$$

which yields a scattering amplitude of the direct plus exchange form as claimed.

Identical Fermion Scattering

One can similarly discuss the scattering of half-integral spin particles — $S = \frac{1}{2}, \frac{3}{2}, \frac{5}{2}, \ldots$ — except that in this case the cross section involves the *difference* between direct and exchange terms. There is an important nuance which must be observed, however, when dealing with particles which have spin — interference occurs *only* in the scattering channel wherein quantum numbers of both scattering particles are the same.

In order to make this point clear, consider scattering of two spin 1/2 particles, *e.g.* electrons, within a wavefunction perspective. Because of the Pauli exclusion principle the "electron"–"electron" wavefunction must be anti-symmetric under interchange of particle identity. Assuming that the interactions involved are *spin-independent*, the wavefunction can be written in the simple factorized form

$$\Psi(1,2) = \psi(\vec{r}_1, \vec{r}_2)|\chi_1\rangle|\chi'_2\rangle - \psi(\vec{r}_2, \vec{r}_1)\rangle|\chi'_1\rangle|\chi_2\rangle \tag{6.11}$$

where $|\chi\rangle$, $|\chi'\rangle$ represent spin states of the electron. First consider the case that both spin configurations are the same, say up with respect to some quantization axis. Then

$$\Psi(1,2) = [\psi(\vec{r}_1, \vec{r}_2) - \psi(\vec{r}_2, \vec{r}_1)]|\uparrow_1\rangle|\uparrow_2\rangle . \tag{6.12}$$

[Note that $\Psi(1,2) = 0$ when $\vec{r}_1 = \vec{r}_2$ in accord with the Pauli exclusion principle.] As in the integral spin case the spatial wavefunction can be written in terms of center of mass coordinate $\vec{R}$ and relative coordinate $\vec{r}$, but now with a minus sign between direct and exchange terms so that

$$\Psi(1,2) \underset{r \to \infty}{\sim} \exp i\vec{P} \cdot \vec{R} \left(e^{ikz} - e^{-ikz} + \frac{1}{r} e^{ikr} (f_k(\theta) - f_k(\pi - \theta)) \right) |\uparrow_1\rangle|\uparrow_2\rangle . \tag{6.13}$$

The differential scattering cross section is then

$$\frac{d\sigma}{d\Omega} = |\text{Direct} - \text{Exchange}|^2 = |f_k(\theta) - f_k(\pi - \theta)|^2 \qquad (6.14)$$
$$= |f_k(\theta)|^2 + |f_k(\pi - \theta)|^2 - 2\operatorname{Re} f_k^*(\theta) f_k(\pi - \theta)$$

so that, for example, at $\theta = \frac{\pi}{2}$ the cross section must vanish! This quantum mechanical interference is verified experimentally.

Now consider the case that the two spin states are orthogonal, e.g.

$$\Psi(1,2) = \psi(\vec{r}_1, \vec{r}_2)|\uparrow_1\rangle|\downarrow_2\rangle - \psi(\vec{r}_2, \vec{r}_1)|\downarrow_1\rangle|\uparrow_2\rangle \ . \qquad (6.15)$$

It is clear that the asymptotic form of the wavefunction is

$$\Psi(1,2) \underset{r\to\infty}{\sim} \exp i\vec{P}\cdot\vec{R}\left[\left(e^{ikz} + \frac{e^{ikr}}{r}f_k(\theta)\right)|\uparrow_1\rangle|\downarrow_2\rangle \right.$$
$$\left. - \left(e^{-ikz} + \frac{e^{ikr}}{r}f_k(\pi - \theta)\right)|\downarrow_1\rangle|\uparrow_2\rangle\right] \ . \qquad (6.16)$$

The differential scattering cross section is given then by

$$\frac{d\sigma}{d\Omega} = \langle\psi_{\text{scat}}|\psi_{\text{scat}}\rangle \qquad (6.17)$$

where

$$|\psi_{\text{scatt}}\rangle = f_k(\theta)|\uparrow_1\rangle|\downarrow_2\rangle - f_k(\pi - \theta)|\downarrow_1\rangle|\uparrow_2\rangle \qquad (6.18)$$

or

$$\frac{d\sigma}{d\Omega} = |f_k(\theta)|^2 + |f_k(\pi - \theta)|^2 \ . \qquad (6.19)$$

That is, there is no interference! The reason for this is clear. One can experimentally distinguish scattering through angle θ from that through $\pi - \theta$ by detection of the final spin state — i.e., $|\uparrow_1\rangle|\downarrow_2\rangle$ implies scattering through angle θ while $|\downarrow_1\rangle|\uparrow_2\rangle$ implies scattering through $\pi - \theta$. Since these routes are in principle distinguishable the rules of quantum mechanics require that we add probabilities rather than amplitudes, so that no interference can result.

In a more general situation where the spin state is given by $|\chi_1\rangle|\rho_2\rangle$ with $\langle\rho|\chi\rangle \neq 0$ we have

$$\frac{d\sigma}{d\Omega} = \langle\psi_{\text{scat}}|\psi_{\text{scat}}\rangle = |f_k(\theta)|^2 + |f_k(\pi - \theta)|^2 - 2\operatorname{Re} f_k(\theta) f_k^*(\pi - \theta)|\langle\rho|\chi\rangle|^2 \qquad (6.20)$$

so that the interference term is proportional to the overlap of the two spin states.

A particularly important case is that of unpolarized scattering which can be considered as a statistical mixture of

$$\begin{aligned}
\frac{1}{4}:|\uparrow_1\rangle|\uparrow_1\rangle \quad &\text{with} \quad \frac{d\sigma_{\uparrow\uparrow}}{d\Omega} = |f_k(\theta) - f_k(\pi - \theta)|^2 \\
\frac{1}{4}:|\uparrow_1\rangle|\downarrow_1\rangle \quad &\text{with} \quad \frac{d\sigma_{\uparrow\downarrow}}{d\Omega} = |f_k(\theta)|^2 + |f_k(\pi - \theta)|^2 \\
\frac{1}{4}:|\downarrow_1\rangle|\uparrow_2\rangle \quad &\text{with} \quad \frac{d\sigma_{\downarrow\uparrow}}{d\Omega} = |f_k(\theta)|^2 + |f_k(\pi - \theta)|^2 \\
\frac{1}{4}:|\downarrow_1\rangle|\downarrow_2\rangle \quad &\text{with} \quad \frac{d\sigma_{\downarrow\downarrow}}{d\Omega} = |f_k(\theta) - f_k(\pi - \theta)|^2 \ .
\end{aligned} \qquad (6.21)$$

The unpolarized cross section then is

$$\frac{d\sigma_{\text{unpol.}}}{d\Omega} = \frac{1}{4}\left(\frac{d\sigma_{\uparrow\uparrow}}{d\Omega} + \frac{d\sigma_{\uparrow\downarrow}}{d\Omega} + \frac{d\sigma_{\downarrow\uparrow}}{d\Omega} + \frac{d\sigma_{\downarrow\downarrow}}{d\Omega}\right) \qquad (6.22)$$
$$= |f_k(\theta)|^2 + |f_k(\pi-\theta)|^2 - \operatorname{Re} f_k^*(\theta) f_k(\pi-\theta)$$

II.7 SCATTERING MATRIX

The assumption of spin independence of the particle-particle interaction made in the above analysis is clearly naive and it is important to generalize the previous formalism to the situation that the interaction is spin dependent. First assume that the particles are distinguishable and that both are spin 1/2. If each begins with spin up along the quantization axis, the incident wavefunction can be taken as

$$\psi_{\text{inc}} = \exp\left(i\vec{P}\cdot\vec{R}\right) e^{ikz} |\uparrow_1\rangle |\uparrow_2\rangle \ . \qquad (7.1)$$

Since the interaction is spin-dependent, the final spin state of the particles can be any of the four combinations — $\uparrow\uparrow$, $\uparrow\downarrow$, $\downarrow\uparrow$, $\downarrow\downarrow$ — and the scattering amplitude can be written in terms of a linear superposition of the four possibilities

$$|\psi_{\text{scat}}\rangle = M_{\uparrow\uparrow,\uparrow\uparrow}|\uparrow_1\rangle|\uparrow_2\rangle + M_{\uparrow\downarrow,\uparrow\uparrow}|\uparrow_1\rangle|\downarrow_2\rangle \qquad (7.2)$$
$$+ M_{\downarrow\uparrow,\uparrow\uparrow}|\downarrow_1\rangle|\uparrow_2\rangle + M_{\downarrow\downarrow,\uparrow\uparrow}|\downarrow_1\rangle|\downarrow_2\rangle \ .$$

In general, for an arbitrary initial spin state the scattering amplitude can be written as a matrix

$$\mathcal{M} = \begin{array}{c} \\ \langle\uparrow_1|\langle\uparrow_2| \\ \langle\uparrow_1|\langle\downarrow_2| \\ \langle\downarrow_1|\langle\uparrow_2| \\ \langle\downarrow_1|\langle\downarrow_2| \end{array} \begin{pmatrix} |\uparrow_1|\uparrow_2\rangle & |\uparrow_1\rangle|\downarrow_2\rangle & |\downarrow_1\rangle|\uparrow_2\rangle & |\downarrow_1\rangle|\downarrow_2\rangle \\ M_{\uparrow\uparrow,\uparrow\uparrow} & M_{\uparrow\uparrow,\uparrow\downarrow} & M_{\uparrow\uparrow,\downarrow\uparrow} & M_{\uparrow\uparrow,\downarrow\downarrow} \\ M_{\uparrow\downarrow,\uparrow\uparrow} & M_{\uparrow\downarrow,\uparrow\downarrow} & M_{\uparrow\downarrow,\downarrow\uparrow} & M_{\uparrow\downarrow,\downarrow\downarrow} \\ M_{\downarrow\uparrow,\uparrow\uparrow} & M_{\downarrow\uparrow,\uparrow\downarrow} & M_{\downarrow\uparrow,\downarrow\uparrow} & M_{\downarrow\uparrow,\downarrow\downarrow} \\ M_{\downarrow\downarrow,\uparrow\uparrow} & M_{\downarrow\downarrow,\uparrow\downarrow} & M_{\downarrow\downarrow,\downarrow\uparrow} & M_{\downarrow\downarrow,\downarrow\downarrow} \end{pmatrix} \qquad (7.3)$$

If the initial spin state is given by the linear superposition

$$c_{\uparrow\uparrow}|\uparrow_1\rangle|\uparrow_2\rangle + c_{\uparrow\downarrow}|\uparrow_1\rangle|\downarrow_2\rangle + c_{\downarrow\uparrow}|\downarrow_1\rangle|\uparrow_2\rangle + c_{\downarrow\downarrow}|\downarrow_1\rangle|\downarrow_2\rangle \qquad (7.4)$$

we may represent this state in terms of the column vector

$$\psi_{\text{inc}} = \begin{pmatrix} c_{\uparrow\uparrow} \\ c_{\uparrow\downarrow} \\ c_{\downarrow\uparrow} \\ c_{\downarrow\downarrow} \end{pmatrix} \cdot e^{ikz} \exp\left(i\vec{P}\cdot\vec{R}\right) \equiv c\, e^{ikz} \exp(i\vec{P}\cdot\vec{R}) \ . \qquad (7.5)$$

The scattered wavefunction becomes

$$|\psi_{\text{scat}}\rangle = \mathcal{M}(\theta,\phi) c \qquad (7.6)$$

in matrix notation and the differential scattering cross section is written as

$$\frac{d\sigma}{d\Omega} = \langle \psi_{\text{scat}} | \psi_{\text{scat}} \rangle = c^\dagger \mathcal{M}^\dagger(\theta,\phi) \mathcal{M}(\theta,\phi) c$$
$$= |c_{\uparrow\uparrow}|^2 |\mathcal{M}_{\uparrow\uparrow,\uparrow\uparrow}(\theta,\phi)|^2 + c^*_{\uparrow\uparrow} c_{\uparrow\downarrow} \mathcal{M}^*_{\uparrow\downarrow,\uparrow\uparrow}(\theta,\phi) \mathcal{M}_{\uparrow\uparrow,\uparrow\downarrow}(\theta,\phi) \quad (7.7)$$
$$+ \ldots \;.$$

In terms of this matrix notation, the rules for identical particles can be easily written

$$\mathcal{M}_{\chi\chi',\chi\chi'}(\theta,\phi) \pm \mathcal{M}_{\chi\chi',\chi'\chi}(\pi-\theta,\pi+\phi) \;. \quad (7.8)$$

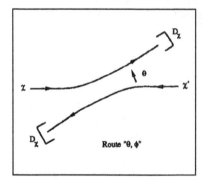
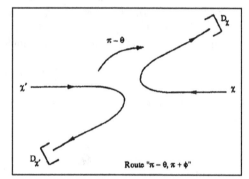

Fig. II.9: *Diagrammatic representation for identical particle scattering of non-spinless particles.*

T-Matrix Connection

The connection of the matrix $\mathcal{M}$ with the T-matrix is easily ascertained. Since when spin is present a state must be characterized in terms of both its spin and momentum, a general scattering amplitude can be written as

$$\text{Amp}_{fi} = \lim_{\substack{t' \to \infty \\ t \to -\infty}} \langle f | \hat{U}(t',t) | i \rangle = \lim_{\substack{t' \to \infty \\ t \to -\infty}} \langle \chi_1 \chi'_2 | \langle \vec{p}'_1 \vec{p}'_2 | \hat{U}(t',t) | \vec{p}_1 \vec{p}_2 \rangle | \lambda_1 \lambda'_2 \rangle \quad (7.9)$$
$$= \delta_{fi} - 2\pi i \delta(E_f - E_i)(2\pi)^3 \delta^3(\vec{P}_f - \vec{P}_i) \langle \chi_1 \chi'_2 | \mathcal{M}(\theta,\phi) | \lambda_1 \lambda_2 \rangle \;.$$

We see that

$$\hat{\mathcal{F}}_{fi}(E_i) = \langle \chi_1 \chi'_2 | \mathcal{M}(\theta,\phi) | \lambda_1 \lambda'_2 \rangle$$
$$= \langle \chi_1 \chi'_2 | \langle \vec{p}'_1 \vec{p}'_2 | \hat{V} + \hat{V} \frac{1}{E_i - \hat{H}_0 + i\epsilon} \hat{V} + \ldots | \vec{p}_1 \vec{p}_2 \rangle | \lambda_1 \lambda'_2 \rangle \;. \quad (7.10)$$

Here while $\hat{V}$ is an operator both in coordinate (or momentum) space and in spin space, the matrix $\mathcal{M}(\theta,\phi)$ operates only on the spin degrees of freedom — coordinate space integration has already been carried out. Often instead of the CM scattering angle θ, ϕ one indicates $\mathcal{M}$ as depending on the initial and final relative momenta $\vec{p}_i, \vec{p}_f$

$$\mathcal{M}(\theta,\phi) \equiv \mathcal{M}(\vec{p}_i, \vec{p}_f) \;. \quad (7.11)$$

(Note here that $\vec{p}_i, \vec{p}_f$ are simply numerical vector quantities, not operators.) In terms of this notation the corresponding amplitude for identical particle scattering becomes

$$\text{Amp} = \langle \chi_1 \chi_2' | \mathcal{M}(\vec{p}_i, \vec{p}_f) | \lambda_1 \lambda_2' \rangle \pm \langle \chi_1 \chi_2' | \mathcal{M}(-\vec{p}_i, \vec{p}_f) | \lambda_1' \lambda_2 \rangle \quad (7.12)$$

which is identical to the indistinguishability requirement written earlier.

Symmetry Restrictions

While one can, in a particular model, explicitly calculate a form for the scattering matrix $\mathcal{M}(\vec{p}_i, \vec{p}_f)$, the result will be in general only approximate since the Schrödinger equation is exactly soluble only for a limited number of potentials, such as the Coulomb, harmonic oscillator, *etc.* However, there are various requirements which symmetry requirements place upon the scattering matrix and these must apply regardless of the detailed form of the interaction potential.

Examples of such symmetries are spatial translation, rotation invariance, spatial inversion, time reversal, *etc.* The first two are familiar and merely express the belief that there exists no special location or direction in space. That is, experiments performed at one point in (free) space should give the same results when translated to another location. Likewise results should not change if the apparatus is rotated to a new orientation. These requirements lead in the usual way to conservation of momentum and angular momentum. The former is clearly realized in that the scattering matrix depends only upon relative momenta and is multiplied by the factor $(2\pi)^3 \delta^3(\vec{P}_i - \vec{P}_f)$ in producing the S-matrix. Similarly rotational invariance leads to angular momentum conservation—*i.e.*, matrix elements only connect states of the same angular momentum. However, we are employing here a representation for the scattering matrix wherein momentum, *not* angular momentum, is diagonal, and we shall discuss below the conditions which prevail on $\mathcal{M}(\vec{p}_i, \vec{p}_f)$.

Symmetry under spatial translation or under rotations is well-documented experimentally and no physicist has serious doubts about their validity. The same is not true of spatial inversion or of time reversal invariance. In fact, *both* are known to be violated! However, there are certain caveats which allow their use in a scattering process. The point is that spatial inversion symmetry, which is associated with parity invariance, is preserved by strong and electromagnetic interactions but violated by the weak force (that which causes the relatively slow decay of various elementary particles such as the neutron or muon). Since the latter is generally at least four orders of magnitude weaker than its strong/electromagnetic counterpart, parity violation can generally be neglected in a strong scattering experiment. Similar comments apply to the case of time reversal symmetry, which is found to be violated only by an interaction even more feeble than the weak force. Here too we can essentially neglect this symmetry violation, leading to conditions on $\mathcal{M}(\vec{p}_i, \vec{p}_f)$ which will be described below.

As an example, we consider the case of scattering of two spin 1/2 particles:

i) Rotational Invariance

The scattering matrix $\mathcal{M}$ can be a function only of the initial and final momenta $\vec{p}_i$ and $\vec{p}_f$ and of the Pauli spin matrices, $\vec{\sigma}_1, \vec{\sigma}_2$, which operate in the space of particles #1 and #2, respectively. It is important to realize that the dependence upon $\vec{\sigma}_1, \vec{\sigma}_2$ can be no more complex than bilinear — i.e., since $\sigma_1^2 = 1$, $\sigma_i \sigma_j = \delta_{ij} + i\epsilon_{ijk}\sigma_k$ one can imagine that $\mathcal{M}$ contains terms such as σ_{1x}, σ_{2z}, etc. or $\sigma_{1x}\sigma_{2z}$, etc. but more complex forms such as $\sigma_{1x}\sigma_{1y}^2$ or $\sigma_{2z}\sigma_{2x}^3$ are not needed.

Consider now the unitary operator $\hat{R}$ associated with some spatial rotation $\mathcal{R}$ and break the operator into its space and spin components

$$\hat{R} = \hat{R}_{\text{space}} \hat{R}_{\text{spin}} \ . \tag{7.13}$$

Since the Hamiltonian is rotationally invariant

$$\hat{R}^\dagger \hat{H} \hat{R} = \hat{H} \tag{7.14}$$

so must the time development operator be:

$$\hat{R}^\dagger \hat{U}(t',t) \hat{R} = \hat{U}(t',t) \ . \tag{7.15}$$

We require then

$$\begin{aligned}\mathcal{M}(\vec{p}_i, \vec{p}_f, \vec{\sigma}_1, \vec{\sigma}_2) &\sim \left\langle \vec{p}_f | \hat{U}(\infty, -\infty) | \vec{p}_i \right\rangle \\ &= \left\langle \vec{p}_f | \hat{R}^\dagger_{\text{space}} \hat{R}^\dagger_{\text{spin}} \hat{U}(\infty, -\infty) \hat{R}_{\text{spin}} \hat{R}_{\text{space}} | \vec{p}_f \right\rangle \ . \end{aligned} \tag{7.16}$$

Here

$$\hat{R}_{\text{space}} | \vec{p}_i \rangle = | \mathcal{R}\vec{p}_i \rangle \quad \text{and} \quad \langle \vec{p}_f | \hat{R}^\dagger_{\text{space}} = \langle \mathcal{R}\vec{p}_f | \tag{7.17}$$

where $\mathcal{R}\vec{p}$ denotes the appropriately rotated vector — e.g. for a rotation by angle ϕ about the z-axis we have[†]

$$\mathcal{R}(p_x, p_y, p_z) = (p_x \cos\phi + p_y \sin\phi, -p_x \sin\phi + p_y \cos\phi, p_z) \ . \tag{7.18}$$

[†] This can be seen from the explicit representation of the spatial rotation operator

$$\hat{R}_{\text{space}} = \exp i\phi \hat{L}_z \ .$$

Considering an infinitesimal rotation by angle $\delta\phi$ we have then

$$\hat{R}_{\text{space}} \cong 1 + i\delta\phi \hat{L}_z = 1 + i\delta\phi (xp_y - yp_x) \ .$$

When applied to the momentum operator $\vec{p}$ we find

$$\begin{aligned}\hat{R}^\dagger_{\text{space}} \vec{p} \hat{R}_{\text{space}} &= (1 - i\delta\phi(xp_y - yp_x)) \vec{p} (1 + i\delta\phi(xp_y - yp_x))) \\ &= \vec{p} - i\delta\phi p_y [x, \vec{p}] + i\delta\phi p_x [y, \vec{p}] \\ &= \vec{p} + \delta\phi (p_y \hat{e}_x - p_x \hat{e}_y) \\ &= \mathcal{R}\vec{p} \end{aligned}$$

where

$$\mathcal{R}\vec{p} = (p_x + \delta\phi p_y, p_y - \delta\phi p_x, p_z) \ .$$

Similarly we have[‡]
$$\hat{R}^\dagger_{\text{spin}} \vec{\sigma} \hat{R}_{\text{spin}} = \mathcal{R}\vec{\sigma} \tag{7.19}$$

— e.g. for rotation by angle ϕ about the z-axis

$$\mathcal{R}(\sigma_x, \sigma_y, \sigma_z) = (\sigma_x \cos\phi + \sigma_y \sin\phi, -\sigma_x \sin\phi + \sigma_y \cos\phi, \sigma_z) \quad. \tag{7.20}$$

The requirement of rotational invariance is then

$$\mathcal{M}(\vec{p}_i, \vec{p}_f, \vec{\sigma}_1, \vec{\sigma}_2) = \mathcal{M}(\mathcal{R}\vec{p}_i, \mathcal{R}\vec{p}_f, \mathcal{R}\vec{\sigma}_1, \mathcal{R}\vec{\sigma}_2) \quad. \tag{7.21}$$

i.e., the scattering matrix $\mathcal{M}$ behaves as a scalar under simultaneous rotation of all participating vectors. Thus, terms like

$$\vec{p}_i \cdot \vec{p}_f \quad , \quad \vec{\sigma} \cdot \vec{p}_i \quad , \quad \vec{\sigma}_1 \cdot \vec{\sigma}_2 \quad , \quad \text{etc.} \tag{7.22}$$

are permitted, but structures such as

$$p_{i_x} \quad , \quad \sigma_{1_x} \quad , \quad \text{etc.} \tag{7.23}$$

are not. In addition, each allowed form can be multiplied by a function of the scalar quantities $\vec{p}_i \cdot \vec{p}_f, p_i^2, p_f^2$.

Equivalently when applied to a plane wavefunction $\exp(i\vec{p} \cdot \vec{r})$

$$\hat{R}_{\text{space}} \exp(i\vec{p} \cdot \vec{r}) = \left(1 + \delta\phi\left(x\frac{\partial}{\partial y} - y\frac{\partial}{\partial x}\right)\right) \exp i\vec{p} \cdot \vec{r}$$
$$= \exp i\mathcal{R}\vec{p} \cdot \vec{r} \quad.$$

[‡] This can be demonstrated from the explicit representation for the rotation operator
$$\hat{R}_{\text{spin}} = \exp\left(i\phi\frac{1}{2}\sigma_z\right) \quad.$$

Considering an infinitesimal rotation by angle $\delta\phi$ we have then

$$\hat{R}^\dagger_{\text{spin}} \vec{\sigma} \hat{R}_{\text{spin}} \cong \left(1 - i\delta\phi\frac{1}{2}\sigma_z\right) \vec{\sigma} \left(1 + i\delta\phi\frac{1}{2}\sigma_z\right)$$
$$\cong \vec{\sigma} - i\delta\phi\frac{1}{2}[\sigma_z, \vec{\sigma}]$$
$$\equiv \vec{\sigma}'$$

where
$$\vec{\sigma}' = \hat{e}_x(\sigma_x + \delta\phi\sigma_y) + \hat{e}_y(\sigma_y - \delta\phi\sigma_x) + \hat{e}_z\sigma_z$$
$$= \mathcal{R}\vec{\sigma}$$

ii) Spatial Inversion

The spatial inversion operation Π corresponds to

$$\begin{aligned} t &\xrightarrow[\Pi]{} t \\ \vec{r} &\xrightarrow[\Pi]{} -\vec{r} \end{aligned} \quad . \tag{7.24}$$

Clearly under such a transformation momentum also changes sign

$$\vec{p} = m\frac{d\vec{r}}{dt} \xrightarrow[\Pi]{} \frac{d(-\vec{r})}{dt} = -\vec{p} \quad . \tag{7.25}$$

However, angular momentum does not

$$\vec{L} = \vec{r} \times \vec{p} \xrightarrow[\Pi]{} (-\vec{r}) \times (-\vec{p}) = +\vec{L} \quad . \tag{7.26}$$

Vectors such as $\vec{r}, \vec{p}$ which change sign under spatial inversion are called "polar" vectors while those such as $\vec{L}$ which do not are termed "axial" vectors. In order that the total angular momentum

$$\vec{J} = \vec{L} + \frac{1}{2}\vec{\sigma} \tag{7.27}$$

behave properly it is clear that $\vec{\sigma}$ should also be an axial vector. Hence the requirement of spatial inversion or parity invariance on the scattering matrix is

$$\mathcal{M}\left(\vec{p}_i, \vec{p}_f, \vec{\sigma}_1, \vec{\sigma}_2\right) = \mathcal{M}(-\vec{p}_i, -\vec{p}_f, \vec{\sigma}_1, \vec{\sigma}_2) \tag{7.28}$$

which rules out forms of the sort $\vec{\sigma}_1 \cdot \vec{p}_i$, $\vec{\sigma}_2 \cdot \vec{p}_f$, etc. but allows terms such as $\vec{\sigma}_1 \cdot \vec{p}_i \vec{\sigma}_2 \cdot \vec{p}_f$, $\vec{\sigma}_1 \cdot (\vec{p}_i \times \vec{p}_f)$, etc.

iii) Time Reversal

The time reversal operation T corresponds to

$$\begin{aligned} t &\xrightarrow[T]{} -t \\ t &\xrightarrow[T]{} \vec{r} \end{aligned} \quad . \tag{7.29}$$

Under such a transformation momentum changes sign

$$\vec{p} = m\frac{d\vec{r}}{dt} \xrightarrow[T]{} m\frac{d\vec{r}}{d(-t)} = -\vec{p} \tag{7.30}$$

as does angular momentum

$$\vec{L} = \vec{r} \times \vec{p} \xrightarrow[T]{} \vec{r} \times (-\vec{p}) = -\vec{L} \quad . \tag{7.31}$$

We require spin to reverse also then under time inversion

$$\vec{\sigma} \xrightarrow[T]{} -\vec{\sigma} \quad . \tag{7.32}$$

However, there is an additional subtlety here. Consider a scattering process as shown below in (a). Under time reversal the scattering appears as in (b). That is, the momentum is indeed reversed but so is the initial and final state of the scattering process. The requirement of time reversal then reads

$$\mathcal{M}\left(\vec{p}_i, \vec{p}_f, \vec{\sigma}_1, \vec{\sigma}_2\right) = \mathcal{M}^\dagger\left(-\vec{p}_f, -\vec{p}_i, -\vec{\sigma}_1, -\vec{\sigma}_2\right) \tag{7.33}$$

The feature that $\mathcal{M} \to \mathcal{M}^\dagger$ is associated with the fact that the time reversal operation is not unitary, but we do not have time to pursue this point further.

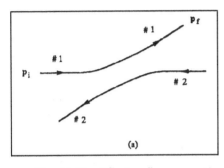 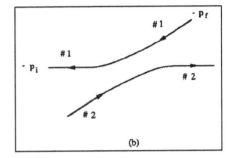

Fig. II.10: "Cartoon" representing processes related by time reversal invariance.

Problem II.7.1

Scattering Matrix for Spin 1/2-Spin 0 Scattering

Two particles, one of spin 1/2 and the other of spin 0 interact according to a potential

$$V = V_A(r) + V_B(r) \vec{\sigma} \cdot \vec{\ell} .$$

Here V_A, V_B are functions only of the distance between the two particles and $\vec{\ell} = \vec{r} \times \vec{p}$ is the relative orbital angular momentum operator.

i) Use the Born approximation to obtain an explicit form for the 2×2 matrix

$$\mathcal{M}(\theta, \phi) = f(\theta) + i g(\theta) \vec{\sigma} \cdot \hat{n}$$

where $f(\theta)$ and $g(\theta)$ are related to the Fourier transforms of $V_A(r)$ and $V_B(r)$:

$$\mathcal{M}(\theta, \phi) = -\frac{\mu}{2\pi}\left[\mathcal{V}_A(q)\mathbf{1} + i\frac{\sin\theta}{q}\frac{d\mathcal{V}_B(q)}{dq}\vec{\sigma}\cdot\hat{n}\right]$$

Here $\vec{q} = \vec{k}_i - \vec{k}_f$ is the momentum transfer, with

$$q = |\vec{k}_i - \vec{k}_f| = 2k_i \sin\frac{1}{2}\theta ,$$

$\mu = m_1 m_2/(m_1 + m_2)$ is the reduced mass for the system of two particles,

$$\mathcal{V}(q) \equiv \int d^3 r\, e^{i\vec{q}\cdot\vec{r}} V(r) = 4\pi \int dr\, r^2 V(r) \frac{\sin qr}{qr}$$

is the Fourier transform of the potential, and

$$\hat{n} = \frac{\vec{k}_i \times \vec{k}_f}{|\vec{k}_i \times \vec{k}_f|}$$

is the unit normal to the scattering plane.

ii) Show that this form is required by parity invariance of the two particle interaction.

Note that if the functions $V_A(r)$ and $V_B(r)$ are real, corresponding to a hermitian interaction $\hat{V}$, then the amplitudes $f(\theta)$ and $g(\theta)$ are likewise real. No interference between these amplitudes and hence no polarization effects would be expected in the scattering as long as the Born approximation is valid. In the real world, however, there are two effects which obviate this expectation. First is the fact that final state interactions generate phases such that the scattering amplitudes $f(\theta), g(\theta)$ are no longer relatively real. Secondly for scattering from a complex nucleus the elastic scattering may be described phenomenologically as though it is due to a potential $\hat{V}$ which is non-hermitian. The non-hermiticity expresses the loss of probability in the elastic channel due to inelastic processes–breakup of the nucleus, nuclear excitation, *etc*. In this case $f(\theta)$ and $g(\theta)$ are no longer both real, develop a relative phase which is in general different from zero, and generate polarization effects even in Born approximation.

II.8 PARTIAL WAVE EXPANSION

It was argued above that the stricture of angular momentum conservation requires that the scattering matrix $\mathcal{M}(\theta, \phi)$ be a scalar quantity under rotations. Actually the angular momentum conservation requirement is even stronger and assures that if we decompose both initial and final states into their respective angular momentum components characterized by J_i, J_f, respectively, then the S-matrix must be diagonal in this basis, *i.e.*,

$$S_{J_f, J_i} \propto \delta_{J_f, J_i} \ . \tag{8.1}$$

The resulting form of the scattering amplitude, when represented in this way, is called the partial wave expansion. In the case of spinless scattering this means that one can represent the scattering amplitude $f_k(\theta)$ as an expansion in Legendre polynomials

$$f_k(\theta) = \sum_{\ell=0}^{\infty} (2\ell+1) a_\ell(k) P_\ell(\cos\theta) \ , \tag{8.2}$$

where θ represents the scattering angle and $a_\ell(k)$ is the (energy dependent) partial wave amplitude, representing the scattering in the channel with angular momentum ℓ. Because of angular momentum conservation the $a_\ell(k)$ are independent of one another. The usefulness of this approach to the scattering process arises due to the feature that at a given energy only a finite number of partial waves need be included. At the classical level this can be understood from the finite range of the potential. If the scattering potential has "radius" r_0 then classically any particle arriving with impact parameter $b > r_0$ will pass by without scattering. Since these correspond to incident angular momenta

$$\ell = kb > kr_0 \tag{8.3}$$

we anticipate that only those partial waves with $\ell < kr_0$ need be included in the summation. We shall see later why this simple argument also obtains quantum mechanically.

Formal Derivation

In order to derive the form of the partial wave scattering amplitude consider first a plane wave e^{ikz} for which we can write

$$e^{ikz} = e^{ikr\cos\theta} = \sum_{\ell=0}^{\infty} (2\ell+1) i^\ell j_\ell(kr) P_\ell(\cos\theta) \tag{8.4}$$

where the $j_\ell(kr)$ are the well-known spherical Bessel functions

$$j_\ell(kr) \equiv \left(\frac{\pi}{2kr}\right)^{1/2} J_{\ell+1/2}(kr) \tag{8.5}$$

whose lowest order forms are

$$\begin{aligned} j_0(x) &= \frac{\sin x}{x} \\ j_1(x) &= \frac{\sin x}{x^2} - \frac{\cos x}{x} \\ j_2(x) &= \left(\frac{3}{x^2} - \frac{1}{x}\right) \sin x - \frac{3}{x^2} \cos x \end{aligned} \tag{8.6}$$

etc. For large x [Me 65]

$$j_\ell(x) \underset{x\to\infty}{\sim} \frac{1}{x} \sin\left(x - \ell\frac{\pi}{2}\right) \tag{8.7}$$

so that

$$\begin{aligned} e^{ikz} &\underset{r\to\infty}{\sim} \frac{1}{2ikr} \sum_{\ell=0}^{\infty} (2\ell+1) i^\ell \left(e^{i(kr-\ell\frac{\pi}{2})} - e^{-i(kr-\ell\frac{\pi}{2})} \right) P_\ell(\cos\theta) \\ &= \frac{1}{2ikr} \sum_{\ell=0}^{\infty} (2\ell+1) \left(e^{ikr} - e^{-i(kr-\ell\pi)} \right) P_\ell(\cos\theta) \ . \end{aligned} \tag{8.8}$$

We observe that this representation is in terms of an incoming (e^{-ikr}/r) and outgoing (e^{ikr}/r) spherical wave with the phase of the latter being shifted by an angle $\ell\pi$ with respect to the former. This shift can be considered to arise because of the presence of the repulsive centrifugal potential.

In the case that there exists an interaction potential $V(r)$ the form of the scattering wavefunction must be slightly different. The relative normalization between incident and outgoing spherical waves should remain the same (since there should be identical probability flow into and out of the scattering region by unitarity) and the form of the incoming wavefunction should be unchanged since it has not yet interacted with the scattering potential. The only modification allowed then is in the *phase* of the outgoing spherical wave, and the scattering wavefunction $\psi^{(+)}(\vec{r})$ must assume the form

$$\psi^{(+)}(\vec{r}) \underset{r \to \infty}{\sim} \frac{1}{2ikr} \sum_{\ell=0}^{\infty} (2\ell+1) \left(e^{i(kr+2\delta_\ell)} - e^{-i(kr-\ell\pi)} \right) P_\ell(\cos\theta) \qquad (8.9)$$

where $\delta_\ell(k)$ is the scattering phase shift in the ℓth partial wave. Writing this in the form

$$\psi^{(+)}(\vec{r}) \underset{r \to \infty}{\sim} e^{ikz} + \frac{e^{ikr}}{r} \sum_{\ell=0}^{\infty} (2\ell+1) \left(\frac{e^{2i\delta_\ell} - 1}{2ik} \right) P_\ell(\cos\theta) \qquad (8.10)$$

and comparing with the asymptotic form

$$\psi^{(+)}(\vec{r}) \underset{r \to \infty}{\sim} e^{ikz} + \frac{e^{ikr}}{r} f_k(\theta) \qquad (8.11)$$

we identify

$$f_k(\theta) = \sum_{\ell=0}^{\infty} (2\ell+1) \left(\frac{e^{2i\delta_\ell} - 1}{2ik} \right) P_\ell(\cos\theta) \qquad (8.12)$$

as the scattering amplitude and

$$a_\ell(k) = \frac{e^{2i\delta_\ell} - 1}{2ik} = e^{i\delta_\ell} \sin\delta_\ell \frac{1}{k} \qquad (8.13)$$

as the corresponding partial wave amplitude.

Connection with the S-Matrix

An alternative way to understand this form is in terms of the unitarity of the S-matrix. In a momentum representation we have

$$\langle \vec{p}_f | S | \vec{p}_i \rangle = \langle \vec{p}_f | 1 | \vec{p}_i \rangle - 2\pi i \delta(E_f - E_i) \langle \vec{p}_f | T | \vec{p}_i \rangle \qquad (8.14)$$

where

$$\langle \vec{p}_f | 1 | \vec{p}_i \rangle = (2\pi)^3 \delta^3(\vec{p}_f - \vec{p}_i) \qquad (8.15)$$

and

$$\langle \vec{p}_f | T | \vec{p}_i \rangle = -\frac{2\pi}{m} f_k(\theta) = -\frac{2\pi}{m} \sum_{\ell=0}^{\infty} (2\ell+1) a_\ell(k) P_\ell(\cos\theta) \ . \qquad (8.16)$$

Using the identity [Me 65]

$$(2\pi)^3 \delta^3(\vec{p}_f - \vec{p}_i) = (2\pi)^3 \delta(p_f - p_i) \frac{1}{4\pi p_i^2} \sum_{\ell=0}^{\infty} (2\ell+1) P_\ell(\cos\theta) \qquad (8.17)$$

where $\cos\theta = \hat{p}_f \cdot \hat{p}_i$ and noting

$$\delta(E_f - E_i) = \frac{dp_i}{dE_i}\delta(p_f - p_i) = \frac{m}{p_i}\delta(p_f - p_i) \tag{8.18}$$

we have

$$\begin{aligned}\langle \vec{p}_f|S|\vec{p}_i\rangle &= (2\pi)^3\delta(p_f-p_i)\frac{1}{4\pi p_i^2}\sum_{\ell=0}^{\infty}(2\ell+1)P_\ell(\cos\theta)\left(1+2ip_i a_\ell(p_i)\right) \\ &\equiv (2\pi)^3\delta(p_f-p_i)\frac{1}{4\pi p_i^2}\sum_{\ell=0}^{\infty}(2\ell+1)P_\ell(\cos\theta)S_\ell(p_i) \end{aligned} \tag{8.19}$$

We find then

$$\begin{aligned}S_\ell(p_i) &= 1 + 2ip_i a_\ell(p_i) \\ &= 1 + 2i\,e^{i\delta_\ell}\sin\delta_\ell = 1 + e^{2i\delta_\ell} - 1 = e^{2i\delta_\ell}\end{aligned} \tag{8.20}$$

which is as expected. That is, since S must be unitary and diagonal in an angular momentum representation it is required that

$$|S_\ell|^2 = 1 \tag{8.21}$$

which is satisfied by the simple exponential form above. In terms of this representation the optical theorem can be easily obtained. Using the orthogonality relation for Legendre polynomials

$$\int_{-1}^{1} dz\, P_\ell(z)P_{\ell'}(z) = \frac{2}{2\ell+1}\delta_{\ell,\ell'} \tag{8.22}$$

we find

$$\begin{aligned}\sigma = \int d\Omega \frac{d\sigma}{d\Omega} &= 2\pi\int_{-1}^{1} d(\cos\theta)\,|f_k(\theta)|^2 \\ &= \frac{4\pi}{k^2}\sum_{\ell=0}^{\infty}(2\ell+1)\sin^2\delta_\ell \end{aligned} \tag{8.23}$$

On the other hand, in the forward direction we have

$$\begin{aligned}\text{Im}\,f_k(\theta=0) &= \text{Im}\,\frac{1}{k}\sum_{\ell=0}^{\infty}(2\ell+1)e^{i\delta_\ell}\sin\delta_\ell P_\ell(1) \\ &= \frac{1}{k}\sum_{\ell=0}^{\infty}(2\ell+1)\sin^2\delta_\ell \end{aligned} \tag{8.24}$$

i.e., $$\text{Im}\,f_k(0) = \frac{k}{4\pi}\sigma \tag{8.25}$$

which is the optical theorem.

Large ℓ Cutoff

The connection between the classical and quantum mechanical arguments for the existence of an $\ell_{\max}$ can now be explored. Writing the scattering solution as

$$\psi^{(+)}(\vec{r}) = \sum_\ell R_\ell(r) P_\ell(\cos\theta) \tag{8.26}$$

the radial function $R_\ell(r)$ obeys the differential equation

$$\left(\frac{d^2}{dr^2} + k^2 - 2mV(r) - \frac{\ell(\ell+1)}{r^2}\right) r R_\ell(r) = 0 \ . \tag{8.27}$$

If $V(r)$ vanishes beyond radius r_0 then we must have

$$r > r_0 \qquad R_\ell(r) = A j_\ell(kr) + B n_\ell(kr) \tag{8.28}$$

where $j_\ell(kr) = \left(\frac{\pi}{2kr}\right)^{\frac{1}{2}} J_{\ell+\frac{1}{2}}(kr)$ are the usual spherical Bessel functions which are well-behaved as $r \to 0$, while $n_\ell(kr)$ are the corresponding irregular solutions, which diverge for small r

$$n_\ell(kr) = (-1)^\ell \left(\frac{\pi}{2kr}\right)^{1/2} J_{-\ell-1/2}(kr) \ . \tag{8.29}$$

Both j_ℓ, n_ℓ satisfy the differential equation

$$\left(\frac{d^2}{dr^2} + k^2 - \frac{\ell(\ell+1)}{r^2}\right) r j_\ell(kr), r n_\ell(kr) = 0 \ . \tag{8.30}$$

From the asymptotic forms

$$j_\ell(kr) \underset{r\to\infty}{\sim} \frac{1}{kr} \sin\left(kr - \frac{1}{2}\ell\pi\right)$$
$$n_\ell(kr) \underset{r\to\infty}{\sim} -\frac{1}{kr} \cos\left(kr - \frac{1}{2}\ell\pi\right) \tag{8.31}$$

and the requirement on the scattering solution that

$$R_\ell(kr) \underset{r\to\infty}{\sim} \frac{1}{kr}\left(e^{i(kr+2\delta_\ell)} - e^{-i(kr-\ell\pi)}\right) \tag{8.32}$$

we find

$$e^{i2\delta_\ell} = \frac{A - iB}{A + iB} \quad i.e., \quad \delta_\ell = -\tan^{-1}\frac{B}{A} \ . \tag{8.33}$$

We can write this alternatively as

$$\delta_\ell = \tan^{-1}\frac{kr_0 j'_\ell(kr_0) - \beta_\ell j_\ell(kr_0)}{kr_0 n'_\ell(kr_0) - \beta_\ell n_\ell(kr_0)} \tag{8.34}$$

where β_ℓ is the logarithmic derivative of the wavefunction at r_0

$$\beta_\ell = x\frac{1}{R_\ell}\frac{dR_\ell}{dx}\bigg|_{x=kr_0} = kr_0\frac{Aj'_\ell(kr_0) + Bn'_\ell(kr_0)}{Aj_\ell(kr_0) + Bn_\ell(kr_0)} \quad (8.35)$$

For $\ell \gg \ell_{\max} = kr_0$ we have

$$j_\ell(kr_0) \sim \frac{1}{(2\ell+1)!!}(kr_0)^\ell$$
$$n_\ell(kr_0) \sim (2\ell-1)!!(kr_0)^{-\ell-1} \quad (8.36)$$

so

$$\delta_\ell \sim -\left(\frac{\ell+\beta_\ell+1}{\ell-\beta_\ell}\right)\frac{(kr_0)^{2\ell+1}}{(2\ell+1)!!(2\ell-1)!!} \ll 1 \quad (8.37)$$

which provides the quantum mechanical cutoff of high partial waves that we were seeking.

Problem II.9.1

Hard Sphere Scattering

It is well known that the classical physics calculation of the differential scattering cross section for the Coulomb potential agrees with an exact quantum mechanical evaluation. In order to study whether this phenomenon is more general we need a second situation which is exactly soluble by both techniques. Such an example is provided by the simple hard sphere potential

$$V(r) = \begin{cases} \infty & r < a \\ 0 & r \geq a. \end{cases}$$

i) Calculate the classical differential scattering cross section.

ii) Evaluate the exact quantum mechanical differential scattering cross section. Suggestion: Use phase shift methods.

iii) Compare the total scattering cross sections calculated in parts i) and ii) in the short wavelength (high energy) limit and demonstrate that

$$\sigma_{\text{tot}}(\text{quantum}) = 2\sigma_{\text{tot}}(\text{classical}) \ .$$

iv) Explain the physical reason behind this inequality.

CHAPTER III

CHARGED PARTICLE INTERACTIONS

III.1 CHARGED PARTICLE LAGRANGIAN

In order to determine the coupling of a charged particle to the electromagnetic field it is useful to marry relativity and the principle of least action (*cf.* Sect. I.2). The latter states that if one constructs the action S

$$S = \int_{t_1}^{t_2} dt\, L(\vec{x}, \dot{\vec{x}}, t) \tag{1.1}$$

as the time integral of the Lagrangian, then the equations of motion are found by demanding that the action be stationary—$\delta S = 0$—for small variations in the classical path $\vec{x}(t)$. [A lovely discussion of this technique exists in Chapter 19 of Volume II of Feynman's *Lectures on Physics*, FeLS 64]. We have then

$$\begin{aligned}\delta S &= \int_{t_1}^{t_2} dt\, \left(L\left(\vec{x}+\delta\vec{x}, \dot{\vec{x}}+\delta\dot{\vec{x}}, t\right) - L\left(\vec{x}, \dot{\vec{x}}, t\right)\right) \\ &= \int_{t_1}^{t_2} dt\, \left(\frac{\partial L}{\partial \vec{x}} \cdot \delta\vec{x} + \frac{\partial L}{\partial \dot{\vec{x}}} \cdot \delta\dot{\vec{x}} \right) = 0 \end{aligned} \tag{1.2}$$

where $\delta\vec{x}$ is by assumption an infinitesimal change in the path $\vec{x}(t)$. It should be noted in making this variation, however, that the endpoints are to be held *fixed*

$$\vec{x}(t_1) = \vec{x}_1\ ,\qquad \vec{x}(t_2) = \vec{x}_2 \tag{1.3}$$

so that we require

$$\delta\vec{x}(t_1) = \delta\vec{x}(t_2) = 0\ . \tag{1.4}$$

We can then integrate by parts and neglect the surface term

$$\begin{aligned}0 = \delta S &= \int_{t_1}^{t_2} dt\, \delta\vec{x} \cdot \left(\frac{\partial L}{\partial \vec{x}} - \frac{d}{dt}\frac{\partial L}{\partial \dot{\vec{x}}} \right) + \delta\vec{x} \cdot \left.\frac{\partial L}{\partial \dot{\vec{x}}}\right|_{t_1}^{t_2} \\ &= \int_{t_1}^{t_2} dt\, \delta\vec{x} \cdot \left(\frac{\partial L}{\partial \vec{x}} - \frac{d}{dt}\frac{\partial L}{\partial \dot{\vec{x}}} \right)\ . \end{aligned} \tag{1.5}$$

If the variation in the action is to vanish for an *arbitrary* change in path $\delta\vec{x}(t)$, we demand

$$0 = \frac{\partial L}{\partial \vec{x}} - \frac{d}{dt}\frac{\partial L}{\partial \dot{\vec{x}}} \tag{1.6}$$

which is Lagrange's equation of motion. For simple potential motion

$$L = T - V = \frac{1}{2}m\dot{\vec{x}}^2 - V(\vec{x}) \tag{1.7}$$

and
$$\frac{\partial L}{\partial \dot{\vec{x}}} = m\dot{\vec{x}} = \vec{p} \ , \qquad \frac{\partial L}{\partial \vec{x}} = -\vec{\nabla}V(\vec{x}) \ . \tag{1.8}$$

The equation of motion becomes
$$\frac{d}{dt}\vec{p} = -\vec{\nabla}V(\vec{x}) \tag{1.9}$$

which is Newton's second law.

Now consider what the form of a relativistic Lagrangian might be [Ja 62]. Since
$$d\tau = \sqrt{dt^2 - d\vec{x}^2} = dt\sqrt{1 - \dot{\vec{x}}^2} \tag{1.10}$$

we have
$$dt = \gamma \, d\tau \tag{1.11}$$

where τ is the proper time and
$$\gamma = \frac{1}{\sqrt{1 - \dot{\vec{x}}^2}} \tag{1.12}$$

is the time dilation factor. The action can be written as
$$S = \int_{t_1}^{t_2} dt\, L = \int_{\tau_1}^{\tau_2} d\tau\, (\gamma L) \ . \tag{1.13}$$

and if the equations of motion are to have the same form in *all* inertial frames γL must be a Lorentz scalar quantity. This scalar must be constructed from the two four-vectors — x^μ and p^μ — which characterize the particle. However, as experiments performed at one point in spacetime x^μ should give identical results to those performed at a different point $x^\mu + a^\mu$ the action must be invariant under translations — $x^\mu \to x^\mu + a^\mu$ — so γL can only be a function of p^μ. Finally, since γL is a scalar, it can only be a function of $p^\mu p_\mu = m^2$ so we conclude
$$\gamma L = f(m) = \text{const.} \ . \tag{1.14}$$

The relativistic action for a freely moving particle is then given by
$$S = f(m)\int_{\tau_1}^{\tau_2} d\tau = f(m)\int_{t_1}^{t_2} dt\,\sqrt{1 - \dot{\vec{x}}^2} \tag{1.15}$$

and the equation of motion is
$$0 = \frac{d}{dt}\frac{\partial L}{\partial \dot{\vec{x}}} - \frac{\partial L}{\partial \vec{x}} = -\frac{d}{dt}f(m)\gamma \dot{\vec{x}} \ . \tag{1.16}$$

Since $m\gamma\dot{\vec{x}}$ is the relativistic momentum, we identify $f(m) = -m$ and
$$L = -\frac{m}{\gamma} = -m\sqrt{1 - \dot{\vec{x}}^2} \ . \tag{1.17}$$

In the non-relativistic limit — $|\dot{\vec{x}}| \ll 1$ — this becomes

$$L = -m + \frac{1}{2}m\dot{\vec{x}}^2 + \ldots \tag{1.18}$$

as usual. In this limit, we know that in the presence of a scalar potential $\phi(x)$ the potential energy is given by

$$V = e\phi(x) \tag{1.19}$$

where e is the charge of the particle. Relativistically ϕ is the scalar component of a four-vector

$$A^\mu = (\phi, \vec{A}) \tag{1.20}$$

whose three-vector piece is the usual vector potential. A Lorentz scalar which reduces to Eq. 1.19 in the nonrelativistic limit is

$$e\frac{p \cdot A}{m} \xrightarrow{v \to 0} e\phi \tag{1.21}$$

where $p^\mu = mdx^\mu/d\tau$ is the relativistic four-momentum, so that the full relativistic Lagrangian must be

$$\gamma L = -m - e\frac{1}{m}p \cdot A \tag{1.22}$$

or

$$L = -m\sqrt{1 - \dot{\vec{x}}^2} - e\phi\frac{dt}{d\tau} + e\frac{d\vec{x}}{d\tau} \cdot \vec{A} \tag{1.23}$$

whose non-relativistic reduction is

$$L = -m + \frac{1}{2}m\dot{\vec{x}}^2 - e\phi + e\dot{\vec{x}} \cdot \vec{A} + \ldots \quad . \tag{1.24}$$

Before proceeding further we check that the Lorentz force law is correctly reproduced. For simplicity, we work in the non-relativistic limit. Recall that the Lagrangian is a function of the trajectory $\vec{x}(t)$ through the position dependence inherent in the vector potential $A^\mu(x)$. Then

$$\frac{d}{dt}\frac{\partial L}{\partial \dot{\vec{x}}} - \frac{\partial L}{\partial \vec{x}} = \frac{d}{dt}\left(m\dot{\vec{x}} + e\vec{A}\right) + e\frac{\partial \phi}{\partial \vec{x}} - e\sum_{i=1}^{3}\dot{x}_i\frac{\partial}{\partial \vec{x}}A_i = 0 \tag{1.25}$$

i.e.,

$$\frac{d}{dt}\vec{p} = -e\vec{\nabla}\phi + e\sum_{i}\dot{x}_i\vec{\nabla}A_i \tag{1.26}$$

where the (canonical) momentum is

$$\vec{p} = \frac{\partial L}{\partial \dot{\vec{x}}} = m\dot{\vec{x}} + e\vec{A} \quad . \tag{1.27}$$

Since by the chain rule

$$\frac{d}{dt}\vec{A} = \frac{\partial \vec{A}}{\partial t} + \sum_{i=1}^{3} \frac{\partial \vec{A}}{\partial x_i}\dot{x}_i = \frac{\partial}{\partial t}\vec{A} + (\dot{\vec{x}} \cdot \vec{\nabla})\vec{A} \qquad (1.28)$$

we find

$$\begin{aligned} m\frac{d\dot{\vec{x}}}{dt} &= e\left(-\vec{\nabla}\phi - \frac{\partial \vec{A}}{\partial t}\right) + e\left(\sum_i \dot{x}_i \vec{\nabla} A_i - \left(\dot{\vec{x}} \cdot \vec{\nabla}\right)\vec{A}\right) \\ &= e\left(-\vec{\nabla}\phi - \frac{\partial \vec{A}}{\partial t}\right) + e\dot{\vec{x}} \times \left(\vec{\nabla} \times \vec{A}\right) \\ &= e\left(\vec{E} + \dot{\vec{x}} \times \vec{B}\right) \end{aligned} \qquad (1.29)$$

i.e., the familiar Lorentz force law.

For quantum mechanical applications we need the form of the Hamiltonian. Performing the usual canonical transformation

$$\begin{aligned} H = \dot{\vec{x}} \cdot \frac{\partial L}{\partial \dot{\vec{x}}} - L = \dot{\vec{x}} \cdot \vec{p} - L &\underset{v \ll 1}{\approx}]m\dot{\vec{x}}^2 + e\dot{\vec{x}} \cdot \vec{A} + m - \frac{1}{2}m\dot{\vec{x}}^2 + e\phi - e\dot{\vec{x}} \cdot \vec{A} \\ &= m + \frac{1}{2}m\dot{\vec{x}}^2 + e\phi = m + \frac{\left(\vec{p} - e\vec{A}\right)^2}{2m} + e\phi \end{aligned}$$
(1.30)

We also require an appropriate Lagrangian or Hamiltonian which reproduces the Maxwell relations. First though we briefly review the usual formulation of electromagnetism.

PROBLEM III.1.1

Relativistic Particle Interactions

We derived in Eq. 1.23 the relativistically correct Lagrangian for the interaction of a particle of mass m and charge e with an external electromagnetic field described by the vector potential A_μ:

$$L = -m\sqrt{1-\dot{\vec{x}}^2} - e\phi + e\dot{\vec{x}} \cdot \vec{A} \ .$$

i) Find the relativistic equation of motion and show that it reduces to the Lorentz force law.

ii) Find the relativistically correct Hamiltonian for the system.

iii) Using this Hamiltonian, show that Hamilton's equations of motion yield the Lorentz force law.

iv) Show that this Hamiltonian reduces, in the non-relativistic limit, to the form

$$H \approx \frac{\left(\vec{p} - e\vec{A}\right)^2}{2m} + e\phi \ .$$

III.2 REVIEW OF MAXWELL EQUATIONS AND GAUGE INVARIANCE

Maxwell Relations: Conventional Form

We begin by reviewing Maxwell's equations which, in rationalized Heaviside-Lorentz units, read

$$\vec{\nabla} \cdot \vec{E} = \rho$$
$$\vec{\nabla} \times \vec{B} - \frac{\partial \vec{E}}{\partial t} = \vec{j}$$
$$\vec{\nabla} \cdot \vec{B} = 0 \tag{2.1}$$
$$\vec{\nabla} \times \vec{E} + \frac{\partial \vec{B}}{\partial t} = 0 \ .$$

The charge on the electron e is related to the fine structure constant α via

$$\alpha = \frac{e^2}{4\pi(\hbar c)} \approx \frac{1}{137} \ . \tag{2.2}$$

However, writing the Maxwell relations in the above form in terms of the three-vector quantities $\vec{E}, \vec{B}$ obscures the relativistic covariance of these equations. In order to bring out this feature, we introduce the Maxwell tensor, whose components are

$$F^{\mu\nu} = \begin{pmatrix} 0 & -E_1 & -E_2 & -E_3 \\ E_1 & 0 & -B_3 & B_2 \\ E_2 & B_3 & 0 & -B_1 \\ E_3 & -B_2 & B_1 & 0 \end{pmatrix} . \tag{2.3}$$

and which is an antisymmetric Lorentz tensor of rank 2. That is, if under a Lorentz transformation

$$x^\mu \to x'^\mu = \sum_{\nu=0}^{3} a^\mu{}_\nu x^\nu \tag{2.4}$$

then under the same transformation $F^{\mu\nu}$ becomes

$$F^{\mu\nu} \to F'^{\mu\nu} = \sum_{\alpha,\beta=0}^{3} a^\mu{}_\alpha a^\nu{}_\beta F^{\alpha\beta} \ . \tag{2.5}$$

Since the charge density ρ and current density $\vec{j}$ constitute components of a four-vector

$$j^\mu = (\rho, \vec{j}) \tag{2.6}$$

we can write the first two Maxwell equations in the concise form

$$\partial_\mu F^{\mu\nu} = j^\nu . \tag{2.7}$$

[Note that here $\partial_\mu = \eta_{\mu\nu}\partial^\nu = \left(\frac{\partial}{\partial t}, \vec{\nabla}\right) = \frac{\partial}{\partial x^\mu}$.] Also, on account of the antisymmetry of the Maxwell tensor

$$0 = \partial_\mu \partial_\nu F^{\mu\nu} = \partial_\nu j^\nu \tag{2.8}$$

the continuity equation for the charge current density is automatic. The physical significance of Eq. 2.8 can be seen by integration over a finite volume V

$$0 = \frac{d}{dt}\int_V \rho\, d^3r + \int_V \vec{\nabla}\cdot\vec{j}\, d^3r\ , \tag{2.9}$$

which, using Gauss' theorem becomes

$$\frac{d}{dt}Q = -\int_{\text{surface}} \vec{j}\cdot d\vec{S} \tag{2.10}$$

where Q is the total charge contained within the volume. Thus any charge which disappears from the volume must flow out through the surface—charge is conserved locally.

The remaining two Maxwell equations can be concisely written in terms of a second relativistically covariant equation as

$$\partial_\mu F^{*\mu\nu} = 0 \tag{2.11}$$

where $F^{*\mu\nu} \equiv \frac{1}{2}\epsilon^{\mu\nu\alpha\beta}F_{\alpha\beta}$ is called the dual tensor and $\epsilon^{\mu\alpha\beta\gamma}$ is the totally antisymmetric tensor in four dimensions

$$\epsilon^{\mu\nu\alpha\beta} = \begin{cases} 1 & \text{if } \mu\nu\alpha\beta \text{ is a symmetric permutation of 0123} \\ -1 & \text{if } \mu\nu\alpha\beta \text{ is an antisymmetric permutation of 0123} \\ 0 & \text{otherwise} \end{cases} . \tag{2.12}$$

Thus the *four* conventional Maxwell relations reduce to just *two* relativistically covariant equations

$$\begin{aligned}\partial_\mu F^{\mu\nu} &= j^\nu \\ \partial_\mu F^{*\mu\nu} &= 0\ .\end{aligned} \tag{2.13}$$

Maxwell Relations: Vector Potential Form

We can simplify things even further by introducing the four-vector potential A^μ

$$A^\mu = (\phi, \vec{A}) \tag{2.14}$$

where ϕ is the scalar potential and $\vec{A}$ is the conventional three-vector potential. In terms of A^μ we have

$$\begin{aligned} F^{\mu\nu} &= \partial^\mu A^\nu - \partial^\nu A^\mu \\ \vec{E} &= -\vec{\nabla}\phi - \frac{\partial \vec{A}}{\partial t} \\ \vec{B} &= \vec{\nabla}\times\vec{A} \end{aligned} \tag{2.15}$$

so that the second covariant Maxwell relation—Eq. 2.11—is *automatically* satisfied on account of the antisymmetry of $\epsilon^{\mu\alpha\beta\gamma}$—i.e., $\epsilon^{\mu\alpha\beta\gamma}\partial_\alpha\partial_\beta \equiv 0$. Equivalently, in three-vector language

$$\nabla \cdot B = \vec{\nabla} \cdot \vec{\nabla} \times \vec{A} = \vec{\nabla} \times \vec{\nabla} \cdot A \equiv 0$$

$$\vec{\nabla} \times \vec{E} = -\vec{\nabla} \times \vec{\nabla}\phi - \frac{\partial}{\partial t}\vec{\nabla} \times \vec{A} \tag{2.16}$$

$$= 0 - \frac{\partial}{\partial t}\vec{\nabla} \times \vec{A} = -\frac{\partial \vec{B}}{\partial t} \quad.$$

Then the first (and now only) Maxwell equation becomes

$$\Box A^\nu - \partial^\nu(\partial_\mu A^\mu) = j^\nu \quad. \tag{2.17}$$

Gauge Invariance

Finally, we can utilize "gauge invariance" to produce a further simplification. We are free to redefine the vector potential by adding the gradient of an arbitrary scalar field χ

$$A^\mu_{\text{new}} = A^\mu_{\text{old}} + \partial^\mu \chi \quad. \tag{2.18}$$

Since

$$F^{\mu\nu}_{\text{new}} = F^{\mu\nu}_{\text{old}} + (\partial^\mu \partial^\nu - \partial^\nu \partial^\mu)\chi = F^{\mu\nu}_{\text{old}} \tag{2.19}$$

the physical quantities $\vec{E}, \vec{B}$ are unchanged. We can then simplify Eq. 2.17 by selecting a χ such that

$$\Box \chi = -\partial_\mu A^\mu_{\text{old}} \quad. \tag{2.20}$$

Then A^μ_{new} satisfies the Lorentz condition

$$\partial_\mu A^\mu_{\text{new}} = \partial_\mu A^\mu_{\text{old}} + \Box \chi = 0 \tag{2.21}$$

and the Maxwell equation becomes

$$\Box A^\mu = j^\mu \quad. \tag{2.22}$$

We are still free to make a further change

$$A^\mu \to A^\mu + \partial^\mu \eta \tag{2.23}$$

as long as

$$\Box \eta = 0 \quad, \tag{2.24}$$

which is called a gauge transformation of the *second* kind.

Classical electrodynamics is then invariant under gauge transformations in that under the redefinition

$$A^\mu \to A^\mu + \partial^\mu \chi \tag{2.25}$$

the physical fields $\vec{E}$ and $\vec{B}$ are unchanged. Let's now see how this invariance is manifested in quantum mechanics. Suppose we have a wavefunction ψ which obeys the Schrödinger equation implied by the Hamiltonian of Eq. 1.30

$$\left(\frac{1}{2m}\left(-i\vec{\nabla} - e\vec{A}\right)^2 + e\phi\right)\psi = i\frac{\partial}{\partial t}\psi \quad. \tag{2.26}$$

Then under the gauge transformation Eq. 2.25, the wavefunction
$$\psi' = e^{-ie\chi}\psi \qquad (2.27)$$
obeys the Schrödinger equation for the gauge-transformed field
$$\left[\frac{1}{2m}\left(-i\vec{\nabla} - e\vec{A} + e\vec{\nabla}\chi\right)^2 + e\left(\phi + \frac{\partial\chi}{\partial t}\right)\right]\psi' = i\frac{\partial}{\partial t}\psi' \ . \qquad (2.28)$$
[Proof:
$$i\frac{\partial}{\partial t}\psi' = e\frac{\partial\chi}{\partial t}\psi' + ie^{-ie\chi}\frac{\partial\psi}{\partial t}$$
$$\left(-i\vec{\nabla} - e\vec{A} + e\vec{\nabla}\chi\right)^2 \psi' = \left(e\vec{A} - e\vec{\nabla}\chi\right)^2 \psi'$$
$$+ 2i\left(e\vec{A} - e\vec{\nabla}\chi\right) \cdot \left(e^{-ie\chi}\vec{\nabla}\psi - ie\vec{\nabla}\chi\psi'\right) + ie^2\vec{\nabla}^2\chi\psi'$$
$$+ e^2\vec{\nabla}\chi \cdot \vec{\nabla}\chi\psi' - e^{-ie\chi}\vec{\nabla}^2\psi + 2ie\vec{\nabla}\chi \cdot e^{-ie\chi}\vec{\nabla}\psi$$
$$= e^{-ie\chi}\left(-i\vec{\nabla} - e\vec{A}\right)^2 \psi$$
$$(2.29)$$
so that Eq. 2.28 becomes
$$\frac{1}{2m}e^{-ie\chi}\left(\left(-i\vec{\nabla} - e\vec{A}\right)^2 + e\phi\right)\psi = e^{-ie\chi}(i\frac{\partial\psi}{\partial t}) \quad \text{q.e.d .]} \qquad (2.30)$$
Under the gauge transformation Eq. 2.25 then, the wavefunction undergoes a change
$$\psi \to e^{-ie\chi}\psi = \psi' \qquad (2.31)$$
but physics remains invariant since
$$|\psi|^2 = |\psi'|^2 \ . \qquad (2.32)$$

The gauge invariance property is an important one and is easily derived within a path integral context. For a vector potential A^μ the propagator is given by
$$D_F(\vec{x}_2, t_2; \vec{x}_1, t_1) = \int \mathcal{D}[\vec{x}(t)] \exp i \int_{t_1}^{t_2} dt\, L\left(\vec{x}, \dot{\vec{x}}, \vec{A}, \phi\right) \qquad (2.33)$$
where
$$L = \frac{1}{2}m\dot{\vec{x}}^2 - e\phi + e\dot{\vec{x}}\cdot\vec{A} \ . \qquad (2.34)$$
After a gauge transformation the propagator becomes
$$D'_F(\vec{x}_2, t_2; \vec{x}_1, t_1) = \int \mathcal{D}[\vec{x}(t)] \exp i \int_{t_1}^{t_2} dt\, L\left(\vec{x}, \dot{\vec{x}}, \vec{A} - \vec{\nabla}\chi, \phi + \frac{\partial\chi}{\partial t}\right)$$
$$= \int \mathcal{D}[\vec{x}(t)] \exp i \int_{t_1}^{t_2} dt\, L\left(\vec{x}, \dot{\vec{x}}, \vec{A}, \phi\right) \qquad (2.35)$$
$$\times \exp -ie \int_{t_1}^{t_2} dt\left(\dot{\vec{x}}\cdot\vec{\nabla}\chi + \frac{\partial\chi}{\partial t}\right) \ .$$

III.2 REVIEW OF MAXWELL EQUATIONS AND GAUGE INVARIANCE

However,
$$\left(\frac{\partial}{\partial t} + \dot{\vec{x}} \cdot \vec{\nabla}\right)\chi = \frac{d}{dt}\chi \tag{2.36}$$

is a total derivative so that
$$\exp\left(-ie\int_{t_1}^{t_2} dt \frac{d}{dt}\chi\right) = \exp(-ie\chi(t_2) + ie\chi(t_1)) . \tag{2.37}$$

This factor can be taken outside the path integration yielding
$$D'_F(\vec{x}_2, t_2; \vec{x}_1, t_1) = e^{-ie\chi(t_2)} D_F(\vec{x}_2, t_2; \vec{x}_1, t_1) e^{ie\chi(t_1)} . \tag{2.38}$$

Finally, expanding Eq. 2.38 in terms of eigenfunctions we have
$$\begin{aligned}\sum_n \psi'_n(\vec{x}_2) e^{-iE_n(t_2-t_1)} \psi'^*_n(\vec{x}_1) \\ = \sum_n e^{-ie\chi(t_2)} \psi_n(\vec{x}_2) e^{-iE_n(t_2-t_1)} \psi^*_n(\vec{x}_1) e^{ie\chi(t_1)}\end{aligned} \tag{2.39}$$

whence we observe that
$$\psi'(\vec{x}, t) = e^{-ie\chi(t)} \psi(\vec{x}, t) \tag{2.40}$$

as found previously (*cf.* Eq. 2.31).

PROBLEM III.2.1

Relativistic Maxwell Equations

Show that the *two* relativistic Maxwell equations
$$\partial_\mu F^{\mu\nu} = j^\nu \qquad \partial_\mu F^{*\mu\nu} = 0$$

are completely equivalent to the *four* Maxwell equations given in Eq. 2.1.

PROBLEM III.2.2

To Gauge or Not to Gauge: That is the Question

Suppose that at time $t = 0$ we have a one-dimensional harmonic oscillator in its ground state
$$\psi(x, 0) = \phi_0(x) .$$

An electric field
$$\vec{E} = E_0 \cos\omega t \,\hat{e}_x$$

is now turned on. What is the amplitude to have made a transition to an excited state $\phi_n(x)$ at later time t? In order to answer this question we use the interaction term
$$V = \frac{e^2}{2m} \vec{A} \cdot \vec{A} - \frac{e}{m} \vec{A} \cdot \vec{p} + e\phi .$$

i) If we choose
$$\vec{A} = 0, \qquad \phi = -E_0 x \cos\omega t$$
show that (to lowest order in e)

$$\mathrm{Amp}_{fi} = ieE_0 e^{-\frac{i}{2}\omega_0 t} \langle \phi_n | x | \phi_0 \rangle$$
$$\times \left(\exp\left(i\frac{\omega - n\omega_0}{2}t\right) \frac{\sin\frac{n\omega_0+\omega}{2}t}{n\omega_0 + \omega} + \exp\left(-i\frac{\omega + n\omega_0}{2}t\right) \frac{\sin\frac{n\omega_0-\omega}{2}}{n\omega_0 - \omega} \right)$$

where ω_0 is the oscillator frequency.

ii) On the other hand if we make a gauge transformation so that
$$\vec{A} = -\frac{E_0}{\omega}\sin\omega t \,\hat{e}_x \qquad \phi = 0$$
show that

$$\mathrm{Amp}_{fi} = ieE_0 e^{-\frac{i}{2}\omega_0 t} \langle \phi_n | x | \phi_0 \rangle \frac{n\omega_0}{\omega}$$
$$\times \left(-\exp\left(i\frac{\omega - n\omega_0}{2}t\right) \frac{\sin\frac{n\omega_0+\omega}{2}t}{n\omega_0 + \omega} + \exp\left(-i\frac{\omega + n\omega_0}{2}t\right) \frac{\sin\frac{n\omega_0-\omega}{2}t}{n\omega_0 - \omega} \right).$$

iii) Construct explicitly the gauge transformation involved.

iv) Why is there a different answer in the two cases? (Isn't quantum mechanics gauge invariant?)

v) Which answer is physically correct?

PROBLEM III.2.3

Resolution of the Gauge Invariance Paradox

In the previous problem we identified a strange situation wherein two apparently equivalent formulations of the same problem appeared to give different predictions for an experimental result. Nevertheless, upon reflection it becomes clear that the problem is not formulated quite as carefully as one might want. The problem is that we calculated the probability of a transition between two *unperturbed* eigenstates of the Hamiltonian. However, the electric field is always present and it is not clear how one would detect such a transition. In order to avoid ambiguity, let's define things more precisely. Imagine turning the electric field on suddenly at some time t_1 and then turning it off again at time t_2. The measurement of the transition probability is now non-ambiguous, as it can be performed in the absence of the perturbing potential — i.e., before time t_1 and after time t_2. Gauge invariance requires the equality of the rates for this process as calculated using either $\vec{p} \cdot \vec{A}$ or $e\phi$ as the interaction potential.

Verify this assertion — assuming an applied electric field
$$\vec{E}(t) = \theta(t - t_1)\theta(t_2 - t)E_0 \cos\omega t \,\hat{e}_x$$

III.2 REVIEW OF MAXWELL EQUATIONS AND GAUGE INVARIANCE

calculate the transition rate from the ground state of a harmonic oscillator (at time $t < t_1$) to the nth excited state (at time $t > t_2$):

i) Using as the interaction potential

$$\phi = -xE(t) \;, \qquad \vec{A} = 0 \;;$$

ii) Using as the interaction potential

$$\phi = 0 \;, \qquad \vec{A}(t) = -\int_{-\infty}^{t} dt' \, \vec{E}(t')$$

and show that the transition probability is the same for either case.

This then is the resolution of the paradox: when the problem is well-posed physically there can be no difference in the results obtained by either technique!

PROBLEM III.2.4

The Weak Equivalence Principle in Quantum Mechanics

According to the weak equivalence principle, an observer in a frame undergoing uniform acceleration $\vec{a}$ must find identical physics to that found by an observer in a uniform gravitational field with $\vec{g} = -\vec{a}$. Thus the wavefunction of a free particle as observed in the accelerated frame — $\psi_a(\vec{r}, t)$ — must be identical (up to a phase) to a solution of the Schrödinger equation in a uniform gravitational field — $\psi'_g(\vec{r}', t)$ — where

$$i\frac{\partial}{\partial t}\psi'_g(\vec{r}', t) = \left(-\frac{1}{2m}\vec{\nabla}'^{\,2} + mgz'\right)\psi'_g(\vec{r}', t) \;.$$

i) In order to explore this assertion, determine the behavior of the propagator under the generalized Galilean transformation

$$\vec{x}' = \vec{x} + \vec{\xi}(t) \;, \qquad t' = t$$

and demonstrate that the transformed propagator corresponds to that for motion under the influence of the potential

$$V(\vec{x}') = -m\ddot{\vec{\xi}} \cdot \vec{x}'$$

provided that the wavefunction picks up a phase factor

$$\exp -im\left(\dot{\vec{\xi}} \cdot \vec{x}' - \frac{1}{2}\int dt\,\dot{\vec{\xi}}^{\,2}\right) \equiv \exp i\phi \;.$$

That is, demonstrate that

$$e^{i\phi(2)}\psi'(\vec{x}'_2, t_2) = \int d^3x'_1 \, D'_F(\vec{x}'_2, t_2; \vec{x}'_1, t_1) e^{i\phi(1)}\psi'(\vec{x}'_1, t_1) \;.$$

Note that for the case that
$$\vec{\xi}(t) = -\frac{1}{2}\vec{g}t^2$$
we find
$$V(\vec{x}') = m\vec{g}\cdot\vec{x}'$$
which is just the quantum mechanical statement of the equivalence principle.

ii) For the case $\vec{\xi}(t) = \vec{u}t$ show that if the wavefunction in the unprimed frame is a plane wave with momentum $\vec{p}$
$$\psi(\vec{x},t) = \exp\left(i\vec{p}\cdot\vec{x} - i\frac{p^2}{2m}t\right)$$
then the corresponding wavefunction in the primed frame is also a plane wave
$$\psi'(\vec{x}',t) = \exp\left(i\vec{p}'\cdot\vec{x}' - i\frac{p'^2}{2m}t\right)$$
where $\vec{p}' = \vec{p} + m\vec{u}$. Thus the Schrödinger equation is Galilean invariant.

iii) Consider the time-independent Schrödinger equation in the primed frame
$$\left(i\frac{1}{2m}\vec{\nabla}'^2 + mgz'\right)\psi'(\vec{x}') = E\psi'(\vec{x}') \ .$$
Show that if $E = 0$ then a solution is
$$\psi(\vec{x}',t) = Ai\left((2m^2 g)^{\frac{1}{3}} z'\right)$$
where $Ai(x)$ is the Airy function.

iv) Using the result of part i) and $\xi(t) = \frac{1}{2}gt^2$ show that
$$\psi(\vec{x},t) = \exp i m g t\left(z - \frac{1}{3}gt^2\right) Ai\left((2m^2 g)^{\frac{1}{3}}\left(z - \frac{1}{2}gt^2\right)\right)$$
obeys the free particle Schrödinger equation
$$i\frac{\partial}{\partial t}\psi(\vec{x},t) = -\frac{1}{2m}\vec{\nabla}^2\psi(\vec{x},t) \ .$$

v) Note that $\psi(\vec{x},t)$ represents a wavepacket which is an exact solution of the free Schrödinger equation but which does not spread. Explain this curious feature.

PROBLEM III.2.5
Charged Particle in a Uniform Field

Consider a particle of mass m and charge e which is embedded in a uniform electric field $\vec{E} = E\hat{e}_x$. This situation may be described in terms of a potential energy
$$V(x) = -eEx \ .$$
From Problem I.3.2 we know that if at $t = 0$ the wavefunction of the particle is of plane wave form
$$\psi(\vec{x}, 0) = e^{i\vec{p}\cdot\vec{x}}$$
then at a later time t, we have
$$\psi_1(\vec{x}, t) = \exp i(\vec{p} + e\vec{E}t) \cdot \vec{x} - i\left(\frac{p^2}{2m}t + \frac{e}{2m}\vec{p}\cdot\vec{E}t^2 - \frac{e^2 E^2 t^3}{6m}\right)$$
so that the problem can be exactly solved.

However, it should also be possible to represent the electric field in terms of a vector potential $\vec{A}$
$$\vec{A} = -\vec{E}t$$
with a vanishing scalar potential
$$\phi = 0$$

i) Find the gauge transformation $\Lambda(\vec{x}, t)$, which accomplishes this change.

ii) Taking as before
$$\psi(\vec{x}, 0) = e^{i\vec{p}\cdot\vec{x}}$$
but now
$$H(\vec{x}, t) = \frac{1}{2m}\left(-i\vec{\nabla} - e\vec{A}(t)\right)^2$$
find the wave function at time t via
$$\psi_2(\vec{x}, t) = \exp\left(-i\int_0^t dt' H(\vec{x}, t')\right) \psi(\vec{x}, 0) \ .$$

iii) Verify that ψ_1 and ψ_2 are related by the simple gauge factor
$$\psi_2(\vec{x}, t) = e^{-ie\Lambda} \psi_1(\vec{x}, t) \ .$$

III.3 THE BOHM–AHARONOV EFFECT

We have emphasized that in classical physics it is sufficient to deal solely with the electromagnetic field tensor $F^{\mu\nu}$ (or equivalently with its components, the electric and magnetic field strengths). The vector potential A^μ is an auxiliary construct from which the fields may be calculated via

$$\vec{B} = \vec{\nabla} \times \vec{A}$$
$$\vec{E} = -\vec{\nabla}A^0 - \frac{\partial \vec{A}}{\partial t} \ . \tag{3.1}$$

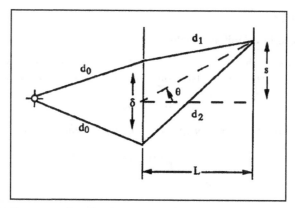

Fig. III.1: Geometry for the discussion of two slit diffraction.

An indication of this is that the vector potential can be redefined via a gauge transformation

$$A^\mu \to A^\mu + \partial^\mu \Lambda \tag{3.2}$$

without affecting any physical laws. The vector potential does have a significance of its own in quantum mechanics, however, as we now demonstrate by discussing the so-called Bohm–Aharonov experiment [AhB 59].

In order to understand this effect, imagine first a two slit diffraction experiment involving an electron beam, as shown in Figure III.1.

Assuming dominance of the classical trajectory, then along path #1 the phase of the electron wavefunction at the screen relative to the source is given by

$$\exp(i\int_{t_1}^{t_2} dt \, \tfrac{1}{2}m\dot{x}^2) \sim \exp\left(i\frac{2\pi d_0}{\lambda} + i\frac{2\pi d_1}{\lambda}\right) \tag{3.3}$$

where

$$\lambda = \frac{2\pi}{p} \approx \frac{2\pi t}{md} \tag{3.4}$$

is the deBroglie wavelength. On the other hand, along path #2 the phase change is given by

$$\exp\left(i\frac{2\pi d_0}{\lambda} + i\frac{2\pi d_2}{\lambda}\right) \; . \tag{3.5}$$

The relative phase difference between the two paths then is given by

$$\exp i\frac{2\pi}{\lambda}(d_2 - d_1) \; . \tag{3.6}$$

so that the wavefunction at a given point at the screen has the form

$$\psi \sim \psi_0 \left(1 + \exp i2\pi \frac{d_2 - d_1}{\lambda}\right) \tag{3.7}$$

III.3 THE BOHM–AHARONOV EFFECT

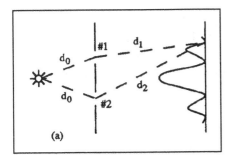
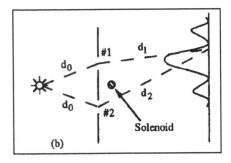

Fig. III.2: Two slit diffraction patterns without (a) and with (b) the presence of an infinite solenoid.

while the intensity is

$$|\psi|^2 \sim |\psi_0|^2 \, 4\cos^2 \pi \frac{d_2 - d_1}{\lambda} \, . \tag{3.8}$$

From the figure above

$$d_2 = \sqrt{L^2 + \left(s + \frac{1}{2}\delta\right)^2} \qquad d_1 = \sqrt{L^2 + \left(s - \frac{1}{2}\delta\right)^2}$$

$$\approx \sqrt{L^2 + s^2} + \frac{s\delta}{2\sqrt{L^2 + s^2}} \qquad \approx \sqrt{L^2 + s^2} - \frac{s\delta}{2\sqrt{L^2 + s^2}} \tag{3.9}$$

$$= \sqrt{L^2 + s^2} + \frac{\delta}{2}\sin\theta \qquad = \sqrt{L^2 + s^2} - \frac{\delta}{2}\sin\theta \, .$$

so we can write the intensity pattern in terms of the angle θ and slit separation δ as

$$I(\theta) = 4I_0 \cos^2 \pi\delta \frac{\sin\theta}{\lambda} \tag{3.10}$$

where I_0 is the intensity assuming only a single slit were open. This is the usual picture of electron diffraction and results in an intensity pattern as shown in Figure III.2a.

Now, however, consider the changes which ensue if an infinite solenoid is introduced into the configuration, as shown in Figure III.2.b. Since the solenoid is infinite, there exists no magnetic field outside the solenoidal volume itself. However, there *does* exist a non-vanishing vector potential $\vec{A}$ in this region. If we assume the solenoid (of radius R and uniform field strength B) to be aligned along the z-direction then, working in cylindrical coordinates r, ϕ, z, a possible choice for the vector potential is

$$\vec{A} = \begin{cases} \hat{\phi}\frac{1}{2}Br & r < R \\ \hat{\phi}\frac{1}{2}B\frac{R^2}{r} & r > R \end{cases} \tag{3.11}$$

since
$$\vec{B} = \vec{\nabla} \times \vec{A} = \hat{k}\frac{1}{r}\frac{\partial}{\partial r}(rA_\phi) = \begin{cases} B\hat{k} & r < R \\ 0 & r > R \end{cases} . \tag{3.12}$$

In order to determine the effect upon the electron wavefunction, we recall that the Lagrangian for a charged particle in the presence of a vector potential $\vec{A}$ becomes

$$L = \frac{1}{2}m\dot{\vec{x}}^2 + e\dot{\vec{x}} \cdot \vec{A} . \tag{3.13}$$

The path integral for path #1 then involves

$$\exp i \int \left(\frac{1}{2}m\dot{x}^2 + e\vec{A} \cdot \frac{d\vec{x}}{dt}\right) dt \simeq \exp\left(i\frac{2\pi}{\lambda}(d_0 + d_1) + ie \int_1 \vec{A} \cdot d\vec{x}\right) \tag{3.14}$$

while that for path #2 involves the phase

$$\exp\left(i\frac{2\pi}{\lambda}(d_0 + d_2) + ie \int_2 \vec{A} \cdot d\vec{x}\right) . \tag{3.15}$$

The phase difference at the screen between the two paths becomes

$$\exp\left(i2\pi\frac{d_2 - d_1}{\lambda} + ie \int_2 - \int_1 \vec{A} \cdot d\vec{x}\right) = \exp\left(i2\pi\frac{d_2 - d_1}{\lambda} + ie \oint \vec{A} \cdot d\vec{x}\right) , \tag{3.16}$$

so that there exists a change in phase over and above that which arises from the path difference. By Stokes' theorem

$$\oint \vec{A} \cdot d\vec{x} = \int \vec{\nabla} \times \vec{A} \cdot d\vec{S} = \int \vec{B} \cdot d\vec{S} = B\pi R^2 , \tag{3.17}$$

and even though the magnetic field *vanishes* in the vicinity of the classical trajectories, its effect can still be seen in that the intensity pattern is shifted to become

$$I(\theta) = 4I_0 \cos^2\left(\pi\frac{\delta}{\lambda}\sin\theta + e\frac{1}{2}B\pi R^2\right) . \tag{3.18}$$

This modification in the intensity pattern due to the existence of a non-vanishing vector potential was suggested first by Bohm and Aharonov [AhB 52] and has since been verified experimentally [To 82]. We conclude that the vector potential (or more precisely its line integral) *does* have a measurable quantum mechanical significance. (Of course, one can write things completely in terms of a magnetic field, without any reference to a vector potential, but only at the cost of making the theory nonlocal.) That the result is gauge-independent is clear from the feature that under a gauge transformation

$$\vec{A} \to \vec{A} + \vec{\nabla}\Lambda, \tag{3.19}$$

since

$$\oint \vec{\nabla}\Lambda \cdot d\vec{x} = 0 \tag{3.20}$$

we find that $\oint \vec{A} \cdot d\vec{x}$ is unchanged.

PROBLEM III.3.1

Rotating Superconductors and The A-B Effect

An interesting variation on the A-B effect can be found in a rotating superconductor. If $x^\mu, x^{\mu'}$ denote the co-ordinates within an inertial, rotating frame respectively then they are related by

$$\begin{aligned} t' &= t & t &= t' \\ z' &= z & z &= z' \\ x' &= x\cos\omega t + y\sin\omega t & x &= x'\cos\omega t' - y'\sin\omega t' \\ y' &= -x\sin\omega t + y\cos\omega t & y &= x'\sin\omega t' + y'\cos\omega t' \end{aligned}$$

for the case of rotation with angular velocity $\vec{\omega} = \omega \hat{e}_z$.

i) Assuming the inertial frame to be Minkowski, i.e.,

$$g_{\mu\nu} = \eta_{\mu\nu},$$

show using the fact that $g_{\mu\nu}$ is a second rank tensor that in the rotating frame

$$g'_{\mu\nu}(x') = \frac{\partial x^\alpha}{\partial x^{\mu'}} \frac{\partial x^\beta}{\partial x^{\nu'}} \eta_{\alpha\beta}$$

i.e.,

$$g'_{00} = 1 - (\vec{\omega} \times \vec{r}') \cdot (\vec{\omega} \times \vec{r}')$$
$$g'_{0i} = (\vec{r}' \times \vec{\omega})_i$$
$$g'_{ij} = -\delta_{ij}$$

ii) Use this new metric to find the Lagrangian for motion of a charged particle in the rotating frame. Show that

$$L = -m(g'_{\mu\nu}\dot{x}^{\mu'}\dot{x}^{\nu'})^{\frac{1}{2}} - eA_\mu \dot{x}^{\mu'}$$

$$\approx -m + \frac{1}{2}m\dot{\vec{r}}'^2 - e\text{``}\phi\text{''} + e\dot{\vec{r}}' \cdot \text{``}\vec{A}\text{''}$$

where

$$\text{``}\phi\text{''} = \phi - \frac{m}{2e}\vec{\omega} \times \vec{r}' \cdot \vec{\omega} \times \vec{r}'$$

$$\text{``}\vec{A}\text{''} = \vec{A} - \frac{m}{e}\vec{r}' \times \vec{\omega}$$

are effective scalar and vector potentials.

iii) Use Lagrange's equation to determine the equation of motion in the rotating frame and show that it can be written as

$$m\frac{d^2\vec{r}'}{dt^2} = e(\vec{E} + \dot{\vec{r}}' \times \vec{B}) + 2m\dot{\vec{r}}' \times \vec{\omega} + m\vec{\omega} \times (\vec{\omega} \times \vec{r}') \ .$$

Thus the effective scalar, vector potential terms lead to the centrifugal and Coriolis terms in the force law for the rotating frame.

iv) Since in an inertial frame it is required that $\vec{B} = 0$ within a superconductor, then within a rotating superconductor show that we must have "$\vec{B}$" $= 0$ and hence a real magnetic field

$$\vec{B} = -\frac{2m}{e}\vec{\omega}$$

must exist.

v) Demonstrate that this leads to an A-B effect within a rotating superconductor surrounding a nonsuperconducting region of area S

$$\Phi = \oint \vec{A} \cdot d\vec{\ell} = -\frac{2m}{e}\omega S \ .$$

Such experiments have actually been performed and lead to results in precise agreement with the above prediction [Se 82, ZiM 65].

III.4 THE MAXWELL FIELD LAGRANGIAN

We embark at this point upon a procedure termed "quantization of the radiation field." The idea is that just as quantum mechanics dictates interesting and important modifications of the classical way of examining particle motion, replacing, for example, the Newtonian idea of a prescribed trajectory by Feynman's prescription involving integration over *all* possible paths, similar quantum mechanical implications should exist for the field degrees of freedom represented by the vector potential $A^\mu(\vec{x}, t)$. One such implication is well-known — that of wave-particle duality. That is, in the classical picture the electric and magnetic fields associated with radiation are represented in terms of an electromagnetic wave. However, at the quantum level, for example, in dealing with the photoelectric effect, it is necessary to ascribe particle attributes to the radiation. One speaks of the radiation in terms of photons—massless particles which carry the energy-momentum of the field. It is this picture in terms of photons which arises from field quantization, and our first step in deriving this description is to represent the interaction of the fields with each other and with the energy-momentum density.

Maxwell Lagrangian

We seek an action which will reproduce the form of the Maxwell equations. Since we are no longer dealing with a simple trajectory $\vec{x}(t)$ but rather with a field $\vec{A}^\mu(\vec{x}, t)$, which has values at all points of spacetime, the procedure is similar but not identical to that used earlier for particles. In the present case one defines a Lagrange *density* $\mathcal{L}(x)$ whose integral over spacetime gives the action S

$$S = \int d^4x\, \mathcal{L}(x) \ . \tag{4.1}$$

III.4 THE MAXWELL FIELD LAGRANGIAN

Since $\mathcal{L}$ is in general a function of the field $\phi(x)$ (using generic notation) and its derivatives $\partial_\mu \phi(x)$, then when we vary the action we find

$$\delta S = \int d^4x \left(\frac{\partial \mathcal{L}}{\partial \phi} \delta \phi(x) + \frac{\partial \mathcal{L}}{\partial(\partial_\mu \phi)} \delta \partial_\mu \phi(x) \right) \quad . \tag{4.2}$$

Integrating by parts and using

$$\delta \partial_\mu \phi(x) = \partial_\mu \delta \phi(x) \tag{4.3}$$

we find (neglecting surface terms as before)

$$\delta S = \int d^4x \, \delta\phi(x) \left(\frac{\partial \mathcal{L}}{\partial \phi} - \partial_\mu \frac{\partial \mathcal{L}}{\partial(\partial_\mu \phi)} \right) \quad . \tag{4.4}$$

Requiring $\delta S = 0$ for arbitrary field variations $\delta \phi(x)$ yields the Euler-Lagrange equation

$$\frac{\partial \mathcal{L}}{\partial \phi} - \partial_\mu \frac{\partial \mathcal{L}}{\partial(\partial_\mu \phi)} = 0 \quad . \tag{4.5}$$

Our problem then reduces to finding an appropriate Lagrange density. Note that under a Lorentz transformation

$$\begin{aligned} dx'^0 &= \gamma \, dx^0 \big|_{d\vec{x}=0} & \text{Time dilation} \\ dx'_\parallel &= \gamma^{-1} dx_\parallel \big|_{dx^0=0} & \text{Lorentz contraction} \\ dx'_\perp &= dx_\perp & \end{aligned} \tag{4.6}$$

so that the integration measure is invariant

$$d^4x' = d^4x \quad . \tag{4.7}$$

We then require S to be invariant in order that the equations of motion should have identical form in all frames— the Lagrange density should be a Lorentz scalar and the correct form is

$$\mathcal{L} = -\frac{1}{4} F_{\mu\nu} F^{\mu\nu} - j^\nu A_\nu \quad . \tag{4.8}$$

Let's verify this. Treating each component of A_μ as an independent field we have

$$\begin{aligned} \frac{\partial \mathcal{L}}{\partial A_\nu} &= -j^\nu \\ \frac{\partial \mathcal{L}}{\partial(\partial_\mu A_\nu)} &= \frac{\partial}{\partial(\partial_\mu A_\nu)} \left(-\frac{1}{4} (\partial_\alpha A_\beta - \partial_\beta A_\alpha)^2 \right) \\ &= \frac{\partial}{\partial(\partial_\mu A_\nu)} \left(-\frac{1}{2} \partial_\alpha A_\beta \left(\partial^\alpha A^\beta - \partial^\beta A^\alpha \right) \right) \\ &= -F^{\mu\nu} \quad . \end{aligned} \tag{4.9}$$

The Euler–Lagrange equation reads then

$$\partial_\mu F^{\mu\nu} = j^\nu ,\qquad(4.10)$$

which is the Maxwell relation. The corresponding Hamiltonian density is found via the canonical transformation

$$\begin{aligned}\mathcal{H} &= \frac{\partial A_\mu}{\partial t}\frac{\partial \mathcal{L}}{\partial\left(\frac{\partial A_\mu}{\partial t}\right)} - \mathcal{L} \\ &= -\left(F_{0\mu} + \partial_\mu A_0\right)F^{0\mu} + \frac{1}{4}F^{\mu\nu}F_{\mu\nu} + j_\mu A^\mu \\ &= \vec{E}\cdot\vec{\nabla}\phi + \frac{1}{2}\left(\vec{E}^2 + \vec{B}^2\right) + j_\mu A^\mu .\end{aligned}\qquad(4.11)$$

We recognize at least one familiar term — the usual electromagnetic field energy density

$$u = \frac{1}{2}\left(\vec{E}^2 + \vec{B}^2\right) .\qquad(4.12)$$

The significance of the remaining terms will be discussed shortly.

Coulomb Gauge

Before proceeding with the quantization program we show that one can write our Hamiltonian in terms of a vector potential which satisfies the Coulomb gauge condition

$$\vec{\nabla}\cdot\vec{A} = 0 .\qquad(4.13)$$

As mentioned above, given a current, charge density distribution $j^\mu(x)$ and an electromagnetic field configuration described by $A^\mu(x)$ we can always choose a gauge such that $\vec{\nabla}\cdot\vec{A}$ vanishes. Suppose that $\vec{\nabla}\cdot\vec{A}^{\text{old}} \neq 0$. If we pick

$$\chi(x) = -\int d^3x'\,\frac{1}{4\pi}\,\frac{1}{|\vec{x}-\vec{x}'|}\vec{\nabla}'\cdot\vec{A}^{\text{old}}(\vec{x}',t) .\qquad(4.14)$$

then

$$\vec{A}^{\text{new}} = \vec{A}^{\text{old}} - \vec{\nabla}\chi\qquad(4.15)$$

and

$$\begin{aligned}\vec{\nabla}\cdot\vec{A}^{\text{new}} &= \vec{\nabla}\cdot\vec{A}^{\text{old}} + \int d^3x'\,\vec{\nabla}^2\frac{1}{4\pi|\vec{x}-\vec{x}'|}\vec{\nabla}'\cdot\vec{A}^{\text{old}}(\vec{x}',t) \\ &= \vec{\nabla}\cdot\vec{A}^{\text{old}} - \int d^3x'\delta^3(\vec{x}-\vec{x}')\vec{\nabla}'\cdot\vec{A}(\vec{x}',t) = 0.\end{aligned}\qquad(4.16)$$

Because of the Lorentz condition

$$\partial_\mu A^\mu = \frac{\partial A^0}{\partial t} + \vec{\nabla}\cdot\vec{A} = 0\qquad(4.17)$$

we have also

$$\frac{\partial}{\partial t}A^0_{\text{new}} = \frac{\partial}{\partial t}A^0_{\text{old}}(x) + \frac{\partial\chi}{\partial t} = 0 ,\qquad(4.18)$$

III.4 THE MAXWELL FIELD LAGRANGIAN

and using the Maxwell equation, Eq. 2.22, dropping henceforth the superscript "new"

$$\left(\frac{\partial^2}{\partial t^2} - \vec{\nabla}^2\right) A^0(\vec{x}, t) = \rho \ . \tag{4.19}$$

But, using Eq. 4.18 this reduces to

$$\vec{\nabla}^2 A^0(\vec{x}, t) = -\rho \tag{4.20}$$

whose solution is

$$A^0(\vec{x}, t) = \int d^3x' \frac{1}{4\pi|\vec{x} - \vec{x}'|} \rho(\vec{x}', t) \ . \tag{4.21}$$

so that the scalar potential is completely determined in terms of the charge density.

Now return to the Hamiltonian density

$$\begin{aligned}\mathcal{H} &= \frac{1}{2}\left(\vec{E}^2 + \vec{B}^2\right) + \vec{E} \cdot \vec{\nabla}\phi + j_\mu A^\mu \\ &= \frac{1}{2}\left(\vec{E}^2 + \vec{B}^2\right) + \vec{\nabla} \cdot (\phi\vec{E}) - \phi\vec{\nabla} \cdot \vec{E} + j_\mu A^\mu \ .\end{aligned} \tag{4.22}$$

The term $\vec{\nabla} \cdot (\phi\vec{E})$ may be dropped, as it is a total divergence and, for local fields, vanishes upon integration over d^3x, leaving

$$\mathcal{H} = \frac{1}{2}\left(\vec{E}^2 + \vec{B}^2\right) - \vec{j} \cdot \vec{A} \ . \tag{4.23}$$

Finally, Eq. 4.23 may be simplified by writing

$$\vec{E} = \vec{E}_\parallel + \vec{E}_\perp \tag{4.24}$$

with

$$\vec{\nabla} \cdot \vec{E}_\perp = 0 \quad \text{and} \quad \vec{\nabla} \times \vec{E}_\parallel = 0 \ . \tag{4.25}$$

From Eq. 2.15

$$\vec{E}_\parallel = -\vec{\nabla}\phi \qquad \vec{E}_\perp = -\frac{\partial \vec{A}_\perp}{\partial t} \tag{4.26}$$

and $\vec{A} = \vec{A}_\perp$ with

$$\vec{A}_\parallel = 0 \tag{4.27}$$

Since

$$\begin{aligned}\int |\vec{E}|^2 d^3x &= \int \left(\left|\vec{E}_\perp\right|^2 + \left|\vec{E}_\parallel\right|^2 + 2\vec{\nabla}\phi \cdot \frac{\partial \vec{A}_\perp}{\partial t}\right) d^3x \\ &= \int \left|\vec{E}_\perp\right|^2 d^3x + \int \vec{\nabla}\phi \cdot \vec{\nabla}\phi \, d^3x \\ &= \int d^3x \left|\vec{E}_\perp\right|^2 - \int d^3x \, \phi \vec{\nabla}^2 \phi \\ &= \int d^3x \left|\vec{E}_\perp\right|^2 + \int d^3x \, \rho\phi \ .\end{aligned} \tag{4.28}$$

the Hamiltonian can be written in final form as

$$H = \int d^3x \frac{1}{2}\left(\vec{B}^2 + \vec{E}_\perp^2 + \rho\phi - \vec{j}\cdot\vec{A}\right)$$

$$= \int d^3x \frac{1}{2}\left((\vec{\nabla}\times\vec{A}_\perp)^2 + \left(\frac{\partial\vec{A}_\perp}{\partial t}\right)^2\right) - \vec{j}\cdot\vec{A}_\perp \qquad (4.29)$$

$$+ \frac{1}{2}\int \frac{\rho(x)\rho(x')d^3x\,d^3x'}{4\pi|\vec{x}-\vec{x}'|}$$

We see that in "Coulomb gauge" — $\vec{\nabla}\cdot\vec{A} = 0$ — the Hamiltonian can be totally written in terms of $\vec{A}_\perp$. The degrees of freedom associated with ϕ and $\vec{A}_\parallel$ have been eliminated, and we can now address the features of the remaining transverse components of the vector potential which are required by quantum mechanics.

PROBLEM III.4.1

The Magnetic Monopole and Charge Quantization

You may have wondered at the asymmetry inherent in the Maxwell equations. That is, we have

$$\vec{\nabla}\cdot\vec{E} = \rho$$

but

$$\vec{\nabla}\cdot\vec{B} = 0 \ .$$

Equivalently, in the relativistic formulation of electrodynamics

$$\partial_\mu F^{\mu\nu} = j^\nu$$

but

$$\partial_\mu F^{*\mu\nu} = 0$$

where

$$F^{*\mu\nu} \equiv \frac{1}{2}\epsilon^{\mu\nu\alpha\beta}F_{\alpha\beta}$$

is the "dual" tensor. Why is there no corresponding magnetic charge density ρ_M such that

$$\vec{\nabla}\cdot\vec{B} = \rho_M$$

or a magnetic four-current density j^*_ν such that

$$\partial_\mu F^{*\mu\nu} = j^{*\nu} \ ?$$

Quantum mechanics does not have an answer to this question. Nevertheless, quantum theory does have something important to say in this regard. In a remarkable paper—[Di 31]—Dirac demonstrated that if a so-called "magnetic monopole" does exist then its charge g_M is quantized in units of $\frac{1}{2e}$. Equivalently, running the argument backwards, if a single magnetic monopole exists anywhere in the universe then charge must be quantized in units of $\frac{1}{2g_M}$. In the absence of any present explanation for the fact that all electric charges are known to be multiples of e, Dirac's result is very tantalizing and could be the answer to the mystery of charge quantization! Although Dirac's paper was somewhat formal, it is also possible to understand this result in an alternative and simpler fashion, as outlined in this problem: A magnetic monopole of "charge" g_M must have the field strength

$$\vec{B} = \frac{g_M}{r^3}\vec{r} \ .$$

i) Show that this field pattern can be calculated from the curl of the vector potential

$$\vec{A}^{(1)} = g_M \hat{\phi} \frac{1 - \cos\theta}{r \sin\theta}$$

or equivalently from an alternate form

$$\vec{A}^{(2)} = -g_M \hat{\phi} \frac{1 + \cos\theta}{r \sin\theta} \ .$$

However, something is wrong here.

ii) Demonstrate that if one takes a surface S enclosing the origin then

$$\int_S \vec{B} \cdot d\vec{S} = \int \vec{\nabla} \cdot \vec{B} d^3 x = 0$$

if $\vec{B}$ is represented everywhere by $\vec{\nabla} \times \vec{A}$ and $\vec{A}$ is non-singular. However, this disagrees with what we expect from the magnetic analog of Gauss' law

$$\int_S \vec{B} \cdot d\vec{S} = \sum_i g_{M_i} \ .$$

What is the solution to this apparent paradox?

iii) Show that vector potential $\vec{A}^{(1)}$ is singular at $\theta = \pi$ while $\vec{A}^{(2)}$ is singular at $\theta = 0$. Thus $\vec{A}^{(2)}$ can be used for all θ except within the wedge $\theta < \epsilon$ and $\vec{A}^{(1)}$ for all θ except within the wedge $\pi - \theta < \epsilon$ where ϵ is a positive infinitesimal. However, neither form can be utilized for all θ. Nevertheless it is possible to employ either $\vec{A}^{(1)}$ or $\vec{A}^{(2)}$ in the angular region $\epsilon < \theta < \pi - \epsilon$. But if these two vector potentials give identical field strengths then they must be related by a gauge transformation.

iv) Show that the gauge function $\Lambda = -2g_M\phi$ will do the trick — that is

$$\vec{A}^{(1)} - \vec{A}^{(2)} = -\vec{\nabla}\Lambda \ .$$

Now consider the wavefunction of an electrically charged particle of charge e which is in the vicinity of the monopole. In the overlap region we can use, of course, either $\vec{A}^{(1)}$ or $\vec{A}^{(2)}$.

v) Show that the corresponding charged particle wavefunctions are related by
$$\psi^{(2)} = \exp\left(-i2eg_M\phi\right)\psi^{(1)} .$$

vi) Demonstrate that the requirement that the wavefunction is single-valued leads to
$$g_M = \frac{n}{2e}, \qquad n = 0, \pm 1, \pm 2, \ldots$$
which is Dirac's result.

PROBLEM III.4.2

The Magnetic Monopole: Another Look

It is also possible to get a feel for the physics which the existence of a monopole would require in a more classical fashion, as pointed out by M. Fierz [Fi 44]. Thus suppose that a monopole of charge g_M is situated at the origin of coordinates so that there is a magnetic field set up
$$\vec{B} = g_M \frac{\vec{r}}{r^3} .$$
Now imagine an electric charge e to be placed on the z-axis at location $z = d$. [Note that if both g_M and e are at rest then there exists no net force on either particle since
$$\vec{F}_e = e\left(\vec{E} + \vec{v} \times \vec{B}\right) = 0$$
$$\vec{F}_{g_M} = g_M\left(\vec{B} - \vec{v} \times \vec{E}\right) = 0 .]$$

i) Show that in the resultant electromagnetic field configuration there is a non-vanishing angular momentum of magnitude
$$\vec{J} = \int d^3r\, \vec{r} \times \left(\vec{E} \times \vec{B}\right) = -eg_M \hat{e}_z .$$
Now quantum mechanically we know that angular momentum is quantized in units of $\hbar/2$ so we expect that
$$g_M e = n\frac{\hbar}{2}, \qquad n = 0, 1, 2, \ldots .$$
Since our result is *independent* of the distance between charges, we see the basis of Dirac's observation that electric charge must be quantized in units of
$$\frac{\hbar}{2g_M} .$$

ii) A second approach to this problem is to examine the equation of motion of a charge e and mass m in the field of a magnetic monopole of strength g located at the origin
$$m\frac{d^2\vec{r}}{dt^2} = eg_M \vec{v} \times \vec{r}\frac{1}{r^3} .$$
Demonstrate that the vector
$$\vec{L} = \vec{r} \times m\vec{v} - eg_M \hat{r}$$
is conserved and interpret this result.

III.5 QUANTIZATION OF THE RADIATION FIELD

If we consider radiation confined to a cavity, we can show that it can be decomposed into a sum of *independent* normal modes, each of which oscillates harmonically.

Electromagnetic Mode Sum

Pick a simple cavity — say a cube of volume $V = 1$ — and represent the radiation field in terms of a vector potential

$$\vec{A}(\vec{x},t) = \sum_{\vec{k}} \vec{A}_{\vec{k}}(t)e^{i\vec{k}\cdot\vec{r}} + \vec{A}_{\vec{k}}^*(t)e^{-i\vec{k}\cdot\vec{r}} \quad . \tag{5.1}$$

The field $\vec{A}(\vec{x},t)$ is real and the summation is over all wavenumbers $\vec{k}$ which are consistent with the boundary conditions. Since we are in Coulomb gauge

$$\vec{\nabla} \cdot \vec{A} = 0 \tag{5.2}$$

where $\vec{A}$ satisfies the wave equation

$$\left(\frac{\partial^2}{\partial t^2} - \vec{\nabla}^2\right)\vec{A} = 0 \quad . \tag{5.3}$$

In terms of the Fourier coefficients $\vec{A}_{\vec{k}}(t)$ these conditions read

$$\begin{aligned}&\text{i)} \quad \vec{k}\cdot\vec{A}_{\vec{k}}(t) = 0 \\ &\text{ii)} \quad \left(\frac{\partial^2}{\partial t^2} + \vec{k}^2\right)\vec{A}_{\vec{k}}(t) = 0\end{aligned} \tag{5.4}$$

and the solution is

$$\vec{A}_{\vec{k}}(t) = \sum_{\lambda=1}^{2} K_{\vec{k},\lambda} \hat{\epsilon}_{\vec{k},\lambda} e^{-i\omega_k t} \equiv \sum_{\lambda=1}^{2} \vec{A}_{\vec{k},\lambda}(t) \tag{5.5}$$

where $\omega_k = |\vec{k}|$ and $\hat{\epsilon}_{\vec{k},\lambda}$ ($\lambda = 1,2$) are two linearly independent unit vectors satisfying $\hat{\epsilon}_{\vec{k},\lambda} \cdot \vec{k} = 0$.

Thus far we are simply doing classical field theory. Now begin the transition to quantum mechanics by defining amplitudes $a_{\vec{k},\lambda}(t)$ such that

$$\vec{A}_{\vec{k},\lambda}(t) \equiv c_{\vec{k},\lambda} \hat{\epsilon}_{\vec{k},\lambda} a_{\vec{k},\lambda}(t) \quad . \tag{5.6}$$

If we pick $c_{\vec{k},\lambda} = \frac{1}{\sqrt{2\omega_k}}$ then

$$\begin{aligned}H_{\text{RAD}} &= \frac{1}{2}\int d^3x \left(\vec{E}^2 + \vec{B}^2\right) \\ &= \frac{1}{2}\int d^3x \sum_{\vec{k},\lambda}\sum_{\vec{k}',\lambda'}\left[\left(i\omega_k \vec{A}_{\vec{k},\lambda}e^{i\vec{k}\cdot\vec{x}} - i\omega_k \vec{A}^*_{\vec{k},\lambda}e^{-i\vec{k}\cdot\vec{x}}\right)\right. \\ &\quad \cdot \left(i\omega_{k'}\vec{A}_{\vec{k}',\lambda'}e^{i\vec{k}'\cdot\vec{x}} - i\omega_{k'}\vec{A}^*_{\vec{k}',\lambda'}e^{-i\vec{k}'\cdot\vec{x}}\right) \\ &\quad + \left(i\vec{k}\times\vec{A}_{\vec{k},\lambda}e^{i\vec{k}\cdot\vec{x}} - i\vec{k}\times\vec{A}^*_{\vec{k},\lambda}e^{-i\vec{k}\cdot\vec{x}}\right) \\ &\quad \left.\cdot \left(i\vec{k}'\times\vec{A}_{\vec{k}',\lambda'}e^{i\vec{k}'\cdot\vec{x}} - i\vec{k}'\times\vec{A}^*_{\vec{k}',\lambda'}e^{-i\vec{k}'\cdot\vec{x}}\right)\right] \ . \end{aligned} \quad (5.7)$$

Using

$$\int_V d^3x \, e^{i(\vec{k}-\vec{k}')\cdot\vec{x}} = \delta_{\vec{k},\vec{k}'} \quad (5.8)$$

this becomes

$$\begin{aligned}H_{\text{RAD}} &= \frac{1}{2}\sum_{\vec{k},\lambda,\lambda'}\left(-\omega_k^2 \vec{A}_{\vec{k},\lambda}\cdot\vec{A}_{-\vec{k},\lambda'} - \omega_k^2 \vec{A}^*_{\vec{k},\lambda}\cdot\vec{A}^*_{-\vec{k},\lambda'}\right. \\ &\quad + 2\omega_k^2 \vec{A}_{\vec{k},\lambda}\cdot\vec{A}^*_{\vec{k},\lambda'} + \vec{k}\times\vec{A}_{\vec{k},\lambda}\cdot\vec{k}\times\vec{A}_{-\vec{k},\lambda'} \\ &\quad \left. + \vec{k}\times\vec{A}^*_{\vec{k},\lambda}\cdot\vec{k}\times\vec{A}^*_{-\vec{k},\lambda'} + 2\vec{k}\times\vec{A}_{\vec{k},\lambda}\cdot\vec{k}\times\vec{A}^*_{\vec{k},\lambda'}\right) \\ &= \sum_{\vec{k},\lambda,\lambda'} 2\omega_k^2 \vec{A}_{\vec{k},\lambda}\cdot\vec{A}^*_{\vec{k},\lambda'} = \sum_{\vec{k},\lambda}\omega_k a^*_{\vec{k},\lambda}a_{\vec{k},\lambda} \ . \end{aligned} \quad (5.9)$$

Defining

$$Q_{\vec{k},\lambda} \equiv \frac{1}{\sqrt{2\omega_k}}(a_{\vec{k},\lambda} + a^*_{\vec{k},\lambda}) \qquad P_{\vec{k},\lambda} = -i\sqrt{\frac{\omega_k}{2}}\left(a_{\vec{k},\lambda} - a^*_{\vec{k},\lambda}\right) \quad (5.10)$$

we can write

$$H_{\text{RAD}} = \sum_{\vec{k},\lambda}\frac{1}{2}\left(P^2_{\vec{k},\lambda} + \omega_k^2 Q^2_{\vec{k},\lambda}\right) \quad (5.11)$$

which is a simple sum of *independent* harmonic oscillator Hamiltonians.

Quantization

We can quantize the theory by making the change from classical variables P, Q to quantum mechanical operators $\hat{P}, \hat{Q}$ which satisfy the canonical commutation relations

$$\begin{aligned}\left[\hat{Q}_{\vec{k},\lambda}, \hat{P}_{\vec{k}',\lambda'}\right] &= i\delta_{\vec{k},\vec{k}'}\delta_{\lambda,\lambda'} \\ \left[\hat{Q}_{\vec{k},\lambda}, \hat{Q}_{\vec{k}',\lambda'}\right] &= 0 \\ \left[\hat{P}_{\vec{k},\lambda}, \hat{P}_{\vec{k}',\lambda'}\right] &= 0 \ . \end{aligned} \quad (5.12)$$

Comparison with Dirac's solution of the harmonic oscillator discussed in section I.3 shows that this is equivalent to the substitution

$$a_{\vec{k},\lambda} \to \hat{a}_{\vec{k},\lambda} \qquad a^*_{\vec{k},\lambda} \to \hat{a}^\dagger_{\vec{k},\lambda} \tag{5.13}$$

where $\hat{a}_{\vec{k},\lambda}, \hat{a}^\dagger_{\vec{k},\lambda}$ are operators which satisfy the commutation relations

$$\left[\hat{a}_{\vec{k},\lambda}, \hat{a}^\dagger_{\vec{k}',\lambda'}\right] = \delta_{\vec{k},\vec{k}'} \delta_{\lambda,\lambda'}$$
$$\left[\hat{a}_{\vec{k},\lambda}, \hat{a}_{\vec{k}',\lambda'}\right] = 0 \tag{5.14}$$
$$\left[\hat{a}^\dagger_{\vec{k},\lambda}, \hat{a}^\dagger_{\vec{k}',\lambda'}\right] = 0$$

The Hamiltonian becomes

$$\hat{H}_{\text{RAD}} = \sum_{\vec{k},\lambda} \omega_k \frac{1}{2} \left(\hat{a}_{\vec{k},\lambda} \hat{a}^\dagger_{\vec{k},\lambda} + \hat{a}^\dagger_{\vec{k},\lambda} \hat{a}_{\vec{k},\lambda}\right)$$
$$= \sum_{\vec{k},\lambda} \omega_k \left(\hat{N}_{\vec{k},\lambda} + \frac{1}{2}\right) \tag{5.15}$$

where

$$\hat{N}_{\vec{k},\lambda} = \hat{a}^\dagger_{\vec{k},\lambda} \hat{a}_{\vec{k},\lambda} \tag{5.16}$$

is the number operator for the mode specified by $\vec{k}, \lambda$. The eigenvalues of $\hat{N}_{\vec{k},\lambda}$ are integers $0, 1, 2, \ldots$ and we can specify eigenstates of the Hamiltonian $\hat{H}_{\text{RAD}}$ by an infinite-dimensional column vector which lists the integer eigenvalue $n_{\vec{k},\lambda}$ associated with each mode

$$\begin{pmatrix} n_{\vec{k}_1,\lambda_1} \\ n_{\vec{k}_2,\lambda_2} \\ \vdots \\ \vdots \end{pmatrix}. \tag{5.17}$$

The eigenstate $|n_{\vec{k},\lambda}\rangle$ which corresponds to a given eigenvalue $n_{\vec{k},\lambda}$ is, of course,

$$|n_{\vec{k},\lambda}\rangle = \frac{\left(\hat{a}^\dagger_{\vec{k},\lambda}\right)^n}{\sqrt{n!}} |0_{\vec{k},\lambda}\rangle \tag{5.18}$$

and when this eigenvalue obtains we say that there exist n photons in this particular mode. Using the commutation relations it is easy to show that

$$\hat{a}_{\vec{k},\lambda} |n_{\vec{k}',\lambda'}\rangle = \sqrt{n_{\vec{k},\lambda}}\, \delta_{\vec{k},\vec{k}'} \delta_{\lambda,\lambda'} |(n-1)_{\vec{k},\lambda}\rangle$$
$$\hat{a}^\dagger_{\vec{k},\lambda} |n_{\vec{k}',\lambda'}\rangle = \sqrt{n_{\vec{k},\lambda}+1}\, \delta_{\vec{k},\vec{k}'} \delta_{\lambda,\lambda'} |(n+1)_{\vec{k},\lambda}\rangle \ . \tag{5.19}$$

Since $\hat{a}^\dagger$ applied to a state of n photons yields a state containing one *more* photon, $\hat{a}^\dagger$ is called a photon "creation" operator. Similarly, since $\hat{a}|n\rangle$ yields a state with one *less* photon $\hat{a}$ is termed a photon "annihilation" operator.

PROBLEM III.5.1
The Photoelectric Effect

Consider the photoelectric effect in a hydrogen-like atom ground state.

i) Using the quantized radiation field write the transition matrix element to lowest order in $\hat{V}$.

ii) Show that if the incident photon direction is along the z-axis and its polarization is $\hat{\epsilon}$ then the cross section is given by

$$\frac{d\sigma}{d\Omega} = 32 \frac{\alpha}{m\omega} \frac{Z^5}{a_0^5} (\hat{\epsilon} \cdot \hat{n})^2 k_f^3 \frac{1}{\left(\frac{Z^2}{a_0^2} + \left(\vec{k}_f - \vec{k}_i\right)^2\right)^4}$$

is the photon energy is large enough that the final state wave function of the ejected electron can be approximated by a plane wave. Here $\hat{n}$ is a unit vector in the θ, ϕ direction and a_0 is the Bohr radius.

iii) What differences would you expect in the circumstances that the plane wave approximation is no longer valid?

III.6 THE VACUUM ENERGY*

In the previous section we have derived expressions for the vector potential

$$\vec{A}(\vec{x}) = \int \frac{d^3k}{(2\pi)^3} \frac{1}{\sqrt{2\omega_k}} \sum_\lambda \left(\hat{\epsilon}_{\vec{k},\lambda} \hat{a}_{\vec{k},\lambda} e^{i\vec{k}\cdot\vec{x}} + \hat{\epsilon}^*_{\vec{k},\lambda} \hat{a}^\dagger_{\vec{k},\lambda} e^{-i\vec{k}\cdot\vec{x}} \right) \qquad (6.1)$$

and for the radiation Hamiltonian

$$\hat{H}_{\text{RAD}} = \int \frac{d^3k}{(2\pi)^3} \sum_\lambda \omega_k \left(\hat{a}^\dagger_{\vec{k},\lambda} \hat{a}_{\vec{k},\lambda} + \frac{1}{2} \right) \ . \qquad (6.2)$$

If we apply $\hat{H}_{\text{RAD}}$ to the vacuum state — i.e., the state with no photons so that $\hat{a}^\dagger_{\vec{k},\lambda} \hat{a}_{\vec{k},\lambda}|0\rangle = 0$ — we have

$$\hat{H}_{\text{RAD}}|0\rangle = \int \frac{d^3k}{(2\pi)^3} \sum_\lambda \frac{1}{2}\omega_k |0\rangle = \infty \ . \qquad (6.3)$$

The vacuum state, because of the zero point energy $\frac{1}{2}\omega_k$ associated with each mode, appears to contain infinite energy! However, at least in the limit of classical physics there does not appear to be any evidence for such an effect. For example, assume that at very high energies, which we have yet to reach, the physics is somehow different than in the familiar low energy domain — there is no $\frac{1}{2}\omega_k$ per mode. Then the vacuum energy density would become

$$\rho = 2\int_0^\Lambda \frac{k^2 dk}{(2\pi)^3} \int d\Omega \frac{1}{2}\omega_k$$
$$= \frac{\Lambda^4}{8\pi^2} \sim 200 \text{ gcm}^{-3} \text{ for } \Lambda \sim m_e \ .$$
(6.4)

which is enormous! Indeed the critical density required in order to close the universe is only $\sim 2\times 10^{-30}$g/cm^3. We conclude that as far as one can determine, since in nonrelativistic physics the zero of energy is arbitrary, this vacuum energy has no classical significance and may be subtracted off. Henceforth then we shall utilize the renormalized radiation Hamiltonian

$$\hat{H}_{\text{RAD}}^{(r)} = \int \frac{d^3k}{(2\pi)^3} \sum_\Lambda \omega_k \hat{a}_{\vec{k},\lambda}^\dagger \hat{a}_{\vec{k},\lambda}$$
(6.5)

which has the property that $\hat{H}_{\text{RAD}}^{(r)}|0\rangle = 0$. Having subtracted off this (infinite) zero point energy and ascribed it to a shift in the energy scale, one may get the impression that there is no significant consequence to this quantum mechanical phenomenon. This is *not* the case. As we shall see later the zero point energy plays an essential role in causing spontaneous decay of a radiating system, producing a measurable energy shift of an atomic state due to coupling to the radiation field, *etc.* However, there exists an even more direct consequence of this phenomenon, which we now explore — the Casimir effect [ElR 91].

The Casimir Effect

In order to understand the physics of the Casimir process, imagine a pair of parallel capacitor plates with a separation a as shown below and consider the modes permitted by the boundary condition that $\vec{E}_\parallel$ must vanish on the surface of the plates. If the normal to the surface defines the z-direction, then for propagation in this direction the wavelength varies from zero to $2a$. For *each* allowed mode there exists a zero point energy $\frac{1}{2}\omega_k$. Consequently the total energy between the plates is given by

$$U(a) = \sum_{\vec{k},\lambda} \frac{1}{2}\omega_k \ .$$
(6.6)

When the plate separation is increased, more modes are permitted so that the energy is an increasing function of a. The energy can be lowered by reducing the distance between plates, suggesting an attractive force of magnitude

$$F = -\frac{\partial E(a)}{\partial a} \ .$$
(6.7)

This force has been detected experimentally by M. Spaarnay—[Sp 58]— and represents *macroscopic* manifestation of quantum field theory.

Now let's get quantitative and evaluate the size of this force. We know that in the x, y directions the allowed wavenumbers k_x, k_y are continuous from $-\infty$ to $+\infty$ with a density of states given by

$$A \int \frac{d^2k}{(2\pi)^2} \tag{6.8}$$

where A is the area of the plates. In the z-direction, on the other hand, the boundary condition $\vec{E}(0) = \vec{E}(a) = 0$ requires

$$E \propto \sin k_z z \ . \tag{6.9}$$

with

$$k_z = \frac{n\pi}{a} \qquad n = 1, 2, \ldots \tag{6.10}$$

Since the mode energy is

$$\omega_k = \sqrt{k_x^2 + k_y^2 + \left(\frac{n\pi}{a}\right)^2} \tag{6.11}$$

the total energy associated with the zero point motion

$$E(a) = 2 \sum_{n=1}^{\infty} A \int \frac{d^2k}{(2\pi)^2} \frac{1}{2}\omega_k \tag{6.12}$$

where the factor of two is associated with the two orthogonal polarizations for each mode. Defining

$$k = \sqrt{k_x^2 + k_y^2} \tag{6.13}$$

we have

$$k\,dk = \omega\,d\omega \tag{6.14}$$

and

$$U(a) = A \sum_{n=1}^{\infty} \frac{1}{2\pi} \int_{\frac{n\pi}{a}}^{\infty} d\omega\, \omega^2 \ . \tag{6.15}$$

As it stands $U(a)$ is divergent. Hence we define the energy by inserting a cutoff factor $\exp(-\epsilon\omega)$ into the integration, yielding

$$\begin{aligned}E(a) &= \frac{A}{2\pi} \sum_{n=1}^{\infty} \int_{\frac{n\pi}{a}}^{\infty} d\omega\, \omega^2\, e^{-\epsilon\omega} = \frac{A}{2\pi} \frac{d^2}{d\epsilon^2} \sum_{n=1}^{\infty} \int_{\frac{n\pi}{a}}^{\infty} d\omega\, e^{-\omega\epsilon} \\ &= \frac{A}{2\pi} \frac{d^2}{d\epsilon^2} \sum_{n=1}^{\infty} \frac{1}{\epsilon} \exp\left(-\frac{n\pi\epsilon}{a}\right) \ . \end{aligned} \tag{6.16}$$

III.6 THE VACUUM ENERGY* 129

Performing the summation, we find

$$E(a) = \frac{A}{2\pi} \frac{d^2}{d\epsilon^2} \frac{1}{\epsilon} \left(\frac{1}{1 - e^{-\frac{\pi\epsilon}{a}}} - 1 \right) . \tag{6.17}$$

We wish to examine this result in the limit of small ϵ. Hence use the expansion [GrR 65]

$$\frac{1}{1 - e^t} = -\sum_{n=0}^{\infty} B_n \frac{t^{n-1}}{n!} \tag{6.18}$$

where B_n are the Bernoulli numbers.

The energy per unit area becomes

$$\begin{aligned}\frac{1}{A} U(a) &= -\frac{1}{2\pi} \frac{d^2}{d\epsilon^2} \frac{1}{\epsilon} \left(1 + \sum_{n=0}^{\infty} B_n \frac{\left(-\frac{\pi\epsilon}{a}\right)^{n-1}}{n!} \right) \\ &= -\frac{1}{2\pi} \frac{d^2}{d\epsilon^2} \big(\frac{1}{\epsilon} - B_0 \frac{a}{\pi\epsilon^2} + B_1 \frac{1}{\epsilon} - B_2 \frac{\pi}{2a} \\ &\quad + B_3 \frac{\pi^2 \epsilon}{6a^2} - B_4 \frac{\pi^3 \epsilon^2}{24 a^3} + \ldots \big) ,\end{aligned} \tag{6.19}$$

and, performing the requisite differentiations, we have

$$\frac{1}{A} U(a) = 3 B_0 \frac{a}{\pi^2 \epsilon^4} - (1 + B_1) \frac{1}{\pi \epsilon^3} + B_4 \frac{\pi^2}{24 a^3} + \ldots . \tag{6.20}$$

The cutoff-dependent components are not relevant to the force calculation. In order to see this imagine that the capacitor is inserted between an additional pair of plates as shown. Defining

$$\begin{aligned}\frac{3 B_0}{\pi^2 \epsilon^4} &\equiv C_0 \\ -\frac{1 + B_1}{\pi \epsilon^3} &\equiv C_1 \\ B_4 \frac{\pi^2}{24} &\equiv C_2\end{aligned} \tag{6.21}$$

we have

$$\begin{aligned}\frac{1}{A} U(a) &= 2 \left(C_0 \left(L - \frac{a}{2} \right) + C_1 + \frac{C_2}{(L - \frac{a}{2})^3} \right) \\ &\quad + C_0 a + C_1 + \frac{C_2}{a^3} \\ &= 2 C_0 L + 3 C_1 + C_2 \left(\frac{1}{a^3} + \frac{2}{(L - \frac{a}{2})^3} \right) .\end{aligned} \tag{6.22}$$

so that in the limit as $L \to \infty$

$$\frac{1}{A} U(a) = \text{Const.} + \frac{C_2}{a^3} . \tag{6.23}$$

Thus the cutoff-dependent (and divergent) pieces serve only to renormalize the overall energy scale and do not contribute to the force between the plates, which is given by

$$\frac{1}{A}F = -\frac{\partial}{\partial a}\frac{1}{A}U(a) = \frac{3C_2}{a^4} \ . \tag{6.24}$$

Since $B_4 = -\frac{1}{30}$ this yields a finite and cutoff-independent prediction

$$\frac{1}{A}F = -\frac{\pi^2}{240a^4} \ , \tag{6.25}$$

with the negative sign indicating that the force is attractive as expected. This prediction was confirmed in the experiment of Spaarnay. The quantum mechanical aspect of the radiation field — the vacuum energy — is an experimentally verifiable fact!

A curious feature here is that the force is independent of the fine structure constant α. Were there no charge on the electron or proton, there would be no electric or magnetic forces and the Casimir energy would have to vanish. Thus there seems to be a paradox. The resolution lies in the fact that the imposition of the boundary condition at the surface of the conducting plates requires a non-zero value of α. For a real conductor there exists some frequency ω_0 — dependent upon α — above which the notion of the surface of a perfect conductor at which the electric field vanishes is no longer valid so that the field penetrates the medium. As α decreases, the critical frequency becomes smaller and in the limit as $\alpha \to 0$ the concept of a perfect conductor is no longer even approximately true. Thus the Casimir force must vanish.

We have seen that quantization of the electromagnetic field is *required* — indeed the Casimir effect is a specific ramification of the vacuum energy $\frac{1}{2}\omega_k$ for each mode $\vec{k}, \lambda$. The energy in each mode in general is

$$E_{\vec{k},\lambda} = \omega_k\left(n_{\vec{k},\lambda} + \frac{1}{2}\right) \tag{6.26}$$

where $n_{\vec{k},\lambda}$ is the number of photons present in that mode. The energy $n_{\vec{k},\lambda}\omega_k$ is just the "classical" energy in the field — $n_{\vec{k},\lambda}$ photons, with energy ω_k for each photon— so that the classical picture is valid provided

$$n_{\vec{k},\lambda} \gg \frac{1}{2} \ . \tag{6.27}$$

For example, consider the region of space at a distance one hundred miles from a clear channel 50,000 watt radio station. The volume contained in the spherical shell between r and $r + dr$ is

$$dV = 4\pi r^2\, dr \ , \tag{6.28}$$

and the energy contained within this volume was emitted between time

$$t = \frac{r}{c} \quad \text{and} \quad t - dt = \frac{r + dr}{c} \tag{6.29}$$

III.6 THE VACUUM ENERGY* 131

and has the value
$$dE = -P\,dt = P\frac{dr}{c}. \tag{6.30}$$
where P represents the power. The energy per unit volume is
$$\frac{dE}{dV} = \frac{P\frac{dr}{c}}{4\pi r^2\,dr} = \frac{P}{4\pi c\,r^2} \tag{6.31}$$
and the classical photon density is
$$\frac{dn}{dV} = \frac{P}{4\pi\omega c\,r^2} \approx 10^{12} m^{-3} \tag{6.32}$$
for a 1000 KHz photon and the parameters cited above. For a realistic macroscopic volume—say $\Delta V \sim 1\text{cm}^3$ —we have
$$n = \frac{dn}{dV}\cdot\Delta V \sim 10^6 \ggg 1 \tag{6.33}$$
so that quantum mechanics does not need to be considered in the operation of conventional electronics. This is *not* the case, however, for a single hydrogen atom emitting (or absorbing) a single photon, as we shall study in the succeeding chapter.

CHAPTER IV

CHARGED PARTICLE INTERACTIONS: APPLICATIONS

IV.1 RADIATIVE DECAY: FORMAL

We are now equipped to calculate the decay rate of an atom which finds itself in an excited state $|B\rangle$ and spontaneously decays to the ground state $|A\rangle$. For simplicity we shall assume that these are the only states in the system and that we are dealing with a one electron atom. The answer is given by Fermi's golden rule

$$\Gamma = \int \frac{d^3k}{(2\pi)^3} \sum_\lambda 2\pi\delta\left(E_A^{(0)} + \omega_k - E_B^{(0)}\right) \left|\langle f|\hat{V}|i\rangle\right|^2 \tag{1.1}$$

where the initial state $|i\rangle$ is

$$|i\rangle = |B\rangle|0\rangle \tag{1.2}$$

(*i.e.*, atom in state $|B\rangle$ with no photons present) while the final state $|f\rangle$ is

$$|f\rangle = |A\rangle|1_{\vec{k},\lambda}\rangle \tag{1.3}$$

(*i.e.*, atom in state $|A\rangle$ plus a photon in state $|\vec{k},\lambda\rangle$). In order to find the potential energy operator $\hat{V}$ to be used we write the Hamiltonian as

$$\hat{H} = \frac{\left(\vec{p} - e\vec{A}\right)^2}{2m} + e\phi + \hat{H}_{\text{RAD}}^{(r)} \equiv \hat{H}_0 + \hat{V} \tag{1.4}$$

where we have divided $\hat{H}$ into a free part

$$\hat{H}_0 = \hat{H}_{\text{RAD}}^{(r)} + \frac{\vec{p}^{\,2}}{2m} + e\phi \equiv \hat{H}_{\text{RAD}}^{(r)} + \hat{H}_{\text{ATOMIC}} \tag{1.5}$$

and an interaction component

$$\hat{V} = -\frac{e}{2m}\left(\vec{p}\cdot\vec{A} + \vec{A}\cdot\vec{p}\right) + \frac{e^2}{2m}\vec{A}\cdot\vec{A} \ . \tag{1.6}$$

[Note: Because of the transversality condition $\vec{\nabla}\cdot\vec{A} = 0$ the operator $\vec{p}$ commutes with $\vec{A}$ so that henceforth we shall write

$$\hat{V} = -\frac{e}{m}\vec{A}\cdot\vec{p} + \frac{e^2}{2m}\vec{A}\cdot\vec{A} \ .] \tag{1.7}$$

The initial and final states are eigenstates of $\hat{H}_0$ with

$$\begin{aligned}\hat{H}_0|B\rangle|0\rangle &= E_B^{(0)}|B\rangle|0\rangle \\ \hat{H}_0|A\rangle|1_{\vec{k},\lambda}\rangle &= \left(E_A^{(0)} + \omega_k\right)|A\rangle|1_{\vec{k},\lambda}\rangle \ .\end{aligned} \tag{1.8}$$

We need now to evaluate the interaction matrix element

$$\left\langle A \middle| \left\langle 1_{\vec{k},\lambda} \middle| \hat{V} \middle| 0 \right\rangle \middle| B \right\rangle . \tag{1.9}$$

Note that the piece of the potential $\hat{V}$ involving $\vec{A} \cdot \vec{A}$ connects only states having the same number of photons or which differ by a pair of photons. Thus $\vec{A} \cdot \vec{A}$ cannot connect the vacuum to a single photon state and we shall discard it, keeping instead only the $-\frac{e}{m} \vec{A} \cdot \vec{p}$ term for which[†]

$$\begin{aligned}\left\langle f \middle| \hat{V} \middle| i \right\rangle &= -\frac{e}{m} \left\langle A \middle| \left\langle 1_{\vec{k},\lambda} \middle| \vec{A} \cdot \vec{p} \middle| 0 \right\rangle \middle| B \right\rangle \\ &= -\frac{e}{m} \frac{1}{\sqrt{2\omega_k}} \left\langle A \middle| \hat{\epsilon}^*_{\vec{k},\lambda} e^{-i\vec{k}\cdot\vec{r}} \cdot \vec{p} \middle| B \right\rangle .\end{aligned} \tag{1.10}$$

The spontaneous decay rate is then given by

$$\frac{d\Gamma}{d\Omega} = \frac{\omega_k^2}{(2\pi)^2} \sum_\lambda \left| \left\langle f \middle| \hat{V} \middle| i \right\rangle \right|^2 = \frac{\alpha \omega_k}{2\pi m^2} \sum_\lambda \left| \left\langle A \middle| e^{-i\vec{k}\cdot\vec{r}} \hat{\epsilon}^*_{\vec{k},\lambda} \cdot \vec{p} \middle| B \right\rangle \right|^2 , \tag{1.11}$$

or

$$\Gamma = \frac{\alpha}{2\pi m^2} \omega_k \int d\Omega \sum_\lambda \left| \left\langle A \middle| \hat{\epsilon}^*_{\vec{k},\lambda} e^{-i\vec{k}\cdot\vec{r}} \cdot \vec{p} \middle| B \right\rangle \right|^2 . \tag{1.12}$$

[Note: For a many electron atom the result is the same but with the matrix element given by

$$\left\langle A \middle| \sum_i e^{-i\vec{k}\cdot\vec{r}_i} \hat{\epsilon}^*_{\vec{k},\lambda} \cdot \vec{p}_i \middle| B \right\rangle , \tag{1.13}$$

where the sum is over all atomic electrons.] A typical wavenumber for an atomic transition is in the eV range while atomic radii are of order Angstroms. Then

$$k \cdot r \sim 1\text{eV} \times 1\text{Å} \frac{1}{\hbar c} \sim \frac{1}{2000} \ll 1 . \tag{1.14}$$

so that it is reasonable to replace

$$e^{-i\vec{k}\cdot\vec{r}} \approx 1 . \tag{1.15}$$

This is called the electric dipole approximation. The resulting matrix element is generally rewritten using a trick. Note that

$$\left[\hat{H}_{\text{ATOMIC}}, \vec{r} \right] = -i \frac{1}{m} \vec{p} \tag{1.16}$$

[†] Note here that $\vec{p}, \vec{r}$ operate on the electron co-ordinates, while $\hat{a}_{\vec{k},\lambda}, \hat{a}^\dagger_{\vec{k},\lambda}$ operate on the photon space. Also, although the quantized vector potential—Eq. III.6.1—involves a sum over *all* possible photon momenta and polarizations, only the component involving $\hat{a}^\dagger_{\vec{k},\lambda}$ survives when the matrix element is taken.

so that

$$\langle A|\vec{p}|B\rangle = im\left\langle A\left|\left[\hat{H}_{\text{ATOMIC}}, \vec{r}\right]\right|B\right\rangle$$
$$= -im\left(E_B^{(0)} - E_A^{(0)}\right)\langle A|\vec{r}|B\rangle \quad (1.17)$$

and we have used the fact that $|A\rangle$ and $|B\rangle$ are eigenstates of $\hat{H}_{\text{ATOMIC}}$. Finally, employing energy conservation we have $E_B^{(0)} - E_A^{(0)} = \omega_k$ and the electric dipole transition rate becomes

$$\frac{d\Gamma}{d\Omega_k} \approx \frac{\alpha\omega_k^3}{2\pi}\sum_\lambda \left|\left\langle A\left|\vec{r}\cdot\vec{\epsilon}_{k,\lambda}^{\,*}\right|B\right\rangle\right|^2 \quad (1.18)$$

which is its traditional form.

PROBLEM IV.1.1

Hydrogen Atom and the Electric Dipole Approximation

Consider a hydrogen atom which is in an unpolarized 2p state. Use the formula for electric dipole emission–Eq. 1.18

i) to show that the angular distribution of the emitted radiation is isotropic, and

ii) to demonstrate that the lifetime of this state is given by

$$\tau_{2p} = 1.6 \times 10^{-9} \text{sec}.$$

Now suppose that the initial state is completely polarized in substate

$$|\ell = 1, m = 1\rangle \ .$$

iii) How would the results found in parts i) and ii) above be changed?

IV.2 RADIATIVE DECAY: INTUITIVE

While the use of the quantized radiation field yields the correct answer for the electromagnetic transition rate, the use of creation and annihilation operators is somewhat obscure. In order to get a bit more insight into what is going on, we consider an alternative derivation of the decay rate from the point of view of the vacuum energy.

Using Fermi's golden rule we write the transition rate from excited state $|B\rangle$ to ground state $|A\rangle$ as

$$\Gamma = \int \frac{d^3k}{(2\pi)^3}\sum_\lambda 2\pi\delta\left(E_A^{(0)} + \omega_k - E_B^{(0)}\right)\left|\left\langle A|\hat{V}|B\right\rangle\right|^2 \quad (2.1)$$

where the interaction potential $\hat{V}$ is as before given by

$$\hat{V} = -\frac{e}{m}\vec{p}\cdot\vec{A} \ . \quad (2.2)$$

[Also we are assuming this to be a single electron atom, so that we do not have to sum over electrons.] In this expression the momentum $\vec{p}$ symbolizes the usual quantum mechanical replacement

$$\vec{p} = -i\vec{\nabla} \tag{2.3}$$

with the gradient taken with respect to the electron coordinate. However, $\vec{A}$ is to be considered as a *classical* field quantity. This procedure of treating the electron quantum mechanically but the electromagnetic field classically is called the "semiclassical" method.

The matrix element in Eq. 2.1 can be put into an alternate form by use of the commutation relation

$$\left[\hat{H}_{ATOMIC}, \vec{r}\right] = -i\frac{1}{m}\vec{p} \tag{2.4}$$

where

$$\hat{H}_{ATOMIC} = \frac{\vec{p}^2}{2m} - \frac{\alpha}{r} \tag{2.5}$$

is the hydrogen atom Hamiltonian. Then[†]

$$\left\langle A \left| \hat{V} \right| B \right\rangle = \left\langle A \left| -\frac{e}{m}\vec{p} \right| B \right\rangle \cdot \vec{A} = -ie \left\langle A \left| \left[\hat{H}_{ATOMIC}, \vec{r}\right] \right| B \right\rangle \cdot \vec{A}$$
$$= ie \left(E_B^{(0)} - E_A^{(0)} \right) \langle A|\vec{r}|B\rangle \cdot \vec{A} \tag{2.6}$$

which by energy conservation may be put into the form

$$\left\langle A \left| \hat{V} \right| B \right\rangle = ie\omega_k \langle B|\vec{r}|A\rangle \cdot \vec{A}$$
$$= e \langle B|\vec{r}|A\rangle \cdot \frac{\partial \vec{A}}{\partial t} = -e \langle B|\vec{r}|A\rangle \cdot \vec{E} \;, \tag{2.7}$$

where $\vec{E}$ is the electric field associated with the photon. We can represent the transition rate as

$$\Gamma = e^2 \int \frac{d^3k}{(2\pi)^3} \sum_\lambda 2\pi\delta\left(E_A^{(0)} + \omega_k - E_B^{(0)}\right) \langle B|r_i|A\rangle \langle A|r_j|B\rangle$$
$$\times \langle E_i E_j \rangle \tag{2.8}$$

where $\langle E_i E_j \rangle$ signifies the vacuum expectation value of this quantity. Now although before the transition takes place there is no electromagnetic radiation present so that

$$\langle E_i \rangle = 0 \tag{2.9}$$

[†] Note that we assume that the vector potential $\vec{A}$ can be taken outside the matrix element, *i.e.*, that $\vec{A}$ is effectively independent of the electron degrees of freedom. This is the electric dipole approximation.

IV.2 RADIATIVE DECAY: INTUITIVE

there *does* exist a non-zero value of $\langle E_i^2 \rangle$ since for each mode—$\vec{k}, \lambda$— of the radiation field there exists an associated energy $\frac{1}{2}\omega_k$. As the electromagnetic field energy density is given by

$$u = \frac{1}{2}\left(\vec{E}^2 + \vec{B}^2\right) \tag{2.10}$$

and for an electromagnetic wave we have $|\vec{E}| = |\vec{B}|$, we can write

$$u = \vec{E}^2 \ . \tag{2.11}$$

The total field energy is then

$$U_{\text{Tot}} = \int_V d^3r\, u = \int_V d^3r\, \vec{E}^2(\vec{r}) = \int \frac{d^3k}{(2\pi)^3} \sum_\lambda \vec{E}_{\vec{k},\lambda}^2 \tag{2.12}$$

where we have decomposed the electric field into its Fourier components $\vec{E}_{\vec{k},\lambda}$. Since in momentum space each mode carries energy $\frac{\omega_k}{2}$, we can represent

$$U_{\text{Tot}} = \int \frac{d^3k}{(2\pi)^3} \sum_\lambda \frac{\omega_k}{2} = \int \frac{d^3k}{(2\pi)^3} \sum_\lambda \frac{\omega_k}{2} \hat{\epsilon}^*_{\vec{k},\lambda} \cdot \hat{\epsilon}_{\vec{k},\lambda} \tag{2.13}$$

which suggests the result

$$\left\langle (E_{\vec{k},\lambda})_j (E_{\vec{k},\lambda})_i \right\rangle = \frac{\omega_k}{2}(\hat{\epsilon}^*_{\vec{k},\lambda})_j(\hat{\epsilon}_{\vec{k},\lambda})_i \ . \tag{2.14}$$

[Note: For a more formal derivation use

$$\left\langle (E_{\vec{k},\lambda})_i (E_{\vec{k},\lambda})_j \right\rangle = (-i\omega_k)^2 \langle 0 | \frac{1}{\sqrt{2\omega_k}}\left((\hat{\epsilon}_{\vec{k},\lambda})_i \hat{a}_{\vec{k},\lambda} - (\hat{\epsilon}^*_{\vec{k},\lambda})_i \hat{a}^\dagger_{\vec{k},\lambda}\right)$$
$$\times \frac{1}{\sqrt{2\omega_k}}\left((\hat{\epsilon}_{\vec{k},\lambda})_j \hat{a}_{\vec{k},\lambda} - (\hat{\epsilon}^*_{\vec{k},\lambda})_j \hat{a}^\dagger_{\vec{k},\lambda}\right)|0\rangle \tag{2.15}$$
$$= \frac{\omega_k}{2}(\hat{\epsilon}^*_{\vec{k},\lambda})_j(\hat{\epsilon}_{\vec{k},\lambda})_i \ .]$$

We find for the transition rate

$$\Gamma = e^2 \int \frac{d^3k}{(2\pi)^3} \sum_\lambda 2\pi\delta\left(E_A^{(0)} + \omega_k - E_B^{(0)}\right) \frac{\omega_k}{2} \left|\left\langle A \left| \vec{r} \cdot \hat{\epsilon}^*_{\vec{k},\lambda} \right| B \right\rangle\right|^2$$
$$= \alpha \int d\Omega \sum_\lambda \frac{1}{2\pi} \omega_k^3 \left|\left\langle A \left| \vec{r} \cdot \hat{\epsilon}^*_{\vec{k},\lambda} \right| B \right\rangle\right|^2 \tag{2.16}$$

which corresponds to the classical dipole formula

$$\frac{d\Gamma}{d\Omega} = \frac{\alpha}{2\pi}\omega_k^3 \sum_\lambda \left|\left\langle A \left| \vec{r} \cdot \hat{\epsilon}^*_{\vec{k},\lambda} \right| B \right\rangle\right|^2 \tag{2.17}$$

in complete agreement with Eq. 1.18 found via quantizing the radiation field.

In this approach then the mystery of what causes the transition is solved. Although there is no external electromagnetic field present — $\langle \vec{E} \rangle = 0$ — since the transition has not yet occurred, there *does* exist a fluctuating field due to the vacuum energy — $\langle \vec{E}^2 \rangle \neq 0$ — and it is the interaction of the electron with the vacuum field which gives rise to the transition!

PROBLEM IV.2.1

Yet Another Derivation of the Dipole Radiation Formula

We have seen both by use of the quantized radiation field and via interaction of an excited atom with vacuum fluctuations that the decay rate for electric dipole emission is given by

$$\frac{d\Gamma}{d\Omega} = \frac{\alpha}{2\pi}\omega_k^3 \sum_\lambda |\langle A|\vec{r}\cdot\hat{\epsilon}_\lambda^*|B\rangle|^2 \ .$$

It is possible to derive this expression by still another route — a clever argument due to Einstein. Consider a two state system — an excited state B and a ground state A — and let $\omega_{AB} = E_B - E_A$. If the atom is bathed in unpolarized radiation with frequency distribution $I(\omega)$, the transition rate from state A to state B is given by

$$R_{A\to B} \equiv N_A u(\omega_{AB}) B_{BA}$$

where N_A is the number of atoms in state A and $u(\omega_{AB})$ is the energy density of radiation at frequency ω_{AB}. The symbol B_{BA} represents a constant of proportionality which we now calculate.

Let the electric field vector of the radiation be

$$\vec{E}_\lambda(t) = \int_{-\infty}^{\infty} d\omega\, e^{i\omega t} \vec{a}_\lambda(\omega) \ .$$

i) Using the fact that $\vec{E}(t)$ is real, show that

$$\vec{a}_\lambda^*(-\omega) = \vec{a}_\lambda(\omega) \ .$$

ii) Employ time-dependent perturbation theory to calculate the transition amplitude from state A to state B. Use the electric dipole approximation

$$V = -e\vec{r}\cdot\vec{E}(t)$$

and assume the radiation pulse begins at time $t = 0$ to show that for t large

$$\text{Amp}_{BA}^\lambda(t) = ie2\pi\vec{a}_\lambda^*(\omega_{AB})\cdot\langle B|\vec{r}|A\rangle \ .$$

Show that the average energy per unit volume in each mode λ is given by

$$u_\lambda^{\text{Tot}} = \frac{1}{t}\int_0^t dt'\, E_\lambda^2(t')$$
$$= \frac{4\pi}{t}\int_0^\infty d\omega|a_\lambda(\omega)|^2 \equiv \int_0^\infty d\omega\, u_\lambda(\omega) \ .$$

iii) Calculate the transition rate

$$\equiv B_{BA} u(\omega_{AB}) = \sum_\lambda \frac{1}{t}\left|\text{Amp}_{BA}^\lambda(t)\right|^2 = 4\pi^2 \alpha \sum_\lambda u_\lambda(\omega_{BA}) |\langle B|\vec{r}\cdot\hat{\epsilon}_\lambda|A\rangle|^2 \ .$$

Hence

$$B_{BA} = 4\pi^2 \alpha \sum_\lambda |\langle B|\vec{r}\cdot\hat{\epsilon}_\lambda|A\rangle|^2 \ .$$

Now suppose that the system A, B is in equilibrium with the radiation at temperature T. Then the rate of transitions from A to B is

$$R_{A\to B} = u(\omega_{AB}) B_{BA} N_A$$

while that from B to A is

$$R_{B\to A} = u(\omega_{AB}) B_{AB} N_B + A_{AB} N_B$$

where B_{AB} is another constant of proportionality and A_{AB} is the spontaneous decay rate. Since we are in thermal equilibrium

$$R_{A\to B} = R_{B\to A} \quad \text{and} \quad \frac{N_B}{N_A} = \exp{-\omega_{AB}\beta}$$

where $\beta = \frac{1}{kT}$ is the inverse temperature.

iv) Solve for $u(\omega_{AB})$ and show

$$u(\omega_{AB}) = \frac{A_{AB}}{B_{BA} e^{\beta \omega_{AB}} - B_{AB}}$$

v) Compare with the famous Planck distribution law [McG 71]

$$u_\lambda(\omega) = \frac{\omega^3}{2\pi^2} \frac{1}{e^{\beta\omega} - 1}$$

and demonstrate that

$$B_{BA} = B_{AB} \quad \text{and} \quad A_{AB} = B_{BA}\frac{\omega_k^3}{2\pi^2}$$

vi) Show then that

$$\Gamma = \frac{\alpha}{2\pi}\omega_k^3 \int d\Omega \sum_\lambda \left|\langle A|\vec{r}\cdot\hat{\epsilon}_{k,\lambda}^*|B\rangle\right|^2$$

in agreement with the expression derived previously by other means.

IV.3 ANGULAR DISTRIBUTION OF RADIATIVE DECAY

Having obtained the formal expression for the differential transition rate for electric dipole radiation, we can now determine the specific angular distribution of the emitted radiation.

E1 Decay Distribution: Wigner-Eckart Theorem Analysis

In order to carry out this evaluation, we represent the transition operator $\vec{r} = x\hat{e}_x + y\hat{e}_y + z\hat{e}_z$ in a spherical rather than a Cartesian basis

$$\vec{r} = r^{(1)}\hat{e}^{(1)*} + r^{(0)}\hat{e}^{(0)*} + r^{(-1)}\hat{e}^{(-1)*} \tag{3.1}$$

where

$$r^{(1)} = -\frac{1}{\sqrt{2}}(x+iy) \qquad r^{(0)} = z \qquad r^{(-1)} = \frac{1}{\sqrt{2}}(x-iy)$$

$$\hat{e}^{(1)} = -\frac{1}{\sqrt{2}}(\hat{e}_x + i\hat{e}_y) \qquad \hat{e}^{(0)} = \hat{e}_z \qquad \hat{e}^{(-1)} = \frac{1}{\sqrt{2}}(\hat{e}_x - i\hat{e}_y) \ . \tag{3.2}$$

The $r^{(m)}$, ($m = -1, 0, 1$) form the components of a spherical tensor of rank one, so that the matrix elements may be evaluated using the Wigner–Eckart theorem

$$\left\langle A; J_A M_A \left| r^{(k)} \right| B; J_B M_B \right\rangle = C_{J_B 1; J_A}^{M_B k; M_A} \langle A\|r\|B\rangle \tag{3.3}$$

where $C_{J_B 1; J_A}^{M_B k; M_A}$ is a Clebsch–Gordan coefficient and $\langle A\|r\|B\rangle$—which is *independent* of M_A, M_B, k—is the reduced matrix element. Before proceeding further we note that the electric dipole matrix element obeys certain "selection rules." Because of the properties of the Clebsch–Gordan coefficient the E1 amplitude must vanish unless

$$J_B - J_A = 0, \pm 1$$
$$M_B - M_A = -k \ . \tag{3.4}$$

Also, if $\hat{\Pi}$ is the spatial inversion or parity operator, we note that

$$\left\{\hat{\Pi}, \vec{r}\right\} = 0 \ . \tag{3.5}$$

Then

$$0 = \left\langle A \left| \left\{\hat{\Pi}, \vec{r}\right\} \right| B \right\rangle = (\pi_B + \pi_A)\langle A|\vec{r}|B\rangle \tag{3.6}$$

where π_A, π_B is the intrinsic parity of state B, A respectively so that

$$\langle A|\vec{r}|B\rangle = 0 \quad \text{unless} \quad \pi_B = -\pi_A \ . \tag{3.7}$$

The selection rules for E1 radiation are then

$$\Delta J = 0, \pm 1 \ , \quad \text{parity change} \ . \tag{3.8}$$

The dominant decay mechanism will be E1 if these criteria are satisfied. Generally the photon polarization is not observed so that the two orthogonal photon polarizations must be summed over. Noting that

$$\sum_{\lambda=1}^{2}(\hat{\epsilon}_{\vec{k},\lambda}^{*})_{i}(\hat{\epsilon}_{\vec{k},\lambda})_{j} = \delta_{ij} - \hat{k}_{i}\hat{k}_{j} \qquad (3.9)$$

(where $\hat{k}$ is a unit vector in the direction of the outgoing photon momentum) we have for the differential transition rate

$$\frac{d\Gamma}{d\Omega} = \frac{\alpha}{2\pi}\omega_{k}^{3}\left(M^{*}\cdot M - M^{*}\cdot\hat{k}M\cdot\hat{k}\right) \qquad (3.10)$$

where

$$\vec{M} \equiv \langle A|\vec{r}|B\rangle \quad . \qquad (3.11)$$

Usually the polarization state of the daughter state $|A\rangle$ is also not observed so we must sum over the $2J_A + 1$ values of M_A. However, we suppose that the parent state $|B\rangle$ has been prepared with probability $p(M_B)$ to have spin projection M_B. We have then

$$\frac{d\Gamma}{d\Omega} = \frac{\alpha}{2\pi}\omega_{k}^{3}|\langle A\|r\|B\rangle|^{2}\sum_{M_A,M_B}p(M_B)$$

$$\times\left[\sum_{k=-1}^{1}\left(C_{J_B 1; J_A}^{M_B k; M_A}\right)^{2} - \cos^{2}\theta\left(C_{J_B 1; J_A}^{M_B 0; M_A}\right)^{2} - \frac{1}{2}\sin^{2}\theta\left(C_{J_B 1; J_A}^{M_B 1; M_A}\right)^{2}\right.$$

$$\left.- \frac{1}{2}\sin^{2}\theta\left(C_{J_B 1; J_A}^{M_B -1; M_A}\right)^{2}\right] \qquad (3.12)$$

$$= \frac{\alpha}{2\pi}\omega_{k}^{3}|\langle A\|r\|B\rangle|^{2}\sum_{M_A,M_B}p(M_B)$$

$$\times\left[\left(1 - \frac{1}{2}\sin^{2}\theta\right)\sum_{k=-1}^{1}\left(C_{J_B 1; J_A}^{M_B k; M_A}\right)^{2} - \left(1 - \frac{3}{2}\sin^{2}\theta\right)\left(C_{J_B 1; J_A}^{M_B 0; M_A}\right)^{2}\right]$$

Since [Me 65]

$$\sum_{M_A,k}\left(C_{J_B 1; J_A}^{M_B k; M_A}\right)^{2} = \frac{2J_A + 1}{2J_B + 1}$$

$$\left(C_{J_B 1; J_A}^{M_B 0; M_A}\right)^{2} = \delta_{M_A,M_B}\begin{cases}\frac{(J_B+1)^{2}-M_B^{2}}{(J_B+1)(2J_B+1)} & J_A = J_B + 1 \\ \frac{M_B^{2}}{J_B(J_B+1)} & J_A = J_B \\ \frac{J_B^{2}-M_B^{2}}{J_B(2J_B+1)} & J_A = J_B - 1\end{cases} \qquad (3.13)$$

the result can be written in the form

$$\frac{d\Gamma}{d\Omega} = \frac{\alpha}{2\pi}\omega_k^3 |\langle A\|r\|B\rangle|^2$$
$$\times \left[\frac{2}{3}\frac{2J_A+1}{2J_B+1} - \frac{1}{3}\Lambda_{J_B,J_A}\left(1 - \frac{3\langle M_B^2\rangle}{J_B(J_B+1)}\right)\left(1 - \frac{3}{2}\sin^2\theta\right)\right] \quad (3.14)$$

where

$$\Lambda_{J_B,J_A} = \begin{cases} \frac{J_B}{2J_B+1} & J_A = J_B+1 \\ -1 & J_A = J_B \\ \frac{J_B+1}{2J_B+1} & J_A + J_B - 1 \end{cases} \quad (3.15)$$

and

$$\langle M_B^2 \rangle = \sum_{M_B} p(M_B) M_B^2 \; . \quad (3.16)$$

[Note: We have used the condition that the probability for having *some* value of M_B adds to unity

$$\sum_{M_B} p(M_B) = 1 \; .] \quad (3.17)$$

If the parent state is unpolarized, the probability distribution is uniform —

$$p(M_B) = \frac{1}{2J_B+1} \; . \quad (3.18)$$

We have then

$$\langle M_B^2 \rangle = \sum_{M_B} \frac{M_B^2}{2J_B+1} = \frac{1}{3}J_B(J_B+1) \quad (3.19)$$

so that the angular distribution becomes

$$\left.\frac{d\Gamma}{d\Omega}\right|_{\text{unpolarized}} = \frac{\alpha}{2\pi}\omega_k^3 \frac{2}{3}\frac{2J_A+1}{2J_B+1}|\langle A\|r\|B\rangle|^2 \; . \quad (3.20)$$

Note that the decay distribution is isotropic, as one would expect—if the initial state is unpolarized there is no "magic" direction in space.

On the other hand suppose the system has been prepared in a state such that

$$\langle M_B^2 \rangle \neq \frac{1}{3}J_B(J_B+1). \quad (3.21)$$

The system is then said to be "aligned" along the axis of quantization with

$$\mathcal{A}_B = 1 - \frac{3\langle M_B^2 \rangle}{J_B(J_B+1)} \quad (3.22)$$

being the alignment. [Caution: Do not confuse the "alignment", which involves $\langle M_B^2 \rangle$, with the "polarization" which is defined as

$$\mathcal{P}_B = \frac{\langle M_B \rangle}{J_B} \ .] \qquad (3.23)$$

For an aligned system, there *does* exist a special direction in space (namely the axis along which the atom is aligned) and the decay distribution becomes

$$\frac{d\Gamma}{d\Omega} = \frac{\alpha}{2\pi} \omega_k^3 |\langle A \| r \| B \rangle|^2 \left[\frac{2}{3} \frac{2J_A+1}{2J_B+1} - \frac{1}{3} \mathcal{A}_B \Lambda_{J_B,J_A} \left(1 - \frac{3}{2}\sin^2\theta\right) \right] \ . \qquad (3.24)$$

If we consider a specific case — say $J_B = 1 \to J_A = 0$ — then

$$\Lambda_{1,0} = \frac{J_B+1}{2J_B+1} = \frac{2}{3} \qquad (3.25)$$

and

$$\frac{d\Gamma}{d\Omega} = \frac{\alpha}{2\pi} \omega_k^3 |\langle A \| r \| B \rangle|^2 \frac{2}{9} \left[1 - \mathcal{A}_B \left(1 - \frac{3}{2}\sin^2\theta\right) \right] \ . \qquad (3.26)$$

If maximal alignment obtains— $\mathcal{A} = 1$, which results say if $p(1) = p(-1) = 0$, $p(0) = 1$ — we have

$$\frac{d\Gamma}{d\Omega} = \frac{\alpha}{2\pi} \omega_k^3 |\langle A \| r \| B \rangle|^2 \frac{1}{3} \sin^2\theta \qquad (3.27)$$

which displays the classic $\sin^2\theta$ distribution expected for dipole radiation, as shown in Figure IV.1.

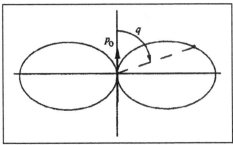

Fig. IV.1: Dipole Radiation Pattern.

E1 Decay Distribution: Intuitive Analysis

The previous calculation of the angular distribution of radiation emitted from an aligned atom, although straightforward, is not particularly transparent. It is thus useful to present an alternative (and more intuitive) way by which to derive the angular dependence [FeLS 65].

For simplicity consider an electric dipole transition from a $J = 1$ atom to a $J = 0$ atomic state. First think about emission along the $+z$ direction of a right-hand circularly polarized photon. By angular momentum conservation, the emission must have been from an initial state with $J = 1$, $J_z = 1$ as shown below. If we apply the spatial inversion or "parity" operator to such a matrix element

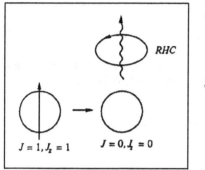

$$\langle \gamma_{\text{RHC}}; J = 0, J_z = 0 | H_{EM} | J = 1, J_z = 1 \rangle$$

(3.28)

then spin, being an axial vector, is unchanged while momentum, being a polar vector, is reversed. This becomes a transition from the same $J = 1$, $J_z = 1$ state to a left-hand circularly polarized photon travelling along the $-z$ axis. Finally a rotation by 180° about the y axis produces the configuration shown in the diagram below wherein an atom with $J = 1$, $J_z = -1$ decays via emission of a left-hand circularly polarized photon along the $+z$ direction.

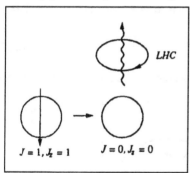

$$\langle \gamma_{\text{LHC}}; J = 0, J_z = 0 | H_{EM} | J = 1, J_z = -1 \rangle$$

(3.29)

By parity invariance the amplitudes for the two processes shown above must be identical up to a phase. For E1 emission, since the parities of initial and final states are opposite, this overall phase must be $e^{i\pi} = -1$. Thus define

$$\begin{aligned} a &\equiv \langle \gamma_{\text{RHC}}; J = 0, J_z = 0 | H_{EM} | J = 1, J_z = 1 \rangle \\ &= -\langle \gamma_{\text{LHC}}; J = 0, J_z = 0 | H_{EM} | J = 1, J_z = -1 \rangle \end{aligned}$$

(3.30)

Next consider the amplitude for a right-hand circularly polarized photon to be emitted into the solid angle around θ, ϕ. In order to calculate this quantity consider the operator

$$\hat{R}(\theta, \phi) = \exp(i\hat{J}_y \theta) \exp(i\hat{J}_z \phi) \equiv \hat{R}_y(\theta) \hat{R}_z(\phi)$$

(3.31)

which takes a state aligned along the $+z$ axis and rotates it into a state aligned along the θ, ϕ direction. Thus

$$\langle JM | \hat{R}(\theta, \phi) | JM' \rangle = \exp(iM'\phi) \langle JM | \hat{R}_y(\theta) | JM' \rangle$$

(3.32)

IV.3 ANGULAR DISTRIBUTION OF RADIATIVE DECAY

is the amplitude for a state with spin J and projection M' along the direction θ, ϕ to have projection M along the $+z$ direction. We have then

$$\langle \gamma(\theta, \phi, \text{RHC}); J = 0, J_z = 0 | H_{EM} | J = 1, J_z = M \rangle = a \left\langle 1, 1 | \hat{R}(\theta, \phi) | 1, M \right\rangle . \quad (3.33)$$

Similarly the amplitude for the emission of a left-handed circularly polarized photon along the θ, ϕ direction is

$$\langle \gamma(\theta, \phi; \text{LHC}); J = 0, J_z = 0 | H_{EM} | J = 1, J_z = M \rangle = -a \left\langle 1, -1 | \hat{R}(\theta, \phi) | 1, M \right\rangle . \quad (3.34)$$

Assuming that the photon polarization is undetected and that the decaying atom is described by probability $p(M)$ of being in state $|J = 1, J_z = M\rangle$, we find the angular distribution

$$f(\theta, \phi) \propto \sum_{M=-1}^{+1} \sum_{\lambda=\text{LHC}}^{\text{RHC}} p(M) |\langle \gamma(\theta, \phi; \lambda); J = 0, J_z = 0 | H_{EM} | J = 1, J_z = M \rangle|^2$$

$$= |a|^2 \sum_{M=-1}^{+1} p(M) \left[\left| \left\langle 1, 1 | \hat{R}_y(\theta) | 1, M \right\rangle \right|^2 + \left| \left\langle 1, -1 | \hat{R}_y(\theta) | 1, M \right\rangle \right|^2 \right] . \quad (3.35)$$

Although we could find these coefficients $\left\langle 1, M' | \hat{R}_y(\theta) | 1, M \right\rangle$ in a table [they are denoted by $d^{(1)}_{M',M}(\theta)$] it is easy to calculate them. Thus for a spin $1/2$ system the rotation operator is

$$\hat{R}_y(\theta) = \exp i \frac{\sigma_y}{2} \theta = \cos \frac{\theta}{2} + i \sigma_y \sin \frac{\theta}{2} . \quad (3.36)$$

so that

$$\left\langle \frac{1}{2}, \frac{1}{2} \middle| \hat{R}_y(\theta) \middle| \frac{1}{2}, \frac{1}{2} \right\rangle = \cos \frac{\theta}{2} \quad \left\langle \frac{1}{2}, -\frac{1}{2} \middle| \hat{R}_y(\theta) \middle| \frac{1}{2}, \frac{1}{2} \right\rangle = -\sin \frac{\theta}{2}$$

$$\left\langle \frac{1}{2}, -\frac{1}{2} \middle| \hat{R}_y(\theta) \middle| \frac{1}{2}, -\frac{1}{2} \right\rangle = \cos \frac{\theta}{2} \quad \left\langle \frac{1}{2}, \frac{1}{2} \middle| \hat{R}_y(\theta) \middle| \frac{1}{2}, -\frac{1}{2} \right\rangle = \sin \frac{\theta}{2} . \quad (3.37)$$

Then since

$$|1, 1\rangle = \left| \frac{1}{2}, \frac{1}{2} \right\rangle \left| \frac{1}{2}, \frac{1}{2} \right\rangle$$

$$|1, 0\rangle = \sqrt{\frac{1}{2}} \left(\left| \frac{1}{2}, \frac{1}{2} \right\rangle \left| \frac{1}{2}, -\frac{1}{2} \right\rangle + \left| \frac{1}{2}, -\frac{1}{2} \right\rangle \left| \frac{1}{2}, \frac{1}{2} \right\rangle \right) \quad (3.38)$$

$$|1, -1\rangle = \left| \frac{1}{2}, -\frac{1}{2} \right\rangle \left| \frac{1}{2}, -\frac{1}{2} \right\rangle$$

we find

$$\left\langle 1,1 \left| \hat{R}_y(\theta) \right| 1,1 \right\rangle = \left\langle 1,-1 \left| \hat{R}_y(\theta) \right| 1,-1 \right\rangle = \cos^2 \frac{\theta}{2} = \frac{1}{2}(1+\cos\theta)$$

$$\left\langle 1,1 \left| \hat{R}_y(\theta) \right| 1,-1 \right\rangle = \left\langle 1,-1 \left| \hat{R}_y(\theta) \right| 1,1 \right\rangle = \sin^2 \frac{\theta}{2} = \frac{1}{2}(1-\cos\theta)$$

$$\left\langle 1,\pm 1 \left| \hat{R}_y(\theta) \right| 1,0 \right\rangle = -\left\langle 1,0 \left| \hat{R}_y(\theta) \right| 1,\pm 1 \right\rangle = \pm\sqrt{2}\cos\frac{\theta}{2}\sin\frac{\theta}{2} = \pm\sqrt{\frac{1}{2}}\sin\theta \ ,$$
(3.39)

and the angular distribution becomes

$$f(\theta,\phi) \propto |a|^2 \left[(p(1)+p(-1))\left(\frac{1}{4}(1+\cos\theta)^2 + \frac{1}{4}(1-\cos\theta)^2\right) + p(0)\sin^2\theta \right] .$$
(3.40)

Since
$$p(1)+p(-1) = \langle M^2 \rangle \quad \text{and} \quad p(1)+p(0)+p(-1) = 1 \ , \tag{3.41}$$

we find
$$f(\theta,\phi) \propto |a|^2 \left[\sin^2\theta + \langle M^2 \rangle \left(\frac{1}{2}(1+\cos^2\theta) - \sin^2\theta\right) \right]$$
$$= |a|^2 \left[\sin^2\theta + \langle M^2 \rangle \left(1 - \frac{3}{2}\sin^2\theta\right) \right]$$
(3.42)

which agrees precisely with Eq. 3.26 derived previously.

Higher Order Multipoles

We have seen that provided the selection rules

$$\Delta J = 0, \pm 1 \ ; \quad \text{parity change} \tag{3.43}$$

are satisfied, then electric dipole radiation will in general be the mechanism for the transition between two states of an atom and we have evaluated explicit forms for the angular distribution. However, what occurs if the selection rules are *not* satisfied — if, e.g. $\Delta J = 2$ or there is no change in parity, or both? In this case the approximation

$$e^{-i\vec{k}\cdot\vec{r}} \approx 1 \tag{3.44}$$

in evaluation of the transition amplitude is not sufficient and one must go to higher order in the expansion. If we take

$$e^{-i\vec{k}\cdot\vec{r}} \approx 1 - i\vec{k}\cdot\vec{r} \tag{3.45}$$

then we can write

$$e^{-i\vec{k}\cdot\vec{r}}\hat{\epsilon}^*_{\vec{k},\lambda}\cdot\vec{p} \cong \hat{\epsilon}^*_{\vec{k},\lambda}\cdot\vec{p} - i\vec{k}\cdot\vec{r}\hat{\epsilon}^*_{\vec{k},\lambda}\cdot\vec{p} + \ldots$$
$$= \hat{\epsilon}^*_{\vec{k},\lambda}\cdot\vec{p} - \frac{i}{2}\left(\vec{k}\cdot\vec{r}\hat{\epsilon}^*_{\vec{k},\lambda}\cdot\vec{p} - \hat{\epsilon}^*_{\vec{k},\lambda}\cdot\vec{r}\vec{k}\cdot\vec{p}\right) - \frac{i}{2}\left(\vec{k}\cdot\vec{r}\hat{\epsilon}^*_{\vec{k},\lambda}\cdot\vec{p} + \hat{\epsilon}^*_{\vec{k},\lambda}\cdot\vec{r}\vec{k}\cdot\vec{p}\right) + \ldots$$
$$= \hat{\epsilon}^*_{\vec{k},\lambda}\cdot\vec{p} + \frac{i}{2}\hat{\epsilon}^*_{\vec{k},\lambda}\times\vec{k}\cdot\vec{r}\times\vec{p} - \frac{i}{2}\left(\vec{k}\cdot\vec{r}\hat{\epsilon}^*_{\vec{k},\lambda}\cdot\vec{p} + \hat{\epsilon}^*_{\vec{k},\lambda}\cdot\vec{r}\vec{k}\cdot\vec{p}\right) + \ldots$$
(3.46)

The first term here — $\hat{\epsilon}^*_{k,\lambda} \cdot \vec{p}$ — is the electric dipole operator, which has already been discussed. Focusing on the second term, which we write as

$$\frac{i}{2}\hat{\epsilon}^*_{k,\lambda} \times \vec{k} \cdot \vec{L} \tag{3.47}$$

where $\vec{L}$ is the angular momentum operator, we see that the basic form is similar to that for E1 radiation, but with the changes

$$\vec{p} \to \vec{r} \times \vec{p} = \vec{L} \qquad \hat{\epsilon}^*_{k,\lambda} \to \frac{i}{2}\hat{\epsilon}^*_{k,\lambda} \times \vec{k} \ . \tag{3.48}$$

Since $\vec{L}$ is a spherical tensor of rank one, the Wigner-Eckart theorem theorem gives

$$\left\langle A; J_A, M_A \left| L^{(k)} \right| B; J_B, M_B \right\rangle = C^{M_B k; M_A}_{J_B 1; J_A} \langle A \| L \| B \rangle \ . \tag{3.49}$$

Also, if we choose

$$\hat{k} = \hat{e}_z \ , \quad \hat{\epsilon}_1 = \hat{e}_x \ , \quad \hat{\epsilon}_2 = \hat{e}_y \tag{3.50}$$

then

$$\begin{aligned} \hat{\epsilon}_1 \times \hat{k} &= -\hat{e}_y \ , & \hat{\epsilon}_2 \times \hat{k} &= \hat{e}_x \\ &= -\hat{\epsilon}_2 & &= +\hat{\epsilon}_1 \end{aligned} \tag{3.51}$$

Hence, the polarization states are simply interchanged, and since we are summing over photon polarizations the decay distribution must be identical to that for E1 radiation. In fact this is magnetic dipole radiation — M1 — and we have

$$\left. \frac{d\Gamma}{d\Omega} \right|_{M1} = \frac{\alpha}{8\pi m^2} \omega_k^3 |\langle A \| L \| B \rangle|^2 \\ \times \left[\frac{2}{3} \frac{2J_A + 1}{2J_B + 1} - \frac{1}{3} \mathcal{A}_B \Lambda_{J_B, J_A} \left(1 - \frac{3}{2} \sin^2 \theta \right) \right] \ . \tag{3.52}$$

The selection rules arising from the Clebsch–Gordon coefficient are the same as before

$$\Delta J = 0, \pm 1 \ . \tag{3.53}$$

However, since $\vec{L} = \vec{r} \times \vec{p}$ is an axial vector and is therefore even under parity, we have

$$[\hat{\Pi}, \vec{L}] = 0 \tag{3.54}$$

whereupon

$$0 = \left\langle A \left| [\hat{\Pi}, \vec{L}] \right| B \right\rangle = (\pi_A - \pi_B) \left\langle A \left| \vec{L} \right| B \right\rangle \ . \tag{3.55}$$

Thus $\left\langle A | \vec{L} | B \right\rangle = 0$ unless $\pi_B = \pi_A$ and the M1 selection rules are

$$\Delta J = 0, \pm 1 \ ; \qquad \text{no parity change} \ . \tag{3.56}$$

Magnetic transitions tend to be slower than corresponding E1 decays by a factor

$$\frac{\Gamma(M1)}{\Gamma(E1)} \sim (kr)^2 \ll 1 \ . \tag{3.57}$$

However, the rates do not "compete" with one another, since the selection rules for parity differ.

Finally, we consider the remaining matrix element

$$\begin{aligned}&-\frac{i}{2}\left\langle A\left|\vec{k}\cdot\vec{r}\hat{\epsilon}^*_{\vec{k},\lambda}\cdot\vec{p}+\hat{\epsilon}^*_{\vec{k},\lambda}\cdot\vec{r}\vec{k}\cdot\vec{p}\right|B\right\rangle \\ &= -\frac{i}{2}k_m(\hat{\epsilon}^*_{\vec{k},\lambda})_n \left\langle A\left|(r_m p_n + r_n p_m)\right|B\right\rangle \ .\end{aligned} \tag{3.58}$$

We may simplify this term by noting that

$$\begin{aligned}\left[\hat{H}_{\text{ATOMIC}}, r_m r_n\right] &= -\frac{i}{m}\left(p_m r_n + r_m p_n\right) \\ &= -\frac{i}{m}\left(r_m p_n + r_n p_m - i\delta_{nm}\right) \ .\end{aligned} \tag{3.59}$$

However, the term in δ_{nm} may be dropped since $\hat{\epsilon}^*_{\vec{k},\lambda} \cdot \vec{k} = 0$. In fact we may write

$$\begin{aligned}-\frac{i}{2}\left\langle A\left|\vec{k}\cdot\vec{r}\hat{\epsilon}^*_{\vec{k},\lambda}\cdot\vec{p}+\hat{\epsilon}^*_{\vec{k},\lambda}\cdot\vec{r}\vec{k}\cdot\vec{p}\right|B\right\rangle &= \frac{m}{2}k_m(\hat{\epsilon}^*_{\vec{k},\lambda})_n \left\langle A\left|\left[\hat{H}_{\text{ATOMIC}}, Q_{mn}\right]\right|B\right\rangle \\ &= -\frac{m\omega_k}{2}k_m(\hat{\epsilon}^*_{\vec{k},\lambda})_n \left\langle A\left|Q_{mn}\right|B\right\rangle\end{aligned} \tag{3.60}$$

where

$$Q_{mn} = r_m r_n - \frac{1}{3}\delta_{mn} r^2 \tag{3.61}$$

is the electric quadrupole moment operator. Note that Q_{mn} is symmetric and traceless

$$Q_{mn} = Q_{nm} \qquad \text{Tr}\, Q = \delta^{mn} Q_{mn} = 0 \ . \tag{3.62}$$

There exist then only five independent components to Q_{mn} and it is straightforward to show that Q_{mn} is equivalent to a spherical tensor operator of rank two. Also since

$$\left[\hat{\Pi}, Q_{mn}\right] = Q_{mn} \tag{3.63}$$

we see that the selection rules associated with this process, called electric quadrupole emission are

$$\Delta J = 0, \pm 1, \pm 2 \ ; \qquad \text{no parity change} \ . \tag{3.64}$$

In this case the angular distribution is more complex, characteristic of quadrupole radiation but a detailed analysis would take us too far afield.

PROBLEM IV.3.1

Dipole Radiation Patterns

Above we derived the angular distribution for dipole radiation emitted from an oriented $J = 1$ parent nucleus to a spinless daughter. The result was obtained in two equivalent ways:

i) from the formal expression for the differential decay rate using the Wigner-Eckart theorem;

ii) from angular momentum considerations using properties of the rotation matrix $\langle JM|\hat{R}_y(\theta)|JM'\rangle$.

Repeat this derivation for dipole transitions a)$(J = 1) \to (J' = 2) + \gamma$ and b) $(J = 1) \to (J' = 1) + \gamma$ and compare the resulting angular distributions as obtained via methods i) and ii) above.

IV.4 LINE SHAPE PROBLEM: WIGNER–WEISSKOPF APPROACH

We now take up the problem of the electromagnetic line shape. For simplicity we again consider the simplified model of a one electron atom with an excited state $|B\rangle$ which can decay to the ground state $|A\rangle$. In our use of Fermi's golden rule in order to calculate the decay rate, we employed the Dirac delta function

$$\delta\left(E_B^{(0)} - E_A^{(0)} - \omega_k\right) \tag{4.1}$$

which suggests that the photon frequency is definite and equal to $E_B^{(0)} - E_A^{(0)}$. However, a little thought reveals that this picture is too simplistic. Since the excited state has a finite lifetime

$$\tau_B = \frac{1}{\Gamma_B} \tag{4.2}$$

by the Heisenberg uncertainty principle we expect an uncertainty in the energy of state $|B\rangle$

$$\Delta E_B \sim \frac{1}{\tau_B} = \Gamma_B \ . \tag{4.3}$$

We anticipate that $\omega_k = E_B - E_A^{(0)}$ will be uncertain by the same amount[†]

$$\Delta \omega_k \sim \Gamma_B \tag{4.4}$$

and that therefore the line shape of the emitted radiation should appear as shown in Figure IV.2.

[†] Since the ground state is stable we have $\tau_A = \infty$ so $\Delta E_A = 0$.

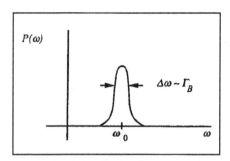

Fig. IV.2: *Line shape expected from uncertainty principle arguments.*

Wigner-Weisskopf Procedure

The classic way of treating this problem is by use of time-dependent perturbation theory and the Wigner–Weisskopf self-consistency approximation. As usual we begin by partitioning the Hamiltonian $\hat{H}$ into a free and interaction piece

$$\hat{H} = \hat{H}_0 + \hat{V} \tag{4.5}$$

where

$$\hat{H}_0 = \hat{H}_{\text{RAD}}^{(r)} + \hat{H}_{\text{ATOMIC}} \tag{4.6}$$

and

$$\hat{V} = -\frac{e}{m}\vec{A}\cdot\vec{p} + \frac{e^2}{2m}\vec{A}\cdot\vec{A} \tag{4.7}$$

couples the atom and radiation field. The initial condition is

$$|\psi(t=0)\rangle = |B,0\rangle \equiv |B\rangle|0\rangle \tag{4.8}$$

i.e., the atom resides in an excited state $|B\rangle$ and no photons are present, with

$$\hat{H}_0|B,0\rangle = E_B^{(0)}|B,0\rangle \; . \tag{4.9}$$

For simplicity, denote a general atom-photon state by $|n\rangle$ with

$$\hat{H}_0|n\rangle = E_n^{(0)}|n\rangle \; . \tag{4.10}$$

Such states are orthonormal and complete

$$\mathbf{1} = \sum_n |n\rangle\langle n| \tag{4.11}$$

so that at $t=0$ we can expand

$$|\psi(t=0)\rangle = \sum_n c_n(0)|n\rangle \tag{4.12}$$

where
$$c_n(0) = \langle n|\psi(t=0)\rangle \ . \tag{4.13}$$

Of course, if $\hat{V} = 0$ the time evolution of the system is known precisely

$$|\psi(t)\rangle = \sum_n c_n(0) e^{-iE_n^{(0)}t}|n\rangle, \tag{4.14}$$

while if $\hat{V} \neq 0$ the wavefunction cannot be exactly determined. We can nevertheless obtain an approximate solution provided $\hat{V}$ is in some sense small compared to $\hat{H}_0$. We can write in general

$$|\psi(t)\rangle = \sum_n c_n(t) e^{-iE_n^{(0)}t}|n\rangle \ . \tag{4.15}$$

where c_n is no longer constant but is a (hopefully slowly varying) function of time, obeying the differential equation

$$i\frac{d}{dt}c_n(t) = \sum_m e^{i\left(E_n^{(0)} - E_m^{(0)}\right)t} \left\langle n|\hat{V}|m\right\rangle c_m(t) \ . \tag{4.16}$$

This (infinite) set of coupled equations is exact and impossible to solve. However, following Wigner and Weisskopf we can make a reasonable approximation. At $t = 0$ only one of the c_n is non-zero — namely c_B. We assume that we can write at *all* times

$$\begin{aligned}
i\frac{d}{dt}c_n &= e^{i\left(E_n^{(0)} - E_B^{(0)}\right)t} \left\langle n|\hat{V}|B,0\right\rangle c_B(t) & |n\rangle \neq |B,0\rangle \\
i\frac{d}{dt}c_B &= \left\langle B,0|\hat{V}|B,0\right\rangle c_B(t) + \sum_n e^{i\left(E_B^{(0)} - E_n^{(0)}\right)t} \left\langle B,0|\hat{V}|n\right\rangle c_n(t)
\end{aligned} \tag{4.17}$$

where only c_B is included in the first equation and only single photon states are included in the sum in the second. We can simplify the second equation by defining

$$c_B(t) \equiv \tilde{c}_B(t) \exp(-i\Lambda t) \tag{4.18}$$

where
$$\Lambda = \left\langle B, 0|\hat{V}|B, 0\right\rangle \ . \tag{4.19}$$

Then using the definition
$$\tilde{E}_B = E_B^{(0)} + \Lambda \tag{4.20}$$

Eq. 4.17 becomes

$$\begin{aligned}
i\frac{d}{dt}c_n &= e^{i(E_n^{(0)} - \tilde{E}_B)t} \left\langle n|\hat{V}|B,0\right\rangle \tilde{c}_B \\
i\frac{d}{dt}\tilde{c}_B &= \sum_n e^{i(\tilde{E}_B - E_n^{(0)})t} \left\langle B,0|\hat{V}|n\right\rangle c_n \ .
\end{aligned} \tag{4.21}$$

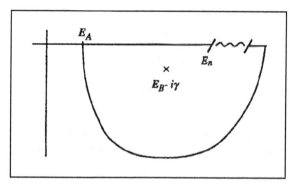

Fig. IV.3: Contour for the continuum integration in Eq. 4.26.

Wigner and Weisskopf proposed the ansatz

$$\tilde{c}_B(t) = e^{-\gamma t} \quad \text{with} \quad \text{Re}\,\gamma > 0 \qquad (4.22)$$

since we anticipate exponential decay. Then

$$i\frac{d}{dt}c_n = e^{i\left(E_n^{(0)}-\tilde{E}_B+i\gamma\right)t}\left\langle n|\hat{V}|B,0\right\rangle . \qquad (4.23)$$

Using the boundary condition $c_n(t=0) = 0$ we have

$$\begin{aligned} c_n(t) &= -i\int_0^t dt'\, e^{i\left(E_n^{(0)}-\tilde{E}_B+i\gamma\right)t'}\left\langle n|\hat{V}|B,0\right\rangle \\ &= \frac{1-e^{i\left(E_n^{(0)}-\tilde{E}_B+i\gamma\right)t}}{E_n^{(0)}-\tilde{E}_B+i\gamma}\left\langle n|\hat{V}|B,0\right\rangle \end{aligned} \qquad (4.24)$$

and

$$i\frac{d\tilde{c}_B}{dt} = -i\gamma e^{-\gamma t} = \sum_n \left|\left\langle n|\hat{V}|B,0\right\rangle\right|^2 \frac{e^{-i\left(E_n^{(0)}-\tilde{E}_B\right)t}-e^{-\gamma t}}{E_n^{(0)}-\tilde{E}_B+i\gamma} . \qquad (4.25)$$

Since the states $|n\rangle$ form a continuum, we need to perform an integral of the form

$$\int_{E_A}^{\infty} dE_n^{(0)}\, f(E_n^{(0)})\,\frac{\exp-i\left(E_n^{(0)}-\tilde{E}_B\right)t}{E_n^{(0)}-\tilde{E}_B+i\gamma} \qquad (4.26)$$

where we have defined

$$f(E_n^{(0)}) = \left|\left\langle n|\hat{V}|B,0\right\rangle\right|^2 . \qquad (4.27)$$

We perform the integration by closing the contour with a huge semicircle in the lower half $E_n^{(0)}$ plane, as shown in Figure IV.3. On the semicircle we have

IV.4 LINE SHAPE PROBLEM: WIGNER–WEISSKOPF APPROACH

$\left|e^{-i(E_n^{(0)} - \tilde{E}_B)t}\right| \sim e^{-\operatorname{Im} E_n^{(0)} t}$ which is highly damped except for points very near E_A. The Wigner–Weisskopf approach is to neglect this contribution. [This nonvanishing component represents a transient effect associated with the fact that we started with an excited atom at $t = 0$ and then suddenly turned on the interaction. We should more correctly have used

$$\tilde{c}_B(t) \underset{t \to \infty}{\sim} e^{-\gamma t} \quad . \tag{4.28}$$

In this approximation we have

$$\int_{E_A}^{\infty} dE_n^{(0)} f(E_n^{(0)}) e^{-i(E_n^{(0)} - \tilde{E}_B)t} \frac{1}{E_n^{(0)} - \tilde{E}_B + i\gamma} \cong -2\pi i f\left(\tilde{E}_B - i\gamma\right) e^{-\gamma t} \quad . \tag{4.29}$$

Generally γ is quite small with respect to $\tilde{E}_B$, so $f\left(\tilde{E}_B - i\gamma\right) \cong f(\tilde{E}_B)$ and

$$\int_{E_A}^{\infty} dE_n^{(0)} f(E_n^{(0)}) e^{-i(E_n^{(0)} - \tilde{E}_B)t} \frac{1}{E_n^{(0)} - \tilde{E}_B + i\gamma}$$
$$\approx -2\pi i\, e^{-\gamma t} \int_{E_A}^{\infty} dE_n^{(0)} \delta\left(E_n^{(0)} - \tilde{E}_B\right) f(E_n^{(0)}) \quad . \tag{4.30}$$

Eq. 4.25 then reads

$$-i\gamma\, e^{-\gamma t} = -e^{-\gamma t} \sum_n \left|\langle n|\hat{V}|B, 0\rangle\right|^2$$
$$\times \left[2\pi i \delta\left(E_n^{(0)} - \tilde{E}_B\right) + \frac{1}{E_n^{(0)} - \tilde{E}_B + i\gamma}\right] \quad . \tag{4.31}$$

Now in section II.1 we have shown that under an integration (summation) we may replace

$$\frac{1}{E_n^{(0)} - \tilde{E}_B + i\gamma} \sim P\frac{1}{E_n^{(0)} - \tilde{E}_B} - i\pi\delta\left(\tilde{E}_B - E_n^{(0)}\right) \quad . \tag{4.32}$$

Hence

$$\frac{1}{E_n^{(0)} - \tilde{E}_B + i\gamma} + 2\pi i\delta\left(\tilde{E}_B - E_n^{(0)}\right) \sim \frac{1}{E_n^{(0)} - \tilde{E}_B - i\gamma} \tag{4.33}$$

so that

$$-i\gamma = \sum_n \frac{-\left|\langle n|\hat{V}|B, 0\rangle\right|^2}{E_n^{(0)} - \tilde{E}_B - i\gamma}$$
$$= -\sum_n \left|\langle n|\hat{V}|B, 0\rangle\right|^2 \left(P\frac{1}{E_n^{(0)} - \tilde{E}_B} + i\pi\delta\left(\tilde{E}_B - E_n^{(0)}\right)\right) \quad . \tag{4.34}$$

We identify

$$\operatorname{Re}\gamma = \sum_n \pi\delta\left(\tilde{E}_B - E_n^{(0)}\right) \left|\langle n|\hat{V}|B, 0\rangle\right|^2 = \frac{1}{2}\Gamma_B \tag{4.35}$$

where
$$\Gamma_B = \sum_n 2\pi\delta\left(\tilde{F}_B - E_n^{(0)}\right)\left|\langle n|\hat{V}|B,0\rangle\right|^2 \qquad (4.36)$$

is the decay rate of state $|B\rangle$ as given by Fermi's golden rule and

$$\text{Im}\,\gamma = \sum_n \left|\langle n|\hat{V}|B,0\rangle\right|^2 P\frac{1}{\tilde{E}_B - E_n^{(0)}} \qquad (4.37)$$

is the energy shift of state $|B\rangle$ as given by second order perturbation theory due to its interaction with the radiation field

$$\text{Im}\,\gamma = \Delta E_B^{(2)} \;. \qquad (4.38)$$

The full energy shift to this order is

$$\Delta E_B = \langle B,0|\hat{V}|B,0\rangle + \Delta E_B^{(2)} = \Lambda + \Delta E_B^{(2)} \qquad (4.39)$$

so that
$$c_B(t) = e^{-i\left(\Delta E_B - \frac{i}{2}\Gamma_B\right)t} \;. \qquad (4.40)$$

The time development of the $|B,0\rangle$ component of the wavefunction is then

$$|\psi(t)\rangle = e^{-i\left(E_B^{(0)} + \Delta E_B\right)t} e^{-\frac{1}{2}\Gamma_B t}|B,0\rangle + \ldots \qquad (4.41)$$

so that
$$|\langle B,0|\psi(t)\rangle|^2 = \exp{-\Gamma_B t}, \qquad (4.42)$$

i.e., the probability to remain in the initial state decays exponentially with time, as expected.

The Line Shape

We finally have the ammunition to study the line shape—we examine the amplitude to have made a transition to the state $|A,\vec{k}\rangle \equiv |A\rangle|1_{\vec{k},\lambda}\rangle$. From Eq. 4.24 this is given by

$$c_{A,\vec{k}}(t) = \frac{1 - e^{i\left(E_A^{(0)} + \omega_k - E_B^{(0)} - \Delta E_B\right)t} e^{-\frac{1}{2}\Gamma_B t}}{E_A^{(0)} + \omega_k - E_B^{(0)} - \Delta E_B + i\frac{1}{2}\Gamma_B} \langle A,\vec{k}|\hat{V}|B,0\rangle$$

$$\underset{\Gamma_B t \gg 1}{\sim} \frac{\langle A,\vec{k}|\hat{V}|B,0\rangle}{E_A^{(0)} + \omega_k - E_B^{(0)} - \Delta E_B + i\frac{1}{2}\Gamma_B} \;. \qquad (4.43)$$

Defining
$$\omega_{AB} = E_B^{(0)} + \Delta E_B - E_A^{(0)} \qquad (4.44)$$

we have
$$|c_{A,\vec{k}}(t)|^2 \underset{\Gamma_B t \gg 1}{\sim} \frac{\left|\langle A,\vec{k}|\hat{V}|B,0\rangle\right|^2}{(\omega_k - \omega_{AB})^2 + \frac{1}{4}\Gamma_B^2} \qquad (4.45)$$

IV.4 LINE SHAPE PROBLEM: WIGNER–WEISSKOPF APPROACH

Summing over the density of photon states then yields the line shape

$$P(\omega_k)d\omega_k = \frac{\omega_k^2 d\omega_k}{(2\pi)^3} \sum_\lambda \int d\Omega \frac{\left|\langle A, \vec{k}|\hat{V}|B, 0\rangle\right|^2}{(\omega_k - \omega_{AB})^2 + \frac{1}{4}\Gamma_B^2} \qquad (4.46)$$

$$\cong \frac{\Gamma_B}{2\pi} \frac{d\omega_k}{(\omega_k - \omega_{AB})^2 + \frac{1}{4}\Gamma_B^2} \; .$$

This shape is called "Lorentzian" and is the line shape which results from a damped classical oscillator. Note that

$$\Delta\omega_k \sim \Gamma_B \qquad (4.47)$$

as expected from uncertainty principle arguments. This result is called the "natural" line shape.

In order to get a sense for orders of magnitude, consider hydrogen in the $2p$ state. Then (cf. Problem IV.1.1)

$$\tau = 1.6 \times 10^{-9} \text{ sec} \; , \qquad \Gamma = \frac{\hbar}{\tau} = \frac{6 \times 10^{-22} \text{ MeV sec}}{1.6 \times 10^{-9} \text{ sec}} \simeq 4 \times 10^{-7} \text{ eV} \; . \qquad (4.48)$$

Of course, the center of the frequency distribution is at

$$\hbar\omega_k = 13.6 \text{ eV}\left(1 - \frac{1}{2^2}\right) = 10.4 \text{ eV} \; . \qquad (4.49)$$

The ratio

$$\frac{\hbar\omega_k}{\Gamma} \qquad (4.50)$$

is called the "quality" or Q-factor, which in this case is

$$Q = \frac{\hbar\omega_k}{\Gamma} = \frac{10 \text{ eV}}{4 \times 10^{-7} \text{ eV}} \cong 2.5 \times 10^7 \; . \qquad (4.51)$$

As a second example, consider the ^{57}Fe transition shown below for which

$$\tau = 1.4 \times 10^{-7} \text{ sec}$$

$$\Gamma = \frac{\hbar}{\tau} = \frac{6 \times 10^{-22} \text{ MeV sec}}{1.4 \times 10^{-7} \text{ sec}} \qquad (4.52)$$

$$\cong 4 \times 10^{-12} \text{ KeV} \; .$$

Then

$$Q = \frac{14.4 \text{ KeV}}{4 \times 10^{-12} \text{ KeV}} \cong 3 \times 10^{12} \; . \qquad (4.53)$$

The line here is fantastically narrow. It was found by Mössbauer that under appropriate circumstances the ^{57}Fe nucleus does not recoil when the photon is emitted and the sharpness of the line shape is preserved. This study of recoiless emission has had wide impact in lattice dynamics, metallurgy, chemistry, and other areas [We 64].

PROBLEM IV.4.1

Wigner Weisskopf Method for a Three Level System

Consider the radiative transitions as shown in the diagram below. Assume A, B, C are all three non-degenerate. Express the state vector $|\psi(t)\rangle$ in terms of its most important components:

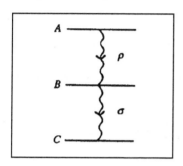

$$\psi(t)\rangle = a(t)e^{-iE_A t}|A\rangle$$
$$+ \sum_\rho b_\rho(t)e^{-i(E_B+\omega_\rho)t}|B,\rho\rangle$$
$$+ \sum_{\rho\sigma} c_{\rho\sigma}(t)e^{-i(E_C+\omega_\rho+\omega_\sigma)t}|C,\rho,\sigma\rangle$$

Impose the boundary conditions: $a(0) = 1$, $b_\rho(0) = c_{\rho\sigma}(0) = 0$ for all ρ, σ.

Now extend the definition of $a(t), b_\rho(t), c_{\rho\sigma}(t)$ to negative times and require

$$a(t) = b_\rho(t) = c_{\rho\sigma}(t) \equiv 0 \quad \text{if} \quad t < 0.$$

Thus $a(t)$ takes a jump from zero to unity as t passes through zero from below to above. Show that this may be accommodated in the differential equations for $a, b_\rho, c_{\rho,\sigma}$ by including a delta function $\delta(t)$ in the equation for $a(t)$:

$$i\frac{da(t)}{dt} = \sum_\rho e^{i(E_A-E_B-\omega_\rho)t}\langle A|\hat{V}|B,\rho\rangle b_\rho(t) + i\delta(t)$$

$$i\frac{db_\rho}{dt} = e^{i(E_B+\omega_\rho-E_A)t}\langle B,\rho|\hat{V}|A\rangle a(t)$$
$$+ \sum_\sigma e^{i(E_B+\omega_\rho-E_C-\omega_\sigma-\omega_\rho)t}\langle B,\rho|\hat{V}|C,\rho,\sigma\rangle c_{\rho\sigma}(t)$$

$$i\frac{dc_{\rho\sigma}}{dt} = e^{i(E_C+\omega_\rho+\omega_\sigma-E_B-\omega_\rho)t}\langle C,\rho,\sigma|\hat{V}|B,\rho\rangle b_\rho(t).$$

Solve these equations by taking Fourier transforms. Define

$$a(t) = \frac{-1}{2\pi i}\int_{-\infty}^{\infty} dw\, e^{i(E_A-w)t} A(w)$$

$$b_\rho(t) = \frac{-1}{2\pi i}\int_{-\infty}^{\infty} dw\, e^{i(E_B+\omega_\rho-w)t} B_\rho(w)$$

IV.4 LINE SHAPE PROBLEM: WIGNER–WEISSKOPF APPROACH

$$c_{\rho\sigma}(t) = \frac{-1}{2\pi i} \int_{-\infty}^{\infty} dw\, e^{i(E_C+\omega_\rho+\omega_\sigma-\omega)t} C_{\rho\sigma}(\omega)$$

Show that the equations for $A(\omega), B_\rho(\omega), C_{\rho\sigma}(\omega)$ become

$$(\omega - E_A + i\epsilon)A(\omega) = 1 + \sum_\rho \langle A|\hat{B},\rho\rangle B_\rho(\omega)$$

$$(\omega - E_B - \omega_\rho + i\epsilon)B_\rho(\omega) = \langle B,\rho|\hat{V}|A\rangle A(\omega) + \sum_\sigma \langle B,\rho|\hat{V}|C,\rho,\sigma\rangle C_{\rho\sigma}(\omega)$$

$$(\omega - E_C - \omega_\rho - \omega_\sigma + i\epsilon)C_{\rho\sigma}(\omega) = \langle C,\rho,\sigma|\hat{V}|B,\rho\rangle B_\rho(\omega)$$

where the $+i\epsilon$ is inserted to guarantee that $a(t), b_\rho(t), c_{\rho\sigma}(t)$ vanish identically for negative times.

We solve these coupled equations by working from the last equation to the first. Thus the bottom equation becomes

$$C_{\rho\sigma}(\omega) = \frac{\langle C,\rho,\sigma|\hat{V}|B,\rho\rangle}{\omega - E_C - \omega_\rho - \omega_\sigma + i\epsilon} B_\rho(\omega) .$$

At this point we need the sum

$$\sum_\sigma \langle B,\rho,\sigma|\hat{V}|C,\rho,\sigma\rangle C_{\rho\sigma}(\omega) = \sum_\sigma \frac{|\langle C,\rho,\sigma|\hat{V}|B,\rho\rangle|^2}{\omega - E_C - \omega_\rho - \omega_\sigma + i\epsilon} B_\rho(\omega) .$$

In the equation for $B_\rho(\omega)$ the most important value of ω is $\omega = E_B + \omega_\rho$. Why? Now make an approximation: in $\sum_\sigma \ldots$ replace ω by $E_B + \omega_\rho$. Show that

$$\sum_\sigma \langle B,\rho|\hat{V}|C,\rho,\sigma\rangle C_{\rho\sigma}(\omega) = (\Delta E_B - i\frac{1}{2}\Gamma_B)B_\rho(\omega)$$

where

$$\Gamma_B = 2\pi \sum_\sigma \delta(\omega_\sigma - E_B + E_C)|\langle C,\rho,\sigma|\hat{V}|B,\rho\rangle|^2$$

$$\Delta E_B = \sum_\sigma P\frac{|\langle C,\rho,\sigma|\hat{V}|B,\rho\rangle|^2}{E_B - E_C - \omega_\sigma}$$

and

$$B_\rho(\omega) = \frac{\langle B,\rho|\hat{V}|A\rangle}{\omega - E_B - \Delta E_B - \omega_\rho + i\frac{1}{2}\Gamma_B} A(\omega).$$

The approximation is justified a posteriori if $\Delta E_B, \Gamma_B \ll E_B + \omega_\rho$. This is generally true of Γ_B but not of ΔE_B. Forge ahead anyway. We know how to renormalize, and the experimental level shifts are indeed small. Proceed in the same fashion to show that approximately

$$\sum_\rho \langle A|\hat{V}|B,\rho\rangle B_\rho(\omega) = (\Delta E_A - i\frac{1}{2}\Gamma_A)A(\omega)$$

with
$$\Gamma_A = 2\pi \sum_\rho \delta(\omega_\rho - E_A + E_B + \Delta E_B)|\langle B, \rho|\hat{V}|A\rangle|^2$$

$$\Delta E_A = \sum_\rho P \frac{|\langle B, \rho|\hat{V}|A\rangle|^2}{E_A - E_B - \Delta E_B - \omega_\rho}$$

Therefore
$$A(\omega) = \frac{1}{\omega - E_A - \Delta E_A + i\frac{1}{2}\Gamma_A}.$$

Calculate the Fourier integrals and find $a(t), b_\rho(t), c_{\rho\sigma}(t)$.

IV.5 COMPTON SCATTERING VIA FEYNMAN DIAGRAMS

The line shape problem is also easily treated via Feynman diagram techniques, as we shall see. However, before doing so it is useful to discuss a somewhat different but related problem — that of photon-atom scattering.

Photon-Atom Scattering: the Kramers-Heisenberg Amplitude

Again we write
$$\hat{H} = \hat{H}_0 + \hat{V} \tag{5.1}$$

with
$$\hat{H}_0 = \hat{H}_{\text{RAD}}^{(r)} + \hat{H}_{\text{ATOMIC}} \tag{5.2}$$

and
$$\hat{V} = -\frac{e}{m}\vec{A}\cdot\vec{p} + \frac{e^2}{2m}\vec{A}\cdot\vec{A} \tag{5.3}$$

Suppose that the initial state is

$$|\phi_i\rangle = |A\rangle|\vec{k}_i, \hat{\epsilon}_i\rangle \qquad \tag{5.4}$$

i.e., atom in state $|A\rangle$ with a single photon of momentum $\vec{k}_i$ and polarization $\hat{\epsilon}_i$. This state is an eigenstate of $\hat{H}_0$ with eigenvalue $E_A^{(0)} + \omega_i = E_i$

$$\hat{H}_0|\phi_i\rangle = E_i|\phi_i\rangle \quad \text{with} \quad E_i = E_A^{(0)} + \omega_i \; . \tag{5.5}$$

Suppose that the final state is

$$|\phi_f\rangle = |B\rangle|\vec{k}_f, \hat{\epsilon}_f\rangle \qquad \tag{5.6}$$

i.e., atom in state $|B\rangle$ with a single outgoing photon of momentum $\vec{k}_f$ and polarization $\hat{\epsilon}_f$. This state is also an eigenstate of $\hat{H}_0$ with eigenvalue $E_f = E_B^{(0)} + \omega_f$.

We wish to calculate the transition amplitude

$$\text{Amp}_{fi} = \lim_{\substack{t' \to \infty \\ t \to -\infty}} e^{iE_f t'} \left\langle \vec{k}_f, \hat{\epsilon}_f \left| \left\langle B \left| \hat{U}(t',t) \right| A \right\rangle \right| \vec{k}_i, \hat{\epsilon}_i \right\rangle e^{-iE_i t} . \tag{5.7}$$

As discussed in Chapter II, we can write Eq. 5.7 as

$$\text{Amp}_{fi} = \delta_{fi} - 2\pi i \delta(E_f - E_i)\mathcal{F}(E_i) \tag{5.8}$$

where $\mathcal{F}(E_i)$ is given by the usual perturbation series in ascending powers of e. Since $\vec{A}(\vec{x})$ only connects states which differ by one in the occupation number of photons, there exists no contribution to $\mathcal{F}(E_i)$ of order e. However, to second order there exist three nonzero terms. There is a single term of first order in $\hat{V}$

$$\mathcal{F}^{(1)}(E_i) = \left\langle \phi_f | \hat{V} | \phi_i \right\rangle = \langle k_f, \hat{\epsilon}_f | \langle B | \frac{e^2}{2m} \frac{e^{i(\vec{k}_i - \vec{k}_f) \cdot r}}{\sqrt{2\omega_f \cdot 2\omega_i}} \hat{\epsilon}_f^* \cdot \hat{\epsilon}_i$$
$$\times \left(\hat{a}_{\vec{k}_i, \hat{\epsilon}_i} \hat{a}^\dagger_{\vec{k}_f, \hat{\epsilon}_f} + \hat{a}^\dagger_{\vec{k}_f, \hat{\epsilon}_f} \hat{a}_{\vec{k}_i, \hat{\epsilon}_i} \right) |A\rangle |\vec{k}_i, \hat{\epsilon}_i\rangle . \tag{5.9}$$

As usual we take the long wavelength limit and replace the exponential by unity. Then

$$\left\langle \phi_f | \hat{V} | \phi_i \right\rangle \approx \frac{e^2}{2m} 2\hat{\epsilon}_f^* \cdot \hat{\epsilon}_i \frac{1}{\sqrt{2\omega_f 2\omega_i}} \langle B | A \rangle$$
$$= \delta_{BA} \frac{e^2}{2m} 2\hat{\epsilon}_f^* \cdot \hat{\epsilon}_i \frac{1}{2\omega_i} \tag{5.10}$$

which corresponds to the Feynman diagram shown in Figure IV.4

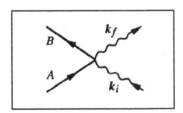

Fig. IV.4: The "seagull" diagram.

and is sometimes called the "seagull" term. [Note that for simplicity we shall assume the atom to have, as usual, but a single electron. Generalization to the multi-electron case is obvious.]

To second order in the interaction we have

$$\mathcal{F}^{(2)}(E_i) = \sum_k \frac{\left\langle \phi_f | \hat{V} | \phi_k \right\rangle \left\langle \phi_k | \hat{V} | \phi_i \right\rangle}{E_i - E_k + i\epsilon} \tag{5.11}$$

which contains two contributing diagrams, as illustrated in Figures IV.5,6. One has an intermediate state consisting of the atom in state $|I\rangle$ and no photons present

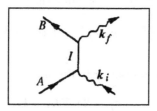

Fig. IV.5: *Pole contribution to photon-atom scattering.*

with amplitude

$$\sum_I \left(\frac{e}{m}\right)^2 \frac{1}{\sqrt{2\omega_f 2\omega_i}} \frac{\left\langle B \left| e^{-i\vec{k}_f \cdot \vec{r}} \hat{\epsilon}_f^* \cdot \vec{p} \right| I \right\rangle \left\langle I \left| e^{i\vec{k}_i \cdot \vec{r}} \hat{\epsilon}_i \cdot \vec{p} \right| A \right\rangle}{E_A^{(0)} + \omega_i - E_I^{(0)} + i\epsilon} \qquad (5.12)$$

which becomes, in the long wavelength limit

$$\sum_I \left(\frac{e}{m}\right)^2 \frac{1}{\sqrt{2\omega_f 2\omega_i}} \frac{\left\langle B \left| \hat{\epsilon}_f^* \cdot \vec{p} \right| I \right\rangle \left\langle I \left| \hat{\epsilon}_i \cdot \vec{p} \right| A \right\rangle}{E_A^{(0)} + \omega_i - E_I^{(0)} + i\epsilon} . \qquad (5.13)$$

The second diagram contains an intermediate state consisting of the atom in state $|I\rangle$ and *both* initial and final photons present

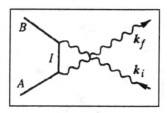

Fig. IV.6: *A second pole diagram for photon-atom scattering.*

with amplitude

$$\sum_I \left(\frac{e}{m}\right)^2 \frac{1}{\sqrt{2\omega_f 2\omega_i}} \frac{\left\langle B \left| e^{i\vec{k}_i \cdot \vec{r}} \hat{\epsilon}_i \cdot \vec{p} \right| I \right\rangle \left\langle I \left| e^{-i\vec{k}_f \cdot \vec{r}} \hat{\epsilon}_f^* \cdot \vec{p} \right| A \right\rangle}{E_A^{(0)} - E_I^{(0)} - \omega_f + i\epsilon} \qquad (5.14)$$

which becomes in the long wavelength limit

$$\sum_I \left(\frac{e}{m}\right)^2 \frac{1}{\sqrt{2\omega_f 2\omega_i}} \frac{\left\langle B \left| \hat{\epsilon}_i \cdot \vec{p} \right| I \right\rangle \left\langle I \left| \hat{\epsilon}_f^* \cdot \vec{p} \right| A \right\rangle}{E_A^{(0)} - E_I^{(0)} - \omega_f + i\epsilon} . \qquad (5.15)$$

There are many other diagrams which contribute to the transition amplitude, as shown in Figure IV.7

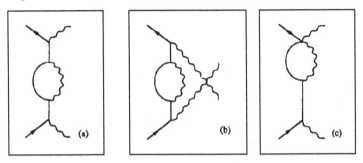

Fig. IV.7: Higher order contributions to photon-atom scattering.

However, these are all of order e^4 or higher and we shall (temporarily) neglect them.

The scattering cross section is calculated in the usual way. Since the photon velocity is unity we have

$$d\sigma = \int |\mathcal{F}(E_i)|^2 \, 2\pi\delta(E_f - E_i) \frac{d^3 k_f}{(2\pi)^3} \tag{5.16}$$

where

$$\mathcal{F}(E_i) = \frac{e^2}{m\sqrt{2\omega_f 2\omega_i}} \left[\hat{\epsilon}_f^* \cdot \hat{\epsilon}_i \langle B|A\rangle \right.$$

$$\left. + \frac{1}{m} \sum_I \frac{\hat{\epsilon}_f^* \cdot \vec{p}_{BI} \hat{\epsilon}_i \cdot \vec{p}_{IA}}{E_A^{(0)} - E_I^{(0)} + \omega_i + i\epsilon} + \frac{1}{m} \sum_I \frac{\hat{\epsilon}_i \cdot \vec{p}_{BI} \hat{\epsilon}_f^* \cdot \vec{p}_{IA}}{E_A^{(0)} - E_I^{(0)} - \omega_f + i\epsilon} \right] . \tag{5.17}$$

The differential scattering section then becomes

$$\frac{d\sigma}{d\Omega_f} = r_0^2 \frac{\omega_f}{\omega_i} \left| \hat{\epsilon}_f^* \cdot \hat{\epsilon}_i \langle B|A\rangle + \frac{1}{m} \sum_I \frac{\hat{\epsilon}_f^* \cdot \vec{p}_{BI} \hat{\epsilon}_i \cdot \vec{p}_{IA}}{E_A^{(0)} - E_I^{(0)} + \omega_i} + \frac{\hat{\epsilon}_i \cdot \vec{p}_{BI} \hat{\epsilon}_f^* \cdot \vec{p}_{IA}}{E_A^{(0)} - E_I^{(0)} - \omega_f} \right|^2 \tag{5.18}$$

where

$$r_0 = \frac{\alpha}{m} = 2.8 \times 10^{-13} \text{ cm} \tag{5.19}$$

is the classical radius of the electron. This result is known as the Kramers-Heisenberg dispersion formula. [The term "dispersion" is used since, as we shall see, there exists an intimate connection between this result and the refractive index for light.] It was derived by Kramers and Heisenberg using arguments based upon Bohr's correspondence principle even before quantum mechanics was fully developed.

Special Cases

There exist several important special cases of this result. Elastic or Rayleigh scattering occurs when the initial and final atomic states are the same — $|B\rangle = |A\rangle$. Then by energy conservation we have also

$$\omega_f = \omega_i . \tag{5.20}$$

Raman scattering is the situation wherein the initial and final atomic states are not the same. If the incident photon carries sufficient energy the atom may be excited to a higher energy state and the final photon has energy

$$\omega_f = \omega_i - (E_B - E_A) \ . \tag{5.21}$$

This is called a "Stokes" line. On the other hand if the atom begins in an excited state and is deexcited via the photon interaction then the final state photon carries away greater energy

$$\omega_f = \omega_i + |E_B - E_A| \ . \tag{5.22}$$

This is called an "anti-Stokes" line.

In the situation that the photon energy is much greater than atomic energy level differences

$$\omega_i \gg E_I - E_A \tag{5.23}$$

the problem must be equivalent to scattering from a free electron. Then we have elastic scattering and the cross section becomes

$$\frac{d\sigma}{d\Omega_f}\bigg|_{\omega_i \gg \widetilde{E_I - E_A}} r_0^2 \left| \hat{\epsilon}_f^* \cdot \hat{\epsilon}_i + \frac{1}{m\omega_i} \sum_I \left(\hat{\epsilon}_f^* \cdot \vec{p}_{AI} \hat{\epsilon}_i \cdot \vec{p}_{IA} - \hat{\epsilon}_i \cdot \vec{p}_{AI} \hat{\epsilon}_f^* \cdot \vec{p}_{IA} \right) \right|^2$$
$$\cong r_0^2 \left| \hat{\epsilon}_f^* \cdot \hat{\epsilon}_i \right|^2 \ . \tag{5.24}$$

This expression was first obtained classically by J. J. Thomson and is called the Thomson cross section. [Note: Eq. 5.24 is valid provided $m \gg \omega_i \gg E_I - E_A$ so that electron recoil may be neglected. If $m \lesssim \omega_i$ we must use relativistic techniques, leading to the so-called Klein–Nishina formula (cf. Section VII.10).]

Suppose the incident photon arrives along the z-direction while the final photon departs into the solid angle $d\Omega$ described by polar angles θ, ϕ, i.e.,

$$\hat{k}_i = (0, 0, 1)$$
$$\hat{k}_f = (\sin\theta \cos\phi, \sin\theta \sin\phi, \cos\theta) \ . \tag{5.25}$$

We may select the associated polarization vectors to be

$$\hat{\epsilon}_i^{(1)} = (1, 0, 0) \qquad \hat{\epsilon}_f^{(1)} = (\sin\phi, -\cos\phi, 0)$$
$$\hat{\epsilon}_i^{(2)} = (0, 1, 0) \qquad \hat{\epsilon}_f^{(2)} = (\cos\theta \cos\phi, \cos\theta \sin\phi, -\sin\theta) \ . \tag{5.26}$$

If we take the incident radiation to be unpolarized we must average over $\hat{\epsilon}_i^{(1)}$ and $\hat{\epsilon}_i^{(2)}$. The cross section for scattering into polarization state $\hat{\epsilon}_f^{(1)}$ is then

$$\frac{d\sigma^{(1)}}{d\Omega} = \frac{1}{2} r_0^2 \left(\sin^2\phi + \cos^2\phi \right) = \frac{1}{2} r_0^2 \ . \tag{5.27}$$

while the cross section for scattering into polarization state $\hat{\epsilon}_f^{(2)}$ is

$$\frac{d\sigma^{(2)}}{d\Omega} = \frac{1}{2}r_0^2\left(\cos^2\theta\cos^2\phi + \cos^2\theta\sin^2\phi\right)$$
$$= \frac{1}{2}r_0^2\cos^2\theta \ . \tag{5.28}$$

[Note that for 90° scattering — $\cos\theta = 0$ — we have

$$\frac{d\sigma^{(2)}}{d\Omega} = 0 \tag{5.29}$$

and the scattered photon is 100% polarized in a direction normal to the plane defined by $\vec{k}_i, \vec{k}_f$.]

Typically the final photon polarization is not detected and the cross section becomes

$$\frac{d\sigma^{\text{tot}}}{d\Omega} = \frac{d\sigma^{(1)}}{d\Omega} + \frac{d\sigma^{(2)}}{d\Omega} = \frac{1}{2}r_0^2(1 + \cos^2\theta) \ . \tag{5.30}$$

The total cross section is easily found

$$\sigma^{\text{tot}} = \int d\Omega \frac{d\sigma^{\text{tot}}}{d\Omega} = \int d\Omega \frac{1}{2}r_0^2(1 + \cos^2\theta)$$
$$= \frac{1}{2}r_0^2 \times 2\pi \int_{-1}^{1} dz\,(1 + z^2) = \frac{8\pi}{3}r_0^2 \tag{5.31}$$

and is identical to the familiar result obtained classically.

PROBLEM IV.5.1

Why is the Sky Blue?

Quantum mechanics should be able to give a simple account of two very obvious facts

a) The sky is blue (and the sunset red)

b) skylight is highly polarized.

(Some of you may not have had direct experience with the polarization. If not, borrow a pair of polaroid sunglasses and study photons coming from the sun scattered nearly perpendicularly.) For the classical explanation of these phenomena read the *Feynman Lectures On Physics*, Vol. I, pp. 32–6 to 32–9 [FeLS 65].

i) Frequency Dependence of the Scattering Cross Section: As a model for the sky consider a gas of one-electron atoms in the ground state $|A\rangle$. Consider elastic (Rayleigh) scattering of photons by one of the atoms in the case that the photon energy is much smaller than the energy separation between $|A\rangle$ and any excited states $|I\rangle$. (Note: $\omega_i \approx \frac{1}{2}$ eV for visible light compared to

($E_I - E_A$)$_{\min}$ = 10.4 eV for hydrogen.) Show that in the long wavelength approximation the cross section is given by

$$\frac{d\sigma}{d\Omega} = m^2 r_0^2 \omega^4 \left| \sum_I \left[(\hat{\epsilon}_f^* \cdot \vec{r})_{AI} (\hat{\epsilon}_i \cdot \vec{r})_{IA} + (\hat{\epsilon}_i \cdot \vec{r})_{AI} (\hat{\epsilon}_f^* \cdot \vec{r})_{IA} \right] \frac{1}{\omega_{IA}} \right|^2 .$$

The strong increase in cross section with frequency ($\sim \omega^4$) gives the blue color of the sky and the red color of the sunset. Rayleigh was the first to understand this many years ago.

Hint: Start from the Kramers–Heisenberg formula. Take $|B\rangle = |A\rangle$ and $\omega_f = \omega_i = \omega$. Expand the energy denominators in powers of ω/ω_{IA}

$$-\frac{1}{E_A - E_I - \omega} = \frac{1}{\omega_{IA}} \left(1 + \frac{\omega}{\omega_{IA}} + \frac{\omega^2}{\omega_{IA}^2} + \cdots \right)$$

keeping terms up to second order. Show that the lowest order terms combine to give $-\hat{\epsilon}_f^* \cdot \hat{\epsilon}_i$ and exactly cancel the seagull term arising from $\vec{A} \cdot \vec{A}$. Get rid of the $\frac{1}{\omega_{IA}}$ by noting that

$$\langle \vec{p} \rangle_{AI} = -im\omega_{IA} \langle \vec{r} \rangle_{AI} , \qquad \langle \vec{p} \rangle_{IA} = im\omega_{IA} \langle \vec{r} \rangle_{IA} \qquad \text{Why?}$$

Use completeness to sum over states $|I\rangle$ and note the commutation role of operators $\vec{r}$ and $\vec{p}$. Show that the first order terms give zero (by repeating the previous trick). Finally, demonstrate that the second order terms combine to give the desired result, neglecting higher powers of ω.

ii) Polarization of Scattered Light: Study the case $\hat{k}_i = \hat{e}_z, \hat{k}_f = \hat{e}_y$. If $\hat{e}_i = \hat{e}_x$ show that
$$\text{Amp}(\hat{\epsilon}_f = \hat{e}_z) = 0 \qquad \text{Amp}(\hat{\epsilon}_f = \hat{e}_x) \neq 0 .$$
If $\hat{\epsilon}_i = \hat{e}_y$ show that there is no scattering in the y direction.

What can you say about the polarization of photons scattered in the y-direction if the incident photon beam is completely unpolarized? What is the relevance of this result to the effectiveness of polaroid sunglasses?

iii) Why isn't the sky violet, since even more violet light is scattered than blue?

IV.6 RESONANT SCATTERING

Suppose that the energy of the incident photon — ω_i — is nearly equal to the excitation energy for some excited state $|B\rangle$ of the atom

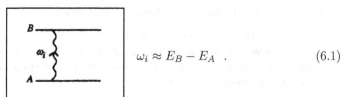

$$\omega_i \approx E_B - E_A . \qquad (6.1)$$

One term in the Kramers–Heisenberg amplitude then becomes extremely large — infinite in fact if $\omega_i = E_B - E_A$. This situation is called resonant scattering, in analogy to the situation wherein a classical harmonic oscillator is driven by an exciting force whose frequency is nearly equal to the natural frequency of the free oscillator. In the classical case the amplitude of the driven oscillator does not become infinite because there always exists some damping mechanism, converting the energy to another form. For example, in an LC circuit, there is always resistance which transforms electrical energy to heat. The classical differential equation

$$\ddot{x} + \gamma \dot{x} + \omega_0^2 x = a \cos \omega t \tag{6.2}$$

has the solution — valid for $\gamma t \gg 1$ —

$$x(t) = \frac{a}{\sqrt{(\omega_0^2 - \omega^2)^2 + \omega^2 \gamma^2}} \cos\left(\omega t - \tan^{-1} \frac{\omega \gamma}{\omega_0^2 - \omega^2}\right) . \tag{6.3}$$

Then "on resonance" (*i.e.*, when $\omega = \omega_o$)

$$x(t) = \frac{a}{\omega_o \gamma} \sin \omega t \tag{6.4}$$

the amplitude is 90° out of phase with the driving force and has magnitude $a/\omega_i \gamma$ so that $x(t)$ would indeed be infinite if there were no damping — $\gamma = 0$. For a realistic situation, however, γ is nonvanishing and results are finite. If we plot a typical amplitude of oscillation vs. the driving frequency one finds the shape shown in Figure IV.8, with the width of the peak being proportional to the damping constant γ.

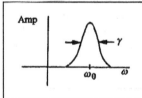

Fig. IV.8: *Oscillation amplitude vs. driving frequency for a classical oscillator.*

For our quantum mechanical problem of a photon incident upon an atom with frequency equal to the excitation energy of some excited state, there also exists a damping mechanism — the excited state $|B\rangle$ is coupled to the radiation field. That is, for the case that the atom is in state $|B\rangle$ with no photons present originally, we would expect on physical grounds that the amplitude to remain in this state — $|B\rangle|0\rangle$ — with no photons present should decay exponentially with time. The longer we wait the larger is the probability that the atom will make a transition out of state $|B\rangle$ into some lower state with emission of a photon. There exists "damping" therefore of the original amplitude, radiation damping, which we anticipate will have the form

$$\exp(-t/2\tau_B) \tag{6.5}$$

where τ_B is the lifetime of state $|B\rangle$. We shall see that the photon scattering cross section near a resonance shows a behavior very similar to the classical resonance curve with $\frac{1}{2\tau_B}$ playing the role of the damping coefficient γ. The phenomenon of resonances is widespread in physics, occurring in molecular, solid state, nuclear, elementary particle physics problems, *etc.* Thus it is worthwhile to examine this phenomenon in the context of photon-atom scattering in some detail.

Suppose the atom begins in the ground state $|A\rangle$, and consider elastic scattering. The resonance in the Kramers–Heisenberg formula is associated with the Feynman diagram shown below. The problem is that the propagator for the intermediate state: $|B\rangle|0\rangle$

$$\frac{1}{E_A^{(0)} + \omega_i - E_B^{(0)} + i\epsilon} \qquad (6.6)$$

becomes very large for incident photon energies ω_i such that

$$\omega_i \approx E_B^{(0)} - E_A^{(0)} \; . \qquad (6.7)$$

However, there exist higher order corrections to this propagator which we have thus far neglected. If we look only at the line connecting the initial and final vertices, then in addition to the simple free propagator Eq. 6.6 for the atom in state $|B\rangle$ with no photons present there also exist an infinite series of diagrams connecting the state $|B\rangle|0\rangle$ to itself but involving increasing orders of the fine structure constant α and the emission and absorption of various numbers of virtual photons, as shown in Figure IV.9.

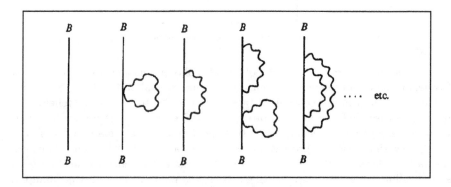

Fig. IV.9: Contributions to the "full" $|B\rangle|0\rangle$ propagator.

The meaning of these diagrams will be explained shortly.

IV.6 RESONANT SCATTERING

The $|B\rangle|0\rangle$ Propagator

It is useful at this point to return to our discussion of the Fourier transform of the full propagator for the state $|B\rangle|0\rangle$.[†] (For generality call the Fourier variable ω. later on, when we take the limit $t' \to \infty$, $t \to -\infty$ we shall replace $\omega \to E_i = E_A^{(0)} + \omega_i$.) We have then

$$-i\left\langle 0\left|\left\langle B|\hat{\mathcal{K}}(\omega)|B\right\rangle\right|0\right\rangle = \frac{1}{\omega - E_B^{(0)} + i\epsilon}$$
$$+ \frac{1}{\omega - E_B^{(0)} + i\epsilon}\left\langle 0\left|\left\langle B|\hat{V}|B\right\rangle\right|0\right\rangle \frac{1}{\omega - E_B^{(0)} + i\epsilon} \quad (6.8)$$
$$+ \frac{1}{\omega - E_B^{(0)} + i\epsilon}\left\langle 0\left|\left\langle B\left|\hat{V}\frac{1}{\omega - \hat{H}_0 + i\epsilon}\hat{V}\right|B\right\rangle\right|0\right\rangle \frac{1}{\omega - E_B^{(0)} + i\epsilon}$$
$$+ \ldots$$

which can be associated with the series of Feynman diagrams above. In order to make this connection, consider first the term in $\hat{V}$ linear in the vector potential

$$\hat{V}_1 = -\frac{e}{m}\vec{A}\cdot\vec{p} \quad (6.9)$$

which creates or destroys a single photon at a vertex. Obviously then

$$\left\langle 0\left|\left\langle B\left|-\frac{e}{m}\vec{A}\cdot\vec{p}\right|B\right\rangle\right|0\right\rangle = 0 \quad (6.10)$$

so that the second term in the propagator expansion—Eq. 6.8—vanishes. On the other hand, the third term is non-zero and corresponds to the Feynman diagram shown below. There are two vertices in this "bubble" diagram — the first

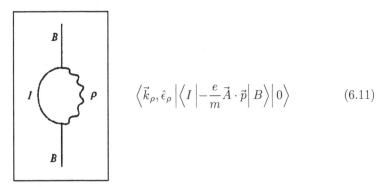

$$\left\langle \vec{k}_\rho, \hat{\epsilon}_\rho\left|\left\langle I\left|-\frac{e}{m}\vec{A}\cdot\vec{p}\right|B\right\rangle\right|0\right\rangle \quad (6.11)$$

involves a transition from $|B\rangle|0\rangle$ to an atom in state $|I\rangle$ and a single photon state $|\vec{k}_\rho, \hat{\epsilon}_\rho\rangle$ while the second

$$\left\langle 0\left|\left\langle B\left|-\frac{e}{m}\vec{A}\cdot\vec{p}\right|I\right\rangle\right|\vec{k}_\rho, \hat{\epsilon}_\rho\right\rangle \quad (6.12)$$

[†] By full propagator we mean the full amplitude to begin and end in the state $|B\rangle|0\rangle$.

involves the return to the state $|B\rangle|0\rangle$ via the absorption of the virtual photon in state $|\vec{k}_\rho, \hat{\epsilon}_\rho\rangle$. Propagation in the state $|I\rangle|\vec{k}_\rho, \hat{\epsilon}_\rho\rangle$ brings down the factor

$$\frac{1}{\omega - E_I^{(0)} - \omega_\rho + i\epsilon} \tag{6.13}$$

since

$$\hat{H}_0 |I\rangle|\vec{k}_\rho, \hat{\epsilon}_\rho\rangle = \left(E_I^{(0)} + \omega_\rho\right)|I\rangle|\vec{k}_\rho, \hat{\epsilon}_\rho\rangle \ . \tag{6.14}$$

We must, of course, sum over all atomic states $|I\rangle$ and all photon modes $|\vec{k}_\rho, \hat{\epsilon}_\rho\rangle$. In order to simplify notation we use

$$\begin{array}{ll} |I, \rho\rangle & \text{meaning} \quad |I\rangle|\vec{k}_\rho, \hat{\epsilon}_\rho\rangle \\ \sum_{I,\rho} & \text{meaning} \quad \sum_I \int \frac{d^3 k_\rho}{(2\pi)^3} \sum_{\lambda_\rho} \cdots \end{array} \tag{6.15}$$

Then define

$$\begin{aligned} f(\omega) &\equiv \left\langle B, 0 \left| \hat{V}_1 \frac{1}{\omega - \hat{H}_0 + i\epsilon} \hat{V}_1 \right| B, 0 \right\rangle \\ &= \sum_{I,\rho} \left\langle B, 0 | \hat{V}_1 | I, \rho \right\rangle \frac{1}{\omega - E_I^{(0)} - \omega_\rho + i\epsilon} \left\langle I, \rho | \hat{V}_1 | B, 0 \right\rangle \end{aligned} \tag{6.16}$$

Now suppose that we sum over a restricted (but infinite) set of Feynman diagrams, wherein between "bubbles" we return to the original state $|B\rangle|0\rangle$, as shown in Figure IV.10.

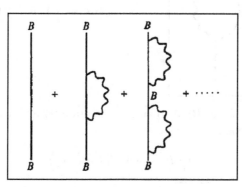

Fig. IV.10: *Geometric series of "bubble" diagrams.*

This is simply a geometric series and can be summed straightforwardly:

$$-i\left\langle B,0\left|\hat{\mathcal{K}}(\omega)\right|B,0\right\rangle_{\text{subset}} = \frac{1}{\omega-E_B^{(0)}+i\epsilon} + \frac{1}{\omega-E_B^{(0)}+i\epsilon}f(\omega)\frac{1}{\omega-E_B^{(0)}+i\epsilon}$$

$$+ \frac{1}{\omega-E_B^{(0)}+i\epsilon}f(\omega)\frac{1}{\omega-E_B^{(0)}+i\epsilon}f(\omega)\frac{1}{\omega-E_B^{(0)}+i\epsilon} + \ldots$$

$$= \frac{1}{\omega-E_B^{(0)}+i\epsilon}\left(1 + \frac{f(\omega)}{\omega-E_B^{(0)}+i\epsilon} + \left(\frac{f(\omega)}{\omega-E_B^{(0)}+i\epsilon}\right)^2 + \ldots\right)$$

$$= \frac{1}{\omega-E_B^{(0)}+i\epsilon} \times \frac{1}{1-\frac{f(\omega)}{\omega-E_B^{(0)}+i\epsilon}} = \frac{1}{\omega-E_B^{(0)}-f(\omega)+i\epsilon}$$

(6.17)

Note that $f(\omega)$ is of order e^2.

There is a second way of getting contributions of order e^2 — from the term $\hat{V}_2$ quadratic in the vector potential

$$\hat{V}_2 = \frac{e^2}{2m}\vec{A}\cdot\vec{A}.$$

(6.18)

Thus define

$$\Lambda \equiv \left\langle B,0\left|\frac{e^2}{2m}\vec{A}\cdot\vec{A}\right|B,0\right\rangle$$

(6.19)

$$= \frac{e^2}{2m}\sum_\rho \langle B,0|a_\rho a_\rho^\dagger|B,0\rangle \frac{1}{2\omega_\rho}$$

$$= \frac{e^2}{2m}\sum_\rho \frac{1}{2\omega_\rho}$$

which corresponds to a Feynman diagram wherein the virtual photon is created and absorbed at the same point. Observe that Λ is *independent* of the state $|B\rangle$ (and is infinite.) As discussed previously, this contribution may have gravitational consequences but is irrelevant as far as energy *differences* are concerned. We shall find that $f(\omega)$ is infinite also! However, $f(\omega)$ is *not* state-independent and cannot therefore be neglected. The procedure whereby such formally divergent quantities are turned into sensible physical results is called "renormalization" and will be discussed shortly. For the present, we shall proceed as if the mathematics made sense.

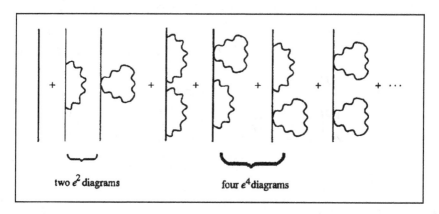

Fig. IV.11: *Full geometric series including all self-energy corrections to $\mathcal{O}(e^2)$.*

Consider an extended subset of diagrams involving both Λ and $f(\omega)$, repeated any number of times, with successive returns to the original state $|B\rangle|0\rangle$, as shown in Figure IV.11. A moment's thought shows that the corresponding contribution to the propagator is

$$-i\left\langle B,0\left|\hat{\mathcal{K}}(\omega)\right|B,0\right\rangle_{\text{subset}} = \frac{1}{\omega - E_B^{(0)} + i\epsilon} + \frac{1}{\omega - E_B^{(0)} + i\epsilon}\left(f(\omega) + \Lambda\right)\frac{1}{\omega - E_B^{(0)} + i\epsilon}$$

$$+\frac{1}{\omega - E_B^{(0)} + i\epsilon}\left(f(\omega) + \Lambda\right)\frac{1}{\omega - E_B^{(0)} + i\epsilon}\left(f(\omega) + \Lambda\right)\frac{1}{\omega - E_B^{(0)} + i\epsilon} + \ldots \quad (6.20)$$

This is a geometric series as in our previous discussion, except for the replacement $f(\omega) \to f(\omega) + \Lambda$. Hence

$$-i\left\langle B,0\left|\hat{\mathcal{K}}(\omega)\right|B,0\right\rangle_{\text{subset}} = \frac{1}{\omega - E_B^{(0)} - f(\omega) - \Lambda + i\epsilon}. \quad (6.21)$$

It is now obvious how to include other diagrams. For example, the diagram below gives

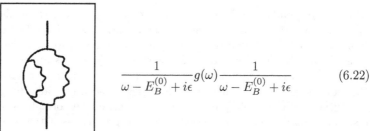

$$\frac{1}{\omega - E_B^{(0)} + i\epsilon}g(\omega)\frac{1}{\omega - E_B^{(0)} + i\epsilon} \quad (6.22)$$

where $g(\omega)$ is order e^4. If we allow $f(\omega)$, $g(\omega)$, Λ to be repeated any number of times, but always returning to $|B\rangle|0\rangle$ we evidently get, as a better approximation to the exact propagator

$$\frac{1}{\omega - E_B^{(0)} - f(\omega) - g(\omega) - \Lambda + i\epsilon} . \qquad (6.23)$$

This can go on, as we build up a series of terms in the denominator

$$f(\omega) + g(\omega) + h(\omega) + \ldots + \Lambda \qquad (6.24)$$

where the successive terms increase in complexity (*i.e.*, involve higher powers of e^2). Each term corresponds to a diagram which defines it

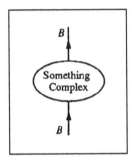

Fig. IV.12: *Cartoon representing a sum over a complete set of self-energy diagrams.*

The full result then is given by the "diagram" shown in Figure IV.12, where a B line leads into "something complicated" and a B line leads out. By construction of the series, "something complicated" cannot be split into two parts separated only by a B line. The technical name for such a diagram is a "proper self-energy graph." Following this reasoning then we see that the full propagator has the general form

$$-i\left\langle B,0 \left|\hat{\mathcal{K}}(\omega)\right| B,0 \right\rangle_{\text{all diagrams}} = \frac{1}{\omega - E_B^{(0)} - \Phi(\omega) + i\epsilon} \qquad (6.25)$$

where

$$\Phi(\omega) = \text{sum over all proper self-energy graphs} . \qquad (6.26)$$

However, to lowest order

$$\Phi_{e^2}(\omega) = f(\omega) + \Lambda \qquad (6.27)$$

since all other proper self-energy diagrams give contributions which are $\mathcal{O}(e^4)$ or higher.

Resonant Photon Atom Scattering

Now return to the problem at hand—elastic scattering—and consider only the resonant diagram, but replace the *free* propagator in state $|B,0\rangle$ by the *full* propagator we have just found. Keeping only $\Phi_{e^2}(\omega)$ in the diagram below, we have

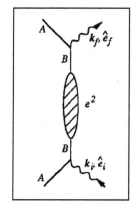

$$\mathcal{F}(\omega) = \frac{\langle A, \vec{k}_f \hat{e}_f | \hat{V} | B, 0 \rangle \langle B, 0 | \hat{V} | A, \vec{k}_i \hat{e}_i \rangle}{\omega - E_B^{(0)} - f(\omega) - \Lambda + i\epsilon} . \quad (6.28)$$

(Do not bother at this point to modify the propagators for the initial and final states $|A\rangle$). Now follow the usual procedure, taking the inverse Fourier transform, to get $\mathrm{Amp}(t',t)$. In the limit $t' \to \infty$, $t \to -\infty$ we find, as shown earlier

$$S_{fi} = \lim_{\substack{t' \to \infty \\ t \to -\infty}} \frac{i}{2\pi} \int_{-\infty}^{\infty} d\omega \frac{e^{i(E_f - \omega)t'}}{\omega - E_f + i\epsilon} \mathcal{F}(\omega) \frac{e^{i(\omega - E_i)t}}{\omega - E_i + i\epsilon} \quad (6.29)$$

$$= -2\pi i \delta(E_f - E_i) \mathcal{F}(E_i)$$

where $E_i = E_A^{(0)} + \omega_i$, $E_f = E_A^{(0)} + \omega_f$. The differential scattering cross section near resonance is then

$$\frac{d\sigma}{d\Omega} \cong \frac{r_0^2}{m^2} \left| \frac{\hat{e}_f^* \cdot \vec{p}_{AB} \hat{e}_i \cdot \vec{p}_{BA}}{\omega_i - (E_B^{(0)} - E_A^{(0)}) - f(E_i) - \Lambda + i\epsilon} \right|^2 \quad (6.30)$$

Now examine the quantity $f(E_i)$, which we write as

$$f(E_i) = \sum_{I,\rho} \frac{\langle B, 0 | \hat{V} | I, \rho \rangle \langle I, \rho | \hat{V} | B, 0 \rangle}{E_A^{(0)} + \omega_i - E_I^{(0)} - \omega_\rho + i\epsilon}$$

$$\approx \sum_{I,\rho} \frac{\langle B, 0 | \hat{V} | I, \rho \rangle \langle I, \rho | \hat{V} | B, 0 \rangle}{E_B^{(0)} - E_I^{(0)} - \omega_\rho + i\epsilon} \quad (6.31)$$

where we have replaced $E_A^{(0)} + \omega_i$ by $E_B^{(0)}$ since we are near resonance. Because the sum $\sum_\rho$ involves an integration over $d\omega_\rho$ we can write

$$f(E_i) \approx \sum_{I,\rho} \left| \langle I, \rho | \hat{V} | B, 0 \rangle \right|^2 \left(P \frac{1}{E_B^{(0)} - E_I^{(0)} - \omega_\rho} - i\pi \delta(E_B^{(0)} - E_I^{(0)} - \omega_\rho) \right) . \quad (6.32)$$

Thus we have

$$\text{Re} f(E_i) = P \sum_{I,\rho} \frac{\left|\langle I, \rho | \hat{V} | B, 0 \rangle\right|^2}{E_B^{(0)} - E_I^{(0)} - \omega_\rho}$$

$$\text{Im} f(E_i) = -\pi \sum_{I,\rho} \left|\langle I, \rho | \hat{V} | B, 0 \rangle\right|^2 \delta(E_B^{(0)} - E_I^{(0)} - \omega_\rho) \ . \tag{6.33}$$

We see that $\text{Re} f(E_i)$ has the form of an energy shift of the state $|B, 0\rangle$ due to second order effects of the interaction Hamiltonian $\hat{V}$, while the first order shift is precisely what we have previously called Λ: $\langle B, 0 | \hat{V} | B, 0 \rangle$. We write then

$$\Delta E_B = \text{Re} f(E_i) + \Lambda \tag{6.34}$$

with ΔE_B being the $\mathcal{O}(e^2)$ shift in energy of the state $|B, 0\rangle$ due to its coupling to the radiation field. It was a great triumph of theoretical physics to show how to be able to calculate these small shifts precisely, as we shall address shortly.

In order to interpret $\text{Im} f(E_i)$ we recall Fermi's golden rule—the transition rate of photon decay from the state $|B, 0\rangle$ is

$$\Gamma_B = \sum_{I,\rho} 2\pi \delta(E_B^{(0)} - E_I^{(0)} - \omega_\rho) \left|\langle I, \rho | \hat{V} | B, 0 \rangle\right|^2 \tag{6.35}$$

so that we identify

$$\text{Im} f(E_i) = -\frac{1}{2}\Gamma_B \ . \tag{6.36}$$

Finally, in order to present the formula for resonant scattering in its usual form, we suppress the ΔE_B which appears in the denominator as well as a shift ΔE_A which would have been present had we used the exact propagator for state $|A\rangle$ in the Feynman diagram. (Note that for the ground state the corresponding function $f(E_A)$ is purely real, giving only an energy shift. There exists no imaginary part since the ground state cannot radiate.) Thus we write

$$\begin{aligned} E_A^{(0)} + \Delta E_A &\equiv E_A^{\text{exp}} \\ E_B^{(0)} + \Delta E_B &\equiv E_B^{\text{exp}} \end{aligned} \tag{6.37}$$

respectively, meaning the experimental values for the energies of these states. We find then

$$\begin{aligned} \frac{d\sigma}{d\Omega} &\cong r_0^2 \frac{1}{m^2} \left| \frac{\hat{\epsilon}_f^* \cdot \vec{p}_{AB} \hat{\epsilon}_i \cdot \vec{p}_{BA}}{\omega_i - (E_B^{\text{exp}} - E_A^{\text{exp}}) + i\frac{1}{2}\Gamma_B} \right|^2 \\ &= \frac{r_0^2}{m^2} \frac{\left|\hat{\epsilon}_f^* \cdot \vec{p}_{AB} \hat{\epsilon}_i \cdot \vec{p}_{BA}\right|^2}{(\omega_i - \omega_{BA})^2 + \frac{1}{4}\Gamma_B^2} \end{aligned} \tag{6.38}$$

with $\omega_{BA} = E_B^{\text{exp}} - E_A^{\text{exp}}$ for the cross section near resonance. Note the Lorentzian energy dependence, with the width being equal to Γ_B, as expected.

[Note: We have implicitly assumed in the above discussion that only a single state $|B\rangle$ is included in the resonant scattering. If there exists a degeneracy or if two states are very close together in energy, then both states can participate in the resonant scattering process and interesting interference effects will arise. The coincidence in energy of the two states $|B_1\rangle, |B_2\rangle$ can often be brought about by varying some external parameter (*e.g.* a magnetic field) affecting the energy levels. The angular distribution of scattered resonant radiation will change drastically when the two states "cross" in energy and this can be used as a probe of their structure. The discussion of such level-crossing experiments uses the obvious generalization of the Kramers-Heisenberg formula

$$\frac{d\sigma}{d\Omega} = r_0^2 \left| \hat{\epsilon}_f^* \cdot \hat{\epsilon}_i + \frac{1}{m} \sum_I \left(\frac{\hat{\epsilon}_f^* \cdot \vec{p}_{AI} \hat{\epsilon}_i \cdot \vec{p}_{IA}}{\omega_i - (E_I^{\exp} - E_A^{\exp}) + \frac{i}{2}\Gamma_I} - \frac{\hat{\epsilon}_i \cdot \vec{p}_{AI} \hat{\epsilon}_f^* \cdot \vec{p}_{IA}}{\omega_f + E_I^{\exp} - E_A^{\exp}} \right) \right|^2 \tag{6.39}$$

which takes account of the resonant denominator in each virtual transition.]

IV.7 LINE SHAPE VIA FEYNMAN DIAGRAMS

We are now prepared to attack the line shape problem within the context of Feynman "diagrammar."

Exponential Decay

First we analyze the related problem of exponential decay. Note that the amplitude to begin in state $|B, 0\rangle$ at time $t = 0$ and remain in this state at time t is easily found in terms of the Fourier transform of the full propagator.

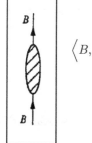

$$\begin{aligned} \langle B, 0 | e^{-i\hat{H}t} | B, 0 \rangle &= \int_{-\infty}^{\infty} \frac{d\omega}{2\pi} e^{-i\omega t} \langle B, 0 | \hat{\mathcal{K}}(\omega) | B, 0 \rangle \\ &= \int_{-\infty}^{\infty} \frac{d\omega}{2\pi} e^{-i\omega t} \frac{i}{\omega - E_B^{(0)} - \Lambda - f(\omega) + i\epsilon} \end{aligned} \tag{7.1}$$

Since the dominant values of ω are those near $\omega = E_B^{(0)}$ we can replace

$$\Lambda + f(\omega) \approx \Lambda + f(E_B^{(0)}) = \Delta E_B - i\frac{1}{2}\Gamma_B \ . \tag{7.2}$$

The integration is easily performed by closing the contour in the lower half plane and using the residue theorem:

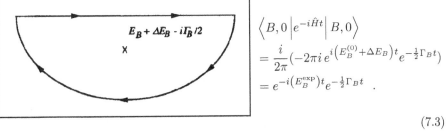

$$\langle B,0|e^{-i\hat{H}t}|B,0\rangle$$
$$= \frac{i}{2\pi}(-2\pi i\,e^{i\left(E_B^{(0)}+\Delta E_B\right)t}e^{-\frac{1}{2}\Gamma_B t})$$
$$= e^{-i(E_B^{\text{exp}})t}e^{-\frac{1}{2}\Gamma_B t} \quad .$$

(7.3)

The probability to remain in state $|B,0\rangle$

$$P_B(t) = \left|\langle B,0|e^{-i\hat{H}t}|B,0\rangle\right|^2 = e^{-\Gamma_B t} \equiv e^{-\frac{t}{\tau_B}} \tag{7.4}$$

thus decays with time exponentially, with mean life

$$\tau_B = \frac{1}{\Gamma_B} \quad . \tag{7.5}$$

The Line Shape

Now consider the case that $|B\rangle$ is the first atomic excited state, which can decay down to the ground state $|A\rangle$ as shown in Figure IV.13.

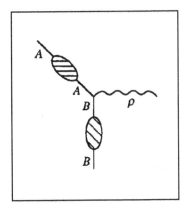

Fig. IV.13: Radiatively corrected photon decay diagram.

The time-dependent amplitude for this process is given by

$$\langle A,\rho|e^{-i\hat{H}t}|B,0\rangle \cong i\int_{-\infty}^{\infty}\frac{d\omega}{2\pi}e^{-i\omega t}\frac{1}{\omega - E_A^{(0)} - \omega_\rho - \Lambda - f_A(\omega) + i\epsilon}$$
$$\times \langle A,\rho|V|B,0\rangle \frac{1}{\omega - E_B^{(0)} - f_B(\omega) - \Lambda + i\epsilon} \quad .$$

(7.6)

The only new point is the function

$$f_A(\omega) = \sum_{I,\sigma} \frac{\langle A,\rho|\hat{V}|I,\rho,\sigma\rangle \langle I,\rho,\sigma|\hat{V}|A,\rho\rangle}{\omega - E_I^{(0)} - \omega_\rho - \omega_\sigma + i\epsilon} \quad . \tag{7.7}$$

Here we are using the "full" propagator for the state $|A,\rho\rangle$, where on the A-line we have virtual emission and reabsorption of photons with the photon ρ playing the role of a "spectator." Of course, $f_A(\omega)$ has only a real part — the imaginary part vanishes since the ground state is stable. From the form of the integral, we see that the dominant values of ω are those for which $\omega \approx E_A^{(0)} + \omega_\rho$. Hence

$$f_A(\omega) \approx \sum_{I,\sigma} \frac{\langle A,\rho|\hat{V}|I,\rho,\sigma\rangle \langle I,\rho,\sigma|\hat{V}|A,\rho\rangle}{E_A^{(0)} - E_I^{(0)} - \omega_\sigma + i\epsilon} = f_A(E_A + \omega_\rho) \tag{7.8}$$

so that[†]

$$\Lambda + f_A(\omega) \approx \Delta E_A \quad . \tag{7.9}$$

Finally then

$$\langle A,\rho|e^{-i\hat{H}t}|B,0\rangle = i \int_{-\infty}^{\infty} \frac{d\omega}{2\pi} e^{-i\omega t} \frac{1}{\omega - E_A^{(0)} - \Delta E_A - \omega_\rho + i\epsilon}$$
$$\times \langle A,\rho|\hat{V}|B,0\rangle \frac{1}{\omega - E_B^{(0)} - \Delta E_B + i\frac{1}{2}\Gamma_B} \tag{7.10}$$

which may be evaluated by closure in the lower half ω plane and use of the residue theorem:

$$\langle A,\rho|e^{-i\hat{H}t}|B,0\rangle = \frac{\langle A,\rho|\hat{V}|B,0\rangle}{\omega_\rho - \left(E_B^{(0)} + \Delta E_B - E_A^{(0)} - \Delta E_A\right) + i\frac{1}{2}\Gamma_B}$$
$$\times \left(e^{-i\left(E_A^{(0)}+\Delta E_A+\omega_\rho\right)t} - e^{-i\left(E_B^{(0)}+\Delta E_B\right)t} e^{-\frac{1}{2}\Gamma_B t}\right) \quad . \tag{7.11}$$

We see that both time-dependent amplitudes

$$\langle A,\rho|e^{-i\hat{H}t}|B,0\rangle \quad , \quad \langle B,0|e^{-i\hat{H}t}|B,0\rangle \tag{7.12}$$

agree with the corresponding expressions derived via the Wigner–Weisskopf procedure. However, the Feynman diagram method is far more straightforward and intuitive.

PROBLEM IV.7.1

[†] Note: $\langle I,\rho,\sigma|V|A,\rho\rangle \approx \langle I,\sigma|V|A,0\rangle$ except for the case that $\rho = \sigma$ but this is an isolated point in the σ integration.

Recoil Effects in Photon Emission

A radiating system, initially at rest, emits a photon with wave vector $\vec{k}$. We expect on physical grounds that the system (atom, molecule, nucleus, or whatever) will recoil following the transition, so that its final total momentum $\vec{P}_f$ is equal and opposite to the momentum $\vec{k}$ of the emitted photon. We wish to demonstrate that the quantum theory of photon emission automatically yields this recoil effect.

i) Consider two charged particles of masses m_1 and m_2, charges e and -e bound together by a potential $V(\vec{r})$. Present a careful discussion of the photon emission rate which goes far enough to demonstrate the recoil effect.

Hint: Write the wavefunction for the system as

$$\psi(\vec{r}_1, \vec{r}_2) = \exp i\vec{P}_i \cdot \vec{R}\psi(\vec{r})$$

where $\vec{r} = \vec{r}_1 - \vec{r}_2$ is the relative co-ordinate vector, $\vec{R} = (m_1\vec{r}_1 + m_2\vec{r}_2)/(m_1 + m_2)$ is the co-ordinate vector of the center of mass, $\vec{P}_i$ is the total momentum of the system, and $\psi(\vec{r})$ represents an eigenstate for the atomic motion of a fictitious particle of mass $\mu = m_1 m_2/(m_1 + m_2)$ in the presence of potential $V(\vec{r})$. During the emission process the wavefunction changes from some $\psi_B(\vec{r})$ to $\psi_A(\vec{r})$. Show that in the electric dipole approximation the differential transition rate can be written in the form

$$d\Gamma = \int \frac{d^3 P_f}{(2\pi)^3} \frac{d^3 k}{(2\pi)^3} (2\pi)^4 \delta^4(P_f + k - P_i) \frac{e^2 \omega_{AB}^2}{2\omega_k} \sum_\lambda |(\hat{\epsilon}_{k,\lambda}^* + \hat{\epsilon}_{k,\lambda}^* \cdot \vec{v}_i \hat{k}) \cdot \langle B|\vec{r}|A\rangle|^2$$

ii) Calculate the differential decay spectrum for photon emission as a function of the angle θ. Choose the z-axis along $\hat{P}_i$ and consider E1 emission from an initial state with $J_B = 1, m_B = 0$ to a final state with $J_A = 0, m_A = 0$. Show that

$$\frac{d\Gamma}{d\Omega} = \frac{\alpha}{2\pi} \omega_{AB}^3 \frac{(1 + \vec{v}_i \cdot \hat{k})^3}{1 - \vec{v}_i \cdot \hat{k}} \frac{1}{3} \sin^2\theta |\langle B||r||A\rangle|^2$$

$$\approx \frac{\alpha}{6\pi} \omega_{AB}^3 |\langle B||r||A\rangle|^2 \frac{\sin^2\theta}{(1 - v_i \cos\theta)^4}$$

iii) Show how the spectrum when $\vec{P}_i \neq 0$ can be obtained from that when $\vec{P}_i = 0$ by means of a simple co-ordinate transformation.

Hint: Using the Lorentz transformation demonstrate the relation

$$\cos\theta = \frac{\cos\theta^* + v_i}{1 + v_i \cos\theta^*}$$

between laboratory and center of mass (*) variables and therefore that

$$\sin^2\theta^* d\Omega^* \approx \frac{\sin^2\theta}{(1 - v_i \cos\theta)^4} d\Omega$$

PROBLEM IV.7.2

The Three Level System

Assume that levels A,B,C are all non-degenerate and undergo radiative transitions as shown in the diagram below. We expand $|\psi(t)\rangle$ in terms of its most important components:

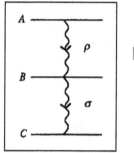

$$|\psi(t)\rangle = a(t)e^{-iE_A^{(0)}t}|A\rangle + \sum_\rho b_\rho(t)e^{-i(E_B^{(0)}+\omega_\rho)t}|B,\rho\rangle$$
$$+ \sum_{\rho\sigma} c_{\rho\sigma}(t)e^{-i(E_C^{(0)}+\omega_\rho+\omega_\sigma)t}|C,\rho,\sigma\rangle$$

with the boundary conditions at t=0: $a(0) = 1$; $b_\rho(0) = 0$; $c_{\rho\sigma}(0) = 0$ for all ρ, σ.

i) Using Feynman diagrams calculate $a(t)$, $b_\rho(t)$, and $c_{\rho\sigma}(t)$. Show that

$$a(t) = e^{-i\Delta E_A t}e^{-\frac{1}{2}\Gamma_A t} \qquad t > 0$$

$$|c_{\rho\sigma}(t)|^2 \underset{t\to\infty}{\sim} \frac{|\langle C,\rho,\sigma|\hat{V}|B,\rho\rangle\langle B,\rho|\hat{V}|A\rangle|^2}{\left[(\omega_\rho + \omega_\sigma - (E_A - E_C))^2 + \frac{1}{4}\Gamma_A^2\right]\left[(\omega_\sigma - (E_B - E_C))^2 + \frac{1}{4}\Gamma_B^2\right]}$$

where we have written E_i for $E_i^{(0)} + \Delta E_i$, with $i = A, B, C$.

ii) Give a physical explanation for the frequency correlation of the ρ, σ in the ρ, σ photon cascade.

iii) Find expressions for the frequency distribution of photon ρ alone (integrating over all σ) and of σ alone (integrating over all ρ) and interpret your results.

iv) Compare your results to those obtained via the Wigner-Weisskopf technique in Problem IV.4.1.

IV.8 THE LAMB SHIFT

Imagine attempting an actual calculation of the energy shift due to coupling of the radiation field

$$\Delta E_B = \Lambda + P \sum_{I,\rho} \frac{\left|\langle I,\rho|\hat{V}|B,0\rangle\right|^2}{E_B^{(0)} - E_I^{(0)} - \omega_\rho} . \qquad (8.1)$$

The term Λ, although badly divergent, is independent of state and hence may be neglected if only energy differences *between* states are considered. [This is equivalent to changing the reference point from which energies are measured. Recall that we previously dropped the zero point energy of the photon vacuum state for the same reason.]

Lamb Shift: Formal Derivation

We shall therefore confine our attention to the principle value integral and note first that the sum on directions and polarizations of the virtual photon ρ may easily be performed. Making the long wavelength approximation, as usual, we have

$$
\begin{aligned}
\Delta E_B &= \sum_I \int \frac{k^2 dk}{(2\pi)^3} d\Omega_{\hat{k}} \sum_\lambda \left(\frac{e}{m}\right)^2 \frac{\left|\hat{\epsilon}_{\vec{k},\lambda} \cdot \vec{p}_{IB}\right|^2}{E_B^{(0)} - E_I^{(0)} - k} \frac{1}{2k} \\
&= \sum_I \int \frac{k\, dk}{(2\pi)^3} d\Omega_{\hat{k}} \frac{e^2}{2m^2} \frac{\vec{p}_{BI} \cdot \vec{p}_{IB} - \vec{p}_{BI} \cdot \hat{k}\vec{p}_{IB} \cdot \hat{k}}{E_B^{(0)} - E_I^{(0)} - k} \\
&= \sum_I \int \frac{k\, dk}{(2\pi)^3} \frac{e^2}{2m^2} \cdot 4\pi \cdot \frac{2}{3} \frac{|\vec{p}_{BI}|^2}{E_B^{(0)} - E_I^{(0)} - k} \\
&= -\frac{2}{3\pi} \frac{\alpha}{m^2} \int_0^\infty dk\, k \sum_I \frac{|\vec{p}_{BI}|^2}{k - E_B^{(0)} + E_I^{(0)}} .
\end{aligned}
\tag{8.2}
$$

This is a divergent integral — $\Delta E_B = -\infty$. The divergence is linear in k — $\int_0^\infty dk$ for large k. [In the relativistic theory, using the Dirac equation and taking account of the existence of positrons as well as electrons, the divergence will be found to be logarithmic — $\int^\infty \frac{dk}{k}$ for large k]. This divergent contribution to the energy caused a good deal of consternation but was largely ignored until the experimental measurement of a small energy shift in the $2S_{\frac{1}{2}} - 2P_{\frac{1}{2}}$ states of hydrogen by Lamb and Retherford compelled attention to this problem [LaR 47, 50].

The basic suggestion for the removal of the divergence was made by Kramers, who noticed that for a free electron of momentum $\vec{p}$ there is also an infinite energy shift—ΔE_p—obtainable from the previous expression if we replace atomic states by plane waves. Since plane waves are eigenstates of the momentum operator, off-diagonal matrix elements vanish and the sum over I reduces to a single term with $\vec{p}_{BB}$ representing the free electron momentum $\vec{p}$. Then

$$
\Delta E_p = -\frac{2}{3\pi} \frac{\alpha}{m^2} \vec{p}^{\,2} \int_0^\infty dk .
\tag{8.3}
$$

The theory is, of course, incorrect for very large photon momenta. However, we may suppose that due to additional effects the integral over k has an effective cutoff $K \sim m$, since for $k \gg K$ e.g. relativistic effects become important. Then

$$
\frac{\Delta E_p}{E_p} \approx -\frac{4}{3\pi}\alpha \ll 1
\tag{8.4}
$$

so that this represents only a tiny correction to the electron kinetic energy.

Since $\Delta E_p \propto \vec{p}^{\,2}$ we may consider the energy shift to be associated with a change δm in the electron rest mass

$$
\Delta E_p = \frac{\vec{p}^{\,2}}{2(m + \delta m)} - \frac{\vec{p}^{\,2}}{2m} \approx -\frac{\vec{p}^{\,2}}{2m^2}\delta m
\tag{8.5}
$$

due to the interaction with the radiation field. We thus identify

$$\delta m = \frac{4\alpha}{3\pi} \int_0^K dk \ . \tag{8.6}$$

If this suggestion is correct, then the quantity m which, up to now, we have employed in the Hamiltonian is *not* the *experimental* electron mass, but is rather a *fictitious* mass which the electron would possess if somehow interaction with the radiation field could be turned off. The physical — experimental — electron mass is given by

$$m_{\text{exp}} = m + \delta m \tag{8.7}$$

which suggests that we should rewrite our Hamiltonian in terms of this measurable quantity

$$\begin{aligned}
\hat{H} &= \frac{\vec{p}^2}{2m} + e\phi(\vec{r}) + \hat{H}_{\text{RAD}}^{(r)} + \hat{V} \\
&= \frac{\vec{p}^2}{2m_{\text{exp}}} + e\phi(\vec{r}) + \hat{H}_{\text{RAD}}^{(r)} + \hat{V} + \left(\frac{\vec{p}^2}{2m} - \frac{\vec{p}^2}{2(m+\delta m)} \right) \\
&\equiv \hat{H}_0' + \hat{V}' \ .
\end{aligned} \tag{8.8}$$

Here $\hat{H}_0'$ is simply the usual Hamiltonian $\hat{H}_0$ but with m_{exp} substituted for the electron mass. However, $\hat{V}'$ now consists of two pieces

$$\hat{V}' = \hat{V} + \frac{\vec{p}^2}{2m^2}\delta m \equiv \hat{V}_1 + \hat{V}_2 \tag{8.9}$$

and *both* must be included in the energy shift calculation, as first done by Bethe [Be 47]. From $\hat{V}_1$ we find as before

$$\Delta E_1 = -\frac{2}{3\pi}\frac{\alpha}{m^2} \int_0^K dk\, k \sum_I \frac{\langle B|\vec{p}|I\rangle \cdot \langle I|\vec{p}|B\rangle}{k + E_I^{(0)} - E_B^{(0)}} \tag{8.10}$$

while for the piece $\hat{V}_2$

$$\Delta E_2 = \frac{1}{2m^2}\delta m \langle B|\vec{p}\cdot\vec{p}|B\rangle$$

(using completeness of atomic states)

$$\begin{aligned}
&= \frac{4\alpha}{3\pi} \int_0^K dk \frac{1}{2m^2} \sum_I \langle B|\vec{p}|I\rangle \cdot \langle I|\vec{p}|B\rangle \\
&= \frac{2\alpha}{3\pi m^2} \int_0^K dk \sum_I \left(k + E_I^{(0)} - E_B^{(0)}\right) \frac{\langle B|\vec{p}|I\rangle \cdot \langle I|\vec{p}|B\rangle}{k + E_I^{(0)} - E_B^{(0)}} \ .
\end{aligned} \tag{8.11}$$

Adding the two contributions, we find

$$\begin{aligned}
\Delta E_{\text{tot}} &= \Delta E_1 + \Delta E_2 \\
&= \frac{2\alpha}{3\pi m^2} \int_0^K dk \sum_I \frac{\left(E_I^{(0)} - E_B^{(0)}\right) \langle B|\vec{p}|I\rangle \cdot \langle I|\vec{p}|B\rangle}{k + E_I^{(0)} - E_B^{(0)}} \ .
\end{aligned} \tag{8.12}$$

The integral now diverges only logarithmically for large $k - \int \frac{dk}{k}$ — and we may be encouraged to hope that in a proper relativistic treatment there will exist an effective cutoff for wavenumbers larger than some $K \approx m$. [This does in fact occur and the corresponding integral in the relativistic theory is finite, as we shall show in Chapter VIII.]

In evaluating the integral recall that we are to take the principle value. Then

$$P \int_0^K \frac{dk}{k-a} \begin{cases} = \int_0^{a-\epsilon} + \int_{a+\epsilon}^K \frac{dk}{k-a} \\ = \ln(k-a)\Big|_0^{a-\epsilon} + \Big|_{a+\epsilon}^K = \ln\frac{\epsilon}{a} + \ln\frac{K-a}{\epsilon} = \ln\frac{K-a}{a} & a > 0 \\ \\ \int_0^K \frac{dk}{k+|a|} = \ln(k+|a|)\Big|_0^K = \ln\frac{K+|a|}{|a|} & a < 0 \end{cases}$$

Hence in general

$$P \int_0^K \frac{dk}{k-a} \simeq \ln\frac{K}{|a|} \qquad (8.13)$$

so that

$$\Delta E = \frac{2\alpha}{3\pi m^2} \sum_I |\langle I|\vec{p}|B\rangle|^2 \left(E_I^{(0)} - E_B^{(0)}\right) \ln \frac{K}{\left|E_I^{(0)} - E_B^{(0)}\right|} \qquad (8.14)$$

Since the logarithm is slowly varying with energy, it makes sense to define

$$\sum_I \left(E_I^{(0)} - E_B^{(0)}\right) |\langle I|\vec{p}|B\rangle|^2 \ln\left|E_I^{(0)} - E_B^{(0)}\right|$$
$$\equiv \ln|E - E_B|_{\text{AV}} \sum_I \left(E_I^{(0)} - E_B^{(0)}\right) |\langle I|\vec{p}|B\rangle|^2 \qquad (8.15)$$

(The average value of the logarithm must be obtained numerically by performing the indicated exact summation on the left-hand side of the equation.)

Now use the completeness property to write

$$\sum_I \left(E_I^{(0)} - E_B^{(0)}\right) \langle B|\vec{p}|I\rangle \cdot \langle I|\vec{p}|B\rangle = \sum_I \langle B|\vec{p}|I\rangle \cdot \left\langle I\left|\left[\hat{H}_{\text{ATOMIC}}, \vec{p}\right]\right|B\right\rangle$$
$$= \left\langle B\left|\vec{p}\cdot\left[\hat{H}_{\text{ATOMIC}}, \vec{p}\right]\right|B\right\rangle . \qquad (8.16)$$

Here

$$\hat{H}_{\text{ATOMIC}} = \frac{1}{2m}\vec{p}^{\,2} - \frac{\alpha}{r} \qquad (8.17)$$

is the atomic Hamiltonian. Then

$$\left[\hat{H}_{\text{ATOMIC}}, \vec{p}\right] = -i\vec{\nabla}\frac{\alpha}{r} \qquad (8.18)$$

and

$$\left\langle B \left| \vec{p} \cdot \left[\hat{H}_{\text{ATOMIC}}, \vec{p} \right] \right| B \right\rangle = -\int d^3r \, \psi_B^*(r) \vec{\nabla} \cdot \left(\psi_B(r) \vec{\nabla} \frac{\alpha}{r} \right) \; . \tag{8.19}$$

Since the integral is a real number, we can write

$$\left\langle B \left| \vec{p} \cdot \left[\hat{H}_{\text{ATOMIC}}, \vec{p} \right] \right| B \right\rangle = -\frac{1}{2} \int d^3r \, \psi_B^*(r) \vec{\nabla} \cdot \left(\psi_B(r) \vec{\nabla} \frac{\alpha}{r} \right)$$
$$- \frac{1}{2} \int d^3r \, \psi_B(r) \vec{\nabla} \cdot \left(\psi_B^*(r) \vec{\nabla} \frac{\alpha}{r} \right)$$

(integrating by parts)

$$= -\frac{1}{2} \int d^3r \, \psi_B^*(r) \left[\vec{\nabla} \cdot \left(\psi_B(r) \vec{\nabla} \frac{\alpha}{r} \right) - \vec{\nabla} \psi_B(r) \cdot \vec{\nabla} \frac{\alpha}{r} \right]$$
$$= -\frac{\alpha}{2} \int d^3r \, |\psi_B(r)|^2 \vec{\nabla}^2 \left(\frac{1}{r} \right) \; . \tag{8.20}$$

However, we know that

$$\vec{\nabla}^2 \frac{1}{r} = -4\pi \delta^3(r) \tag{8.21}$$

so

$$\left\langle B \left| \vec{p} \cdot \left[\hat{H}_{\text{ATOMIC}}, \vec{p} \right] \right| B \right\rangle = 2\pi \alpha |\psi_B(0)|^2 \; . \tag{8.22}$$

Since the result is proportional to the value of the atomic wavefunction at the origin so that only S-waves are shifted. Using

$$\left| \psi_{nS_{\frac{1}{2}}}(0) \right|^2 = \frac{1}{\pi n^3 a_0^3} \tag{8.23}$$

where

$$a_0 = \frac{1}{\alpha m} \tag{8.24}$$

is the Bohr radius, we find

$$\Delta E = \frac{4}{3\pi} \alpha^5 m \ln \frac{K}{\left| E - E_B^{(0)} \right|_{\text{AV}}} \times \begin{cases} \frac{1}{n^3} & \text{for } S\text{-states} \\ 0 & \text{otherwise} \end{cases} \; . \tag{8.25}$$

For the $2S_{\frac{1}{2}}$ state of hydrogen $\left| E - E_B^{(0)} \right|_{\text{av}}$ is calculated by Bethe, Brown and Stehn [BeBS 50] as

$$\left| E - E_B^{(0)} \right|_{\text{AV}} = 16.6 \, \text{Rydberg} = 8.3 \alpha^2 m \; . \tag{8.26}$$

In the sum over intermediate states, the contributions from the continuum are most important, giving the surprisingly large value for $\left| E - E_B^{(0)} \right|_{\text{AV}}$.

As we shall see later, the Dirac theory of the hydrogen atom predicts an exact degeneracy for the $2S_{1/2}$ and $2P_{1/2}$ states of hydrogen. However, because of the coupling to the radiation field, we expect the degeneracy to be lifted, with

$$E_{2S_{1/2}} - E_{2P_{1/2}} = -\frac{4}{3\pi}\alpha^5 m \ln(8.3\alpha^2) \cdot \frac{1}{8} \approx 1048\,\text{MHz} \; . \tag{8.27}$$

This splitting was first measured by Lamb and Retherford [LaR 47,LaR 50], who found

$$E_{2S_{1/2}} - E_{2P_{1/2}} = 1057.8 \pm 0.1\,\text{MHz} \; . \tag{8.28}$$

The measurement preceded the theoretical calculations and stimulated the remarkable theoretical developments which followed. Bethe's calculation, outlined above, was the first of these and, while incomplete, was clearly on the right track. More elaborate relativistically correct calculations of the Lamb shift by Feynman, by Schwinger and by Tomanaga found a result in precise agreement with the measured value, winning a Nobel prize for these theorists.

Lamb Shift: Intuitive Derivation

The derivation of the Lamb shift which has just been presented, while correct, is somewhat lacking in physical intuition. A more intuitive approach is provided by T. A. Welton [We 48], who looks at the Lamb shift as arising from the interaction of a bound electron with the zero point fluctuations of the electromagnetic field.

Consider an electron in some state of the hydrogen atom $\psi_n(\vec{r})$ with no photons present. That is, the state vector is

$$|\psi_n\rangle|0\rangle \; . \tag{8.29}$$

The meaning of $|0\rangle$ is that the radiation field is in its ground state — i.e., each of the equivalent radiation oscillators is in its ground state. As we have previously argued, in this configuration the expectation value of the electric and magnetic fields must vanish

$$\left\langle 0\left|\vec{E}\right|0\right\rangle = \left\langle 0\left|\vec{B}\right|0\right\rangle = 0 \tag{8.30}$$

at each point in spacetime since the vector potential $\vec{A}$ either creates or destroys photons and therefore has a vanishing vacuum expectation value

$$\left\langle 0\left|\vec{A}\right|0\right\rangle = 0 \; . \tag{8.31}$$

On the other hand, the field energy is *non*-zero, namely

$$\left\langle 0\left|\frac{1}{2}\left(\vec{E}^2 + \vec{B}^2\right)\right|0\right\rangle = \sum_{\text{modes}} \frac{1}{2}\omega_k = \int \frac{d^3k}{(2\pi)^3}\omega_k$$
$$= \frac{1}{2\pi^2}\int_0^\infty d\omega_k\, \omega_k^3 \; . \tag{8.32}$$

Since for the radiation field

$$\left\langle 0|\vec{E}^2|0\right\rangle = \left\langle 0|\vec{B}^2|0\right\rangle \tag{8.33}$$

we have
$$\langle 0|\vec{E}^2|0\rangle = \frac{1}{2\pi^2}\int_0^\infty d\omega_k\, \omega_k^3 \ . \tag{8.34}$$

There exist therefore fluctuating field strengths in empty space and such fields will have an effect on systems containing charged particles. In particular, consider an electron bound in a hydrogen atom. When subjected to a fluctuating electric field of frequency larger than the corresponding frequency of its bound state motion (speaking now somewhat loosely and classically) the electron will follow the field and fluctuate in its position. This in turn will affect the potential energy of interaction with the proton. (The position of the proton will also fluctuate but these fluctuations can be neglected since the proton mass is so large.) The kinetic energy of the electron will also be affected by this fluctuating motion, but we shall ignore this on the grounds that it will be similar for all states and can be neglected in taking differences.

For the high frequency components of the field fluctuations we shall ignore the binding energy and will treat the electron motion classically. Let $\Delta\vec{r}$ be the displacement of the electron from its equilibrium "orbit" in response to the rapidly fluctuating electric field $\vec{E}(t)$. Then

$$m\frac{d^2}{dt^2}\Delta\vec{r} = e\vec{E} \ . \tag{8.35}$$

Now perform a Fourier analysis. Consider the x-component of the motion and write

$$\Delta x(t) = \int_{-\infty}^\infty \frac{d\omega}{2\pi} e^{-i\omega t}\Delta x(\omega) \tag{8.36}$$

where

$$\Delta x(\omega) = \Delta x^*(-\omega) \tag{8.37}$$

since Δx is real. Also

$$E_x(t) = \int_{-\infty}^\infty \frac{d\omega}{2\pi} e^{-i\omega t} E_x(\omega) \tag{8.38}$$

with

$$E_x(\omega) = E_x^*(-\omega) \tag{8.39}$$

since $E_x(t)$ is real. From the equation of motion, Eq. 8.35, we have

$$\Delta x(\omega) = -\frac{e}{m\omega^2}E_x(\omega) \ . \tag{8.40}$$

Consider the average value of $(\Delta x)^2$ over some long time interval T

$$\begin{aligned}\langle (\Delta x)^2\rangle &= \frac{1}{T}\int_{-\frac{T}{2}}^{\frac{T}{2}} dt\, \Delta x(t)\Delta x^*(t) \\ &= \frac{1}{T}\int_{-\frac{T}{2}}^{\frac{T}{2}} dt \int_{-\infty}^\infty \frac{d\omega}{2\pi} e^{-i\omega t}\Delta x(\omega)\int_{-\infty}^\infty \frac{d\omega'}{2\pi} e^{i\omega' t}\Delta x^*(\omega') \\ &\simeq \frac{1}{T}\int_{-\infty}^\infty \frac{d\omega}{2\pi}\int_{-\infty}^\infty \frac{d\omega'}{2\pi} 2\pi\delta(\omega'-\omega)\Delta x(\omega)\Delta x^*(\omega') \\ &= \frac{1}{T}\int_{-\infty}^\infty \frac{d\omega}{2\pi}\Delta x(\omega)\Delta x^*(\omega) \ ,\end{aligned} \tag{8.41}$$

i.e.,

$$\left\langle (\Delta x)^2 \right\rangle = \frac{1}{T} \frac{e^2}{m^2} \int_{-\infty}^{\infty} \frac{d\omega}{2\pi} \frac{1}{\omega^4} E_x(\omega) E_x^*(\omega) \ . \tag{8.42}$$

However,

$$\begin{aligned}
\left\langle E_x^2 \right\rangle &= \frac{1}{3} \left\langle E^2 \right\rangle = \frac{1}{T} \int_{-\infty}^{\infty} \frac{d\omega}{2\pi} E_x(\omega) E_x^*(\omega) \\
&= \frac{2}{T} \int_0^{\infty} \frac{d\omega}{2\pi} E_x(\omega) E_x^*(\omega) = \frac{1}{6\pi^2} \int_0^{\infty} d\omega \, \omega^3
\end{aligned} \tag{8.43}$$

or

$$\frac{1}{T} E_x^*(\omega) E_x(\omega) = \frac{1}{6\pi} \omega^3 \ . \tag{8.44}$$

Thus we find

$$\left\langle (\Delta \vec{r})^2 \right\rangle = 3 \left\langle (\Delta x)^2 \right\rangle = \frac{2\alpha}{\pi m^2} \int_0^{\infty} \frac{d\omega}{\omega} \ . \tag{8.45}$$

The frequency integral diverges logarithmically at both limits. We have argued, however, that the electron cannot respond to the fluctuating field if the frequency is much less than the atomic binding frequency. Thus take

$$\omega_{\min} \sim E_{\text{Rydberg}} = \frac{\alpha^2 m}{2} \ . \tag{8.46}$$

At the high frequency end we should also expect an effective cutoff at

$$\omega_{\max} \sim m \tag{8.47}$$

since we have neglected relativistic effects. We thus have

$$\left\langle (\Delta \vec{r})^2 \right\rangle = \frac{2\alpha}{\pi m^2} \ln \frac{\omega_{\max}}{\omega_{\min}} \ . \tag{8.48}$$

In order finally to compute the effect of these fluctuations on the potential energy, we write

$$V(\vec{r}(t) + \Delta \vec{r}(t)) = V(\vec{r}) + \Delta \vec{r} \cdot \vec{\nabla} V(\vec{r}) + \frac{1}{2} \left(\Delta \vec{r} \cdot \vec{\nabla} \right)^2 V(\vec{r}) + \ldots \tag{8.49}$$

where $\vec{r}(t)$ is the (smooth) classical trajectory and

$$V(r) = -\frac{\alpha}{r} \tag{8.50}$$

is the Coulomb potential energy. Since

$$\langle \Delta \vec{r} \rangle = 0 \tag{8.51}$$

the second term in the expansion of the potential drops out. For the third, when we take the angular average we have

$$\int \frac{d\Omega}{4\pi} \Delta r_i \, \Delta r_j \, (\nabla_i \nabla_j) = \frac{1}{3} (\Delta \vec{r})^2 \, \vec{\nabla}^2 \ . \tag{8.52}$$

Then
$$\Delta V = \langle V(\vec{r}+\Delta\vec{r}) - V(\vec{r})\rangle = \frac{1}{6}\langle(\Delta\vec{r})^2\rangle \vec{\nabla}^2 V \qquad (8.53)$$
$$= 4\pi\delta^3(\vec{r})\alpha\frac{1}{6}\langle(\Delta\vec{r})^2\rangle \ .$$

This may be regarded as a perturbation in the Hamiltonian for the atom. For a state with wave function $\psi(\vec{r})$ this leads to an energy shift

$$\begin{aligned}\Delta E &= \frac{4\pi\alpha}{6}\langle(\Delta\vec{r})^2\rangle \int d^3r\,\psi(\vec{r})\,\delta^3(\vec{r})\,\psi(\vec{r}) \\ &= \frac{4\pi\alpha}{6}\langle(\Delta\vec{r})^2\rangle |\psi(0)|^2 \\ &= \frac{4\alpha^2}{3m^2}|\psi(0)|^2 \ln\frac{\omega_{\max}}{\omega_{\min}} \\ &= \frac{4\alpha^5}{3\pi}m\left\{\begin{array}{cc}\frac{1}{n^3} & \text{for } S\text{-states} \\ 0 & \text{otherwise}\end{array}\right\}\ln\frac{\omega_{\max}}{\omega_{\min}}\end{aligned} \qquad (8.54)$$

which is very similar to Bethe's result. For a numerical estimate, we have

$$\begin{aligned}\Delta E_{2S_{\frac{1}{2}}-2P_{\frac{1}{2}}} &= m\frac{\alpha^5}{6\pi}\ln\frac{2}{\alpha^2} = \frac{\alpha^3}{3\pi}\left(\frac{\alpha}{2a_0}\right)\ln\frac{2}{\alpha^2} \\ &= \frac{\alpha^3}{3\pi} \times 13.6\,\text{eV} \times 11.9 \approx 1.6 \times 10^9\,\text{Hz}\end{aligned} \qquad (8.55)$$

which is the correct order of magnitude and sign!

IV.9 DISPERSION RELATIONS

The arguments presented in Section II.2 concerning the interaction of an incident plane wave on a thin slab of material can easily be extended to the case of photons.

The Index of Refraction

Assume for simplicity that the slab shown below is thin compared to the mean free path for interaction.[†] Then we showed in section II.2 that the amplitude after passing through a thickness L is material is given by

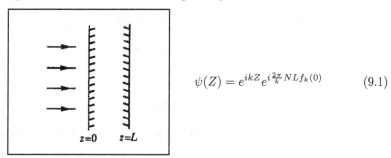

$$\psi(Z) = e^{ikZ}e^{i\frac{2\pi}{k}NLf_k(0)} \qquad (9.1)$$

[†] This is generally the case if the slab is a dilute gas, so that multiple scattering can be neglected.

where N is the number density of target particles and $f_k(0)$ is the forward scattering amplitude. This result is usually written in terms of the index of refraction of the medium—$n(\omega)$

$$\psi(Z) = e^{ik(Z-L)}e^{in(\omega)kL} \qquad (9.2)$$

whereupon we identify

$$n(\omega) = 1 + \frac{2\pi}{k^2}Nf_k(0) \qquad (9.3)$$

Here $f_k(0)$ is the amplitude for forward scattering with no change in polarization and no change in energy

$$f_k(0) = -r_o\left[1 + \frac{1}{m}\sum_I\left(\frac{|\hat{\epsilon}\cdot\vec{p}_{IA}|^2}{\omega - (E_I - E_A) + i\frac{\Gamma_I}{2}} - \frac{|\hat{\epsilon}\cdot\vec{p}_{IA}|^2}{\omega + (E_I - E_A)}\right)\right] \qquad (9.4)$$

where we have written $\hat{\epsilon}_i = \hat{\epsilon}_f = \hat{\epsilon}$ and $|A\rangle$ is the ground state.

Now separate the index into real and imaginary parts with

$$\operatorname{Re} n(\omega) = 1 - \frac{2\pi}{k^2}Nr_o\left[1 + \frac{1}{m}\sum_I|\hat{\epsilon}\cdot\vec{p}_{IA}|^2\left(\frac{\omega - \omega_{IA}}{(\omega - \omega_{IA})^2 + \frac{1}{4}\Gamma_I^2} - \frac{1}{\omega + \omega_{IA}}\right)\right]$$

$$\operatorname{Im} n(\omega) = \frac{\pi}{k^2}Nr_o\frac{1}{m}\sum_I|\hat{\epsilon}\cdot\vec{p}_{IA}|^2\frac{\Gamma_I}{(\omega - \omega_{IA})^2 + \frac{1}{4}\Gamma_I^2} \qquad (9.5)$$

where $\omega_{IA} = E_I - E_A$. Since resonances are in general very narrow, then except in the vicinity of a resonance we may approximate

$$\operatorname{Re} n(\omega) \approx 1 - \frac{2\pi}{k^2}Nr_o\left[1 + \frac{2}{m}\sum_I|\hat{\epsilon}\cdot\vec{p}_{IA}|^2\frac{\omega_{IA}}{\omega^2 - \omega_{IA}^2}\right]$$

$$\operatorname{Im} n(\omega) \approx 0 \ . \qquad (9.6)$$

Notice that for large ω — ω much greater than any of the important resonant frequencies ω_{IA} — $\operatorname{Re} n(\omega) \to 1$ approaching unity from below. Thus for X-rays the index of refraction is slightly less than one so that the phase velocity

$$v_{ph} = \frac{\omega}{nk} = \frac{c}{n} \qquad (9.7)$$

actually exceeds the speed of light! (Of course, the group velocity

$$v_{gr} = \frac{d\omega}{dk} \qquad (9.8)$$

remains less than c.)

On the other hand, at optical frequencies, as ω approaches a resonant frequency ω_{IA} from below $\operatorname{Re} n(\omega)$ is an increasing function of ω — $\frac{d}{d\omega}\operatorname{Re} n(\omega) > 0$. This is "normal dispersion." A typical plot of the refractive index is then as shown in Figure IV.14.

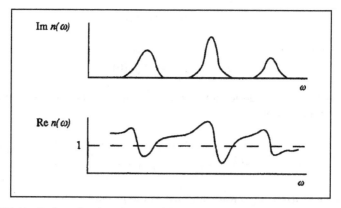

Fig. IV.14: Index of refraction in the resonance region.

In the immediate vicinity of a resonance there is a narrow frequency range over which $\operatorname{Re} n(\omega)$ has negative slope, which is "anomalous dispersion."

Of course, $\operatorname{Im} n(\omega)$ is related to the attenuation factor for light transmitted through the slab and thus to the total photon interaction cross section σ. This relation is just the optical theorem

$$\operatorname{Im} n(\omega) = \frac{N\sigma}{2k} \ . \tag{9.9}$$

From the graph above we see that $\operatorname{Im} n(\omega)$ and hence σ is significantly different from zero only in the immediate neighborhood of a resonance.

As we shall see, there exists an intimate connection between $\operatorname{Re} n(\omega)$ and $\operatorname{Im} n(\omega)$. This is well-known in classical electromagnetic theory, where one has

$$\begin{aligned}\operatorname{Re} n(\omega) &= 1 + \frac{1}{\pi} P \int_0^\infty d\omega' \frac{2\omega' \operatorname{Im} n(\omega')}{\omega'^2 - \omega^2} \\ &= 1 + \frac{1}{\pi} P \int_0^\infty d\omega' \operatorname{Im} n(\omega) \left(\frac{1}{\omega' - \omega} + \frac{1}{\omega' + \omega} \right) \ .\end{aligned} \tag{9.10}$$

[See, e.g. the discussion in Feynman, Vol. II, Sec. 32-1 for the classical expression for $n(\omega)$ using as a model for the atom an electron bound in a system which has many characteristic modes of oscillation with resonant frequencies ω. One finds an expression for $n(\omega)$ remarkably similar to the quantum mechanical result given above.] The tie implied by Eq. 9.10 between $\operatorname{Re} n(\omega)$ and $\operatorname{Im} n(\omega)$ is called a dispersion relation and follows from the property of $n(\omega)$ as a function of the complex variable ω. It is a consequence of the fact that $n(\omega)$ is an analytic function of ω with no singularities located in the *upper* half of the complex ω plane.

Dispersion Relations: Physical Basis

The physical basis of dispersion integrals was realized by Kramers and Kronig who deduced these relations from the requirement that signals in an optical medium should not exceed the speed of light. Similar connections are known in other branches of physics. In electrical circuit theory, for example, the output voltage across the terminals of an arbitrary linear network can differ from zero only *after* the incident signal has been applied at the input terminals. For signals at frequency ω this implies that the complex impedance as a function of ω is analytic in the upper half of the complex ω plane and hence that its real and imaginary parts are connected in a characteristic fashion. Similarly in magnetic phenomena, the requirement that we find no induced magnetization before an external magnetic field is applied implies a corresponding requirement on the analytic structure of the magnetic susceptibility $\chi(\omega)$. Also in elementary particle physics, the condition that no scattering can occur before the incident beam has struck the target puts restrictions on the analytic form of the scattering amplitude.

All of these relations between real and imaginary components of complex response functions are known as dispersion relations because their first application was to optical phenomena and that is the context in which we shall continue our discussion. A very nice article in this regard is that of R. Hagedorn [Sh 66] wherein he imagined a conversation between an inventor and a physicist. The inventor has a "great idea". Imagine that in a darkened room a flash bulb goes off at time $t = 0$. The light signal is then described by a delta function in time — $\delta(t)$ — or in frequency space in terms of a constant response function $f(\omega)$

$$g(t) = \delta(t) = \int \frac{d\omega}{2\pi} e^{-i\omega t} f(\omega) \Rightarrow f(\omega) = 1 \ . \tag{9.11}$$

The inventor imagines wearing a pair of sunglasses which restrict the frequency spectrum, e.g.

$$f(\omega) = \begin{cases} 1 & \text{if } |\omega| < \omega_0 \\ 0 & \text{otherwise} \end{cases} \tag{9.12}$$

Then the light signal that will be seen through the sunglasses is

$$f(t) = \int_{-\omega_0}^{\omega_0} \frac{d\omega}{2\pi} e^{-i\omega t} = \frac{1}{\pi t} \sin \omega_0 t, \tag{9.13}$$

which is non-vanishing even *before* the flashbulb went off. This obviously violates causality and the physicist patiently explains to the inventor what is wrong—any realistic pair of sunglasses must have a response function $f(\omega)$ which obeys a dispersion relation and which therefore, as we shall show, does not violate causality.

The argument follows from a mathematical theorem put forth in Titchmarsh's treatise on Fourier Integrals [Ti 48]. Let $g(t)$ be a bounded and square integrable function of the real variable t which vanishes identically for $t < 0$. Let

$$f(\omega) = \int_0^\infty g(t) e^{i\omega t} dt = \int_{-\infty}^\infty g(t) e^{i\omega t} dt \tag{9.14}$$

be the corresponding Fourier transform. Then this integral definition for $f(\omega)$ can be extended to complex values of ω and defines a function of the complex variable ω which is analytic in the upper half of the ω plane.

The theorem is very plausible since, for $t > 0$ and $\operatorname{Im}\omega > 0$ the quantity $e^{i\omega t}$ is exponentially damped. Thus $\left|e^{i\omega t}\right|$ is bounded no matter how large t becomes. The function $f(\omega)$ is then a bounded superposition of analytic functions $e^{i\omega t}$ and is itself analytic. Taking the inverse Fourier transform

$$g(t) = \int_{-\infty}^{\infty} \frac{d\omega}{2\pi} e^{-i\omega t} f(\omega), \tag{9.15}$$

we see that for $t < 0$, we can complete the contour in the upper half ω plane. Since $f(\omega)$ has no singularities we find

$$g(t) \equiv 0 \quad \text{for} \quad t < 0 \tag{9.16}$$

which is the causality condition which we require.

Dispersion Relations: Formal Derivation

Consider transmission of an optical wave in a dispersive medium with index of refraction $n(\omega)$. The medium extends into the half plane $z > 0$ and is bounded by the plane $z = 0$. Suppose that an incident pulse does not arrive at the boundary plane before the time $t = 0$. We could anticipate that no disturbance would reach a point z within the medium prior to the time $t = \frac{z}{c}$. The amplitude at z at time t is of the form

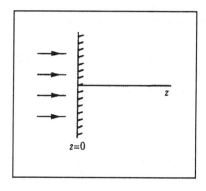

$$g(z,t) = \int_{-\infty}^{\infty} \frac{d\omega}{2\pi} e^{-i\omega\left(t - n(\omega)\frac{z}{c}\right)} f(\omega)$$

$$= \int_{-\infty}^{\infty} \frac{d\omega}{2\pi} e^{-i\omega\left(t - \frac{z}{c}\right)} f(\omega) e^{i\omega(n(\omega)-1)\frac{z}{c}} \tag{9.17}$$

Now define

$$\tau = t - \frac{z}{c} \quad \text{and} \quad \Phi(\omega,z) \equiv f(\omega) e^{i\omega(n(\omega)-1)\frac{z}{c}} \tag{9.18}$$

in terms of which

$$g(z,\tau) = \int_{-\infty}^{\infty} \frac{d\omega}{2\pi} e^{-i\omega\tau} \Phi(\omega,z)$$

or

$$\Phi(\omega,z) = \int_{-\infty}^{\infty} d\tau\, e^{i\omega\tau} g(z,\tau) \ . \tag{9.19}$$

Physics requires that $g(z,\tau) = 0$ for $\tau < 0$. Hence

$$\Phi(\omega, z) = \int_0^\infty d\tau\, e^{i\omega\tau} g(z,\tau) \qquad (9.20)$$

and we expect that $\Phi(\omega, z)$ is analytic in the upper half ω plane. In particular for $z = 0$, $\Phi(\omega, z = 0) = f(\omega)$ has this property and for finite z the same is true of $\exp i\omega\, (n(\omega) - 1)\, \frac{z}{c}$. Since $n(\omega) - 1$ cannot have a pole at $\omega = 0$ we conclude that $n(\omega) - 1$ is analytic in the upper half ω plane.

Now since the pulse amplitude $g(z,\tau)$ is real

$$g^*(z,\tau) = g(z,\tau) \qquad (9.21)$$

we have

$$\begin{aligned} g^*(z,\tau) &= \int_{-\infty}^\infty \frac{d\omega}{2\pi} e^{i\omega\left(t-n^*(\omega)\frac{z}{c}\right)} f^*(\omega) \\ &= \int_{-\infty}^\infty \frac{d\omega}{2\pi} e^{-i\omega\left(t-n^*(-\omega)\frac{z}{c}\right)} f^*(-\omega) \qquad (9.22) \\ &= \int_{-\infty}^\infty \frac{d\omega}{2\pi} e^{-i\omega\left(t-n(\omega)\frac{z}{c}\right)} f(\omega) = g(z,\tau) \ . \end{aligned}$$

Hence

$$n^*(-\omega) = n(\omega) \quad \text{and} \quad f^*(-\omega) = f(\omega) \ , \qquad (9.23)$$

and, using these conditions, we can study the consequence of the analyticity of $n(\omega) - 1$.

Let ω be some fixed point in the upper half plane. Then from Cauchy's formula we have

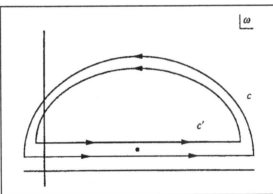

Fig. IV.15: *Integration contours C, C' for application of Cauchy's theorem.*

$$n(\omega) - 1 = \frac{1}{2\pi i} \int_C d\omega' \frac{n(\omega') - 1}{\omega' - \omega}$$

and $\qquad (9.24)$

$$0 = \frac{1}{2\pi i} \int_{C'} d\omega' \frac{n(\omega') - 1}{\omega' - \omega} \ .$$

where the contours C, C' are as shown in Figure IV.15. Then

$$n(\omega) - 1 = \frac{1}{2\pi i}\int_C + \int_{C'} \frac{n(\omega') - 1}{\omega' - \omega} d\omega' \quad . \tag{9.25}$$

On physical grounds we expect that $n(\omega) - 1$ approaches zero for $|\omega| \to \infty$ in the upper half plane and hope that integrals over the semicircles in C and C' converge to zero in the limit of infinite radius. In this case $n(\omega) - 1$ is expressed as the sum of two integrals along lines ℓ and ℓ' parallel to the real axis with point ω sandwiched in between, as indicated in Figure IV.16.

Fig. IV.16: Contours C, C' in the vicinity of the pole.

Allow the line to converge down upon the real ω' axis keeping point ω within the sandwich. Then in the limit as the radius of the circle becomes infinitesimal the semi-circular contributions are seen to cancel, while the straight line sections define the principal value integral. For real ω

$$n(\omega) - 1 = \frac{1}{i\pi} P \int_{-\infty}^{\infty} d\omega' \frac{n(\omega') - 1}{\omega' - \omega} \quad , \tag{9.26}$$

or, taking the real and imaginary parts

$$\operatorname{Re} n(\omega) - 1 = P\frac{1}{\pi}\int_{-\infty}^{\infty} d\omega' \frac{\operatorname{Im} n(\omega')}{\omega' - \omega}$$
$$\operatorname{Im} n(\omega) = -P\frac{1}{\pi}\int_{-\infty}^{\infty} d\omega' \frac{\operatorname{Re} n(\omega') - 1}{\omega' - \omega} \quad . \tag{9.27}$$

The integrals on ω' may be changed to the range $0 \le \omega' \le \infty$ by using Eq. 9.23

$$\operatorname{Re} n(-\omega) - i \operatorname{Im}(-\omega) = \operatorname{Re} n(\omega) + i \operatorname{Im} n(\omega) \quad , \tag{9.28}$$

so

$$\operatorname{Re} n(\omega) - 1 = P\frac{1}{\pi}\int_0^\infty d\omega' \frac{\operatorname{Im} n(\omega')}{\omega' - \omega} + P\frac{1}{\pi}\int_0^\infty d\omega' \frac{\operatorname{Im} n(-\omega')}{-\omega' - \omega}$$
$$= P\frac{1}{\pi}\int_0^\infty d\omega' \operatorname{Im} n(\omega')\left(\frac{1}{\omega' - \omega} + \frac{1}{\omega' + \omega}\right) \tag{9.29}$$
$$= P\frac{1}{\pi}\int_0^\infty d\omega' \operatorname{Im} n(\omega') \frac{2\omega'}{\omega'^2 - \omega^2} \quad .$$

Similarly

$$\operatorname{Im} n(\omega) = -\frac{2\omega}{\pi} P \int_0^\infty d\omega' \frac{\operatorname{Re} n(\omega') - 1}{\omega'^2 - \omega^2} \tag{9.30}$$

which are the canonical dispersion relations.

PROBLEM IV.9.1

Dispersion Relations and the Kramers-Heisenberg Relation

Consider the amplitude for elastic scattering of photons by atoms in the forward direction and with no change of polarization state. Writing the forward scattering amplitude as $f_k(0)$, such that

$$\frac{d\sigma}{d\Omega}(\theta = 0; \hat{\epsilon}_i = \hat{\epsilon}_f) = |f_k(0)|^2$$

i) Show that the Kramers-Heisenberg formula requires

$$f_k(0) = -r_o \left[1 + \frac{1}{m} \sum_I \left(\frac{|\vec{p}_{IA} \cdot \hat{\epsilon}|^2}{E_A - E_I + \omega + i\frac{1}{2}\Gamma_I} + \frac{|\vec{p}_{IA} \cdot \hat{\epsilon}|^2}{E_A - E_I - \omega} \right) \right]$$

where $r_o = \alpha/m$ is the classical radius of the electron. Why is the sign taken to be negative?

ii) In atomic physics a typical decay width is $\Gamma_I \sim 10^{-7} eV$ while energy spacings are $\mathcal{O}(1\ eV)$. Show then, using the same tricks which were employed in Problem IV.5.1 that

$$\mathrm{Re} f_k(0) \approx \sum_I \frac{2 r_o \omega^2}{m \omega_{IA}} \frac{|\vec{p}_{IA} \cdot \hat{\epsilon}|^2}{\omega_{IA}^2 - \omega^2}$$

$$\mathrm{Im} f_k(0) \approx \frac{r_o}{m} \sum_I \frac{1}{2} \Gamma_I \frac{|\vec{p}_{IA} \cdot \hat{\epsilon}|^2}{(\omega_{IA} - \omega)^2 + \frac{1}{4}\Gamma_I^2}$$

where $\omega_{IA} \equiv E_I - E_A$.

iii) Observe that $\mathrm{Im} f_k(0)$ is essentially negligible except when $\omega \approx \omega_{IA}$. Demonstrate that a reasonable approximation is given by

$$\mathrm{Im} f_k(0) \approx \frac{\pi r_o}{m} \sum_I |\vec{p}_{IA} \cdot \hat{\epsilon}|^2 \delta(\omega_{IA} - \omega)$$

iv) Show that the real and imaginary parts of $f_k(\omega)$ satisfy a dispersion relation

$$\mathrm{Re} f_k(0) = \frac{2\omega^2}{\pi} \int \frac{\mathrm{Im} f_{k'}(0) d\omega'}{\omega'(\omega'^2 - \omega^2)}$$

PROBLEM IV.9.2

Sum Rules and the Optical Theorem

Consider the forward elastic Kramers-Heisenberg amplitude

$$f_k(0) = -r_o \left[1 + \frac{1}{m} \sum_I \left(\frac{|\vec{p}_{IA} \cdot \hat{\epsilon}|^2}{E_A - E_I + \omega + i\frac{1}{2}\Gamma_I} + \frac{|\vec{p}_{IA} \cdot \hat{\epsilon}|^2}{E_A - E_I - \omega} \right) \right]$$

The optical theorem for elastic photon atom scattering requires

$$\sigma_{tot}(\omega) = \frac{4\pi}{\omega} \mathrm{Im} f_k(0)$$

i) Consider resonant scattering from a spinless atomic ground state $|A\rangle$ and involving an excited state $|B\rangle$ of unit spin. Assume that state B decays only to state A. Calculate the total resonant cross section for the scattering of a photon of polarization ϵ and compare with $4\pi \mathrm{Im} f_k(0)/\omega$ where $f_k(0)$ is the resonant forward scattering amplitude for the same photon polarization.

ii) Demonstrate that

$$\int_0^\infty \sigma_{tot}(\omega) d\omega = 2\pi^2 r_o$$

This result is called the Thomas-Reiche-Kuhn sum rule.

Hint: Show that

$$\left\langle A \left| [x, [x, \hat{H}_{ATOMIC}]] \right| A \right\rangle = -\frac{1}{m}$$

and that the integral above can be reduced to this form.

PROBLEM IV.9.3

Radiative Energy Shifts: Another Look

Although we have looked at the energy shift of an atom coupled to the radiation field as being due to the emission and subsequent reabsorption of virtual photons, there exists an alternative picture of this effect which is due to Feynman. The idea is that in the presence of matter the energy of any photons becomes shifted because of the existence of an index of refraction $n(\omega)$, which changes the free space relation between frequency and wave number—$k = \omega_k$—to $k = n(\omega_k)\omega_k$. Now although the state $|B,0\rangle$ with which we have been dealing does not contain any real photons, there does exist an energy $\frac{1}{2}\omega_k$ associated with each mode and this zero point energy is shifted due to the presence of the medium by amount

$$\Delta E = \int \frac{d^3k}{(2\pi)^3} \sum_\lambda \frac{1}{2}\omega_k \left(\frac{1}{n(\omega_k)} - 1 \right) .$$

We have previously derived an expression

$$n(\omega_k) = 1 + \frac{2\pi}{\omega_k^2} N f_k(0)$$

where N is the number density of atoms and $f_k(0)$ is the forward elastic scattering amplitude.

i) Using N=1 and

$$f_k(0) = -\frac{\alpha}{m} \left[\hat{\epsilon}_f^* \cdot \hat{\epsilon}_i + \frac{1}{m} \sum_I \left(\frac{\hat{\epsilon}_f^* \cdot \vec{p}_{BI} \hat{\epsilon}_i \cdot \vec{p}_{IB}}{E_B - E_I - \omega_k} + \frac{\hat{\epsilon}_i \cdot \vec{p}_{BI} \hat{\epsilon}_f^* \cdot \vec{p}_{IB}}{E_B - E_I + \omega_k} \right) \right]$$

from the Kramers-Heisenberg dispersion formula, demonstrate that the Lamb shift associated with this zero point motion is identical to that derived in perturbation theory.

ii) Should or should not the energy calculated above be added to the perturbation theory result? Why?

IV.10 EFFECTIVE LAGRANGIANS*

The concept of an "effective Lagrangian" is a familiar one in physics, although the term itself may be unfamiliar. Suppose one is dealing with a system with $n \geq 2$ degrees of freedom but wishes to examine only $m < n$ of these directly. One writes an m-dimensional Lagrangian which includes implicitly the effects of the remaining $n-m$ degrees of freedom. This effective Lagrangian is consequently often non-local, but offers a useful way to characterize the physical system.

Index of Refraction: Classical Mechanics

A familiar example of this technique involves interaction of an electromagnetic field with matter. Although a detailed microscopic view of such a system involves a careful discussion of the dynamical polarizability of individual atoms, it is common to replace this $N \sim 10^{23}$-dimensional picture by a macroscopic description in terms of a (frequency-dependent) dielectric constant and magnetic susceptibility. This medium-dependent electrodynamics represents the effective Lagrangian approach to the problem.

Before tackling the quantum mechanical analog, we review the classical description of this matter-field system. As a simple model of the atom imagine the electron to be bound to the nucleus by a simple spring of spring constant

$$k \equiv m\omega_0^2 \ . \tag{10.1}$$

(Here ω_0 is basically the excitation frequency of the atomic state.) Now imagine that an oscillating electric field

$$\vec{E}(t) = \vec{E}_0 e^{-i\omega t} \tag{10.2}$$

is applied to the atom. The corresponding motion of the electron is given by Newton's law

$$m\frac{d^2}{dt^2}\vec{x} + m\omega_0^2 \vec{x} = e\vec{E}_0 \, e^{-i\omega t} \tag{10.3}$$

whose solution is

$$\vec{x}(t) = \frac{e\vec{E}_0 \, e^{-i\omega t}}{m(\omega_0^2 - \omega^2)} \tag{10.4}$$

and represents an oscillating electric dipole moment

$$\vec{p} = e\vec{x}(t) = \frac{e^2 \vec{E}_0 e^{-i\omega t}}{m(\omega_0^2 - \omega^2)} \ . \tag{10.5}$$

Consider first the case of a constant electric field — $\omega = 0$. One finds then an induced dipole moment which is proportional to the applied field

$$\vec{p} = \frac{e^2 \vec{E}_0}{m\omega_0^2} \equiv \alpha_E \vec{E}_0 \ . \tag{10.6}$$

where the constant of proportionality $\alpha_E = e^2/\omega_0^2$ is the static electric polarizability of the atom. The energy of such a dipole immersed in an electric field ω is given by [Ja 62]

$$U = -\frac{1}{2}\vec{p}\cdot\vec{E}_0 = -\frac{1}{2}\alpha_E \vec{E}_0^2 \ . \tag{10.7}$$

Now imagine that instead of a single atom we are dealing with a medium, containing N atoms per unit volume. As there exists an energy density

$$u = \frac{1}{2}\vec{E}_0^2 \tag{10.8}$$

associated with the free electric field, the total energy density of the system of field plus matter becomes

$$u_{\text{tot}} = \frac{1}{2}\vec{E}_0^2 - \frac{N}{2}\alpha_E \vec{E}_0^2 \ . \tag{10.9}$$

This effective energy density is usually written in a different fashion, since in the presence of all these dipoles the observed field is no longer $\vec{E}_0$. Imagine the matter to exist between capacitor plates located at $z = 0$ and $z = a$. The electrostatic potential associated with a *single* dipole is given by

$$\phi = \frac{\vec{p}\cdot\vec{r}}{4\pi r^3} \tag{10.10}$$

and the net electrostatic potential at a point z_0 between the plates due to these dipole fields is found to be

$$\phi(z_0) = Np \int_0^a dz \int_0^\infty 2\pi\rho\, d\rho \frac{z_0 - z}{4\pi\left[(z-z_0)^2 + \rho^2\right]^{3/2}}$$
$$= Np\left(z_0 - \frac{a}{2}\right) \ . \tag{10.11}$$

The corresponding electric field is given by

$$\vec{E}_{\text{dipole}} = -\nabla\phi(z_0) = -\hat{e}_z\, Np \equiv -\vec{P} \tag{10.12}$$

where $-\vec{E}_{\text{dipole}}$ is usually represented by the polarization density $\vec{P}$. The net electric field in this region is then[‡]

$$\vec{E} = \vec{E}_0 - \vec{P} = \hat{e}_z(E_0 - Np)$$
$$= \hat{e}_z E_0 (1 - N\alpha_E) \tag{10.13}$$

and in terms of this field one writes the effective energy density as

$$u_E = \frac{1}{2}\vec{E}_0^2(1 - N\alpha_E) \approx \frac{1}{2}\vec{E}^2(1 + N\alpha_E) \equiv \frac{1}{2}\epsilon\vec{E}^2 \tag{10.14}$$

[‡] Note: In conventional notation the "polarizing" field $\vec{E}_0$ is called the displacement field $\vec{D}$.

where
$$\epsilon = 1 + N\alpha_E \qquad (10.15)$$
is the dielectric constant.[†]

Now drop the assumption of a constant field. In the case that the applied field has frequency ω, our result becomes
$$u_E(\omega) = \frac{1}{2}\epsilon(\omega)\vec{E}^2(\omega) \qquad (10.16)$$
with
$$\epsilon(\omega) = 1 + \frac{Ne^2}{m(\omega_0^2 - \omega^2)} \;, \qquad (10.17)$$
while in the case of a spatially uniform but time-dependent applied field $\vec{E}(t)$, the effective energy density can be found by decomposing $\vec{E}(t)$ into its Fourier components
$$\vec{E}(\omega) = \int_{-\infty}^{\infty} dt\, e^{i\omega t}\vec{E}(t) \qquad (10.18)$$
in which case
$$\langle u_{\text{tot}} \rangle = \frac{1}{2T}\int_{-\infty}^{\infty}\frac{d\omega}{2\pi}\epsilon(\omega)\vec{E}(\omega)\cdot\vec{E}(-\omega) \;. \qquad (10.19)$$
Equivalently in the time representation
$$\langle u_{\text{tot}} \rangle = \frac{1}{2T}\int_{-\frac{T}{2}}^{\frac{T}{2}} dt\, \vec{E}^2(t) - \frac{Ne^2}{2m}\int_{-\frac{T}{2}}^{\frac{T}{2}} dt \int_{-\frac{T}{2}}^{\frac{T}{2}} dt'\,\vec{E}(t)\cdot\vec{E}(t')D(t-t') \qquad (10.20)$$
where
$$D(t-t') = \int_{-\infty}^{\infty}\frac{d\omega}{2\pi}e^{-i\omega(t-t')}\frac{1}{\omega^2 - \omega_0^2} \;. \qquad (10.21)$$
We see that $D(t-t')$ is non-local. However, this is a price which is worth paying in order to achieve this simple macroscopic representation.

Index of Refraction: Quantum Mechanics

The above is familiar material from a basic course in electricity and magnetism. However, it is also possible to perform a simple quantum mechanical analysis. This calculation is easiest to perform within the path integral formalism. Suppose we

---

[†] We assume here that $N\alpha_E \ll 1$. If this is not the case Eq. 10.15 is replaced by the Clausius-Mossotti relation [Ja 62]
$$\epsilon = \frac{1 + \frac{2}{3}N\alpha_E}{1 - \frac{1}{3}N\alpha_E} \;.$$

have a propagator D_F for all the matter fields under the influence of an external electromagnetic field described by some vector potential A_μ

$$D_F \sim \int \prod_i \mathcal{D}[\vec{x}_i] \exp i \int_{-\frac{T}{2}}^{\frac{T}{2}} dt \, L_{\text{mat}}[\vec{x}_i(t), A_\mu] \quad . \tag{10.22}$$

Here the product symbol is over all atomic coordinates in the system. The effective Lagrangian is defined to be

$$\exp i \int_{-\frac{T}{2}}^{\frac{T}{2}} dt \, L_{\text{eff}}[A_\mu] = \frac{\int \prod_i \mathcal{D}[\vec{x}_i] \exp i \int dt \, L_{\text{mat}}[\vec{x}_i(t), A_\mu]}{\int \prod_i \mathcal{D}[\vec{x}_i] \exp i \int dt \, L_{\text{mat}}[x_i(t), 0]} \tag{10.23}$$

and accounts fully for the effects of the medium. Specifically

$$i \int_{-T/2}^{T/2} dt \, L_{\text{eff}}[A_\mu] \sim -i \Delta E_g[A_\mu] T \tag{10.24}$$

where $\Delta E_g[A_\mu]$ is the shift in ground state energy of the medium due to its interaction with the field.

As an elementary model, we again employ a harmonic oscillator picture of the atom. Including interaction with the vector potential via the usual $j_\mu A^\mu$ term we find the action

$$S = \int_{-\frac{T}{2}}^{\frac{T}{2}} dt \sum_i \left(\frac{1}{2} m \dot{\vec{x}}_i^2(t) - \frac{1}{2} m \omega_0^2 \vec{x}_i^2(t) - e\phi(\vec{x}_i, t) + e\dot{\vec{x}}_i \cdot \vec{A}(\vec{x}_i, t) \right) \quad . \tag{10.25}$$

Assuming the field to be slowly varying over the size of the oscillator we can make a Taylor expansion

$$A_\mu(\vec{x}, t) = A_\mu(\vec{0}, t) + \vec{x} \cdot \vec{\nabla} A_\mu(\vec{x}, t)\Big|_{\vec{x}=0} + \frac{1}{2!} x_i x_j \nabla_i \nabla_j A_\mu(\vec{x}, t)\Big|_{\vec{x}=0} + \ldots \tag{10.26}$$

and discard all but the first order term. Integrating by parts, the action becomes

$$S \simeq \int_{-\frac{T}{2}}^{\frac{T}{2}} dt \sum_i \left(\frac{1}{2} m \dot{\vec{x}}_i^2(t) - \frac{1}{2} m \omega_0^2 \vec{x}_i^2(t) - e\vec{x}_i \cdot \vec{\nabla} \phi(0, t) - e\vec{x}_i \cdot \frac{\partial \vec{A}}{\partial t}(0, t) \right) + \ldots$$

$$\cong \int_{-\frac{T}{2}}^{\frac{T}{2}} dt \sum_i \left(\frac{1}{2} m \dot{\vec{x}}_i^2(t) - \frac{1}{2} m \omega_0^2 x_i^2(t) + e\vec{x}_i \cdot \vec{E}_0(t) \right)$$

$$\tag{10.27}$$

which has the form of a harmonic oscillator potential plus a simple electric dipole interaction. As the dependence upon $\vec{x}_i$ is quadratic, the functional integration required to yield the effective Lagrangian may be performed directly — this is simply a forced harmonic oscillator with driving term $\vec{F}(t) = e\vec{E}_0(t)$—and in the large time limit we find (*cf.* Section I.6)

$$\frac{\int \mathcal{D}[\vec{x}_i] \exp i \int_{-\frac{T}{2}}^{\frac{T}{2}} dt \left(\frac{1}{2}m\dot{\vec{x}}_i^2(t) - \frac{1}{2}m\omega_0^2\vec{x}_i^2 + e\vec{E}_0(t)\cdot\vec{x}_i\right)}{\int \mathcal{D}[\vec{x}_i] \exp i \int_{-\frac{T}{2}}^{\frac{T}{2}} dt \left(\frac{1}{2}m\dot{\vec{x}}_i^2(t) - \frac{1}{2}m\omega_0^2\vec{x}_i^2\right)} \quad (10.28)$$

$$\cong \exp -i\frac{e^2}{2m}\int_{-\frac{T}{2}}^{\frac{T}{2}} dt \int_{-\frac{T}{2}}^{\frac{T}{2}} ds\, \vec{E}_0(t)\cdot\vec{E}_0(s) D(t-s)$$

where

$$D(t-s) = \int_{-\infty}^{\infty} \frac{d\omega}{2\pi} \frac{e^{-i\omega(t-s)}}{\omega^2 - \omega_0^2 + i\epsilon} = -\frac{i}{2\omega_0} \exp -i\omega_0|t-s| \quad (10.29)$$

is the oscillator Green's function.

Characterizing the medium by a uniform number density of atoms N, the effective action becomes

$$\int_{-T/2}^{T/2} dt\, L_{\text{eff}}[A_\mu] = -\frac{Ne^2}{2m} \int_V d^3x \int_{-T/2}^{T/2} dt \int_{-T/2}^{T/2} ds\, \vec{E}_0(t)\cdot\vec{E}_0(s) D(t-s) \sim -\Delta E_g T \,. \quad (10.30)$$

Combining this result with that for the field energy in the absence of the medium

$$E_g^{(0)} T = \int_{-T/2}^{T/2} dt \int_V d^3x \frac{1}{2} E_0^2(t) \quad (10.31)$$

we find the total energy of the field-medium system to be given by

$$\begin{aligned}
E_g T &\sim \int_{-T/2}^{T/2} dt \int d^3x \frac{1}{2} E_0^2(t) \\
&\quad + \frac{Ne^2}{2m} \int d^3x \int_{-T/2}^{T/2} dt \int_{-T/2}^{T/2} ds\, \vec{E}_0(t)\cdot\vec{E}_0(s) D(t-s) \\
&= \int d^3x \int_{-\infty}^{\infty} \frac{d\omega}{2\pi} \frac{1}{2} \vec{E}_0(\omega)\cdot\vec{E}_0(-\omega)\left(1 + \frac{Ne^2}{m}\frac{1}{\omega^2 - \omega_0^2}\right) \\
&\cong \int d^3x \int_{-\infty}^{\infty} \frac{d\omega}{2\pi} \frac{1}{2} \vec{E}(\omega)\cdot\vec{E}(-\omega)\epsilon(\omega) \,,
\end{aligned} \quad (10.32)$$

which is identical to the result found classically.

The physics of this energy shift can be understood if we deal with a constant applied field—$\omega = 0$. In this case the change in the ground state energy U_0 of each oscillator due to its interaction with the field is easily found using simple time-independent perturbation theory

$$\begin{aligned}
\Delta U_0^{(1)} &= \langle 0|eE_0\hat{x}|0\rangle = 0 \\
\Delta U_0^{(2)} &= \sum_{n\neq 0} \frac{\langle 0|eE_0\hat{x}|n\rangle \langle n|eE_0\hat{x}|0\rangle}{E_0 - E_n} \\
&= -\frac{e^2\vec{E}_0^2}{\omega_0} |\langle 1|\hat{x}|0\rangle|^2 = -\frac{e^2\vec{E}_0^2}{2m\omega_0^2} \,.
\end{aligned} \quad (10.33)$$

With N atoms per unit volume, the total energy shift of the medium is

$$\Delta E_g = -\frac{Ne^2 \vec{E}_0^2}{2m\omega_0^2} \int_V d^3x \qquad (10.34)$$

which agrees with the value calculated above. The effective Lagrangian procedure also allows an elementary evaluation of ΔE_g in the case of a time-dependent electric field, as shown in Eq. 10.23, and represents a powerful way in which to eliminate unwanted degrees of freedom. We shall see additional examples in later sections.

Some readers may be justifiably concerned that the use of a harmonic oscillator picture of the atom is too naive to be believed. However, the connection with real atoms can be made straightforwardly. Since for the harmonic oscillator

$$|\langle 1_{\hat{\epsilon}}|\hat{\epsilon}\cdot\vec{x}|0\rangle|^2 = \frac{1}{2m\omega_0} \qquad (10.35)$$

where $\hat{\epsilon}$ is a unit vector in the direction of the electric field, we can write for the dielectric constant

$$\epsilon(\omega) = 1 + 2Ne^2\omega_0 \frac{1}{\omega_0^2 - \omega^2} |\langle 1_{\hat{\epsilon}}|\hat{\epsilon}\cdot\vec{x}|0\rangle|^2 \ . \qquad (10.36)$$

The corresponding expression for the index of refraction becomes

$$n(\omega) = \sqrt{\epsilon(\omega)\mu(\omega)} \cong \sqrt{\epsilon(\omega)}$$
$$\cong 1 + Ne^2\omega_0 \frac{1}{\omega_0^2 - \omega^2} |\langle 1_{\hat{\epsilon}}|\hat{\epsilon}\cdot\vec{x}|0\rangle|^2 \ . \qquad (10.37)$$

However, one can also write the index of refraction in terms of the forward scattering amplitude (*cf.* Section IV.9)

$$n(\omega) = 1 + \frac{2\pi}{\omega^2} N f_k(0) \ , \qquad (10.38)$$

and in problem IV.9.1 we found, using the Kramers–Heisenberg dispersion formula that

$$f_k(0) = \sum_I \frac{e^2}{2\pi m^2} \frac{\omega^2}{\omega_{I0}} \frac{|\hat{\epsilon}\cdot\vec{p}_{I0}|^2}{\omega_{I0}^2 - \omega^2} \qquad (10.39)$$

for a transition from the ground state $|0\rangle$ to some excited state $|I\rangle$. Using the relation

$$\left[\hat{H}_{\text{ATOMIC}}, \vec{r}\right] = -i\frac{1}{m}\vec{p} \qquad (10.40)$$

we find

$$\frac{1}{m}\vec{p}_{I0} = \frac{1}{m}\langle I|\vec{p}|0\rangle = i\left\langle I\left|\left[\hat{H}_{\text{ATOMIC}}, \vec{r}\right]\right|0\right\rangle$$
$$= i\omega_{I0}\langle I|\vec{r}|0\rangle \qquad (10.41)$$

so that
$$n(\omega) = 1 + \sum_I e^2 N \omega_{I0} \frac{|\hat{\epsilon} \cdot \vec{r}_{I0}|^2}{\omega_{I0}^2 - \omega^2} \quad . \tag{10.42}$$

Comparison with Eq. 10.37 shows that generalization of the harmonic oscillator result to real atoms involves the replacement of the oscillator frequency ω_0 by the excitation energy ω_{I0} and summation over the possible excited states.

PROBLEM IV.10.1

Classical Harmonic Oscillator and the Index of Refraction

The relation
$$n(\omega) = 1 + \frac{2\pi}{\omega^2} N f_k(0)$$
between the forward scattering amplitude and the index of refraction allows an alternative derivation of $n(\omega)$ for a classical harmonic oscillator, provided we know the forward scattering amplitude $f_k(0)$. The latter may be found using formal techniques of E&M, for example by use of the Leonard–Weickert potentials [Ja 62]. However, a simple equivalence principle argument is perhaps more enlightening.

Consider a system of two charges $q_1 q_2$ hung from the ceiling by a pair of string. Let each charge have mass m and suppose the equilibrium separation of the charges is d.

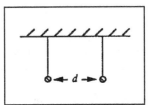

In the absence of electromagnetic effects it is clear that the downward force expected on the ceiling is simply $2mg$.

i) Show that the energy stored in the electromagnetic fields generated by these charges is given by
$$U = \frac{q_1 q_2}{4\pi d}$$
(where we have neglected the self-energy of the charges) and that this implies, by the fact that gravitation applies to all energy, that the actual force exerted on the ceiling is
$$F = 2mg + \frac{q_1 q_2 g}{4\pi d} \quad .$$

The origin of this "extra" force is that, due to the attraction by gravity the field lines which emerge from a point charge as not purely radial but must rather droop somewhat, as shown below.

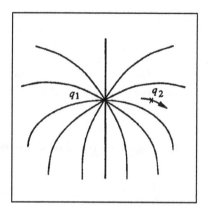

ii) Demonstrate that because of this droop there must exist a force on charge #2 due to #1 (and vice versa) in a downward direction

$$F_{2z} = \frac{q_1 q_2 \vec{g}}{8\pi d}$$

which corresponds to our electric field

$$E_{2z} = \frac{q_1 \vec{g}}{8\pi d} \ .$$

iii) By the equivalence principle this result implies that an identical force must be exerted in the case of a particle having acceleration $\vec{a} = -\vec{g}$. Consider then a test charge q_2 fixed at $x = d$, $y = 0$, $z = 0$ and a second test charge q_1 located at $x = 0$, $y = 0$ $z = \frac{1}{2} g t^2$. Show that the vertical electric field at the location of q_2 associated with the (retarded) position of charge #1 is, from Coulomb's law

$$E_{2z} = -\frac{1}{2} \frac{q_1 \vec{g}}{4\pi d}$$

which is of the same size but opposite *sign* from the answer obtained above. Hence, there must exist an additional component

$$E^r_{2z} = \frac{q_1 \vec{g}}{4\pi d}$$

which is the radiation electric field. It is easy to see that this result agrees precisely with the result of a careful evaluation

$$\vec{E}^r = \frac{q_1}{4\pi r^3} \vec{r} \times (\vec{r} \times \vec{a}) = -\frac{q_1 \vec{a}}{4\pi r} \quad (\vec{a} \cdot \vec{r} = 0)$$

where $\vec{a}$ is the acceleration.

iv) Suppose q_1 is part of a harmonic oscillator of frequency ω to which an electric field
$$\vec{E} = E_0 \hat{e}_z\, e^{-i\omega t}$$
is applied. Demonstrate that the resulting motion generates an electric field
$$E_{2z} = \frac{\omega^2 q_1 E_0 e^{-i\omega t}}{m(\omega_0^2 - \omega^2) 4\pi d}$$
at the location of the second charge and that this corresponds to a forward scattering amplitude
$$f_k(0) = \frac{q_1^2 \omega^2}{4\pi m(\omega_0^2 - \omega^2)}$$
Substitution into the relation between forward scattering and the index of refraction then yields the previously given expression for $n(\omega)$ in the case of a harmonic oscillator.

IV.11 COMPLEX ENERGY AND EFFECTIVE LAGRANGIANS*

We have seen that in order to account for the effect of radiative damping on electromagnetic decay processes, one can to lowest order simply replace all (free) propagators by full propagators

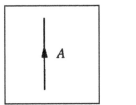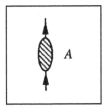

or equivalently replace energies by their complex values
$$E_n^{(0)} \quad \rightarrow \quad E_n^{(0)} + \Delta E_n - i\frac{1}{2}\Gamma_n \ .$$

It is interesting and suggestive that one can also understand this simple substitution from the perspective of effective Lagrangians. That is to say — we wish to evaluate the effective Lagrangian for matter with the electromagnetic field integrated out.

Effective Matter Lagrangian [FeH 65]

In order to perform this integration, we require the Lagrangian as derived in Chapter III for a system of charged particles, say those making up an atom, interacting with a system of fields described by scalar potential ϕ and vector potential $\vec{A}$:

$$L = \sum_{i=1}^{n} \left(\frac{1}{2} m_i \dot{\vec{x}}_i^2(t) + e\dot{\vec{x}}_i \cdot \vec{A}\left(\vec{x}_i(t), t\right)\right) - \frac{1}{2}\sum_{i,j=1}^{n} \frac{e_i e_j}{4\pi |\vec{x}_i(t) - \vec{x}_j(t)|}$$
$$- \frac{1}{2}\int d^3x\, \vec{A}(\vec{x}, t) \left(\frac{\partial^2}{\partial t^2} - \vec{\nabla}^2\right) \cdot \vec{A}(\vec{x}, t) \ . \tag{11.1}$$

It is convenient to work in momentum space, writing

$$\vec{A}(\vec{x},t) = \sum_{\vec{k}} e^{i\vec{k}\cdot\vec{x}} \sum_{\ell=1}^{2} \vec{a}(\vec{k},\ell,t) \tag{11.2}$$

where $\vec{k}\cdot\vec{a}(\vec{k},\ell,t) = 0$ due to the gauge condition $\vec{\nabla}\cdot\vec{A} = 0$. In this case the action can be written as a sum of three terms

$$S = S_{\text{mat}} + S_{\text{rad}} + S_{\text{int}} \tag{11.3}$$

where

$$S_{\text{mat}} = \int_0^T dt \left(\sum_{i=1}^{n} \frac{1}{2} m_i \dot{\vec{x}}_i^2(t) - \frac{1}{2} \sum_{i,j=1}^{n} e_i e_j \frac{1}{4\pi |\vec{x}_i(t) - \vec{x}_j(t)|} \right) \tag{11.4}$$

involves only matter variables,

$$S_{\text{rad}} = \frac{1}{2} \int_0^T dt \sum_{\vec{k}} \left(|\dot{\vec{a}}(\vec{k},1)|^2 - \vec{k}^2 |\vec{a}(\vec{k},1)|^2 + |\dot{\vec{a}}(\vec{k},2)|^2 - \vec{k}^2 |\vec{a}(\vec{k},2)|^2 \right) \tag{11.5}$$

is the action associated with the free radiation field (hereafter we delete the time dependence of $\vec{a}(\vec{k},\ell)$) and

$$S_{\text{int}} = \int_0^T dt \sum_{i=1}^{n} e_i \sum_{\vec{k}} \left(\vec{a}(\vec{k},1) \cdot \dot{\vec{x}}_i(t) + \vec{a}(\vec{k},2) \cdot \dot{\vec{x}}_i(t) \right) e^{i\vec{k}\cdot\vec{x}_i(t)} \tag{11.6}$$

describes the interaction between the matter and radiation.

The full path integral propagator can then be written as

$$D_F(f,t_f;i,t_i) = \int \prod_{i=1}^{n} \mathcal{D}[\vec{x}_i(t)] \prod_{\vec{k},\ell} \mathcal{D}\left[\vec{a}(\vec{k},\ell)\right]$$
$$\times \exp i \left[S_{\text{mat}}(\vec{x}_i) + S_{\text{rad}}\left(\vec{a}(\vec{k},\ell)\right) + S_{\text{int}}\left(\vec{x}_i, \vec{a}(\vec{k},\ell)\right) \right] \tag{11.7}$$

where here the symbols f, i denote the final, initial configurations of matter and field variables. The appearance of the radiation field variables in the integrand is quadratic, so that the path integration may be performed exactly. The action for a particular field degree of freedom $\vec{a}(\vec{k},\ell)$ is given by

$$S_{\vec{k},\ell} = \int_0^T dt \left[\frac{1}{2} \left(|\dot{a}(\vec{k},\ell)|^2 - \vec{k}^2 |a(\vec{k},\ell)|^2 \right) \right.$$
$$\left. + \frac{1}{2} a(\vec{k},\ell) j^*_{\vec{k},\ell}(t) + \frac{1}{2} a^*(\vec{k},\ell) j_{\vec{k},\ell}(t) \right] \tag{11.8}$$

IV.11 COMPLEX ENERGY AND EFFECTIVE LAGRANGIANS*

where we have defined the current

$$j_{\vec{k},\ell}(t) = \sum_{i=1}^{n} e_i \dot{\vec{x}}_i(t) \, e^{i\vec{k}\cdot\vec{x}_i(t)} \cdot \hat{a}(\vec{k},\ell) \; . \tag{11.9}$$

We recognize Eq. 11.8 as the action associated with a simple harmonic oscillator of unit mass and frequency $\omega = |\vec{k}|$ under the influence of a driving term $j_{\vec{k},\ell}(t)$, the path integration of which has already been discussed in Chapter I. For example, in the case that one begins and ends in the oscillator ground state $|0\rangle_{\rm osc}$, the solution is given by

$$\begin{aligned}_{\rm osc}\langle 0| \int \mathcal{D}\left[\vec{a}(\vec{k},\ell)\right] \exp i\left(S_{\rm rad}^{\vec{k},\ell} + S_{\rm int}^{\vec{k},\ell}\right) |0\rangle_{\rm osc} \\ = \exp \frac{i}{4\omega} \int_0^T dt \int_0^T ds \, e^{-i\omega|t-s|} j_{\vec{k},\ell}(t) j_{\vec{k},\ell}^*(s) \; . \end{aligned} \tag{11.10}$$

Hence, the transition amplitude from a given initial atomic state $|\phi_n\rangle$ with no photons present in any of the oscillator modes to the same final state — $|\phi_n\rangle$ is given by

$$\text{Amp}_{n,n}(T) = \left\langle \phi_n \left| \int \mathcal{D}[\vec{x}_i,(t)] \exp iS_{\rm eff}\left[\vec{x}_i(t)\right] \right| \phi_n \right\rangle \tag{11.11}$$

where

$$S_{\rm eff} = S_{\rm mat} + \sum_{\vec{k}} \sum_{\ell=1}^{2} \frac{i}{4\omega} \int_0^T ds \int_0^T dt \, e^{-i\omega|t-s|} j_{\vec{k},\ell}(s) j_{\vec{k},\ell}^*(t) \tag{11.12}$$

is the effective action. Note that all explicit reference to photon degrees of freedom has disappeared — only atomic coordinates remain. Yet all effects of the interactions of photons with the atomic system are included. This is indeed a remarkable and simple expression.

Complex Energy Connection

Having obtained this elegant form for $S_{\rm eff}$, let's now discuss its meaning. First imagine that interaction between the atom and radiation field is turned off so that $S_{\rm eff} = S_{\rm mat}$ and for simplicity suppose that we are dealing with a one electron atom. The propagator becomes

$$D_F^{(0)}(\vec{x}_f,T;\vec{x}_i,0) = \int \mathcal{D}[\vec{x}(t)] \exp i \int_0^T dt \left(\frac{1}{2} m\dot{\vec{x}}^2 + \frac{Z\alpha}{|\vec{x}|} \right) \tag{11.13}$$

and using the representation

$$\begin{aligned} D_F^{(0)}(\vec{x}_f,T;\vec{x}_i,0) &= \left\langle \vec{x}_f \left| \exp -i\hat{H}_{\rm ATOMIC} T \right| \vec{x}_i \right\rangle \\ &= \sum_s \phi_s(\vec{x}_f) e^{-iE_s^{(0)}T} \phi_s^*(\vec{x}_i) \end{aligned} \tag{11.14}$$

we find
$$\text{Amp}_{n,n}^{(0)}(T) = \exp{-iE_n^{(0)}T} \ . \tag{11.15}$$
(Here the Hamiltonian $\hat{H}_{\text{ATOMIC}}$ has the usual Coulombic form
$$H_{\text{ATOMIC}}(\vec{x}) = -\frac{\vec{\nabla}^2}{2m} - \frac{Z\alpha}{|\vec{x}|} \tag{11.16}$$
and $\phi_s(\vec{x})$ denotes an eigenstate with eigenvalue $E_s^{(0)}$
$$H_{\text{ATOMIC}}(\vec{x})\phi_s(\vec{x}) = E_s^{(0)}\phi_s(\vec{x}) \ .) \tag{11.17}$$

Now, however, we restore the interaction with the radiation field and attempt to evaluate the corresponding transition amplitude using the full effective action. An exact answer can no longer be found. However, since $e^2/4\pi \cong 1/137$ is a small quantity, one should be able to find an approximate form by expanding S_{eff} in powers of e^2. To lowest order

$$\exp iS_{\text{eff}} \cong \exp(iS_{\text{mat}}) \times \left[1 - \int \frac{d^3k}{(2\pi)^3} \frac{1}{2k} \sum_{\ell=1}^{2} \frac{1}{2} \int_0^T ds \int_0^T dt \right.$$
$$\left. \times e^{-ik|s-t|} j_{\vec{k},\ell}(x) j_{\vec{k},\ell}^*(t) + \ldots \right] \tag{11.18}$$

The full path integration can now be performed by breaking an arbitrary trajectory into three components — that between time 0 and the earlier of times s and t (say t for definiteness), that between t and s, and that between s and final time T. Noting that

$$\frac{1}{2}\int_0^T ds \int_0^T dt \, e^{-ik|s-t|} j_{\vec{k},\ell}(x) j_{\vec{k},\ell}^*(t)$$
$$= \int_0^T ds \int_0^s dt \, e^{-ik(s-t)} j_{\vec{k},\ell}(s) j_{\vec{k},\ell}^*(t) \tag{11.19}$$

and recalling that intermediate locations $\vec{x}(s), \vec{x}(t)$ must be summed over, we can write

$$\text{Amp}_{n,n}(T) = \int d^3x_f \int d^3x_i \, \phi_n^*(\vec{x}_f) \bigg\{ D_F^{(0)}(\vec{x}_f, T; \vec{x}_i, 0)$$
$$- \int_0^T ds \int_0^s dt \int d^3x_s \int d^3x_t \, D_F^{(0)}(\vec{x}_f, T; \vec{x}_s, s) \int \frac{d^3k}{(2\pi)^3} \frac{1}{2k}$$
$$\times \sum_{\ell=1}^{2} e^{-ik(s-t)} j_{\vec{k},\ell}(s) D_F^{(0)}(\vec{x}_s, s; \vec{x}_t, t) j_{\vec{k},\ell}^*(t) D_F^{(0)}(\vec{x}_t, t; \vec{x}_i, 0)$$
$$- \int_0^T ds \int d^3x_s \, D_F^{(0)}(\vec{x}_f, T; \vec{x}_s, s) \int \frac{d^3k}{(2\pi)^3} \frac{1}{2k} \sum_{\ell=1}^{2} \frac{1}{2} j_{\vec{k},\ell}(s) j_{\vec{k},\ell}^*(s)$$
$$\times D_F^{(0)}(\vec{x}_s, s; \vec{x}_i, 0) + \ldots \bigg\} \phi_n(\vec{x}_i) \ , \tag{11.20}$$

where we have isolated the term wherein the two currents act at the same point, as this must be treated carefully.

This result can be represented in more familiar form by using the eigenstate expansion of the free propagators, whereby we find

$$\text{Amp}_{n,n}(T) = e^{-iE_n^{(0)}T} - \int_0^T ds \int_0^s dt \int d^3x_s \int d^3x_t \, e^{-iE_n^{(0)}(T-s)} \phi_n^*(\vec{x}_s)$$

$$\times \int \frac{d^3k}{(2\pi)^3} \frac{1}{2k} \sum_{\ell=1}^2 e^{-ik(s-t)} j_{\vec{k},\ell}(s) \sum_b \phi_b(\vec{x}_s) e^{-iE_b^{(0)}(s-t)} \phi_b^*(\vec{x}_t)$$

$$\times j_{\vec{k},\ell}^*(t) \phi_n(\vec{x}_t) e^{-iE_n^{(0)}t} - \int_0^T ds \int d^3x_s \int \frac{d^3k}{(2\pi)^3} \frac{1}{2k} \sum_{\ell=1}^2$$

$$\times \frac{1}{2} e^{-iE_n^{(0)}(T-s)} \phi_n^*(\vec{x}_s) j_{\vec{k},\ell}(s) j_{\vec{k},\ell}^*(s) \phi_n(\vec{x}_s) e^{-iE_n^{(0)}s} + \ldots$$

(11.21)

Matrix elements of the current are easily interpreted. Since

$$j_{\vec{k},\ell}(s) = e \, e^{i\vec{k}\cdot\vec{x}_s} \dot{x}(s) \cdot \hat{a}(\vec{k},\ell) \tag{11.22}$$

involves a velocity we must consider integration over position at *two* contiguous time slices — t and $t' = t + \epsilon$. Then, more precisely

$$A_j = \int d^3x_t \, \phi_n^*(\vec{x}_t) j_{\vec{k},\ell}(t) \phi_b(\vec{x}_t)$$

$$\equiv \int d^3x_{t'} \int d^3x_t \, \phi_n^*(\vec{x}_{t'}) \left(\frac{m}{2\pi i \epsilon}\right)^{3/2} \exp i \frac{m}{2} \frac{(\vec{x}_{t'} - \vec{x}_t)^2}{\epsilon} \tag{11.23}$$

$$\times e \exp\left(i\vec{k} \cdot \frac{1}{2}(\vec{x}_{t'} + \vec{x}_t)\right) \frac{1}{\epsilon}(\vec{x}_{t'} - \vec{x}_t) \cdot \hat{a}_{\vec{k},\ell} \phi_b(\vec{x}_t) \, .$$

Now define $\vec{\xi} = \vec{x}_{t'} - \vec{x}_t$ and note the result

$$\int d^3\xi \, \xi_i \xi_j \exp i \frac{m}{2\epsilon} \xi^2 = \delta_{ij} \frac{i\epsilon}{m} \left(\frac{2\pi i \epsilon}{m}\right)^{3/2} . \tag{11.24}$$

We find

$$A_j = \int d^3\xi \int d^3x_t \left(\phi_n^*(\vec{x}_t) + \vec{\xi} \cdot \vec{\nabla} \phi_n^*(\vec{x}_t) + \ldots\right) \left(\frac{m}{2\pi i \epsilon}\right)^{3/2} \exp i \frac{m}{2\epsilon} \xi^2$$

$$\times e \exp i\vec{k} \cdot \vec{x}_t \left(1 + \frac{i}{2}\vec{k}\cdot\vec{\xi} + \ldots\right) \frac{1}{\epsilon} \vec{\xi} \cdot \hat{a}_{\vec{k},\ell} \phi_b(\vec{x}_t) \tag{11.25}$$

$$= \frac{e}{m} \int d^3x_t \, \hat{a}_{\vec{k},\ell} \cdot i\vec{\nabla} \phi_n^*(\vec{x}_t) e^{i\vec{k}\cdot\vec{x}_t} \phi_b(\vec{x}_t)$$

$$\equiv \frac{e}{m} \left\langle b \left| \hat{a}_{\vec{k},\ell} \cdot \vec{p} \, e^{i\vec{k}\cdot\vec{x}} \right| n \right\rangle$$

which is the usual form of the current matrix element.

Similarly, for the isochronous integration over $j_{\vec{k},\ell}(s)j^*_{\vec{k},\ell}(s)$ we must consider contiguous time slices s and $s' = s + \epsilon$. Including a factor ϵ representing dt for this interval we have

$$B_j = \int d^3x_s \, \phi_n^*(\vec{x}_s) j_{\vec{k},\ell}(s) j^*_{\vec{k},\ell}(s) \phi_n(\vec{x}_s) = \epsilon \int d^3x_{s'} \int d^3x_s$$

$$\times \phi_n^*(\vec{x}_{s'}) e^2 \frac{1}{\epsilon^2} \left(\hat{a}(\vec{k},\ell) \cdot (\vec{x}_{s'} - \vec{x}_s) \right)^2 \left(\frac{m}{2\pi i \epsilon} \right)^{3/2} \exp i \frac{m}{2\epsilon} (\vec{x}_{s'} - \vec{x}_s)^2 \phi_n(\vec{x}_s)$$

$$= e^2 \frac{1}{\epsilon} \int d^3\xi \int d^3x_s \left(\phi_n^*(\vec{x}_s) + \vec{\xi} \cdot \vec{\nabla} \phi_n^*(\vec{x}_s) + \ldots \right) \left(\hat{a}(\vec{k},\ell) \cdot \vec{\xi} \right)^2$$

$$\times \left(\frac{m}{2\pi i \epsilon} \right)^{3/2} \exp(i \frac{m}{2\epsilon} \xi^2) \phi_n(\vec{x}_s) = i \frac{e^2}{m} \ .$$

(11.26)

Returning to Eq. 11.21 the integration over s, t can be performed by changing variables to

$$w = \frac{1}{2}(t+s) \qquad z = s - t \ . \tag{11.27}$$

Then for $T \gg \frac{1}{a}$

$$\int_0^T ds \int_0^s dt \, e^{ia(s-t)} \underset{T \gg \frac{1}{a}}{\sim} \left(\int_0^{T/2} dw \int_0^w dz + \int_{T/2}^T dw \int_0^{T-w} dz \right) e^{i(a+i\epsilon)z}$$

$$= \frac{1}{i(a+i\epsilon)} \left(\int_0^{T/2} dw(e^{iaw} - 1) + \int_{T/2}^T dw(e^{ia(T-w)} - 1) \right)$$

$$\underset{T \gg \frac{1}{a}}{\sim} \frac{iT}{a+i\epsilon}$$

(11.28)

whereupon

$$\text{Amp}_{n,n}(T) = e^{-iE_n^{(0)}T} \left(1 - iT \sum_b \int \frac{d^3k}{(2\pi)^3} \frac{1}{2k} \sum_{\ell=1}^2 \frac{\left| \langle b | \frac{e}{m} \hat{a}(\vec{k},\ell) \cdot \vec{p} e^{i\vec{k}\cdot\vec{x}} | n \rangle \right|^2}{E_n^{(0)} - E_b^{(0)} - k + i\epsilon} \right.$$

$$\left. - iT \frac{e^2}{2m} \int \frac{d^3k}{(2\pi)^3} \frac{1}{2k} \sum_{\ell=1}^2 |\hat{a}(\vec{k},\ell)|^2 + \ldots \right)$$

(11.29)

In higher orders the correction to the zeroeth order amplitude is itself exponentiated yielding the full answer

$$\text{Amp}_{n,n}(T) = \exp\left(-i\left(E_n^{(0)} + \Delta E_n\right)T\right) \exp -\frac{1}{2}\Gamma_n T \tag{11.30}$$

where, as before,

$$\Delta E_n = \frac{e^2}{2m} \int \frac{d^3k}{(2\pi)^3} \frac{1}{2k} \sum_{\ell=1}^2 |\hat{a}(\vec{k},\ell)|^2$$

$$+ P \sum_b \int \frac{d^3k}{(2\pi)^3} \frac{1}{2k} \sum_{\ell=1}^2 \frac{\left| \langle b | \frac{e}{m} \hat{a}(\vec{k},\ell) \cdot \vec{p} e^{i\vec{k}\cdot\vec{x}} | n \rangle \right|^2}{E_n^{(0)} - E_b^{(0)} - k}$$

(11.31)

IV.11 COMPLEX ENERGY AND EFFECTIVE LAGRANGIANS

is the energy shift of state ϕ_n due to the interaction with the radiation field and

$$\Gamma_n = \sum_b \int \frac{d^3k}{(2\pi)^3} \frac{1}{2k} \sum_{\ell=1}^{2} 2\pi\delta\left(E_n^{(0)} - E_b^{(0)} - k\right) \left|\left\langle b \left| \frac{e}{m} \hat{a}(\vec{k},\ell) \cdot \vec{p} e^{i\vec{k}\cdot\vec{x}} \right| n \right\rangle\right|^2 \tag{11.32}$$

is the radiative width according to Fermi's golden rule. We observe that the net effect of the change from S_{mat} to S_{eff} is to replace the free energy in $\text{Amp}_{n,n}(T) = \exp -iE_n^{(0)}T$ by its complex value — $\text{Amp}_{n,n}(T) = \exp i\left(E_n^{(0)} + \Delta E_n - i\frac{1}{2}\Gamma_n\right)T$.

Likewise this simple rule obtains for more complex amplitudes. For example, consider the amplitude for photon emission. In this case the path integration over photon degrees of freedom involves an initial vacuum (no photon) state and a final state with a single photon in some mode with momentum $\vec{q}$ and polarization $\hat{a}(\vec{q},\ell)$. The quadratic integration over modes with $\vec{k} \neq \vec{q}$ is performed as before, while if $\vec{k} = \vec{q}$ we have [cf. Problem I.3.1]

$$\left\langle \vec{q}, \hat{a}(\vec{q},\ell) \left| \int \mathcal{D}[a(\vec{q},\ell)] \exp i \left(S_{\text{rad}}^{\vec{k},\ell} + S_{\text{int}}^{\vec{k},\ell}\right) \right| 0 \right\rangle$$
$$= \frac{1}{\sqrt{2\omega_q}} \int_0^T dt\, e^{iqt} j_{\vec{q},\ell}^*(t) \left\langle 0 \left| \int \mathcal{D}[a(\vec{q},\ell)] \exp i \left(S_{\text{rad}}^{\vec{q},\ell} + S_{\text{int}}^{\vec{q},\ell}\right) \right| 0 \right\rangle . \tag{11.33}$$

The transition amplitude between atomic initial state n and atomic final state b is given by the path integral

$$\text{Amp}_{b,n}^{\vec{q},\hat{a}}(T) = \int d^3x_f \int d^3x_i\, \phi_b^*(\vec{x}_f) \int \mathcal{D}[\vec{x}(t)] \exp iS_{\text{eff}}[\vec{x}(t)]$$
$$\times \frac{1}{\sqrt{2\omega_q}} \int_0^T dt\, e^{iqt} j_{\vec{q},\ell}^*(t) \phi_n(\vec{x}_i) \tag{11.34}$$

and comparison with the lowest order amplitude for this transition

$$\text{Amp}_{b,n}^{\vec{q},\hat{a}(0)}(t) = \int d^3x_f \int d^3x_i\, \phi_b^*(\vec{x}_f) \int \mathcal{D}[\vec{x}(t)] \exp iS_{\text{mat}}[\vec{x}(t)]$$
$$\times \frac{1}{\sqrt{2\omega_q}} \int_0^T dt\, e^{iqt} j_{\vec{k},\ell}^*(t) \phi_n(\vec{x}_i) , \tag{11.35}$$

reveals that the only difference in the replacement of L_{mat} by L_{eff}. Therefore, as before, the full amplitude is given essentially by replacement in the lowest order amplitude of bare eigenvalues $E_n^{(0)}$ by their complex counterparts $E_n - i\frac{1}{2}\Gamma_n$. This prescription is quite apparent in the effective Lagrangian formalism.

Alternate Form for S_{eff}

Before leaving this discussion, it is interesting to note that the effective action can be written in a strikingly simple form by use of the identity

$$\exp -ik|s-t| = \int_{-\infty}^{\infty} \frac{d\omega}{2\pi} \frac{2ik\, e^{i\omega(s-t)}}{\omega^2 - k^2 + i\epsilon} . \tag{11.36}$$

Similarly the Coulombic term can be expressed in momentum space as

$$\begin{aligned}L_{\text{Coul}} &= -\frac{1}{2}\int d^3x_1 \int d^3x_2\, \rho(\vec{x}_1,t)\frac{1}{4\pi|\vec{x}_1-\vec{x}_2|}\rho(\vec{x}_2,t) \\ &= -\frac{1}{2}\int \frac{d^3k}{(2\pi)^3}\int \frac{d\omega}{2\pi}\frac{1}{\vec{k}^2}\rho(\vec{k},\omega)\rho^*(\vec{k},\omega) \\ &= -\frac{1}{2}\int \frac{d^3k}{(2\pi)^3}\int \frac{d\omega}{2\pi}\frac{1}{\omega^2-\vec{k}^2}\left(\frac{\omega^2}{\vec{k}^2}-1\right)\rho(\vec{k},\omega)\rho^*(\vec{k},\omega) \;.\end{aligned} \quad (11.37)$$

By current conservation we have

$$\omega\rho(\vec{k},\omega) - \vec{k}\cdot\vec{j}(\vec{k},\omega) = 0 \;. \quad (11.38)$$

Defining $j_\parallel$ as that piece of the current density parallel to $\vec{k}$, then

$$\frac{\omega}{|\vec{k}|}\rho(\vec{k},\omega) = j_\parallel(\vec{k},\omega) \quad (11.39)$$

and

$$L_{\text{Coul}} = -\frac{1}{2}\int \frac{d^3k}{(2\pi)^3}\int \frac{d\omega}{2\pi}\frac{1}{\omega^2-\vec{k}^2}\left(j_\parallel(\vec{k},\omega)j_\parallel^*(\vec{k},\omega) - \rho(\vec{k},\omega)\rho^*(\vec{k},\omega)\right) \;, \quad (11.40)$$

so that the effective action can be written in the form

$$\begin{aligned}S_{\text{eff}} &= \int dt \sum_{i=1}^n \frac{1}{2}m_i\dot{\vec{x}}_i^2(t) + \frac{1}{2}\int \frac{d\omega}{2\pi}\int \frac{d^3k}{(2\pi)^3}\frac{|\rho(\vec{k},\omega)|^2 - |\vec{j}(\vec{k},\omega)|^2}{\omega^2-\vec{k}^2+i\epsilon} \\ &= \int dt \sum_{i=1}^n \frac{1}{2}m_j\dot{\vec{x}}_i^2 + \frac{1}{8\pi^2}\int d^4x_1 \int d^4x_2\, \frac{j_\mu(x_1)j^\mu(x_2)}{(x_1-x_2)^2-i\epsilon}.\end{aligned} \quad (11.41)$$

It is remarkable and suggestive that the real part of the effective matter Lagrangian given in Eq. 11.41 is equivalent to the formulation of classical electrodynamics given by Feynman and Wheeler in terms of use of one-half the retarded wave solution and half the advanced wave solution [FeW 49]. This point is a deep one and we do not have time to explore it further.

CHAPTER V

ALTERNATE APPROXIMATE METHODS

V.1 WKB APPROXIMATION

We have thus far been dealing with perturbation theory — used to find an approximate solution to the Schrödinger equation in the situation that the Hamiltonian can be partitioned into a soluble piece $\hat{H}_0$ and an interaction term $\hat{V}$ which is small compared to $\hat{H}_0$. An alternative approximate technique is available in the situation that the action divided by $\hbar$ is large — the so-called semiclassical method, which we now discuss.

Generally the semiclassical technique is presented in terms of the WKB approximation [We 26, Kr 26, Br 26]. The basic idea is that since, for a constant potential V_0, a stationary state solution can be written as[†]

$$\psi(x,t) \simeq \exp(ikx - iEt) = \exp i \left(\int_0^x k \, dx' - Et \right) \qquad (1.1)$$

where

$$k = \sqrt{2m(E - V_0)} \qquad (1.2)$$

is the (constant) wavenumber, then for a non-constant but slowly varying potential $V(x)$, we should be able to represent an approximate solution by the form

$$\psi(x,t) \sim \exp i \left(\int_0^x k(x') dx' - Et \right) \qquad (1.3)$$

with

$$k(x) = \sqrt{2m(E - V(x))} \qquad (1.4)$$

being the local wavenumber. Of course, a solution of the Schrödinger equation must satisfy local conservation of probability —

$$\frac{\partial \rho}{\partial t} + \frac{\partial}{\partial x}(\rho v) = 0 \qquad (1.5)$$

where $\rho = |\psi(x,t)|^2 = $ const. is the probability density and ρv is the probability current density. Since $\partial \rho / \partial t = 0$ we require $\rho v = $ const., i.e., $\rho \propto $ const./v. Thus

$$\psi(x,t) \sim \frac{1}{\sqrt{k(x)}} \exp i \left(\int_0^x k(x') dx' - Et \right) \qquad (1.6)$$

[†] For notational simplicity we shall deal in this section with one dimensional systems. However, generalization to three dimensions is straightforward.

which is the canonical WKB form. This approximation is valid when refraction dominates over reflection, or equivalently when the change in potential energy over a distance of a de Broglie wavelength is much smaller than the local kinetic energy, i.e.,

$$\lambda \frac{|dV/dx|}{E - V(x)} \ll 1 \ . \tag{1.7}$$

Since

$$\frac{d\lambda}{dx} = \left| \frac{d}{dx} \frac{2\pi}{k} \right| = \frac{2\pi}{k^2} \frac{dk(x)}{dx} = \lambda \frac{dV}{dx} \frac{m}{k^2} \tag{1.8}$$

an equivalent statement is

$$\frac{d\lambda}{dx} \ll 1 \ . \tag{1.9}$$

Thus the approximation clearly breaks down at a classical turning point $x = a$, where $E = V(a)$. The usual procedure to deal with this breakdown is to use the WKB form of the wavefunction except in the immediate vicinity of the turning point, wherein a linear approximation to the potential is used in order to match wavefunction forms to the left and right, respectively. By analyzing the asymptotic forms of the Airy function, we show in Appendix V.1 that the connection formulae are

$$\frac{2}{\sqrt{k(x)}} \cos \left(\int_x^a k(x')dx' - \frac{\pi}{4} \right) \leftrightarrow \frac{1}{\sqrt{\kappa(x)}} \exp \left(-\int_a^x \kappa(x')dx' \right)$$
$$\frac{1}{\sqrt{k(x)}} \sin \left(\int_x^a k(x')dx' - \frac{\pi}{4} \right) \leftrightarrow -\frac{1}{\sqrt{\kappa(x)}} \exp \left(\int_a^x \kappa(x')dx' \right) \tag{1.10}$$

for the situation $dV/dx > 0$ and

$$\frac{1}{\sqrt{\kappa(x)}} \exp \left(-\int_x^a \kappa(x')dx' \right) \leftrightarrow \frac{2}{\sqrt{k(x)}} \cos \left(\int_a^x k(x')dx' - \frac{\pi}{4} \right)$$
$$-\frac{1}{\sqrt{\kappa(x)}} \exp \left(\int_x^a \kappa(x')dx' \right) \leftrightarrow \frac{1}{\sqrt{k(x)}} \sin \left(\int_a^x k(x')dx' - \frac{\pi}{4} \right) \tag{1.11}$$

when $dV/dx < 0$. Here $k(x) = \sqrt{2m(E - V(x))}$ is the local wavenumber while $\kappa(x) = \sqrt{2m(V(x) - E)}$ is the corresponding quantity in the forbidden region.

Appendix V.1

In order to determine the connection formulas, which relate oscillatory and exponential behavior of semiclassical waveforms on opposite sides of a classical turning point, consider the simplest sort of "wall" — the linear potential

$$V(x) = ax \ . \tag{1.12}$$

We can then solve for the stationary state solutions $\psi(x)$ satisfying

$$\left(-\frac{1}{2m} \frac{d^2}{dx^2} + ax \right) \psi(x) = E\psi(x) \ . \tag{1.13}$$

No generality is lost by setting $E = 0$ (we can always redefine the zero of energy) and if we define $z = (2ma)^{1/3} x$ the differential equation becomes

$$\left(\frac{d^2}{dz^2} - z\right) \psi(z) = 0 \ . \tag{1.14}$$

The solutions to this equation are well-known and are the Airy functions

$$Ai(z) \ , \qquad Bi(z) \ . \tag{1.15}$$

The asymptotic behavior of $Ai(z)$ is given by [GrR 80]

$$\begin{aligned} Ai(z) &\underset{z \to \infty}{\sim} \frac{1}{2} \frac{1}{\pi^{1/2} z^{1/4}} \exp\left(-\frac{2}{3} z^{3/2}\right) \\ Ai(-z) &\underset{z \to \infty}{\sim} \frac{1}{\pi^{1/2} z^{1/4}} \sin\left(\frac{2}{3} z^{3/2} + \frac{\pi}{4}\right) \ . \end{aligned} \tag{1.16}$$

For $z > 0$

$$\kappa(x) = \sqrt{2m(V(x) - E)} = \sqrt{2max} \ , \tag{1.17}$$

so that

$$\int_0^x \kappa(x') dx' = \int_0^x dx' \sqrt{2max'} = \sqrt{2ma} \cdot \frac{2}{3} x^{3/2} = \frac{2}{3} z^{3/2} \ . \tag{1.18}$$

Likewise for $z < 0$, we have

$$k(x) = \sqrt{2m(E - V(x))} = \sqrt{-2max} \tag{1.19}$$

and

$$\int_x^0 k(x') \, dx' = \int_0^{-x} dx' \sqrt{2max'} = \sqrt{2ma} \cdot \frac{2}{3} (-x)^{3/2} = \frac{2}{3} (-z)^{3/2} \ . \tag{1.20}$$

Finally, since

$$\sin\left(\phi + \frac{\pi}{4}\right) = \cos\left(\phi - \frac{\pi}{4}\right) \tag{1.21}$$

we can write

$$\frac{1}{\sqrt{k(x)}} \cos\left(\int_x^0 dx' k(x') - \frac{\pi}{4}\right) \leftrightarrow \frac{1}{2\sqrt{\kappa(x)}} \exp\left(-\int_0^x dx' \kappa(x')\right) \ . \tag{1.22}$$

Similarly in the case of $Bi(z)$ we have

$$\begin{aligned} Bi(z) &\underset{z \to \infty}{\sim} \frac{1}{\pi^{1/2} z^{1/4}} \exp\left(\frac{2}{3} z^{3/2}\right) \\ Bi(-z) &\underset{z \to \infty}{\sim} \frac{1}{\pi^{1/2} z^{1/4}} \cos\left(\frac{2}{3} z^{3/2} + \frac{\pi}{4}\right) \ . \end{aligned} \tag{1.23}$$

Since
$$\cos\left(\phi + \frac{\pi}{4}\right) = -\sin\left(\phi - \frac{\pi}{4}\right) \qquad (1.24)$$
we find
$$\frac{1}{\sqrt{k(x)}} \sin\left(\int_x^0 dx' k(x') - \frac{\pi}{4}\right) \leftrightarrow -\frac{1}{\sqrt{\kappa(x)}} \exp\left(\int_0^x dx' \kappa(x')\right) \ . \qquad (1.25)$$

The connection relations for the case of a decreasing potential are verified analogously.

With the connection formulae in hand it is straightforward to determine the factors which arise upon reflection from a barrier, as will be needed in the succeeding section. If the particle is incident on the left from the classically allowed side, the wavefunction must be exponentially decreasing on the opposite side of an increasing potential barrier and we must use Eq. 1.10a. Then representing the factor which arises upon reflection as ϕ_{allowed} we have

$$\frac{1}{\sqrt{k(x)}} \left(\exp i \int_{x_0}^x k(x')dx' + \phi_{\text{allowed}} \exp -i \int_{x_0}^x k(x')dx' \right)$$
$$\sim \frac{2}{\sqrt{k(x)}} \cos\left(\int_x^{x_0} k(x')dx' - \frac{\pi}{4}\right) \qquad (1.26)$$
$$= \frac{e^{+i\frac{\pi}{4}}}{\sqrt{k(x)}} \left(\exp i \int_{x_0}^x k(x')dx' + \exp -i \int_{x_0}^x k(x')dx' \, e^{-i\frac{\pi}{2}} \right)$$

so we identify
$$\phi_{\text{allowed}} = e^{-i\frac{\pi}{2}} = -i \ . \qquad (1.27)$$

On the other hand, if the particle is incident from the classically forbidden side of the barrier the boundary condition on the classically allowed side is that only an outgoing wave should result. Then for the increasing potential barrier case, denoting the factor which arises upon reflection as $\phi_{\text{forbidden}}$, we have

$$\frac{1}{\sqrt{k(x)}} \exp -i \int_{x_0}^x k(x')dx' = e^{-i\frac{\pi}{4}} \frac{1}{\sqrt{k(x)}} \left(\cos\left(\int_x^{x_0} k(x')dx' - \frac{\pi}{4}\right) \right.$$
$$\left. + i \sin\left(\int_x^{x_0} k(x')dx' - \frac{\pi}{4}\right) \right)$$
$$\underset{\substack{\text{via connection} \\ \text{formula Eq. 1.10}}}{\sim} e^{-i\frac{\pi}{4}} \frac{1}{\sqrt{\kappa(x)}} \left(\frac{1}{2} \exp\left(-\int_{x_0}^x \kappa(x')dx'\right) - i \exp\left(\int_{x_0}^x \kappa(x')dx'\right) \right)$$
$$\equiv \frac{e^{-i\frac{3\pi}{4}}}{\sqrt{\kappa(x)}} \left(\exp\left(\int_{x_0}^x \kappa(x')dx'\right) + \phi_{\text{forbidden}} \exp\left(-\int_{x_0}^x \kappa(x')dx'\right) \right) \ .$$
$$(1.28)$$

Thus we identify
$$\phi_{\text{forbidden}} = \frac{i}{2} \ . \qquad (1.29)$$

V.2 SEMICLASSICAL PROPAGATOR

The WKB technique has been applied with great success to a wide variety of quantum mechanical problems:

i) alpha decay [Pr 62]
ii) cold emission of electrons [Bo 51]
iii) mesonic atoms, *etc.* [Ka 77]

However, use of the "mysterious" connection formulae renders this procedure somewhat in the realm of black magic, so that one does not really develop a feel for the physics.

As an alternative approach, we present a version of the semiclassical approximation which is based upon the Feynman path integral. The results obtained in this fashion are totally equivalent to those found via WKB in most cases. However, the path integral viewpoint generates a more intuitive picture of the physics of the process under consideration and offers ways to treat some aspects, such as superbarrier penetration, which lie outside the simple WKB technique [Ho 84].

Quadratic Approximation to the Propagator

In order to understand the semiclassical procedure, consider the path integral form of the propagator, wherein temporarily we restore the $\hbar$ in the phase of the exponential

$$D_F(x_2, t; x_1, 0) = \int \mathcal{D}\left[x(t)\right] \exp \frac{i}{\hbar} S\left[x(t), \dot{x}(t)\right] \ . \tag{2.1}$$

We write, as in Section I.3

$$x(t) = x_{\text{cl}}(t) + \delta x(t), \tag{2.2}$$

where $x_{\text{cl}}(t)$ is a solution of Hamilton's equation

$$\delta S\left[x_{\text{cl}}(t)\right] = 0 \ , \tag{2.3}$$

and expand the action in terms of $\delta x(t)$

$$D_F(x_2, t; x_1, 0) = \int \mathcal{D}\left[\delta x(t)\right] \exp \frac{i}{\hbar} \int_0^t dt' \left(\frac{1}{2} m \dot{x}_{\text{cl}}^2(t') - V(x_{\text{cl}}(t'))\right.$$
$$\left. + m \dot{x}_{\text{cl}}(t') \delta \dot{x}(t') - V'(x_{\text{cl}}(t')) \delta x(t') + \frac{1}{2} m \left(\delta \dot{x}(t')\right)^2 - \frac{1}{2!} V''(x_{\text{cl}}) \left(\delta x(t')\right)^2 + \ldots \right) \ . \tag{2.4}$$

The terms linear in $\delta x, \delta \dot{x}$ vanish after an integration by parts and use of the classical equations of motion, so we find

$$D_F(x_2, t; x_1, 0) \cong \exp \frac{i}{\hbar} S[x_{\text{cl}}(t)] \int \mathcal{D}[\delta x(t)]$$
$$\times \exp \frac{i}{\hbar} \int_0^t dt' \left(\frac{1}{2} m (\delta \dot{x}(t'))^2 - \frac{1}{2} V''(x_{\text{cl}}(t'))(\delta x(t'))^2\right) \tag{2.5}$$

as an approximate representation of the propagator. The exponential phase factor is just the classical action, while the additional multiplicative term is written in terms of the propagator for a particle of mass m moving under the influence of the time-dependent potential $\frac{1}{2}x^2 V''(x_{\rm cl}(t))$.

For a quadratic potential Eq. 2.5 is exact, but for a general potential the semiclassical approximation involves dropping terms of higher order than quadratic in δx, since one can show that the contribution of the quadratic order terms to the path integral is lower order in $\hbar$ than that from cubic or higher order terms. Indeed we saw in Sect. I.2 that only the classical path survives in the limit $\hbar \to 0$. Deviations from this path— $\delta x(t)$— are only important if the change which they induce in $S/\hbar$ is of order $\sim \pi$ or less

$$\frac{1}{\hbar}\delta^2 S[x_{\rm cl}] \lesssim \pi \ . \tag{2.6}$$

since larger deviations imply rapid oscillation of the integrand and result in nearly complete cancellation. We conclude that $\delta x \sim \hbar^{1/2}$ so that

$$\frac{1}{\hbar}S[x_{\rm cl}] \sim \mathcal{O}\left(\hbar^{-1}\right)$$
$$\frac{1}{\hbar}\delta^2 S[x_{\rm cl}] \sim \mathcal{O}(1) \tag{2.7}$$
$$\frac{1}{\hbar}\delta^3 S[x_{\rm cl}] \sim \mathcal{O}\left(\hbar^{1/2}\right) \quad etc.$$

Cubic and higher order terms in δx can be neglected then in the semiclassical limit — $\hbar \to 0$.

The quadratic path integration over $\delta x(t)$ may be performed via a clever analyticity argument due to Coleman, yielding, as shown in Appendix V.2 (restoring $\hbar = 1$)

$$\int \mathcal{D}[\delta x(t)] \exp i \int_0^t dt' \left(\frac{1}{2}m(\delta \dot{x}(t'))^2 - \frac{1}{2}(\delta x(t'))^2 V''(x_{\rm cl}(t'))\right)$$
$$= \left(\frac{m}{2\pi i \dot{x}_{\rm cl}(0)\dot{x}_{\rm cl}(t) \int_{x_1}^{x_2} dx\, \dot{x}_{\rm cl}^{-3}(x)}\right)^{1/2} . \tag{2.8}$$

Eq. 2.8 can be written in a simpler form by noting that if $S \equiv S[x_{\rm cl}(t)]$

$$\frac{\partial S}{\partial x_1} = \frac{\partial}{\partial x_1} \int_0^t dt' \left(\frac{1}{2}m\dot{x}_{\rm cl}^2(t') - V(x_{\rm cl}(t'))\right) = \frac{\partial}{\partial x_1} \int_0^t dt' \left(m\dot{x}_{\rm cl}^2(t') - E_{\rm cl}\right)$$
$$= \frac{\partial}{\partial x_1} \left(\int_{x_1}^{x_2} dx \sqrt{2m(E_{\rm cl} - V(x))} - E_{\rm cl} t\right) \tag{2.9}$$

where $E_{\rm cl}$ is the classical energy and is a function of x_1, x_2, t defined via

$$t = \int_0^t dt' = \int_{x_1}^{x_2} dx \sqrt{\frac{m}{2(E_{\rm cl} - V(x))}} \ . \tag{2.10}$$

Then

$$\frac{\partial S}{\partial x_1} = -\sqrt{2m\left(E_{cl} - V(x_1)\right)} + \frac{\partial E_{cl}}{\partial x_1}\left(\int_{x_1}^{x_2} dx\sqrt{\frac{m}{2\left(E_{cl} - V(x)\right)}} - t\right) \quad (2.11)$$
$$= -\sqrt{2m\left(E_{cl} - V(x_1)\right)}\ .$$

Differentiating once more, we have

$$\frac{\partial^2 S}{\partial x_2 \partial x_1} = -\frac{\partial E_{cl}}{\partial x_2}\sqrt{\frac{m}{2\left(E_{cl} - V(x_1)\right)}}\ . \quad (2.12)$$

However, from Eq. 2.10 for the time interval t (recall that x_1, x_2, t are independent variables)

$$0 = \frac{\partial t}{\partial x_2} = \sqrt{\frac{m}{2\left(E_{cl} - V(x_2)\right)}} - \frac{1}{2}\frac{\partial E_{cl}}{\partial x_2}\int_{x_1}^{x_2} dx\sqrt{\frac{m}{2\left(E_{cl} - V(x)\right)^3}} \quad (2.13)$$

or

$$\frac{\partial E_{cl}}{\partial x_2} = 2\frac{\sqrt{\frac{m}{2\left(E_{cl} - V(x_2)\right)}}}{\int_{x_1}^{x_2} dx\sqrt{\frac{m}{2\left(E_{cl} - V(x)\right)^3}}} = \frac{m}{\dot{x}_{cl}(t)}\frac{1}{\int_{x_1}^{x_2} dx\,\dot{x}_{cl}^{-3}(x)}\ . \quad (2.14)$$

Thus

$$\frac{\partial^2 S}{\partial x_2 \partial x_1} = -\frac{m}{\dot{x}_{cl}(0)\dot{x}_{cl}(t)\int_{x_1}^{x_2} dx\,\dot{x}_{cl}^{-3}(x)} \quad (2.15)$$

so that the prefactor in the quadratic approximation to the propagator can be written in the simple form

$$\left(-\frac{1}{2\pi i}\frac{\partial^2 S}{\partial x_2 \partial x_1}\right)^{1/2}\ . \quad (2.16)$$

This prescription yields an exact result for the propagator if the potential is no higher than quadratic in $x, \dot{x}$. Indeed, for the free particle we have

$$S_{cl}(x_2, t; x_1, 0) = \frac{m(x_2 - x_1)^2}{2t}$$
$$\left(-\frac{1}{2\pi i}\frac{\partial^2 S}{\partial x_2 \partial x_1}\right)^{1/2} = \left(\frac{m}{2\pi i t}\right)^{1/2} \quad (2.17)$$

while for the harmonic oscillator

$$S_{cl}(x_2, t; x_1, 0) = \frac{m\omega}{2\sin\omega t}\left((x_2^2 + x_1^2)\cos\omega t - 2x_1 x_2\right)$$
$$\left(-\frac{1}{2\pi i}\frac{\partial^2 S}{\partial x_2 \partial x_1}\right)^{1/2} = \left(\frac{m\omega}{2\pi i \sin\omega t}\right)^{1/2}\ , \quad (2.18)$$

both of which agree with the exact forms derived previously (*cf.* Eq. I.3.34).

WKB Propagator

In order to make the connection between the quadratic approximation and the conventional WKB technique consider the propagator written in terms of WKB solutions

$$D(x_2, t; x_1, 0) = \int dE \, h(E) \, e^{-iEt} \psi_{\text{WKB}}(x_2, E) \psi^*_{\text{WKB}}(x_1, E) \quad x_2 > x_1 \quad (2.19)$$

where

$$\psi_{\text{WKB}}(x, E) = \frac{1}{\sqrt{k(x)}} \exp i \int_a^x dx' \, k(x') \quad (2.20)$$

with

$$k(x) = \sqrt{2m(E - V(x))} \quad (2.21)$$

being the local wavenumber. Here $h(E)$ is a weighting function which can be determined from the normalization condition

$$\lim_{t \to 0^+} D(x_2, t; x_1, 0) = \delta(x_2 - x_1) \, . \quad (2.22)$$

If we pick x_2, x_1 in a region where the potential is negligible, then $k(x_2) = k(x_1) = \sqrt{2mE}$ and

$$\begin{aligned} D(x_2, t; x_1, 0) &= \int \frac{dE}{k} h(E) \, e^{-iEt} \exp\left(i \int_{x_1}^{x_2} k(x') \, dx'\right) \\ &\xrightarrow[t \to 0^+]{} \int \frac{dE}{k} h(E) \, e^{ik(x_2 - x_1)} \, . \end{aligned} \quad (2.23)$$

Hence

$$\lim_{t \to 0^+} D(x_2, t; x_1, 0) = \delta(x_2 - x_1) = \int \frac{dk}{k} \frac{dE}{dk} h(E) \, e^{ik(x_2 - x_1)} \, . \quad (2.24)$$

We require then

$$\frac{1}{k} \frac{dE}{dk} h(E) = \frac{1}{2\pi} \quad (2.25)$$

or, since $dE/dk = k/m$

$$h(E) = \frac{m}{2\pi} \, . \quad (2.26)$$

We can perform the resulting energy integration in Eq. 2.19

$$D(x_2, t; x_1, 0) = \frac{m}{2\pi} \int dE \, e^{-iEt} \frac{1}{\sqrt{k(x_2) k(x_1)}} \exp\left(i \int_{x_1}^{x_2} dx \, k(x)\right) \quad (2.27)$$

by the stationary phase approximation [MoF 53]. The idea is that if one has an integral of the form

$$J = \int_{-\infty}^{\infty} dx \, e^{if(x)} g(x) \quad (2.28)$$

where $g(x)$ is slowly varying, then contributions to the integral tend to cancel if $f(x)$ rapidly changes as x is varied. Only in the vicinity of points x_0 where $f'(x_0) = 0$ do the phases add coherently and can significant contributions to the integral arise. Assuming that there is a single such point we can expand $f(x)$ in this region

$$f(x) = f(x_0) + \frac{1}{2}(x - x_0)^2 f''(x_0) + \ldots \tag{2.29}$$

and perform the resulting quadratic integral exactly

$$J \cong g(x_0) e^{if(x_0)} \int_{-\infty}^{\infty} dx \exp\left(i\frac{1}{2}(x - x_0)^2 f''(x_0)\right)$$
$$= \sqrt{\frac{2\pi i}{f''(x_0)}} g(x_0) e^{if(x_0)} . \tag{2.30}$$

In our case $x \sim E$ and

$$f(E) = \int_{x_1}^{x_2} dx' \sqrt{2m(E - V(x'))} - Et . \tag{2.31}$$

We determine the stationary phase energy E_0 by requiring

$$0 = \frac{\partial f}{\partial E} = \int_{x_1}^{x_2} dx' \sqrt{\frac{m}{2(E_0 - V(x'))}} - t . \tag{2.32}$$

Since

$$\sqrt{\frac{2(E - V(x))}{m}} = \dot{x}_{\rm cl} \tag{2.33}$$

is the classical velocity, we find

$$t = \int_{x_1}^{x_2} dx' \frac{1}{\dot{x}_{\rm cl}(x')} \tag{2.34}$$

as the defining equation for E_0. However, this means that $E_0 = E_{\rm cl}$, where $E_{\rm cl}$ is the classical energy for the trajectory $x_{\rm cl}(t)$ connecting $(x_1, 0)$ and (x_2, t). Also, since

$$f''(E_{\rm cl}) = -\frac{1}{2}\int_{x_1}^{x_2} dx' \left(\frac{m}{2(E_{\rm cl} - V(x'))^3}\right)^{\frac{1}{2}} = -\frac{1}{m}\int_{x_1}^{x_2} dx\, \dot{x}_{\rm cl}^{-3}(x) , \tag{2.35}$$

the WKB propagator becomes

$$D(x_2, t; x_1, 0) = \left(\frac{m}{2\pi i \dot{x}_{\rm cl}(t)\dot{x}_{\rm cl}(0)\int_{x_1}^{x_2} dx\, \dot{x}_{\rm cl}^{-3}(x)}\right)^{1/2} \exp\left(i\int_{x_1}^{x_2} dx\, k(x) - iE_{\rm cl}t\right) \tag{2.36}$$

which is identical in form to the quadratic approximation to the path integral, since the prefactor is obviously the same while the classical action can be written as

$$S[x_{\rm cl}(t)] = \int_0^t dt' \left(\frac{1}{2}m\dot{x}_{\rm cl}^2(t') - V(x_{\rm cl}(t'))\right) = \int_0^t dt' \left(m\dot{x}_{\rm cl}^2(t')^2 - E_{\rm cl}\right)$$
$$= \int_{x_1}^{x_2} dx\, m\dot{x}_{\rm cl}(x) - E_{\rm cl}t = \int_{x_1}^{x_2} dx \sqrt{2m(E_{\rm cl} - V(x))} - E_{\rm cl}t . \tag{2.37}$$

We conclude that any problem which can be treated via WKB can equally well be handled using path integral techniques in the quadratic approximation.

Barrier Penetration: Semiclassical Approach

An example of such a problem is that of barrier penetration. Consider a potential barrier as shown in V.1

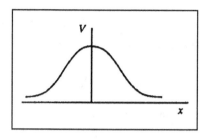

Fig. V.1: A "generic" barrier potential.

and suppose that a wavepacket is incident from that left with central energy E_0 less than the height of the barrier. Classically the particle would be entirely reflected, but quantum mechanically there exists a small probability for transmission. Using the connection formulae, Eqs. 1.10,1.11, the standard WKB expression for the probability is found to be [Me 70]

$$T = 1 - R = \frac{e^{-2\sigma}}{\left(1 + \tfrac{1}{4}e^{-2\sigma}\right)^2} \quad \text{with} \quad \sigma = \int_a^b dx\, \kappa(x) \ . \tag{2.38}$$

where a, b are the classical turning points as shown in Figure V.1 and

$$\kappa(x) = \sqrt{2m\left(V(x) - E\right)} \tag{2.39}$$

is the forbidden region analog of the wavenumber.

In order to see how the same result can be derived via path integral methods it is necessary to generalize the idea of a classical path. In the case of barrier penetration a conventional classical trajectory connecting points x_1, x_2 far to the left, right of the barrier, respectively, and with classical energy *less* than the height of the barrier, does not exist since the particle must be completely reflected. Nevertheless, there do exist trajectories connecting the two spacetime points — $(x_1, 0)$ and (x_2, t) — which satisfy

$$m\ddot{x}_{\text{cl}} = -\left.\frac{dV}{dx}\right|_{x_{\text{cl}}} \tag{2.40}$$

provided we allow the time t to become complex, and these paths can be chosen to have a real classical energy $E < V_{\text{max}}$. In fact, for given x_1, x_2, E there exists a denumerably *infinite* set of such paths which can be labeled by their propagation "times" $t^{(n)}, \quad n = 0, 1, 2, \ldots$

$$t^{(n)} = \int_{x_1}^a + \int_b^{x_2} dx \sqrt{\frac{m}{2(E - V(x))}} - i(2n+1) \int_a^b dx \sqrt{\frac{m}{2(V(x) - E)}} \ . \tag{2.41}$$

The propagation from x_1 to x_2 can be considered as occurring in three successive steps. First the particle travels from the initial point x_1 to the left-hand classical turning point a in the real time interval

$$\Delta t_a = \int_{x_1}^{a} dx \sqrt{\frac{m}{2\left(E - V(x)\right)}} \quad , \tag{2.42}$$

and then traverses the interval between a and b in pure *imaginary* time

$$\Delta t_i^{(n)} = -i(2n+1) \int_{a}^{b} dx \sqrt{\frac{m}{2\left(V(x) - E\right)}} \quad . \tag{2.43}$$

Finally, the particle propagates from the right-hand turning point b to x_2 in the real time interval

$$\Delta t_b = \int_{b}^{x_2} dx \sqrt{\frac{m}{2\left(E - V(x)\right)}} \quad . \tag{2.44}$$

The total temporal interval is then

$$t^{(n)} = \Delta t_a + \Delta t_i^{(n)} + \Delta t_b \tag{2.45}$$

as given in Eq. 2.41.

Physically, the different values $t^{(n)}$ ($n = 0, 1, 2, \ldots$) correspond to the possibility of internal reflections from the walls inside the barrier. This interpretation follows from writing Newton's equations in terms of imaginary time $\tau = it$

$$m \frac{d^2 x}{d\tau^2} = -m \frac{d^2 x}{dt^2} = \left. \frac{dV}{dx} \right|_{x = x(\tau)} \quad . \tag{2.46}$$

Eq. 2.46 describes the motion of a classical particle of mass m moving inside a potential *well* $-V(x)$. The particle can travel from turning point a to turning point b in an infinite number of ways, corresponding to the direct path plus an *arbitrary* number ($n = 1, 2, \ldots$) of complete loops from a to b and back again to a. Thus the "trajectory" labeled by $t^{(0)}$ represents a path which enters the forbidden region at point a, "propagates" directly to turning point b and exits. The path labeled by $t^{(1)}$ passes into the barrier at point a, "propagates" to b where it is reflected, returns to a where it is reflected again, and finally "propagates" to point b where it exits the barrier. Similarly, paths labeled by $t^{(n)}$ correspond to "trajectories" with $2n + 1$ traversals of the forbidden region before final exit at point b.

Obviously these "classical" paths cannot be utilized directly to produce a semiclassical propagator. They become relevant, however, if the propagator is analytically continued through the introduction of its Fourier transform

$$D(x_2, x_1; E) = \int_{0}^{\infty} dt \, e^{iEt} D(x_2, t; x_1, 0) \quad . \tag{2.47}$$

(The lower limit of integration is set by the feature that the Feynman propagator $D(x_2, t; x_1, 0)$ is defined to vanish for $t < 0$ by the $i\epsilon$ prescription.) In fact, it is this energy-space form of the propagator which is relevant for calculation of the barrier reflection/transmission properties, since incident wavepackets are constructed to have fixed energy.

The form of $D(x_2, t; x_1, 0)$ to be used here has already been given

$$D(x_2, t; x_1, 0) \cong \left(\frac{1}{2\pi i\, k(x_2)k(x_1) \int_{x_1}^{x_2} dx\, k^{-3}(x)} \right)^{1/2} \exp\left(i \int_{x_1}^{x_2} k(x)\, dx - i\tilde{E}t \right) . \quad (2.48)$$

The integral $\int_{x_1}^{x_2} k(x)\, dx$ is to be understood as a line integral that can loop around the edges of the barrier $2n$ times ($n = 0, 1, 2, \ldots$). In the immediate vicinity of a turning point, the quadratic approximation to the propagator is no longer valid, and an additional factor λ can be produced each time one passes through such a point. Thus for the "classical" path labeled by integer n, we interpret

$$\int_{x_1}^{x_2} dx\, k(x) = \int_{x_1}^{a} + \int_{b}^{x_2} dx\, k(x) + i(2n+1) \int_{a}^{b} \kappa(x)\, dx - i2n \ln \lambda \quad (2.49)$$

with the energy $\tilde{E}$ defined implicitly in terms of the time $t^{(n)}$ via

$$t^{(n)} = \int_{x_1}^{a} + \int_{b}^{x_2} dx \sqrt{\frac{m}{2\left(\tilde{E} - V(x)\right)}} + i(2n+1) \int_{a}^{b} dx \sqrt{\frac{m}{2\left(V(x) - \tilde{E}\right)}} . \quad (2.50)$$

Here the turning points a, b are functions of $\tilde{E}(x_2, x_1, t^{(n)})$ but λ is independent of energy (and time) and will be evaluated below.

The Fourier transform can be performed by the stationary phase approximation. Writing

$$D(x_2, t; x_1, 0) = \rho(t)\, e^{i\phi(t)} \quad (2.51)$$

(where we have suppressed the dependence upon x_2, x_1) we have

$$D(x_2, x_1; E) = \int_0^\infty dt\, e^{i(Et + \phi(t))} \rho(t) . \quad (2.52)$$

The stationary phase point $\bar{t}$ is determined via

$$\begin{aligned}
0 &= \frac{\partial}{\partial t}(Et + \phi(t)) = \frac{\partial}{\partial t}\left(Et + \int_{x_1}^{x_2} k(x)\, dx - \tilde{E}t \right) \\
&= E - \tilde{E}(t) - t\frac{\partial \tilde{E}}{\partial t} + \int_{x_1}^{x_2} dx\, \frac{\partial k}{\partial \tilde{E}} \frac{\partial \tilde{E}}{\partial t} .
\end{aligned} \quad (2.53)$$

However,

$$t - \int_{x_1}^{x_2} dx\, \frac{\partial k}{\partial \tilde{E}} = 0 \quad (2.54)$$

since this is the equation which defined $\tilde{E}(t)$ in the first place. Thus the stationary phase point $\bar{t}$ is defined via

$$E = \tilde{E}(\bar{t}) . \quad (2.55)$$

Finally, using

$$\left.\frac{\partial^2}{\partial t^2}(Et+\phi(t))\right|_{t=\bar{t}} = -\left.\frac{\partial \tilde{E}}{\partial t}\right|_{t=\bar{t}} = -\frac{1}{\int_{x_1}^{x_2} dx \frac{\partial^2 k}{\partial \tilde{E}^2}}$$
$$= \left(m^2 \int_{x_1}^{x_2} dx\, k^{-3}(x)\right)^{-1} \qquad (2.56)$$

the stationary phase approximation yields

$$D(x_2, x_1; E) = \left(\frac{m^2}{k(x_2)k(x_1)}\right)^{1/2} \exp i \int_{x_1}^{x_2} k(x)\, dx \qquad (2.57)$$

where the integral is to be understood in the sense of Eq. 2.49.

As $\bar{t}^{(n)}$ are complex, the stationary phase procedure requires the path from $t=0$ to $t=+\infty$ to be deformed into the complex plane so as to pass through *each* $\bar{t}^{(n)}$— $n = 0, 1, 2, \ldots$. We find then

$$D(x_2, x_1; E) = \sum_{n=0}^{\infty} D^{(n)}(x_2, x_1; E)$$
$$= \left(\frac{m^2}{k(x_2)k(x_1)}\right)^{1/2} \exp\left(i\int_{x_1}^{a} + i\int_{b}^{x_2} k(x)dx\right) \qquad (2.58)$$
$$\times \sum_{n=0}^{\infty} \lambda^{2n} \exp\left(-(2n+1)\int_{a}^{b} \kappa(x)\, dx\right) .$$

The contribution from paths involving $n \geq 1$ are exponentially suppressed and of questionable accuracy — corrections to the semiclassical approximation to the path integral could well be larger. Nevertheless, it is of interest to include the effects of these "interior bounce" solutions since only then do we have a consistent picture of the transmission and reflection process which conserves probability. Thus in the calculation of the propagator for transmission, as given above, we can perform the summation over n, yielding

$$D(x_2, x_1; E) = \left(\frac{m^2}{k(x_2)\, k(x_1)}\right)^{1/2} \exp\left(i\int_{x_1}^{a} + i\int_{b}^{x_2} k(x)\, dx\right) \frac{e^{-\sigma}}{1 - \lambda^2 e^{-2\sigma}} \qquad (2.59)$$

where

$$\sigma \equiv \int_{a}^{b} \kappa(x)\, dx \qquad (2.60)$$

is the usual WKB barrier penetration coefficient.

We see that this path integral approach to the semiclassical approximation allows a very appealing and graphical picture of the transmission process. Instead of the sum over *all possible* paths required in calculation of the complete path integral, the semiclassical approximation utilizes a sum over all "classical" paths connecting the initial and final points. (Here "classical" is used in the sense defined above, wherein analytic continuation into the complex time plane is permitted.) The form of the propagator can be found by using the simple rules:[†]

i) Propagation from x_1 to x_2 in a classically allowed, forbidden region produces a factor

$$\exp i \int_{x_1}^{x_2} k(x)\,dx \ , \qquad \exp - \int_{x_1}^{x_2} \kappa(x)\,dx \qquad (2.61)$$

respectively;

ii) Reflection from a classical turning point within a classically forbidden region yields a factor λ.

In order to deal with the corresponding reflection coefficient, we require one additional rule

iii) Reflection from a classical turning point within a classically allowed region yields a factor η.

The propagator for the reflection process may be constructed by taking x_1, x_2 both to the left of the barrier and including all possible "classical" trajectories:

i) Propagation from x_1 to the left-hand wall at $x = a$ followed by a reflection and propagation back to x_2;

ii) Propagation from x_1 to the left-hand wall at $x = a$ followed by transmission into the barrier, reflection from the right-hand wall, transmission back to $x = a$ and then propagation from $x = a$ to x_2, etc.

The total contribution to the propagator is then

$$\begin{aligned}
D(x_2, x_1; E) &= \left(\frac{m^2}{k(x_2)\,k(x_1)}\right)^{1/2} \exp\left(i\int_{x_1}^{a} + i\int_{a}^{x_2} k(x)\,dx\right) \\
&\quad \times \left[\eta + \sum_{n=0}^{\infty} \lambda^{2n+1} e^{-(2n+2)\sigma}\right] \\
&= \left(\frac{m^2}{k(x_2)\,k(x_1)}\right)^{1/2} \exp\left(i\int_{x_1}^{a} + i\int_{a}^{x_2} k(x)\,dx\right) \\
&\quad \times \left[\eta + \frac{\lambda e^{-2\sigma}}{1 - \lambda^2 e^{-2\sigma}}\right] \ .
\end{aligned} \qquad (2.62)$$

Note that when propagation is to the left the wavenumber is to be interpreted as $k(x) = -\sqrt{2m(E - V(x))}$ so that if $a > x_2$

$$\int_{a}^{x_2} k(x)\,dx = \int_{x_2}^{a} \sqrt{2m(E - V(x))}\,dx > 0 \ . \qquad (2.63)$$

[†] For completeness, it should be noted that there exist corresponding phases $\exp(\pm i\frac{\pi}{4})$ which arise upon *transmission* through a barrier. However, these do not play a role when the transmission or reflection probabilities are determined and therefore, for simplicity, will be omitted here.

The connection with the transmission and reflection coefficients

$$R = |r(E)|^2 \quad \text{and} \quad T = |t(E)|^2 \qquad (2.64)$$

can now be made via the identification

$$D(x_2, x_1; E) = \left(\frac{m^2}{k(x_2)\,k(x_1)}\right)^{1/2} t(E) \qquad x_1 \ll a \;,\quad b \ll x_2$$

$$D(x_2, x_1; E) = \left(\frac{m^2}{k(x_2)\,k(x_1)}\right)^{1/2} r(E) \qquad x_1, x_2 \ll a \;. \qquad (2.65)$$

If E is above the maximum height of the barrier then

$$r(E) = 0 \quad \text{and} \quad t(E) = \exp\left(i\int_{x_1}^{x_2} k(x)\,dx\right) \;. \qquad (2.66)$$

with

$$R = 0 \quad \text{and} \quad T = 1 \qquad (2.67)$$

as expected. However, in the case of barrier transmission we find using Eq. 2.59 and 2.62

$$T = |t(E)|^2 = \frac{e^{-2\sigma}}{|1 - \lambda^2 e^{-2\sigma}|^2}$$

$$R = |r(E)|^2 = \left|\eta + \frac{\lambda e^{-2\sigma}}{1 - \lambda^2 e^{-2\sigma}}\right|^2 \;. \qquad (2.68)$$

The unitarity requirement

$$R + T = 1 \qquad (2.69)$$

yields

$$|\eta|^2 = 1 \;,\quad 2\operatorname{Re}\eta^*\lambda = -1 \;,\quad \left|\frac{1}{\lambda} - \eta\right|^2 = 1 \qquad (2.70)$$

whose solution is

$$\eta = e^{\pm i\frac{\pi}{2}} \;,\quad \lambda = \frac{1}{2}e^{\mp i\frac{\pi}{2}} \;. \qquad (2.71)$$

The correct phase may be found either i) by comparison with the analytic WKB result which, as shown in Appendix V.1, gives

$$\eta = e^{-i\frac{\pi}{2}} \;,\quad \lambda = \frac{1}{2}e^{i\frac{\pi}{2}} \qquad (2.72)$$

or ii) by a careful consideration of the behavior of the propagator in the vicinity of a classical turning point [Sc 81,McL 72]. Thus the reflection and transmission factors are found to be

$$R = \frac{e^{-2\sigma}}{\left(1 + \tfrac{1}{4}e^{-2\sigma}\right)^2}$$

$$T = \left(\frac{1 - \tfrac{1}{4}e^{-2\sigma}}{1 + \tfrac{1}{4}e^{-2\sigma}}\right)^2 \qquad (2.73)$$

in complete agreement with the corresponding WKB expressions, Eq. 2.38.

Multiple Scattering and Classical Optics

The procedure developed above is purely algebraic, yet produces identical results to those obtained via WKB methods, which involve (approximate) solutions of a differential equation. This might appear surprising at first but can be made understandable via a familiar optical analog—the multiple scattering series. Consider a light signal of frequency ω incident normally on a piece of glass with unit magnetic permeability, index of refraction $n = \sqrt{\epsilon}$ and thickness d. In order to calculate the transmission, reflection probabilities, the standard procedure is to solve the Maxwell equations subject to the boundary conditions

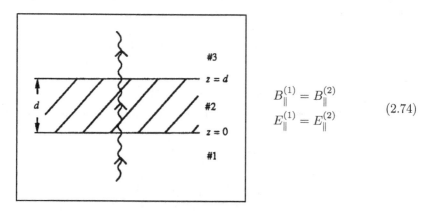

$$B_\parallel^{(1)} = B_\parallel^{(2)}$$
$$E_\parallel^{(1)} = E_\parallel^{(2)} \qquad (2.74)$$

In region #1 ($\omega/k = \frac{1}{n_1}$), we represent the electric and magnetic fields as

$$E = \left(E_i\, e^{ikz} + E_r\, e^{-ikz}\right) e^{-i\omega t}$$
$$B = n_1 \left(E_i\, e^{ikz} - E_r\, e^{-ikz}\right) e^{-i\omega t} \quad . \qquad (2.75)$$

In region #2 ($\omega/k' = \frac{1}{n_2}$) we write

$$E = \left(E_1\, e^{ik'z} + E_2 e^{-ik'z}\right) e^{-i\omega t}$$
$$B = n_2 \left(E_1\, e^{ik'z} - E_2\, e^{-ik'z}\right) e^{-i\omega t} \quad . \qquad (2.76)$$

Finally, in region #3 where again $\omega/k = \frac{1}{n_1}$ we define

$$E = E_t e^{ikz - i\omega t}$$
$$B = n_1 E_t e^{ikz - i\omega t} \quad . \qquad (2.77)$$

At the first boundary — $z = 0$ — we require

$$E_i + E_r = E_1 + E_2$$
$$E_i - E_r = \frac{n_2}{n_1}(E_1 - E_2) \quad , \qquad (2.78)$$

while at the second — $z = d$ — we have
$$E_1 e^{ik'd} + E_2 e^{-ik'd} = E_t e^{ikd}$$
$$E_1 e^{ik'd} - E_2 e^{-ik'd} = \frac{n_1}{n_2} E_t e^{ikd} \ . \tag{2.79}$$

Solving for the reflected, transmitted waves E_r, E_t we find

$$r(k) = \frac{E_r}{E_i} = \frac{\frac{-i}{2} \sin k'd \left(\frac{n_1}{n_2} - \frac{n_2}{n_1} \right)}{\cos k'd - \frac{i}{2} \sin k'd \left(\frac{n_1}{n_2} + \frac{n_2}{n_1} \right)}$$

$$t(k) = \frac{E_t}{E_i} = \frac{e^{-ikd}}{\cos k'd - \frac{i}{2} \sin k'd \left(\frac{n_1}{n_2} + \frac{n_2}{n_1} \right)} \ . \tag{2.80}$$

We easily verify that
$$|r(k)|^2 + |t(k)|^2 = 1 \tag{2.81}$$
as required by conservation of probability.

The above derivation is the standard method of solution for such problems — in terms of the differential equation and its associated boundary conditions. It is "macroscopic" in that it deals with the full electric and magnetic fields in a given region. An equivalent way to attack the same problem, however, is microscopic and algebraic. Thus consider an incident beam with electric field E_i. This beam can be transmitted through the glass in an infinite number of ways:

i) entering into region #2 at $z = 0$, propagating to $z = d$ and exiting,
$$\text{Amp} = t_1 t_2 \, e^{ik'd} \tag{2.82}$$
where t_1, t_2 are transmission factors from air to glass, glass to air, respectively.

ii) entering into region #2 at $z = 0$, propagating to $z = d$, reflecting back to $z = 0$, reflecting back to $z = d$, and exiting
$$\text{Amp} = e^{i3k'd} t_1 t_2 r_2^2 \tag{2.83}$$
where r_2 is the factor which arises upon reflection within the glass, *etc.*

The net transmitted amplitude is then
$$t(k) = t_1 t_2 \, e^{ik'd} \sum_{n=0}^{\infty} \left(r_2^2 \, e^{i2k'd} \right)^n = \frac{t_1 t_2 \, e^{ik'd}}{1 - r_2^2 \, e^{i2k'd}} \ . \tag{2.84}$$

Here the factors r_1, r_2, t_1, t_2 are determined by careful analysis of the situation at the *single* boundary. For example, using the boundary conditions we find for a beam incident from region #1

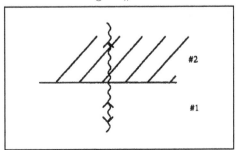

$$E_i + E_r = E_t$$
$$E_i - E_r = \frac{n_2}{n_1} E_t \ , \tag{2.85}$$

so that

$$t_1 = \frac{E_t}{E_i} = \frac{2}{1+\frac{n_2}{n_1}}$$
$$r_1 = \frac{E_r}{E_i} = \frac{1-\frac{n_1}{n_2}}{1+\frac{n_1}{n_2}}.$$
(2.86)

Similarly, for a beam incident from region #2,

$$t_2 = \frac{2}{1+\frac{n_1}{n_2}}$$
$$r_2 = \frac{1-\frac{n_2}{n_1}}{1+\frac{n_2}{n_1}}.$$
(2.87)

Then for the transmission coefficient $t(k)$ we find

$$t(k) = \frac{4e^{ik'd}}{\left(1+\frac{n_1}{n_2}\right)\left(1+\frac{n_2}{n_1}\right)} \times \frac{1}{1-\left(\frac{1-\frac{n_2}{n_1}}{1+\frac{n_2}{n_1}}\right)^2 e^{i2k'd}}$$

$$= \frac{1}{\cos k'd - \frac{i}{2}\sin k'd \left(\frac{n_2}{n_1}+\frac{n_1}{n_2}\right)}$$
(2.88)

which is (up to an unimportant phase) identical to that derived previously.

Similarly, for reflection of the incident beam from the surface there exists an infinite number of trajectories,

i) direct reflection at $z=0$

$$\text{Amp} = r_1$$
(2.89)

ii) transmission at $z=0$, reflection at $z=d$, propagation back to $z=0$ where transmission occurs

$$\text{Amp} = e^{i2k'd} r_2 t_1 t_2 .$$
(2.90)

etc. The overall reflection amplitude is the sum of all possible contributions

$$r(k) = r_1 + t_1 t_2 r_2\, e^{i2k'd} \sum_{n=0}^{\infty} r_2^{2n}\, e^{2ik'd}$$

$$= r_1 + t_1 t_2 r_2 \frac{e^{i2k'd}}{1-r_2^2\, e^{i2k'd}} .$$
(2.91)

We find then

$$r(k) = t(k)\left(\frac{r_1}{t(k)} + r_2\, e^{ik'd}\right)$$
$$= \frac{i}{2} t(k) \sin k'd \left(\frac{n_2}{n_1}-\frac{n_1}{n_2}\right)$$
(2.92)

which again agrees with the expression derived macroscopically.

We see then that the "microscopic" procedure is completely equivalent to the conventional "macroscopic" technique, and in addition offers a clearer intuitive picture for what is taking place. It is apparent that the path integral rules set up above correspond directly to a multiple scattering approach to semiclassical quantum mechanics and yield results completely equivalent to those obtained via the conventional "macroscopic" WKB techniques. Applications to more complex problems than treated here, such as the double well or alpha decay are straightforward [Ho 88].

Appendix V.2

There exist at least two methods for evaluating the prefactor Φ for the semiclassical path integral. The first approach is self-contained but is lengthy [Ma 80]. We present instead a procedure which is quicker but more formal [Co 85]. We employ the result that

$$\Phi \propto \left[\det\left(m\frac{d^2}{dt^2} + V''(x_{\text{cl}}(t))\right)\right]^{-1/2} \equiv D^{-1/2} \qquad (2.93)$$

and evaluate the determinant by means of a clever argument due to Coleman who showed that up to a normalization constant N

$$D = N\psi(t_f - t_i) \qquad (2.94)$$

where $\psi(t)$ satisfies the differential equation

$$m\ddot{\psi}(t) = -V''(x_{\text{cl}}(t))\psi(t) \qquad (2.95)$$

and is subject to the boundary conditions

$$\psi(t_i) = 0 \quad \text{and} \quad \left.\frac{d\psi(t)}{dt}\right|_{t=t_i} = 1 \ . \qquad (2.96)$$

Coleman's argument is deceptively simple. Consider the equation

$$\left(\frac{\partial^2}{\partial t^2} + V_1\right)\psi_\lambda(t) = \lambda\psi_\lambda(t) \qquad (2.97)$$

where $\psi(t)$ will be an eigenfunction and λ an eigenvalue provided

$$\psi_\lambda(t_f) = 0 \ . \qquad (2.98)$$

Now consider a different potential function $V_2(t)$ and consider the ratio

$$f(\lambda) = \frac{\psi_\lambda^{(1)}(t_f)}{\psi_\lambda^{(2)}(t_f)} \ . \qquad (2.99)$$

As a function of λ, f has a simple zero at each eigenvalue of V_1 and a simple pole at each eigenvalue of V_2. But these properties are shared by the function

$$g(\lambda) = \det \left(\frac{\frac{\partial^2}{\partial t^2} + V_1(t) - \lambda}{\frac{\partial^2}{\partial t^2} + V_2(t) - \lambda} \right) . \tag{2.100}$$

It is easy to see that

$$\begin{aligned} g(\lambda) &\underset{\lambda \to \infty}{\sim} 1 \\ f(\lambda) &\underset{\lambda \to \infty}{\sim} 1 \end{aligned} \tag{2.101}$$

as long as λ is not positive and real. Hence, by simple analyticity arguments

$$f(\lambda) = g(\lambda) \tag{2.102}$$

and we find

$$\frac{\det \left(\frac{\partial^2}{\partial t^2} + V(t) \right)}{\psi_{\lambda=0}(t_f)} = \text{const.} \tag{2.103}$$

which is Coleman's result.

Applying this to the problem at hand we introduce $v(t) = \dot{x}_{\rm cl}(t)$ and note that $v(t)$ satisfies the differential equation

$$m\ddot{v}(t) = -V''(x_{\rm cl}(t))\, v(t) \tag{2.104}$$

but not the boundary conditions, Eq. 2.96. However,

$$\frac{d}{dt}\left(v\dot{\psi} - \psi\dot{v} \right) = 0 \tag{2.105}$$

or

$$v^2 \frac{d}{dt}\left(\frac{\psi}{v} \right) = v\dot{\psi} - \psi\dot{v} = \text{const.} = v(t_i) \tag{2.106}$$

where the integration constant is fixed by the boundary conditions. Solving Eq. 2.106 for $\psi(t_f)$ we find

$$\begin{aligned} \psi(t_f) &= v(t_f)v(t_i) \int_{t_i}^{t_f} dt\, \frac{1}{v^2(t)} \\ &= mv(t_f)v(t_i) \int_{x_i}^{x_f} dx\, \frac{1}{(2m(E - V(x)))^{3/2}} \end{aligned} \tag{2.107}$$

The normalization constant may be determined by comparison with the free particle propagator, yielding

$$\Phi = \left(\frac{1}{2\pi i m v(t_f) m v(t_i) \int_{x_i}^{x_f} \frac{dx}{[2m(E-V(x))]^{3/2}}} \right)^{1/2} \tag{2.108}$$

PROBLEM V.2.1

Barrier Penetration via WKB Methods

Use the WKB connection formulae Eqs. 1.10,1.11 to evaluate the wavefunction appropriate for the situation of penetration through a generic barrier potential, e.g. Figure V.1, and demonstrate that the result is as given in Eq. 2.38.

PROBLEM V.2.2

The Forced Harmonic Oscillator: Semiclassical Approach

We have seen in problem I.3.3 that the forced harmonic oscillator possesses an exact solution. The content of this result is most succinctly expressed in terms of the probabilities P_{mn} for the oscillator which began in free oscillator eigenstate $|n\rangle$ at time t_1 (before the force begins to act) to end up in free oscillator eigenstate $|m\rangle$ at time t_2 (after the force has ended)–cf. Appendix V.3:

$$P_{mn} = \begin{cases} \frac{m!}{n!} z^{n-m} e^{-z} \left(L_m^{n-m}(z)\right)^2 & n \geq m \\ P_{nm} & n < m \end{cases}$$

where

$$L_m^{n-m}(z) = \sum_{\ell=0}^{m} \binom{n}{m-\ell} \frac{(-z)^\ell}{\ell!}$$

represents a generalized Laguerre polynomial and

$$z = \frac{1}{2m\omega_0} \left| \int_{-\infty}^{\infty} dt\, F(t) e^{-i\omega_0 t} \right|^2 .$$

It is also possible to understand the physical content of this solution via semiclassical arguments. Thus, although it is conventional to express the path integral in terms of its co-ordinate space form (cf. Problem I.2.1)

$$D_F(x_2, T; x_1, 0) = \int \mathcal{D}[x(t)] \mathcal{D}[p(t)] \exp i \int dt\, (p\dot{x} - H(p, x, t))$$

one can alternatively transform to an equivalent set of canonical co-ordinates– the action-angle variables n, ϕ:

$$n = \frac{p^2}{2m\omega_0} + \frac{m\omega_0 x^2}{2} \qquad \phi = \tan^{-1} \frac{m\omega_0 x}{p}$$

in terms of which

$$D_F(n_2, T; n_1, 0) = \mathcal{D}[n(t)] \mathcal{D}[\phi(t)] \exp -i \int dt\, \left(n\dot{\phi} + H(n, \phi, t)\right) .$$

Consider a forced oscillator with the Hamiltonian

$$H = \frac{p^2}{2m} + \frac{1}{2}m\omega_0^2 x^2 - xF(t) \ .$$

i) If at time $t = 0$

$$\sqrt{n_1}\sin\phi_1 = \sqrt{\frac{m\omega_0}{2}}x_1 \qquad \sqrt{n_1}\cos\phi_1 = \frac{1}{\sqrt{2m\omega_0}}p_1$$

show that at time T

$$\sqrt{n_2}\sin\phi_2 = \sqrt{\frac{m\omega_0}{2}}\left(x_1\cos\omega_0 T + \frac{p_1}{m\omega_0}\sin\omega_0 T\right.$$
$$\left.+ \frac{1}{m\omega_0}\int_0^T dt' F(t')\sin\omega_o(T-t')\right)$$

$$\sqrt{n_2}\cos\phi_2 = \frac{1}{\sqrt{2m\omega_0}}(-m\omega_0 x_1 \sin\omega_0 T + p_1\cos\omega_0 T$$
$$+ \int_0^T dt' F(t')\cos\omega_o(T-t')\Bigg) \ .$$

ii) Squaring, show that

$$n_2 = n_1 + z + 2\sqrt{n_1 z}\cos(\phi_1 - \arg\int_0^T dt' F(t')e^{-i\omega_0 t'})$$

and that for $|\sqrt{n_2} - \sqrt{n_1}| > (<)\sqrt{z}$ there exist two (no) classical trajectories connecting n_1, n_2. In the former case we have

$$D_F(n_2, T; n_1, 0) \sim \exp iS[x_{cl}^{(1)}(t)] + \exp iS[x_{cl}^{(2)}(t)]$$

while for the latter $D_F(n_2, T; n_1, 0)$ is exponentially suppressed.

iii) Sketch then the transition probability P_{n_2, n_1} as a function of n_2 for fixed n_1 and z, indicating the oscillatory (damped) behavior expected for $|\sqrt{n_2} - \sqrt{n_1}| < (>)\sqrt{z}$ and compare the exact result given above with your sketch.

V.3 THE ADIABATIC APPROXIMATION

A second non-perturbative approximation scheme is the adiabatic approximation, which obtains when the Hamiltonian is time-dependent but changes over time scales long compared to typical quantum mechanical oscillation periods $\Delta t \sim \frac{1}{\Delta E}$.

Adiabatic Formalism

We consider the case of a time dependent Hamiltonian $\hat{H}(t)$ of the form

$$\hat{H}(t) = \hat{H}_0 + \hat{V}(t) \tag{3.1}$$

and construct *time-dependent* instantaneous eigenstates, eigenvalues $|\psi_n(t)\rangle$, $E_n(t)$ which satisfy

$$\hat{H}(t)|\psi_n(t)\rangle = E_n(t)|\psi_n(t)\rangle . \tag{3.2}$$

At any time t one can write an arbitrary solution of the Schrödinger equation in terms of these (complete) instantaneous eigenstates

$$|\psi(t)\rangle = \sum_n b_n(t)|\psi_n(t)\rangle . \tag{3.3}$$

The time evolution of the expansion coefficients $b_n(t)$ may be found by requiring

$$i\frac{\partial}{\partial t}|\psi(t)\rangle = \hat{H}(t)|\psi(t)\rangle \tag{3.4}$$

which yields

$$\sum_n \left(i\dot{b}_n(t)|\psi_n(t)\rangle + ib_n(t)|\dot{\psi}_n(t)\rangle\right) = \sum_n b_n(t)E_n(t)|\psi_n(t)\rangle . \tag{3.5}$$

Defining

$$b_n(t) = \exp\left(-i\int_{t_i}^t dt' E_n(t')\right) c_n(t) \tag{3.6}$$

we learn that $c_n(t)$ obeys the equation

$$\dot{c}_n(t) = -\sum_m c_m(t) \exp\left(-i\int_{t_i}^t dt' \left(E_m(t') - E_n(t')\right)\right) \langle\psi_n(t)|\dot{\psi}_m(t)\rangle . \tag{3.7}$$

From the normalization condition

$$\langle\psi_n(t)|\psi_n(t)\rangle = 1 \tag{3.8}$$

we find

$$0 = \langle\psi_n(t)|\dot{\psi}_n(t)\rangle + \langle\dot{\psi}_n(t)|\psi_n(t)\rangle = 2\operatorname{Re}\langle\psi_n(t)|\dot{\psi}_n(t)\rangle \tag{3.9}$$

which merely states that there exists an arbitrary phase in the definition of instantaneous eigenstates. Until recently it had been assumed that the value of this phase could always be adjusted to zero in which case the sum over a complete set of states m in Eq. 3.7 omits the state $m = n$. In the next section we shall see how recent theoretical advances have modified this condition, but for now we assume it to be the case so that

$$\dot{c}_n(t) = -\sum_{m\neq n} c_m(t) \exp\left(-i\int_{t_i}^t dt' \left(E_m(t') - E_n(t')\right)\right) \langle\psi_n(t)|\dot{\psi}_m(t)\rangle . \tag{3.10}$$

Also, differentiating Eq. 3.2

$$\dot{\hat{V}}(t)|\psi_m(t)\rangle + \hat{H}(t)|\dot{\psi}_m(t)\rangle = \dot{E}_m(t)|\psi_m(t)\rangle + E_m(t)|\dot{\psi}_m(t)\rangle \qquad (3.11)$$

so that

$$\langle \psi_n(t)|\dot{\psi}_m(t)\rangle = \frac{\langle \psi_n(t)|\dot{\hat{V}}(t)|\psi_m(t)\rangle}{E_m(t) - E_n(t)} \qquad n \neq m \ . \qquad (3.12)$$

Then

$$\dot{c}_n(t) = \sum_{m \neq n} c_m(t) \exp\left(-i \int_{t_i}^{t} dt' \left(E_m(t') - E_n(t')\right)\right) \frac{\langle \psi_n(t)|\dot{\hat{V}}(t)|\psi_m(t)\rangle}{E_n(t) - E_m(t)} \ . \qquad (3.13)$$

In the case that $\hat{V}$ is time-independent we see that the c_n are constant and Eq. 3.3 becomes just the usual eigenfunction expansion. However, in the situation that $\dot{\hat{V}} \neq 0$, we can write $c_n(t)$ in terms of an integral equation

$$c_n(t) = c_n(t_i) + \sum_{m \neq n} \int_{t_i}^{t} dt' \exp\left(-i \int_{t_i}^{t'} dt'' \left(E_m(t'') - E_n(t'')\right)\right)$$
$$\times \frac{\langle \psi_n(t')|\dot{\hat{V}}(t')|\psi_m(t')\rangle}{E_n(t') - E_m(t')} c_m(t') \qquad (3.14)$$

which can be solved iteratively

$$c_n(t) = c_n(t_i) + \sum_{m \neq n} c_m(t_i) \int_{t_i}^{t} dt' \exp\left(-i \int_{t_i}^{t'} dt'' \left(E_m(t'') - E_n(t'')\right)\right)$$
$$\times \frac{\langle \psi_n(t')|\dot{\hat{V}}(t')|\psi_m(t')\rangle}{E_n(t') - E_m(t')} + \ldots \qquad (3.15)$$

This expansion makes sense provided $\Delta ET \gg 1$ where T is the time scale over which $\hat{V}(t)$ varies, and this is the adiabatic approximation. When $t_f - t_i$ is allowed to become large, the suppression is generally quite substantial, as can be seen if we take a "generic" form for the time-dependence

$$\dot{\hat{V}}(t) \sim \frac{V_0 T}{t^2 + T^2} \qquad (3.16)$$

and neglect the time-dependence of the eigenvalue difference in the argument of the exponential. The first order correction then involves the integral [†]

$$I(T) \sim \int_{-\infty}^{\infty} dt \, \frac{\exp -i(E_m - E_n)t}{t^2 + T^2} \frac{V_0 T}{E_m - E_n} \qquad (3.17)$$

[†] Here the factor $\frac{1}{E_m - E_n}$ has been added in order to make $I(T)$ dimensionless.

which can be evaluated using contour methods, yielding

$$I(T) = \frac{\pi V_0}{|E_m - E_n|} \exp{-|E_m - E_n|T} . \quad (3.18)$$

We find then

$$c_n(t) \approx c_n(t_i) \left(1 + \mathcal{O}\exp(-\Delta E T)\right) . \quad (3.19)$$

Hence if one starts at time t_i, in some instantaneous eigenstate $|\psi_\ell(t_i)\rangle$, i.e.,

$$c_n(t_i) = \delta_{\ell n} \quad (3.20)$$

then the adiabatic theorem states that at later time t the system will be in the state

$$|\psi_\ell(t)\rangle \exp\left(-i \int_{t_i}^{t} dt' E_\ell(t')\right) \quad (3.21)$$

i.e., the *same* instantaneous eigenstate multiplied by a simple dynamical phase factor. Finally, since the propagator represents the kernel which transports an arbitrary wavefunction from t_i to t_f a concise representation of the adiabatic approximation is given by the "adiabatic propagator:" If $\psi_n(\vec{x};t) = \langle \vec{x}|\psi_n(t)\rangle$ is the instantaneous eigenstate wavefunction.

$$D_f(\vec{x}_f, t_f; \vec{x}_i, t_i) = \sum_n \psi_n(\vec{x}_f; t_f) \exp\left(-i \int_{t_i}^{t_f} dt\, E_n(t)\right) \psi_n^*(\vec{x}_i; t_i)$$

$$+ \sum_n \sum_{m \neq n} \psi_n(\vec{x}_f; t_f) \int_{t_i}^{t_f} dt \exp\left(-i \int_{t}^{t_f} dt'\, E_n(t')\right) \frac{\langle \psi_n(t) | \dot{V}(t) | \psi_m(t) \rangle}{E_n(t) - E_m(t)}$$

$$\times \exp\left(-i \int_{t_i}^{t} dt''\, E_m(t'')\right) \psi_m^*(\vec{x}_i; t_i) + \ldots$$

$$(3.22)$$

[Note: Those who find this derivation somewhat complex may appreciate a more direct approach which is available by dividing the time interval $t_f - t_i$ into N steps of size

$$\epsilon = \frac{t_f - t_i}{N} \quad (3.23)$$

Then using completeness of the time-dependent instantaneous eigenstates we have

$$D_F(\vec{x}_f, t_f; \vec{x}_i, t_i) = \left\langle \vec{x}_f \left| T \exp\left(-i \int_{t_i}^{t_f} dt \hat{H}(t)\right) \right| \vec{x}_i \right\rangle$$

$$= \left\langle \vec{x}_f \left| \prod_{k=1}^{N} \exp\left(-i\epsilon \hat{H}(k\epsilon)\right) \right| \vec{x}_i \right\rangle$$

$$= \sum_{p,m,\ldots r} \langle \vec{x}_f | \psi_p(t_f) \rangle \left\langle \psi_p(t_f) \left| \exp\left(-i\epsilon \hat{H}(t_f)\right) \right| \psi_m(t_i + (N-1)\epsilon) \right\rangle \quad (3.24)$$

$$\times \langle \psi_m(t_i + (N-1)\epsilon) | \exp\left(-i\epsilon H(t_i + (N-1)\epsilon)\right) | \psi_q(t_i + (N-2)\epsilon)\rangle$$

$$\times \ldots \times \left\langle \psi_s(t_i + \epsilon) \left| \exp\left(-i\epsilon \hat{H}(t_i + \epsilon)\right) \right| \psi_r(t_i) \right\rangle \langle \psi_r(t_i) | \vec{x}_i \rangle$$

We require then the amplitudes

$$
\begin{aligned}
A_{mn}^k &\equiv \left\langle \psi_m(t_i + k\epsilon) \left| \exp\left(-i\epsilon \hat{H}(t_i + k\epsilon)\right) \right| \psi_n(t_i + (k-1)\epsilon) \right\rangle \\
&= \exp\left(-i\epsilon E_m(t_i + k\epsilon)\right) \langle \psi_m(t_i + k\epsilon) | \psi_n(t_i + (k-1)\epsilon) \rangle \\
&= \text{(from Eq. 3.14)} \quad \exp\left(-i\epsilon E_m(t_i + k\epsilon)\right) \\
&\quad \times \left(\delta_{mn} + \epsilon \frac{\left\langle \psi_m(t_i + (k-1)\epsilon) \left| \dot{V}(t_i + (k-1)\epsilon) \right| \psi_n(t_i + (k-1)\epsilon) \right\rangle}{E_m(t_i + (k-1)\epsilon) - E_n(t_i + (k-1)\epsilon)} + \mathcal{O}(\epsilon^2) \right)
\end{aligned}
$$
(3.25)

Substitution into Eq. 3.24 yields

$$
\begin{aligned}
D_F(\vec{x}_f, t_f; \vec{x}_i, t_i) &= \sum_n \psi_n(\vec{x}_f; t_f) \exp\left(-i\epsilon \sum_{k=1}^N E_n(t_i + k\epsilon)\right) \psi_n(\vec{x}_i; t_i) \\
&+ \sum_n \sum_{m \neq n} \sum_{k=1}^N \psi_n(\vec{x}_f; t_f) \exp\left(-i\epsilon \sum_{\ell=k}^N E_n(t_i + \ell\epsilon)\right) \\
&\times \epsilon \frac{\left\langle \psi_n(t_i + k\epsilon) \left| \dot{V}(t_i + k\epsilon) \right| \psi_m(t_i + k\epsilon) \right\rangle}{E_n(t_i + k\epsilon) - E_m(t_i + k\epsilon)} \\
&\times \exp\left(-i\epsilon \sum_{p=1}^k E_m(t_i + p\epsilon)\right) \psi_m(\vec{x}_i; t_i) + \ldots
\end{aligned}
$$
(3.26)

Taking the limit as $N \to \infty$, we see that this form of the propagator is completely equivalent to Eq. 3.22.]

Example 1: Spin 1/2 Particle in a Magnetic Field

As an example of the use of such techniques consider a spin-1/2 particle which at time t_i is aligned along an external magnetic field $\vec{B}(t)$. The adiabatic theorem asserts that if $\vec{B}(t)$ is allowed to vary, but at a frequency slow compared to the energy splitting between states aligned parallel and antiparallel to the field — $\Delta E = 2\mu B$ — then the particle spin will simply follow the motion of the magnetic field — at later time t_f the spin will still be aligned along the field even though $\vec{B}(t_f)$ may be in a completely different direction than $\vec{B}(t_i)$.

We can verify this assertion explicitly since in certain cases this problem is exactly soluble. Consider, for example, a magnetic field which is of fixed magnitude but whose direction rotates uniformly, with angular velocity ω, about the z-axis and at angle θ to it — i.e., as time evolves the field sweeps out a cone of half angle θ about the z-direction. Using the solution given in Problem V.3.1 we begin the system at $t = 0$ with its spin aligned along $\vec{B}(t = 0)$ by the choice

$$\psi_a(0) = \cos\frac{\theta}{2} \qquad \psi_b(0) = \sin\frac{\theta}{2} \tag{3.27}$$

which corresponds to

$$c_1 = \frac{\omega + \Omega + 2\mu B \cos\theta}{8\Omega\mu B \cos\frac{\theta}{2}} (2\mu B + \Omega - \omega)$$
$$c_2 = \frac{\omega - \Omega + 2\mu B \cos\theta}{8\Omega\mu B \cos\frac{\theta}{2}} (-2\mu B + \Omega + \omega) \ . \tag{3.28}$$

Now evaluate the projection of the exact solution

$$\psi(t) = \begin{pmatrix} \psi_a(t) \\ \psi_b(t) \end{pmatrix} \tag{3.29}$$

onto a spinor

$$\chi_B(t) = \begin{pmatrix} \cos\frac{\theta}{2} e^{-i\frac{1}{2}\omega t} \\ \sin\frac{\theta}{2} e^{i\frac{1}{2}\omega t} \end{pmatrix} \tag{3.30}$$

which is at *all* times aligned along $\vec{B}(t)$

$$\langle \chi_B(t)|\psi(t)\rangle = \frac{(\Omega + 2\mu B)^2 - \omega^2}{8\Omega\mu B} e^{i\frac{1}{2}\Omega t} - \frac{(\Omega - 2\mu B)^2 - \omega^2}{8\Omega\mu B} e^{-i\frac{1}{2}\Omega t} \ . \tag{3.31}$$

The adiabatic limit is that in which the precession frequency ω is much slower than the energy difference $2\mu B$, in which case

$$\Omega \approx 2\mu B + \omega \cos\theta + \ldots \ . \tag{3.32}$$

We find then

$$\langle \chi_b(t)|\psi(t)\rangle = \exp(i\mu Bt) \exp\left(i\frac{\omega}{2}\cos\theta t\right) + \mathcal{O}\left(\frac{\omega^2}{4\mu^2 B^2}\right) \ . \tag{3.33}$$

We observe that, as expected, the exact eigenstate simply follows the direction of the magnetic field accompanied by a dynamical phase

$$\exp(-i \times (-\mu B) \times t) = \exp i\mu Bt \ . \tag{3.34}$$

(There exists an additional phase factor $\exp\left(i\frac{\omega}{2}\cos\theta t\right)$. However, this can generally be neglected (more about that in the succeeding section) since $\omega \ll 2\mu B$ and since it can in principle be absorbed into the overall phase of the state $|\psi(t)\rangle$.) Corrections to the adiabatic prediction are $\mathcal{O}\left(\frac{\omega^2}{4\mu^2 B^2}\right)$ and negligible.

In the opposite limit $\omega \gg 2\mu B$ wherein the field rotates with angular velocity much faster than the magnetic energy splitting we have

$$\Omega \approx \omega + 2\mu B \cos\theta + \ldots \tag{3.35}$$

and

$$\langle \chi_B(t)|\psi(t)\rangle \approx \cos\frac{1}{2}\omega t + i\cos\theta \sin\frac{1}{2}\omega t \ . \tag{3.36}$$

Thus the moment does *not* follow the field precession in this case. In fact, from the probability

$$|\langle \chi_B(t)|\psi(t)\rangle|^2 = \cos^2\frac{1}{2}\omega t + \cos^2\theta \sin^2\frac{1}{2}\omega t \tag{3.37}$$

we observe that the spin stays fixed, while the field rotates.

Example 2: Forced Harmonic Oscillator

As a second example of the adiabatic approximation consider a harmonic oscillator of frequency ω_0 subject to a driving force

$$F(t) = F_0 \exp\left(-\frac{t^2}{\tau^2}\right) . \qquad (3.38)$$

This satisfies the adiabatic condition when $\omega_0 \tau \gg 1$ and we shall begin our discussion by deriving an exact solution of this system. The instantaneous eigenstates and eigenvalues of the problem as easily determined by changing variables from x to $x - \frac{1}{m\omega_0^2} F(t)$, yielding[†]

$$\psi_n(x;t) = \left(\frac{m\omega_0}{\pi}\right)^{1/4} (2^n n!)^{-1/2} H_n\left(\sqrt{m\omega_0}\left(x - \frac{1}{m\omega_0^2} F(t)\right)\right)$$
$$\times \exp -\frac{1}{2} m\omega_0 \left(x - \frac{1}{m\omega_0^2} F(t)\right)^2 \qquad (3.39)$$
$$E_n(t) = \left(n + \frac{1}{2}\right)\omega_0 - \frac{1}{2}\frac{F^2(t)}{m\omega_0^2} .$$

Since $\lim_{t \to \pm\infty} F(t) = 0$ the asymptotic eigenstates are just the usual harmonic oscillator states and the exact solution is completely characterized by the transition amplitudes A_{mn} to begin in a harmonic oscillator state $|n\rangle$ at a time $-T/2 \ll -\tau$ and to end in oscillator state $|m\rangle$ at time $T/2 \gg \tau$. We have then

$$A_{mn} = \exp i(E_m + E_n)\frac{T}{2} \int_{-\infty}^{\infty} dx_2 \int_{-\infty}^{\infty} dx_1\, \phi_m^*(x_2) D_F\left(x_2, \frac{T}{2}; x_1, -\frac{T}{2}\right) \phi_n(x_1) \qquad (3.40)$$

where

$$E_k = \left(k + \frac{1}{2}\right)\omega_0$$
$$\phi_k(x) = \left(\frac{m\omega_0}{\pi}\right)^{1/4} (2^k k!)^{-1/2} H_k\left(\sqrt{m\omega_0}\, x\right) \exp -\frac{1}{2} m\omega_0 x^2 \qquad (3.41)$$

[†] Note that under

$$x \to x' \equiv x - \frac{1}{m\omega_0^2} F(t)$$

the Hamiltonian becomes

$$H(x) = -\frac{1}{2m}\frac{d^2}{dx^2} + \frac{1}{2} m\omega_0^2 x^2 - x F(t)$$
$$\to H(x') = -\frac{1}{2m}\frac{d^2}{dx'^2} + \frac{1}{2} m\omega_0^2 x'^2 - \frac{1}{2}\frac{F^2(t)}{m\omega_0^2} .$$

are oscillator eigenvalues, eigenfunctions, respectively, and

$$D_F\left(x_2, \frac{T}{2}; x_1, -\frac{T}{2}\right) = \left(\frac{m\omega_0}{2\pi i \sin\omega_0 T}\right)^{1/2} \exp i \frac{m\omega_0}{2\sin\omega_0 T}\left[(x_2^2 + x_1^2)\cos\omega_0 T\right.$$
$$- 2x_2 x_1 + 2\frac{x_2}{m\omega_0}\int_{-T/2}^{T/2} dt\, F(t)\sin\omega_0\left(t + \frac{T}{2}\right)$$
$$+ 2\frac{x_1}{m\omega_0}\int_{-T/2}^{T/2} dt\, F(t)\sin\omega_0\left(\frac{T}{2} - t\right)$$
$$\left. - \frac{2}{m^2\omega_0^2}\int_{-\frac{T}{2}}^{\frac{T}{2}} dt \int_{-T/2}^{t} ds\, F(t)F(s)\sin\omega_0\left(\frac{T}{2} - t\right)\sin\omega_0\left(s + \frac{T}{2}\right)\right]$$
(3.42)

is the propagator for the forced oscillator, as calculated previously. For a transition from ground state to ground state, the integration required to calculate A_{mn} is a simple Gaussian and may be done easily. Equivalently we can identify the ground state to ground state amplitude by examining the limit $T \to \infty$, which yields

$$D_F\left(x_2, \frac{T}{2}; x_1, -\frac{T}{2}\right) \sim \phi_0(x_2)\phi_0^*(x_1)e^{-\frac{i}{2}\omega_0 T}$$
$$\times \exp-\frac{1}{2m\omega_0}\int_{-T/2}^{T/2} dt \int_{-T/2}^{t} ds\, F(t)F(s)\, e^{-i\omega_0(t-s)}$$
(3.43)

so that

$$A_{00} = \exp-\frac{1}{2m\omega_0}\int_{-T/2}^{T/2} dt \int_{-T/2}^{t} ds\, F(t)F(s)\, e^{-i\omega_0(t-s)} \ . \quad (3.44)$$

The integration when excited states are involved is somewhat more complex and is performed in appendix V.3. The general result is

$$P_{mn} = \begin{cases} \frac{m!}{n!} z^{n-m} \exp(-z)\left(L_m^{n-m}(z)\right)^2 & n \geq m \\ P_{nm} & n < m \end{cases} \quad (3.45)$$

where P_{mn} denotes the probability to end in oscillator state $|m\rangle$ at $t \to +\infty$ having begun in oscillator state $|n\rangle$ at $t \to -\infty$,

$$L_m^{n-m}(z) = \sum_{\ell=0}^{m} \binom{n}{m-\ell}\frac{(-z)^\ell}{\ell!} \quad (3.46)$$

are the generalized Laguerre polynomials, and

$$z = \frac{1}{2m\omega_0}\left|\int_{-\infty}^{\infty} dt\, F(t)\, e^{-i\omega_0 t}\right|^2 \equiv \frac{|\gamma|^2}{2m\omega_0} \ . \quad (3.47)$$

For the specific form of $F(t)$ given in Eq. 3.38 we have

$$z = \frac{F_0^2}{2m\omega_0}\left|\int_{-\infty}^{\infty} dt\, e^{-\frac{t^2}{\tau^2} - i\omega_0 t}\right|^2 = \frac{\pi F_0^2 \tau^2}{2m\omega_0}\exp\left(-\frac{1}{2}\omega_0^2\tau^2\right) \ . \quad (3.48)$$

The adiabatic condition $\omega_0\tau \gg 1$ then guarantees that the contribution from non-leading terms is exponentially small. We have in general

$$P_{n,n} = e^{-z}\left(1 - nz + \frac{n(n-1)}{4}z^2 - \ldots\right)^2$$
$$= 1 - (2n+1)z + \mathcal{O}(z^2) \tag{3.49}$$

$$P_{m,n} = \frac{m!}{n!\left((m-n)!\right)^2} z^{m-n}\left(1 + \mathcal{O}(z)\right) = \begin{cases} (n+1)z & m = n+1 \\ \mathcal{O}(z^{m-n}) & m > n+1 \end{cases} \tag{3.50}$$

$$P_{m,n} = \frac{n!}{m!\left((n-m)!\right)^2} z^{n-m}\left(1 + \mathcal{O}(z)\right) = \begin{cases} nz & m = n-1 \\ \mathcal{O}(z^{n-m}) & m < n-1 \end{cases} \tag{3.51}$$

Now consider the approximate solution via adiabatic techniques. The adiabatic theorem asserts that when $\omega_0\tau \gg 1$ a state which begins as $|n\rangle$ at time $t \to -\infty$ must end up in the *same* state at time $t \to +\infty$ after the perturbing force is removed, i.e.,

$$P^{\text{ad}}_{m,n} = \delta_{m,n} \tag{3.52}$$

Since $z \ll 1$ in this limit we have

$$P_{m,n} = \begin{cases} 1 + \mathcal{O}(z) \approx 1 & m = n \\ \mathcal{O}(z^{n-m}) \ll 1 & m \neq n \end{cases} \tag{3.53}$$

so that the adiabatic prediction is well-satisfied. However, we can also utilize the formalism developed above in order to evaluate corrections to the adiabatic limit. We find, *e.g.*

$$A_{n-1,n} \cong \langle n-1|x|n\rangle \int_{-\infty}^{\infty} dt \exp\left(i\left(n - \frac{1}{2}\right)\omega_0 t\right) \frac{dF(t)}{dt}$$
$$\times \frac{1}{\omega_0} \exp\left(-i\left(n + \frac{1}{2}\right)\omega_0 t\right) \exp i \int_{-\infty}^{\infty} dt' \frac{F^2(t')}{2m\omega_0^2} \, . \tag{3.54}$$

The integration over dt can be performed via an integration by parts, yielding

$$A_{n-1,n} = i\langle n-1|x|n\rangle \exp i\Phi \int_{-\infty}^{\infty} dt\, e^{-i\omega_0 t} F(t) + \ldots$$
$$= i\sqrt{n}\,\frac{\gamma}{\sqrt{2m\omega_0}}\exp i\Phi + \ldots \tag{3.55}$$

where

$$\Phi = \int_{-\infty}^{\infty} dt' \frac{F^2(t')}{2m\omega_0^2} \tag{3.56}$$

is a phase. Squaring, we find

$$P_{n-1,n} = |A_{n-1,n}|^2 = nz + \mathcal{O}(z^2) \tag{3.57}$$

in agreement, with the exact result quoted in Eq. 3.51. Similarly, it is easy to see that

$$A_{n+1,n} = -i\sqrt{n+1}\frac{\gamma^*}{\sqrt{2m\omega_0}}\exp i\Phi \tag{3.58}$$

so that

$$P_{n+1,n} = (n+1)z + \mathcal{O}(z^2) \tag{3.59}$$

again concurring with Eq. 3.50.

Finally, we can calculate the diagonal amplitude $A_{n,n}$ which to $\mathcal{O}(z)$ involves both the leading and second order pieces of the adiabatic propagator

$$\begin{aligned}A_{n,n} = \exp i\Phi \Bigg\{ 1 - \int_{-\infty}^{\infty} dt \int_{-\infty}^{t} dt' \bigg(\exp i \left(n+\frac{1}{2}\right)\omega_0 t \left\langle n|x|n-1\right\rangle F(t) \\
\times \exp -i\left(n-\frac{1}{2}\right)\omega_0(t-t')\left\langle n-1|x|n\right\rangle F(t')\exp -i\left(n+\frac{1}{2}\right)\omega_0 t' \\
+ \exp i\left(n+\frac{1}{2}\right)\omega_0 t\left\langle n|x|n+1\right\rangle F(t)\exp -i\left(n+\frac{3}{2}\right)\omega_0(t-t') \\
\times \left\langle n+1|x|n\right\rangle F(t')\exp -i\left(n+\frac{1}{2}\right)\omega_0 t' \bigg) + \ldots \Bigg\}\end{aligned} \tag{3.60}$$

where we have integrated by parts, as in Eq. 3.55. We recognize then

$$A_{n,n} = \exp i\Phi \left(1 - \frac{1}{2}\frac{|\gamma|^2}{2m\omega_0}(n+n+1) + \ldots\right) \tag{3.61}$$

so that

$$P_{n,n} = 1 - (2n+1)z + \ldots \tag{3.62}$$

as expected. To this order it is clear that probability is conserved

$$\begin{aligned}\sum_m P_{m,n} &= P_{n,n} + P_{n+1,n} + P_{n-1,n} + \ldots \\
&= 1 - (2n+1)z + (n+1)z + nz + \ldots = 1 \quad .\end{aligned} \tag{3.63}$$

Appendix V.3

The transition probability for a harmonic oscillator, under the influence of a driving force $F(t)$, to make a transition from free oscillator state $|n\rangle$ at $t<0$ to state $|m\rangle$ at $t>T$ can easily be calculated using a trick due to Feynman [FeH 65]. (Here we are assuming that $F(t)$ vanishes outside $0<t<T$ so that asymptotic states are simply free oscillator eigenstates.) We shall utilize the propagator for the forced oscillator previously derived in problem I.3.3, which can be written in the form

$$\begin{aligned}D_F(x_2,T;x_1,0) &= \left(\frac{m\omega_0}{2\pi i \sin\omega_0 T}\right)^{1/2}\exp\bigg(i\frac{m\omega_0}{2\sin\omega_0 T} \\
&\quad \times \left[\cos\omega_0 T\left((x_2-\lambda_2)^2 + (x_1-\lambda_1)^2\right) - 2(x_2-\lambda_2)(x_1-\lambda_1)\right]\bigg)e^{i\sigma} \\
&= D_F^{(0)}(x_2-\lambda_2,T;x_1-\lambda_1,0)\,e^{i\sigma}\end{aligned} \tag{3.64}$$

where here $D_F^{(0)}$ represents the free oscillator propagator and

$$\lambda_1 = \frac{1}{m\omega_0 \sin^2 \omega_0 T} \int_0^T dt' F(t') (\sin \omega t' + \cos \omega_0 T \sin \omega_0 (T - t'))$$

$$= \frac{1}{m\omega_0 \sin \omega_0 T} \int_0^T dt' F(t') \cos \omega_0 (T - t')$$

$$\lambda_2 = \frac{1}{m\omega_0 \sin^2 \omega_0 T} \int_0^T dt' F(t') (\sin \omega_0 (T - t') + \cos \omega_0 T \sin \omega_0 t') \quad (3.65)$$

$$= \frac{1}{m\omega_0 \sin \omega_0 T} \int_0^T dt' F(t') \cos \omega_0 t'$$

$$\sigma = -\left(\lambda_1^2 + \lambda_2^2\right) \cos \omega_0 T + 2\lambda_1 \lambda_2 - \frac{2}{m^2 \omega_0^2} \int_0^T dt' \int_0^{t'} ds' F(t') F(s')$$
$$\times \sin \omega_0 (T - t') \sin \omega_0 s' \ .$$

Now calculate the quantity

$$F(\alpha, \beta) = \int_{-\infty}^{\infty} dx_2 \int_{-\infty}^{\infty} dx_1 \exp -\frac{m\omega_0}{2} (x_2 - \beta)^2$$
$$D_F(x_2, T; x_1, 0) \exp -\frac{m\omega_0}{2} (x_1 - \alpha)^2 \quad (3.66)$$
$$= \int_{-\infty}^{\infty} dx'_2 \int_{-\infty}^{\infty} dx'_1 \exp -\frac{m\omega_0}{2} (x'_2 + \lambda_2 - \beta)^2 D_F^{(0)}(x'_2, T; x'_1, 0)$$
$$\times \exp -\frac{m\omega_0}{2} (x'_1 + \lambda_1 - \alpha)^2 e^{i\sigma}$$

where we have changed variables to

$$\begin{aligned} x'_2 &= x_2 - \lambda_2 \\ x'_1 &= x_1 - \lambda_1 \ . \end{aligned} \quad (3.67)$$

The integration may be easily performed by use of the identity [Me 70]

$$\exp -\frac{m\omega_0}{2} (x - \lambda)^2 = \sum_{\ell=0}^{\infty} \frac{\lambda^\ell}{\sqrt{\ell!}} \exp\left(-\frac{m\omega_0}{4} \lambda^2\right) \left(\frac{m\omega_0}{2}\right)^{\ell/2} \psi_\ell(x) \quad (3.68)$$

where $\psi_\ell(x)$ is a free oscillator wavefunction. Using the representation

$$D_F^{(0)}(x'_2, T; x'_1, 0) = \sum_{n=0}^{\infty} \psi_n(x'_2) \psi_n^*(x'_1) \exp -iE_n T \quad (3.69)$$

and the orthogonality condition

$$\int_{-\infty}^{\infty} dx \, \psi_n^*(x) \psi_\ell(x) = \delta_{n\ell} \quad (3.70)$$

we find

$$\begin{aligned}
F(\alpha,\beta) &= \sum_{\ell=0}^{\infty} \frac{(\lambda_1-\alpha)^\ell (\lambda_2-\beta)^\ell}{\ell!} \left(\frac{m\omega_0}{2}\right)^\ell \exp-\frac{m\omega_0}{4}\left[(\lambda_1-\alpha)^2+(\lambda_2-\beta)^2\right] \\
&\quad \times e^{i\sigma-i\omega_0 T(\ell+\frac{1}{2})} \\
&= \exp\left(\frac{m\omega_0}{2}(\lambda_1-\alpha)(\lambda_2-\beta)e^{-i\omega_0 T}-\frac{m\omega_0}{4}\left[(\lambda_1-\alpha)^2+(\lambda_2-\beta)^2\right]\right) \\
&\quad \times e^{i\sigma-i\frac{1}{2}\omega_0 T} \, .
\end{aligned} \tag{3.71}$$

Defining

$$A_{mn} \equiv \int_{-\infty}^{\infty} dx_2 \int_{-\infty}^{\infty} dx_1\, e^{iE_m T} \psi_m^*(x_2) D_F(x_2,T;x_1,0)\psi_n(x_1) \tag{3.72}$$

as the desired transition amplitude, we have an alternative representation of $F(\alpha,\beta)$ in terms of the A's

$$\begin{aligned}
F(\alpha,\beta) &= \sum_{m,n=0}^{\infty} \frac{\beta^m \alpha^n}{\sqrt{m!n!}} \exp\left(-\frac{1}{4}m\omega_0(\alpha^2+\beta^2)\right) \\
&\quad \times \left(\frac{m\omega_0}{2}\right)^{\frac{1}{2}(m+n)} A_{mn} e^{-i\omega_0 T(m+\frac{1}{2})} \, .
\end{aligned} \tag{3.73}$$

We can read off the form of A_{mn} by comparing these two alternate forms for $F(\alpha,\beta)$. Thus, expanding Eq. 3.71, we find

$$\begin{aligned}
F(\alpha,\beta) &= \exp-\frac{m\omega_0}{4}\left(\lambda_1^2+\lambda_2^2-2\lambda_1\lambda_2 e^{-i\omega_0 T}\right) e^{i\sigma-i\frac{\omega_0 T}{2}}\exp-\frac{m\omega_0}{4}(\alpha^2+\beta^2) \\
&\quad \times \sum_{s,t,u=0}^{\infty} \frac{1}{s!t!u!}\left(-\frac{m\omega_0}{4}\beta\eta_2\right)^s\left(-\frac{m\omega_0}{4}\alpha\eta_1\right)^t\left(\frac{m\omega_0}{2}\alpha\beta\, e^{-i\omega_0 T}\right)^u
\end{aligned} \tag{3.74}$$

where

$$\begin{aligned}
\eta_1 &= -2\lambda_1+2\lambda_2\, e^{-i\omega_0 T} = -i\frac{2}{m\omega_0}\gamma \\
\eta_2 &= -2\lambda_2+2\lambda_1\, e^{-i\omega_0 T} = -i\frac{2}{m\omega_0}\gamma^*\, e^{-i\omega_0 T}
\end{aligned} \tag{3.75}$$

and

$$\gamma = \int_{-\infty}^{\infty} dt'\, F(t')\, e^{-i\omega_0 t'} \, . \tag{3.76}$$

Now resum the series using $m=u+s$, $n=t+u$ yielding

$$\begin{aligned}
F(\alpha,\beta) &= \exp\left[-\frac{m\omega_0}{4}\left(\lambda_1^2+\lambda_2^2-2\lambda_1\lambda_2\, e^{-i\omega_0 T}\right)+i\sigma-i\frac{1}{2}\omega_0 T-\frac{m\omega_0}{4}(\alpha^2+\beta^2)\right] \\
&\quad \times \sum_{m,n=0}^{\infty}\sum_{r=0}^{\min(n,m)} \frac{\alpha^n \beta^m\, e^{-im\omega_0 T}}{r!(n-r)!(m-r)!}\left(\frac{m\omega_0}{2}\right)^r\left(\frac{1}{2}\right)^{n+m-2r}(i\gamma)^{n-r}(i\gamma^*)^{m-r} \, .
\end{aligned} \tag{3.77}$$

Comparison with Eq. 3.73 then yields, assuming $n > m$

$$A_{mn} = A_{00} \sum_{r=0}^{m} \sqrt{m!n!} \, (2m\omega_0)^{r-\frac{1}{2}(m+n)} \frac{(i\gamma)^{n-m}(-|\gamma|^2)^{m-r}}{r!(n-r)!(m-r)!} \tag{3.78}$$

where

$$A_{00} = \exp\left(-\frac{1}{4m\omega_0}|\gamma|^2\right) e^{i\sigma - i\frac{1}{2}m\omega_0\lambda_1\lambda_2 \sin \omega_0 T} \tag{3.79}$$

and we have used the result

$$\lambda_1^2 + \lambda_2^2 - 2\lambda_1\lambda_2 \cos \omega_0 T = \frac{1}{m^2\omega_0^2}|\gamma|^2 \ . \tag{3.80}$$

Finally, defining $j = m - r$ and $z = |\gamma|^2/2m\omega_0$ we have

$$A_{00} = \exp\left(-\frac{1}{2}z\right) e^{i\sigma - i\frac{1}{2}m\omega_0\lambda_1\lambda_2 \sin \omega_0 T}$$

$$A_{mn} = A_{00}\sqrt{m!n!}\left(\frac{i\gamma}{\sqrt{2m\omega_0}}\right)^{n-m} \sum_{j=0}^{m} \frac{(-z)^j}{j!(m-j)!(n+j-m)!} \tag{3.81}$$

$$= A_{00}\left(\frac{m!}{n!}\right)^{1/2}\left(\frac{i\gamma}{\sqrt{2m\omega_0}}\right)^{n-m} L_m^{n-m}(z)$$

which agrees precisely with the result quoted in Eq. 3.45.

PROBLEM V.3.1

Spin 1/2 Particle in a Rotating Magnetic Field [LaL 77]

Consider a spin-1/2 particle of magnetic moment μ in a spatially uniform but time varying magnetic field

$$\vec{B}(t) = B_0 \left(\sin \theta \cos \omega t \, \hat{e}_x + \sin \theta \sin \omega t \, \hat{e}_y + \cos \theta \, \hat{e}_z\right) \ .$$

Clearly this represents a field whose direction maintains constant angle θ with respect to the z-axis but which rotates about $\hat{k}$ with angular velocity ω.

i) Using the Hamiltonian

$$H = -\mu\vec{\sigma} \cdot \vec{B}(t)$$

show that the wavefunction

$$\psi(t) = \begin{pmatrix} \psi_a(t) \\ \psi_b(t) \end{pmatrix}$$

obeys the differential equation

$$i\dot{\psi}_a = -\mu B \left(\cos\theta \psi_a + \sin\theta e^{-i\omega t}\psi_b\right)$$
$$i\dot{\psi}_b = -\mu B \left(\sin\theta e^{i\omega t}\psi_a - \cos\theta \psi_b\right)$$

V.3 THE ADIABATIC APPROXIMATION   245

ii) Employing the substitution $\psi_a = e^{-i\frac{1}{2}\omega t}\chi_a$, $\psi_b = e^{i\frac{1}{2}\omega t}\chi_b$ demonstrate that this set of coupled equations can be solved exactly to yield

$$\psi_a = e^{-i\frac{1}{2}\omega t}\left(c_1 e^{i\frac{1}{2}\Omega t} + c_2 e^{-i\frac{1}{2}\Omega t}\right)$$

$$\psi_b = 2\mu B \sin\theta e^{i\frac{1}{2}\omega t}\left(\frac{c_1 e^{i\frac{1}{2}\Omega t}}{\Omega + \omega + 2\mu B \cos\theta} - \frac{c_2 e^{-i\frac{1}{2}\Omega t}}{\Omega - \omega - 2\mu B \cos\theta}\right)$$

where

$$\Omega = \left[(\omega + 2\mu B \cos\theta)^2 + 4(\mu B \sin\theta)^2\right]^{1/2}$$

and c_1, c_2 are arbitrary constants.

PROBLEM V.3.2

The Forced Oscillator and Functional Techniques

The general result for forced oscillator transition probabilities is quoted in Eq. 3.45 and derived in the appendix. However, it is also possible to calculate such amplitudes directly from the propagator using the tricks developed in Chapter I in our discussion of the generating functional. Thus, defining

$$W[j] = \lim_{\substack{t_2\to\infty \\ t_1\to-\infty}} D_F^j(x_2,t_2;x_1,t_1)$$

i) Show that

a)
$$\langle 0(\text{out})|0(\text{in})\rangle = \frac{W[j]}{W[0]} = \exp -\frac{1}{4m\omega_0}\int_{-\infty}^{\infty}dt\int_{-\infty}^{\infty}dt'\,j(t)j(t')e^{-i\omega_0|t-t'|}$$
$$\equiv \exp -\frac{1}{2}z\ .$$

b)
$$\langle 0(\text{out})|1(\text{in})\rangle = \lim_{t_1\to-\infty}-i\sqrt{\frac{m\omega_0}{2}}\left(1-\frac{i}{\omega_0}\frac{\partial}{\partial t_1}\right)\frac{1}{W[0]}\frac{\delta W[j]}{\delta j(t_1)}$$
$$= -\frac{\gamma}{\sqrt{2m\omega_0}}\exp\left(-\frac{1}{2}z\right)e^{i\omega_0 t_1}$$

c)
$$\langle 1(\text{out})|1(\text{in})\rangle = (-i)^2 \lim_{\substack{t_2\to\infty \\ t_1\to-\infty}}\frac{m\omega_0}{2}\left(1+\frac{i}{\omega_0}\frac{\partial}{\partial t_2}\right)\left(1-\frac{i}{\omega_0}\frac{\partial}{\partial t_1}\right)$$
$$\times \frac{1}{W[0]}\frac{\delta^2 W[j]}{\delta j(t_2)\delta j(t_1)}$$
$$= (-i)^2(1-z)\exp\left(-\frac{1}{2}z\right)e^{-i\omega_0(t_2-t_1)}$$

where
$$\gamma = \int_{-\infty}^{\infty} dt\, e^{i\omega t} j(t)$$
$$z = \frac{|\gamma|^2}{2m\omega_0}$$

ii) Compare these results with those found in the appendix.

PROBLEM V.3.3

Time-Dependent Harmonic Oscillator and the Adiabatic Approximation

The problem of the harmonic oscillator with time-dependent frequency $\omega(t)$ is exactly soluble.

i) Demonstrate that

$$\phi_n(x,t) = \left(\frac{m\dot\gamma}{\pi}\right)^{\frac{1}{4}} \frac{1}{(2^n n!)^{1/2}} \exp\left(-i\left(n+\frac{1}{2}\right)\gamma + \frac{im}{2}\left(i\dot\gamma + \frac{\dot s}{s}\right)x^2\right)$$
$$\times H_n\left((m\dot\gamma)^{1/2} x\right)$$
(3.82)

represents an exact solution to the time-dependent Schrödinger equation, with γ, s defined via
$$\dot\gamma = s^{-2}$$
$$\ddot s - s^{-3} + \omega^2(t)s = 0 \ .$$
(3.83)

The corresponding representation of the propagator is

$$D_F(x_2, t_2; x_1, t_1) = \sum_n \phi_n(x_2, t_2)\phi_n^*(x_1, t_1)$$
$$= \left(\frac{m(\dot\gamma_2\dot\gamma_1)^{1/2}}{2\pi i \sin(\gamma_2 - \gamma_1)}\right)^{1/2} \exp i\frac{m}{2}\left\{\left(\dot\gamma_2 x_2^2 + \dot\gamma_1 x_1^2\right)\ctn(\gamma_2 - \gamma_1)\right.$$
(3.84)
$$\left. - 2x_2 x_1 \csc(\gamma_2 - \gamma_1)(\dot\gamma_2\dot\gamma_1)^{1/2} + \frac{\dot s_2}{s_2}x_2^2 - \frac{\dot s_1}{s_1}x_1^2\right\} \ .$$

ii) An oscillator which at time $t \to \infty$ is in the ground state must then have evolved from a wavefunction

$$\phi_0(x_1, t) = \left(\frac{m\dot\gamma}{\pi}\right)^{1/4} \exp\left(-i\frac{\gamma}{2} + \frac{im}{2}\left(i\dot\gamma + \frac{\dot s}{s}\right)x^2\right)$$
(3.85)

at an earlier instant of time. Derive the identity

$$\int_{-\infty}^{\infty} dx\, e^{-x^2} H_{2n}(ax) = \sqrt{\pi}\frac{(2n)!}{n!}(a^2 - 1)^n$$
(3.86)

and show then that if $a_n(t)$ is the amplitude to be in asymptotic oscillator eigenstate $|n\rangle$, i.e.,

$$a_n(t) = \int_{-\infty}^{\infty} dx\, \psi_n^*(x)\phi_0(x,t) \qquad (3.87)$$

we have

$$a_{2n}(t) = \frac{\sqrt{(2n)!}}{2^n \sigma^{1/2}} (m^2\omega_0\dot\gamma)^{1/4} \left(\frac{m\omega_0}{\sigma} - 1\right)^n \exp\left(-i\frac{\gamma}{2} + i(n+\frac{1}{2})\omega_0 t\right) \qquad (3.88)$$

where

$$\sigma = \frac{m}{2}\left(\dot\gamma + \omega_0 - i\frac{\dot s}{s}\right) \qquad (3.89)$$

iii) Calculate $P_{2n}(t)$ and show that

$$P_{2n}(t) = \frac{(2n)!}{4^n (n!)^2} \frac{2(\omega_0\dot\gamma)^{1/2}}{\left[(\omega_0+\dot\gamma)^2 + \left(\frac{\dot s}{s}\right)^2\right]^{1/2}} \left[\frac{(\omega_0-\dot\gamma)^2 + \left(\frac{\dot s}{s}\right)^2}{(\omega_0+\dot\gamma)^2 + \left(\frac{\dot s}{s}\right)^2}\right]^n \qquad (3.90)$$

iv) Compare this exact result for $P_{2n}(t)$ with that of the adiabatic approximation.

V.4 BERRY'S PHASE

In the previous section we studied the adiabatic limit, wherein the time scale over which the (time-dependent) Hamiltonian varies is long compared to typical quantum mechanical oscillation periods, and showed that the leading piece of the propagator is given by

$$D_F(\vec x_2, t_2; \vec x_1, t_1) = \sum_n \psi_n(\vec x_2; t_2) \exp\left(-i\int_{t_1}^{t_2} dt\, E_n(t)\right) \psi_n^*(\vec x_1; t_1) \qquad (4.1)$$

where $E_n(t)$, $\psi_n(\vec x; t)$ are instantaneous eigenvalues, eigenfunctions of the time-dependent Hamiltonian $h(t)$, satisfying

$$h(t)\psi_n(\vec x; t) = E_n(t)\psi_n(\vec x; t) \ . \qquad (4.2)$$

It was emphasized in our derivation that the result is dependent upon the assumption that one is free to adjust the phase of the instantaneous eigenstates arbitrarily and that such freedom exists in many but not all circumstances. In this section we examine the more general result which asserts that when at time t_2 the Hamiltonian $h(t)$ returns, after an adiabatic excursion, to its form at time t_1 then in addition to the dynamical phase factor

$$\exp\left(-i\int_{t_1}^{t_2} E_n(t)dt\right) \qquad (4.3)$$

associated with instantaneous eigenstate $\psi_n(\vec x; t)$ there exists a topological phase factor $\exp i\gamma_n$ [Be 84].

Berry's Phase: Formalism

Consider a Hamiltonian

$$h(R_i(t)) \qquad i = 1, 2, \ldots k \tag{4.4}$$

which depends on a set of k time-dependent parameters $R_i(t)$. An example could be dependence on a time-dependent vector field, in which case $k = 3$ and the parameters are the three independent components of the field vector. However, we shall deal here with the general case. We suppose that the rate of change of the parameters $R_i(t)$ is much slower than a typical orbital frequency $\Delta E_n(t)$ so that the adiabatic condition obtains. We already know that a system that begins at time t_1 in instantaneous eigenstate $\psi_n(\vec{x}; t_1)$ will evolve almost completely into eigenstate $\psi_n(\vec{x}; t_2)$ at time t_2. Here we wish to look at the phase factor which accompanies $\psi_n(\vec{x}; t_2)$. Writing the general solution to the Schrödinger equation as $\phi(\vec{x}, t)$, where

$$i\frac{\partial}{\partial t}\phi(\vec{x}, t) = h(R_i(t))\phi(\vec{x}, t) \tag{4.5}$$

we define

$$\phi(\vec{x}, t) \cong \psi_n(\vec{x}; t) \exp\left(-i \int_{t_1}^{t} dt' E_n(t')\right) a_n(t) \ . \tag{4.6}$$

Then

$$i\dot{\psi}_n(\vec{x}; t)a_n + i\psi_n(\vec{x}; t)\dot{a}_n = 0 \ . \tag{4.7}$$

From the normalization condition

$$\langle \psi_n(\vec{x}; t)|\psi_n(\vec{x}; t)\rangle = \langle \phi(\vec{x}, t)|\phi(\vec{x}, t)\rangle = 1 \tag{4.8}$$

we find

$$a_n(t)\dot{a}_n^*(t) + a_n^*(t)\dot{a}_n(t) = 0 \tag{4.9}$$

which implies that $a_n(t)$ is a simple phase factor

$$a_n(t) = \exp i\gamma_n(t) \tag{4.10}$$

with

$$\dot{\gamma}_n(t) = i \int d^3x \, \psi_n^*(\vec{x}; t)\dot{\psi}_n(\vec{x}; t) \ . \tag{4.11}$$

Since ψ_n achieves its time dependence only from the existence of the parameters $R_i(t)$ (*i.e.*, $\psi_n(\vec{x}; t)$ would be time-independent were R_i = constant), we may write

$$\psi_n(\vec{x}; t) \equiv \psi_n(\vec{x}, R_i(t)) \tag{4.12}$$

where, using the chain rule

$$\dot{\gamma}_n(t) = i \int d^3x \sum_i \psi_n^*(\vec{x}, R_i(t)) \frac{d}{dR_i} \psi_n(\vec{x}, R_i(t)) \dot{R}_i(t) \ . \tag{4.13}$$

Writing
$$\vec{R}(t) = \begin{pmatrix} R_1(t) \\ R_2(t) \\ \vdots \\ R_k(t) \end{pmatrix} \tag{4.14}$$

as a k-component column vector, we can express Eq. 4.13 in the simplified notation

$$\dot{\gamma}_n(t) = i \left\langle n; \vec{R}(t) | \vec{\nabla}_R | n; \vec{R}(t) \right\rangle \cdot \dot{\vec{R}}(t) \ . \tag{4.15}$$

Thus far, there is nothing new and the existence of this phase $\gamma_n(t)$ in addition to the usual dynamical phase $\int_{t_1}^{t} dt' E_n(R_i(t'))$ has been known for a long time. However, it was always assumed that $\gamma_n(t)$ could be eliminated by redefining the phase of the eigenstate $|n; \vec{R}(t)\rangle$. In 1984 Berry realized that such a phase *is* observable when the time evolution brings the parameter vector $\vec{R}(t)$ back to its starting point, *i.e.*, when

$$\vec{R}(t_2) = \vec{R}(t_1) \tag{4.16}$$

whereby the state vector $|n, \vec{R}(t_2)\rangle$ can be interfered with $|n; \vec{R}(t_1)\rangle$. This quantity, which may be written as

$$\gamma_n = i \int_{t_1}^{t_2} dt \, \dot{\vec{R}}(t) \cdot \left\langle n; \vec{R}(t) | \vec{\nabla}_R \, n; \vec{R}_n(t) \right\rangle$$
$$= i \oint d\vec{R} \cdot \left\langle n; \vec{R}(t) | \vec{\nabla}_R n; \vec{R}(t) \right\rangle \ , \tag{4.17}$$

is called Berry's phase and is an *observable*. That γ_n is a physically meaningful quantity can be further emphasized by expressing this result in the suggestive notation

$$\vec{A}(\vec{R}) = i \left\langle n; \vec{R} | \vec{\nabla}_R n; \vec{R} \right\rangle \tag{4.18}$$

so that

$$\gamma_n = \oint d\vec{R} \cdot \vec{A}(\vec{R}) \tag{4.19}$$

is written in terms of a "vector potential" like field. If we choose to redefine the phase of the eigenstate by

$$|n; \vec{R}\rangle \to \exp i\phi(\vec{R}) |n; \vec{R}\rangle \tag{4.20}$$

then

$$\vec{A}(\vec{R}) \to \vec{A}(\vec{R}) - \vec{\nabla}_R \phi(\vec{R}) \tag{4.21}$$

which is analogous to a gauge transformation. An observable cannot depend upon the choice of gauge and it is clear that Berry's phase has this property since, by Stokes' theorem (suitably generalized if $k \neq 3$) we can write

$$\gamma_n = \oint d\vec{R} \cdot \vec{A}(\vec{R}) = \int_{\text{surf}} d\vec{S} \cdot \left(\vec{\nabla} \times \vec{A}(\vec{R}) \right)$$
$$\to \int_{\text{surf}} d\vec{S} \cdot \vec{\nabla} \times \left(\vec{A}(\vec{R}) - \vec{\nabla}\phi(\vec{R}) \right) = \int_{\text{surf}} d\vec{S} \cdot \vec{\nabla} \times \vec{A}(\vec{R}) \tag{4.22}$$

so that γ_n is unchanged.

A physical significance may be ascribed to this "vector potential" in the circumstance that the parameters R_i are themselves quantized. That is, suppose that the full Hamiltonian for some system is

$$H = \frac{\vec{p}^2}{2m} + \frac{\vec{P}^2}{2M} + V(\vec{R}, \vec{r}) \;. \tag{4.23}$$

in an obvious notation. Writing the full wavefunction as

$$|\psi(\vec{R}, \vec{r})\rangle = |n; \vec{R}\rangle \chi\left(\vec{R}\right) \tag{4.24}$$

we identify

$$h\left(\vec{R}\right) = \frac{\vec{p}^2}{2m} + V\left(\vec{R}, \vec{r}\right) \tag{4.25}$$

as the simple time-dependent Hamiltonian considered previously, with eigenstates $|n; \vec{R}\rangle$ and eigenvalues $E_n(\vec{R})$ which satisfy

$$h\left(\vec{R}\right) |n; \vec{R}\rangle = E_n\left(\vec{R}\right) |n; \vec{R}\rangle \;. \tag{4.26}$$

Then

$$\begin{aligned} H|\psi(\vec{R}, \vec{r})\rangle &= E_n(\vec{R})|n; \vec{R}\rangle \chi(\vec{R}) + |n, \vec{R}\rangle \frac{\vec{P}^2}{2M} \chi(\vec{R}) \\ &\quad - i\vec{\nabla}_R |n; \vec{R}\rangle \cdot \frac{1}{M} \vec{P} \chi(\vec{R}) - \frac{1}{2M} \left(\vec{\nabla}_R^2 |n; \vec{R}\rangle \right) \chi(\vec{R}) \;. \end{aligned} \tag{4.27}$$

Finally, projecting out the $|n; \vec{R}\rangle$ states and neglecting off-diagonal matrix elements — $\langle n'; \vec{R} | \vec{\nabla}_R n; \vec{R}\rangle \propto \delta_{nn'}$ — we find that $\chi(\vec{R})$ obeys a Schrödinger equation with an effective Hamiltonian

$$H_{\text{eff}} = \frac{1}{2M} \left(\vec{P} - \vec{A}(\vec{R}) \right)^2 + U(\vec{R}) \;. \tag{4.28}$$

This is, of course, simply the Born–Oppenheimer approach to a system containing both fast $(\vec{r}, \vec{p})$ and slow $(\vec{R}, \vec{P})$ degrees of freedom [BoO 27]. We observe that the fast system affects the dynamics of its slow counterpart by means of an effective scalar potential

$$U(\vec{R}) = E_n(\vec{R}) - \frac{1}{2M} \left(\langle n; \vec{R} | \vec{\nabla}_R^2 | n; \vec{R} \rangle + \vec{A}^2(\vec{R}) \right) \tag{4.29}$$

and vector potential

$$\vec{A}(\vec{R}) = i \langle n; \vec{R} | \vec{\nabla}_R n; \vec{R} \rangle \;. \tag{4.30}$$

Berry's phase is simply the closed line integral of this vector potential

$$\gamma_n = \oint \vec{A}(\vec{R}) \cdot d\vec{R} \equiv \int \vec{B}(\vec{R}) \cdot d\vec{S} \tag{4.31}$$

V.4 BERRY'S PHASE

where we have used Stokes' theorem to write γ_n as the flux of the field

$$\vec{B}(\vec{R}) = \vec{\nabla}_R \times \vec{A}(\vec{R}) \tag{4.32}$$

through the corresponding surface. The field $\vec{B}(\vec{R})$ will have a non-trivial structure in the presence of sources, which occur when two or more of the "fast" eigenvalues $|n; \vec{R}(t)\rangle$ become degenerate for some value of $\vec{R}(t)$. We can see this since

$$\begin{aligned}\gamma_n &= i \oint d\vec{R} \cdot \left\langle n; \vec{R} | \vec{\nabla}_R n; \vec{R} \right\rangle \\ &= - \operatorname{Im} \int d\vec{S} \cdot \left\langle \vec{\nabla}_R n; \vec{R} | \times \vec{\nabla}_R n; \vec{R} \right\rangle .\end{aligned} \tag{4.33}$$

Now insert a complete set of intermediate states

$$\mathbf{1} = \sum_m |m; \vec{R}\rangle \langle m; \vec{R}| \tag{4.34}$$

and note that the diagonal contribution vanishes since $\left\langle n; \vec{R} | \vec{\nabla}_R n; \vec{R} \right\rangle$ is pure imaginary.[†] From the eigenvalue condition, Eq. 4.26, we find

$$\left(\vec{\nabla}_R h(\vec{R})\right) |n; \vec{R}\rangle + h(\vec{R}) \vec{\nabla}_R |n; R\rangle = E_n(\vec{R}) \vec{\nabla}_R |n; \vec{R}\rangle + \left(\vec{\nabla}_R E_n(R)\right) |n; \vec{R}\rangle \tag{4.35}$$

so that, projecting on to $|\ell; \vec{R}\rangle$

$$\left\langle \ell; \vec{R} | \vec{\nabla}_R h(\vec{R}) | n; \vec{R} \right\rangle = \left(E_n(\vec{R}) - E_\ell(\vec{R})\right) \left\langle \ell; \vec{R} | \vec{\nabla}_R n; \vec{R} \right\rangle \quad (n \neq \ell) . \tag{4.36}$$

We find then

$$\vec{B}(\vec{R}) = - \operatorname{Im} \sum_{\ell \neq n} \frac{\left\langle n; \vec{R} | \vec{\nabla}_R h | \ell; \vec{R} \right\rangle \times \left\langle \ell; \vec{R} | \vec{\nabla}_R h | n; \vec{R} \right\rangle}{\left(E_\ell(\vec{R}) - E_n(\vec{R})\right)^2} . \tag{4.37}$$

The existence of a degeneracy implies an infinity in $\vec{B}(\vec{R})$ and thus the presence of a field source. The Berry phase is the flux associated with such sources.

[†] Note that the normalization condition $\left\langle n; \vec{R} | n; \vec{R} \right\rangle = 1$ requires that

$$\begin{aligned} 0 = \vec{\nabla}_R \left\langle n; \vec{R} | n; \vec{R} \right\rangle &= \left\langle n; \vec{R} | \vec{\nabla}_R n; \vec{R} \right\rangle \\ &+ \left\langle \vec{\nabla}_R n; \vec{R} | n; \vec{R} \right\rangle = 2 \operatorname{Re} \left\langle n; \vec{R} | \vec{\nabla}_R n; \vec{R} \right\rangle .\end{aligned}$$

Example: Spin 1/2 Particle in a Time Dependent Magnetic Field

Now let's see an example of the use of this formalism. The simplest is a spin-1/2 particle in a time dependent external magnetic field — "$\vec{R}(t)$" — for which the relevant Hamiltonian is

$$h\left(\vec{R}(t)\right) = -\frac{\mu}{2}\vec{\sigma}\cdot\vec{R}(t) = -\frac{\mu}{2}\begin{pmatrix} Z(t) & X(t)-iY(t) \\ X(t)+iY(t) & -Z(t) \end{pmatrix}. \qquad (4.38)$$

[We have written the external magnetic field as $\vec{R}(t)$ so as not to be confused with the curl of the vector potential which we have denoted by the symbol $\vec{B}(t)$.] The eigenvalues of h are given by

$$E_+(\vec{R}) = -E_-(\vec{R}) = \frac{\mu}{2}\left(X^2(t)+Y^2(t)+Z^2(t)\right)^{1/2} \qquad (4.39)$$

so that a degeneracy exists when $\vec{R}=0$. Since

$$\vec{\nabla}_R h(\vec{R}) = -\frac{\mu}{2}\vec{\sigma} \qquad (4.40)$$

we observe that, picking $\vec{R}$ to lie along the z-axis

$$\begin{aligned}\vec{B}_\uparrow(R\hat{k}) &= -\operatorname{Im}\langle\uparrow|\vec{\sigma}|\downarrow\rangle\times\langle\downarrow|\vec{\sigma}|\uparrow\rangle\frac{1}{4R^2(t)} \\ &= -\frac{\hat{k}}{4R^2(t)}\operatorname{Im}\left(\langle\uparrow|\sigma_x|\downarrow\rangle\langle\downarrow|\sigma_y|\uparrow\rangle - \langle\uparrow|\sigma_y|\downarrow\rangle\langle\downarrow|\sigma_x|\uparrow\rangle\right) \qquad (4.41) \\ &= -\frac{\hat{k}}{2R^2(t)}\end{aligned}$$

or in general

$$\vec{B}_\uparrow(\vec{R}) = -\frac{\hat{R}}{2R^2(t)} \qquad (4.42)$$

which corresponds to a "magnetic monopole" of strength $-1/2$ located at the origin (i.e., at the place of degeneracy). The Berry phase is the flux associated with this monopole through the surface which is circumnavigated in parameter space and is given by

$$\gamma_\uparrow = \pm\frac{1}{2}\Delta\Omega \qquad (4.43)$$

where $\Delta\Omega$ is the solid angle subtended by the closed path as seen from the origin of parameter space and the $\pm$ refers to the direction in which the line integration is traversed.

Eq. 4.43 may be verified explicitly by use of the representation

$$|\uparrow;\vec{R}\rangle = \begin{pmatrix} \cos\frac{\theta}{2} \\ \sin\frac{\theta}{2}e^{i\phi} \end{pmatrix} \qquad (4.44)$$

for a spinor aligned along $\hat{R}$ and specified by spherical coordinates θ, ϕ. The associated vector potential $|\uparrow; \vec{R}\rangle$ is found to be

$$\vec{A}_\uparrow = i\left\langle \uparrow; \vec{R} | \vec{\nabla}_R \uparrow; \vec{R} \right\rangle = -\hat{a}_\phi \frac{1}{R\sin\theta} \sin^2\frac{\theta}{2}$$
$$= -\hat{a}_\phi \frac{1}{2R} \tan\frac{\theta}{2} \qquad (4.45)$$

which since

$$\vec{\nabla}_R \times \vec{A}_\uparrow(\vec{R}) = \hat{a}_r \frac{1}{R\sin\theta} \frac{\partial}{\partial\theta}(\sin\theta A_\phi) = -\hat{a}_r \frac{1}{2R^2} \qquad (4.46)$$

represents the vector potential of a magnetic monopole of "charge" $-1/2$ located at the origin.

Taking a path in parameter space at fixed θ from θ, ϕ to $\theta, \phi + 2\pi$ we find the Berry phase to be

$$\gamma_\uparrow = \oint \vec{A}_\uparrow(\vec{R}) \cdot d\vec{R} = A_\phi 2\pi R \sin\theta = \pm\pi(1 - \cos\theta) \ . \qquad (4.47)$$

As seen from the origin, the solid angle swept out by this trajectory is

$$\Delta\Omega = \int_0^\theta d\theta' \sin\theta' \int_0^{2\pi} d\phi = 2\pi(1 - \cos\theta) \qquad (4.48)$$

so that

$$\gamma_\uparrow = \pm\frac{1}{2}\Delta\Omega \qquad (4.49)$$

as expected from Eq. 4.43.

Actually, we observed the existence of such an "extra" geometric phase factor in an earlier calculation—(cf. section V.3 wherein we studied the motion of a spin-1/2 particle originally aligned along a magnetic field of the form

$$\vec{R}(t) = R_0\left(\sin\theta\cos\omega t\, \hat{e}_x + \sin\theta\sin\omega t\, \hat{e}_y + \cos\theta\, \hat{e}_z\right) \ . \qquad (4.50)$$

After a time $\Delta t = 2\pi/\omega$ the field has returned to its original configuration and we found that the inner product of initial and final wave functions was given by[†]

$$\left\langle \psi\left(t = \frac{2\pi}{\omega}\right) | \psi(t=0) \right\rangle = \exp\left(i2\pi \frac{\mu R_0}{\omega}\right) \exp{-i\pi(1-\cos\theta)} \qquad (4.51)$$

in the circumstance that the adiabatic condition — $\omega \ll \mu B$ — is satisfied. We observe that $2\pi\mu R_0/\omega$ is the expected dynamical phase

$$\phi_{\rm dyn} = -\int_0^{2\pi/\omega} dt\, E(t) = \mu R_0 \frac{2\pi}{\omega} \ . \qquad (4.52)$$

[†] Note that this result is slightly different from that given in Eq. 3.33. The difference is that $|\chi_B(t)\rangle$ as used there contains a time-dependent phase contribution. The form quoted in Eq. 4.51 corresponds rather to $\langle \chi_B(t=0) | \psi(t=2\pi/\omega)\rangle$.

However, in addition there exists a geometrical phase factor equal to half the solid angle swept out by $\vec{R}(t)$

$$\phi_{\text{geom}} = -\frac{1}{2}\Delta\Omega = -\pi(1-\cos\theta) \qquad (4.53)$$

as predicted in Eq. 4.43.

Another interesting consequence of Berry's phase is that it enables a simple understanding of Dirac's argument that magnetic charge must be quantized in inverse units of electric charge, as discussed in Ch. III. Consider the surface integral shown in Figure V.2

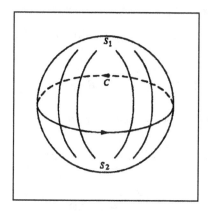

Fig. V.2 : In performing the surface integral one can utilize either surface S_1 or S_2 — the physics must be the same.

One has the choice of employing either the surface S_1 or S_2. But either choice must lead to the same physics, so we conclude

$$\int_{S_1} \vec{B}\cdot d\vec{S} = \int_{S_2} \vec{B}\cdot d\vec{S} + 2\pi p \qquad p = 0, \pm 1, \pm 2, \ldots \qquad (4.54)$$

In the simple spin-1/2 case studied above, we have $p=1$. However, it is straightforward to show that if one begins with a system with spin S and projection S_z that $p = 2S_z$. The Dirac condition follows from use of the conventional electromagnetic Hamiltonian

$$H = \frac{\left(\vec{p}-e\vec{A}\right)^2}{2m} \ . \qquad (4.55)$$

Taking

$$\vec{B} = \frac{g_m}{r^2} \qquad (4.56)$$

as the field of a magnetic monopole with "charge" g_m, we find

$$e\int_{\text{closed}} \vec{B}\cdot d\vec{S} = 4\pi e g_m = 2\pi p \qquad (4.57)$$

or
$$g_m = \frac{p}{2e} \tag{4.58}$$

which is Dirac's constraint that magnetic charge is quantized in units of $1/2e$. Turning the argument around, if there exists a monopole of strength g_m anywhere in the universe then electric charges must be quantized in units

$$e = \frac{p}{2g_m} \tag{4.59}$$

which "explains" the otherwise mysterious charge quantization found in Nature.

There is another interesting aspect here which deserves comment, which is the relationship between the Berry potential and rotational invariance. Specifying an arbitrary rotation by angle $\delta\alpha$ and axis $\hat{n}$ it is well-known that the operator

$$\mathcal{O} = \exp\left(i\vec{L}\cdot\hat{n}\delta\alpha\right) \quad \text{with} \quad \vec{L} = \vec{r}\times\vec{p} \tag{4.60}$$

is the generator of rotations. For example, taking $\hat{n} = \hat{e}_z$ we have

$$\vec{L}\cdot\hat{n} = -i\left(x\frac{\partial}{\partial y} - y\frac{\partial}{\partial x}\right) = -i\frac{\partial}{\partial \phi} \ . \tag{4.61}$$

Then
$$\begin{aligned}\mathcal{O}\psi(r,\theta,\phi) &\cong \left(1 + i\vec{L}\cdot\hat{n}\delta\alpha\right)\psi(r,\theta,\phi) \\ &= \psi(r,\theta,\phi) + \delta\alpha\frac{\partial}{\partial\phi}\psi(r,\theta,\phi) + \ldots \\ &\cong \psi(r,\theta,\phi+\delta\alpha)\end{aligned} \tag{4.62}$$

as expected. If the Hamiltonian of the system is of the simple rotationally invariant form
$$H = \frac{\vec{P}^2}{2M} + U(|\vec{R}|) \tag{4.63}$$

then
$$\mathcal{O}H\mathcal{O}^{-1} = \frac{\left(\vec{P} + \delta\alpha\hat{n}\times\vec{P}\right)^2}{2M} + U(|\vec{R}|) = H + \mathcal{O}(\delta\alpha)^2 \tag{4.64}$$

which implies
$$\left[\vec{L}, H\right] = 0 \ . \tag{4.65}$$

That is, rotational invariance of the Hamiltonian is equivalent to conservation of angular momentum, as required by Noether's theorem.

Now consider the Hamiltonian in the presence of a vector potential $\vec{A}(\vec{R})$

$$H = \frac{1}{2M}\left(\vec{P} - \vec{A}(\vec{R})\right)^2 + U(|\vec{R}|) \ . \tag{4.66}$$

Suppose that $\vec{A}(\vec{R})$ were an ordinary vector field, of the form
$$\vec{A}(\vec{R}) = \hat{R} f(|\vec{R}|) \ . \tag{4.67}$$
Then under the infinitesimal rotation
$$\vec{R} \to \vec{R} + \delta\alpha \hat{n} \times \vec{R} \tag{4.68}$$
we would have
$$\vec{A}(\vec{R}) \to \vec{A}\left(\vec{R} - \delta\alpha\hat{n} \times \vec{R}\right) + \delta\alpha\hat{n} \times \vec{A}(\vec{R}) = \vec{A}(\vec{R}) + \delta\alpha\hat{n} \times \vec{A}(\vec{R}) - \delta\alpha\hat{n} \times \vec{R} \cdot \vec{\nabla}_R \vec{A}(\vec{R}) \ . \tag{4.69}$$
This transformation is achieved by use of the rotation operator
$$\tilde{\mathcal{O}} = \exp\left(i\delta\alpha\hat{n} \cdot \vec{R} \times \left(\vec{P} - \vec{A}(\vec{R})\right)\right) \tag{4.70}$$
which is suggested by the identification
$$\vec{P} - \vec{A}(\vec{R}) = M\dot{\vec{R}} \ . \tag{4.71}$$
Under such a rotation we have
$$\begin{aligned}\tilde{\mathcal{O}}\frac{1}{2M}(\vec{P} - \vec{A}(\vec{R})^2)\tilde{\mathcal{O}}^{-1} &= \frac{1}{2M}\left(\vec{P} - \vec{A}(\vec{R}) + \delta\alpha\hat{n} \times \left(\vec{P} - \vec{A}(\vec{R})\right)\right)^2 \\ &= \frac{1}{2M}\left(\vec{P} - \vec{A}(\vec{R})\right)^2 + \mathcal{O}(\delta\alpha)^2\end{aligned} \tag{4.72}$$
so that the Hamiltonian remains invariant. However, for more general forms of the vector potential we find
$$\begin{aligned}&\tilde{\mathcal{O}}\frac{1}{2M}\left(\vec{P} - \vec{A}(\vec{R})\right)^2 \tilde{\mathcal{O}}^{-1} \\ &= \frac{1}{2M}\left(\vec{P} - \vec{A}(\vec{R}) + \delta\alpha\hat{n} \times (\vec{P} - \vec{A}(\vec{R})) + \delta\alpha\left(\hat{n} \times \vec{R}\right) \times \left(\vec{\nabla} \times \vec{A}(\vec{R})\right)\right)^2 \\ &= \frac{1}{2M}\left(\left(\vec{P} + \vec{A}(\vec{R})\right)^2 + \delta\alpha\sum_{i=1}^{3}\left\{\left(\vec{P} - \vec{A}(\vec{R})\right)_i, \left(\hat{n} \times \vec{R}\right) \times \left(\vec{\nabla} \times \vec{A}(\vec{R})\right)_i\right\}\right) \\ &\neq \frac{1}{2M}\left(\vec{P} - \vec{A}(\vec{R})\right)^2\end{aligned} \tag{4.73}$$
so that the Hamiltonian is *not* invariant. However, invariance of the Hamiltonian is not required in order that the physics remain unchanged. Rather the physics can still be the same provided that the new vector potential is related to the old by a gauge transformation
$$\vec{A}(\vec{R}) \to \vec{A}(\vec{R}) + \vec{\nabla}_R \phi(\vec{R}) \ . \tag{4.74}$$
The condition which we wish to satisfy is
$$\begin{aligned}\vec{A}(\vec{R}) &\to A(\vec{R}) + \delta\alpha\hat{n} \times \vec{A}(\vec{R}) - \delta\alpha\hat{n} \times \vec{R} \cdot \vec{\nabla}_R \vec{A}(\vec{R}) \\ &= \vec{A}(\vec{R}) + \vec{\nabla}_R \phi(\vec{R})\end{aligned} \tag{4.75}$$

or
$$\delta\alpha\left(\hat{n}\times\vec{A}(\vec{R})-\hat{n}\times\vec{R}\cdot\vec{\nabla}_R\vec{A}(\vec{R})\right)=\vec{\nabla}_R\phi(\vec{R}) \ . \tag{4.76}$$

For the case of the monopole vector potential, given in Eq. 4.45 the appropriate scalar function is
$$\phi(\vec{R})=-\delta\alpha\hat{n}\cdot\left(\vec{R}\times\vec{A}(\vec{R})+\frac{1}{2}\hat{R}\right) \ . \tag{4.77}$$

This is clear since
$$\vec{\nabla}_R\phi(\vec{R})=\delta\alpha\left(-\frac{1}{2}\frac{\hat{n}}{R}+\frac{1}{2R}\hat{R}\hat{n}\cdot\hat{R}+\hat{n}\times\vec{A}(\vec{R})-\left(\hat{n}\times\vec{R}\right)_i\vec{\nabla}_R A_i(\vec{R})\right)$$
$$=\delta\alpha\left(-\frac{1}{2R}\hat{R}\times\left(\hat{n}\times\hat{R}\right)+\hat{n}\times\vec{A}(\vec{R})\right. \tag{4.78}$$
$$\left.-\left(\hat{n}\times\vec{R}\right)\times\left(\vec{\nabla}_R\times\vec{A}(\vec{R})\right)-\hat{n}\times\vec{R}\cdot\vec{\nabla}_R\vec{A}(\vec{R})\right) \ .$$

Substitution of $\vec{\nabla}_R\times\vec{A}(\vec{R})=-\frac{1}{2R^2}\hat{R}$ then yields Eq. 4.76.

The Hamiltonian *can* be made invariant by the use of the modified rotation operator
$$\tilde{\mathcal{O}}=\exp i\delta\alpha\hat{n}\cdot\left(\vec{R}\times\left(\vec{P}-\vec{A}(\vec{R})\right)+\frac{1}{2}\hat{R}\right) \tag{4.79}$$

for which we find
$$\tilde{\mathcal{O}}\frac{1}{2M}\left(\vec{P}-\vec{A}(\vec{R})\right)^2\tilde{\mathcal{O}}^{-1}=\frac{1}{2M}\left(\vec{P}-\vec{A}(\vec{R})\right)^2 \ . \tag{4.80}$$

The interpretation must be then that
$$\vec{J}=\vec{R}\times\left(\vec{P}-\vec{A}(\vec{R})\right)+\frac{1}{2}\hat{R} \tag{4.81}$$

is somehow a conserved angular momentum. Writing the first piece as $\vec{R}\times M\dot{\vec{R}}$ we recognize this term as the usual (kinetic) rotational angular momentum. The second piece — $\frac{1}{2}\hat{R}$ — is not as familiar, but it too has an elementary interpretation as the angular momentum contained in the $\vec{E},\vec{B}$ fields of a charge e, monopole g_m system with $eg_m=-\frac{1}{2}$. If we place the monopole at the origin of coordinates and the charge at height a along the z-axis, we have

$$\vec{E}=\frac{e}{4\pi}\frac{\vec{r}-a\hat{k}}{|\vec{r}-a\hat{e}_z|^3}$$
$$\vec{B}=\frac{g_m\vec{r}}{r^3} \tag{4.82}$$

and the field angular momentum is found to be

$$\begin{aligned}
\vec{J}_{\text{field}} &= \int d^3r\, \vec{r} \times \left(\vec{E} \times \vec{B}\right) = \frac{g_m ea}{4\pi} \int \frac{d^3r}{r^3} \frac{r^2 \cos\theta\, \hat{r} - \hat{e}_z r^2}{(r^2 + a^2 - 2ra\cos\theta)^{3/2}} \\
&= \hat{e}_z \frac{1}{2} g_m ea \int_0^\infty dr\, r \int_{-1}^1 d(\cos\theta) \frac{\cos^2\theta - 1}{(r^2 + a^2 - 2ar\cos\theta)^{3/2}} \\
&= -\hat{e}_z \frac{1}{2} g_m e \int_{-1}^1 d(\cos\theta)\,(1 + \cos\theta) = -\hat{e}_z g_m e \ ,
\end{aligned} \qquad (4.83)$$

which, for $g_m e = -\frac{1}{2}$, is the result we were seeking. The symbol $\vec{J}$ then represents the *total* angular momentum of the charge/monopole system — kinetic plus field angular momentum — as might be expected.

CHAPTER VI

THE KLEIN-GORDON EQUATION

VI.1 DERIVATION AND COVARIANCE

The requirements which special relativity imposes upon quantum mechanics are both fascinating and far-reaching[†] We begin our discussion of these effects by considering the wave equation obeyed by particles of zero spin, examples of which are provided by π, K, η mesons, etc.

Relativistic Schrödinger Equation

We review first the heuristic "derivation" of the Schrödinger equation which results from writing the standard non-relativistic relation between energy and momentum

$$E = \frac{\vec{p}^2}{2m} \tag{1.1}$$

and making the correspondence between the energy-momentum four-vector $p^\mu = (E, \vec{p})$ and the four-dimensional gradient operator ∂^μ

$$i\partial^\mu = i\left(\frac{\partial}{\partial t}, -\vec{\nabla}\right) \sim p^\mu = (E, \vec{p}) \quad . \tag{1.2}$$

We have then

$$i\frac{\partial}{\partial t}\psi = -\frac{1}{2m}\vec{\nabla}^2 \psi \tag{1.3}$$

which is the Schrödinger equation for a free particle. [Recall that if we make a Lorentz transformation to a frame S' moving with velocity $v\hat{k}$ with respect to frame S, then

$$t' = \frac{t - vz}{\sqrt{1 - \vec{v}^2}} \quad , \quad z' = \frac{z - vt}{\sqrt{1 - \vec{v}^2}} \quad , \quad x' = x \quad , \quad y' = y \quad . \tag{1.4}$$

According to the chain rule

$$\begin{aligned}
\partial'^0 &= \frac{\partial}{\partial t'} = \frac{\partial t}{\partial t'}\frac{\partial}{\partial t} + \frac{\partial z}{\partial t'}\frac{\partial}{\partial z} = \frac{\partial t}{\partial t'}\partial^0 - \frac{\partial z}{\partial t'}\partial^3 \\
\partial'^3 &= -\frac{\partial}{\partial z'} = -\frac{\partial t}{\partial z'}\frac{\partial}{\partial t} - \frac{\partial z}{\partial z'}\frac{\partial}{\partial z} = -\frac{\partial t}{\partial z'}\partial^0 + \frac{\partial z}{\partial t'}\partial^3 \\
\partial'^1 &= \partial^1 \\
\partial'^2 &= \partial^2 \quad .
\end{aligned} \tag{1.5}$$

[†] Much of our discussion in this section is based on corresponding material found in Gordon Baym's *Lectures on Quantum Mechanics* [Ba69].

From the inverse Lorentz transformation

$$t = \frac{t' + vz'}{\sqrt{1-\vec{v}^2}} \quad, \quad z = \frac{z' + vt'}{\sqrt{1-\vec{v}^2}} \quad, \quad x = x' \quad, \quad y = y' \tag{1.6}$$

we find

$$\frac{\partial t}{\partial t'} = \frac{\partial z}{\partial z'} = \frac{1}{\sqrt{1-\vec{v}^2}} \quad, \quad \frac{\partial t}{\partial z'} = \frac{\partial z}{\partial t'} = \frac{v}{\sqrt{1-\vec{v}^2}} \; . \tag{1.7}$$

Hence

$$\partial'^0 = \frac{\partial^0 - v\partial^3}{\sqrt{1-\vec{v}^2}} \quad, \quad \partial'^3 = \frac{\partial^3 - v\partial^0}{\sqrt{1-\vec{v}^2}} \quad, \quad \partial'^1 = \partial^1 \quad, \quad \partial'^2 = \partial^2 \tag{1.8}$$

so that ∂^μ is indeed a four-vector. Note that $\partial_\mu = \left(\frac{\partial}{\partial t}, \vec{\nabla}\right)$ does *not* have this property. The minus sign in Eq. 1.2 is essential.]

Now consider how Eq. 1.3 might be modified by the strictures of special relativity, wherein the relation between energy and momentum is

$$E = \sqrt{m^2 + \vec{p}^2} \; . \tag{1.9}$$

As a first guess, we might try the wave equation

$$i\frac{\partial}{\partial t}\psi = \sqrt{m^2 - \vec{\nabla}^2}\,\psi \; . \tag{1.10}$$

Seeing the square root with an operator inside is a bit peculiar, but this operation is well-defined if we write ψ as a Fourier transform

$$\psi(\vec{x}, t) = \int \frac{d^3p}{(2\pi)^3} e^{i\vec{p}\cdot\vec{x}} \phi(\vec{p}, t) \; . \tag{1.11}$$

Then

$$\begin{aligned} i\frac{\partial}{\partial t}\psi(\vec{x}, t) &= \int \frac{d^3p}{(2\pi)^3} e^{i\vec{p}\cdot\vec{x}} \sqrt{m^2 + \vec{p}^2}\, \phi(\vec{p}, t) \\ &= \int \frac{d^3p}{(2\pi)^3} e^{i\vec{p}\cdot\vec{x}} \sqrt{m^2 + \vec{p}^2} \int d^3x'\, e^{-i\vec{p}\cdot\vec{x}'} \psi(\vec{x}', t) \; . \end{aligned} \tag{1.12}$$

Interchanging orders of integration, this becomes

$$i\frac{\partial}{\partial t}\psi(\vec{x}, t) = \int d^3x'\, K(\vec{x}, \vec{x}')\psi(\vec{x}', t) \tag{1.13}$$

with

$$K(x, \vec{x}') = \int \frac{d^3p}{(2\pi)^3} e^{i\vec{p}\cdot(\vec{x}-\vec{x}')} \sqrt{m^2 + \vec{p}^2} \; . \tag{1.14}$$

For large $|\vec{x} - \vec{x}'|$ most values of p except for those with $p \lesssim \frac{1}{|x-x'|}$ will lead to rapid oscillation of the exponential and consequently a very small value for the integral. In fact, the integral will be sizable only for $|x - x'| \lesssim \frac{1}{m}$, but this leads to a severe problem. Eq. 1.13 may be used via Taylor's expansion to relate $\psi(x, t+\delta t)$ to values of $\psi\left(x' \sim x \pm \frac{1}{m}, t\right)$.

$$\psi(\vec{x}, t + \delta t) = \psi(\vec{x}, t) + \delta t \frac{\partial \psi(\vec{x}, t)}{\partial t}$$
$$= \psi(\vec{x}, t) - i\delta t \int d^3x' \, K(\vec{x}, \vec{x}')\psi(\vec{x}'t) \ . \tag{1.15}$$

This means that values of $\psi\left(\vec{x} \pm \frac{1}{m}, t\right)$ are affecting $\psi(\vec{x}, t + \delta t)$ even though these two spacetime points are outside the forward light cone, *i.e.*, since δt can be made very small,

$$\left(\delta t^2 - \frac{1}{m^2}\right) < 0 \tag{1.16}$$

which violates causality. We must then abandon Eq. 1.10 as a possible relativistic wave equation.

Klein–Gordon Equation

Next try squaring the energy, momentum relation, yielding

$$E^2 = \vec{p}^2 + m^2 \ . \tag{1.17}$$

The quantum mechanical analog becomes

$$-\frac{\partial^2}{\partial t^2}\phi(\vec{x}, t) = \left(-\vec{\nabla}^2 + m^2\right)\phi(\vec{x}, t) \tag{1.18}$$

or

$$\left(\Box + m^2\right)\phi(\vec{x}, t) = 0 \tag{1.19}$$

where

$$\Box = \frac{\partial^2}{\partial t^2} - \vec{\nabla}^2 \tag{1.20}$$

is the D'Alembertian. We see that Eq. 1.19 is simply the wave equation with the addition of a term involving m^2. This differential equation is properly relativistic and is called the Klein–Gordon equation.

That it is relativistically covariant can be seen by transforming from the original frame S to a new frame S'. Since ∂_μ is a four-vector —

$$\Box = \partial_\mu \partial^\mu = \partial'_\mu \partial^{\mu'} = \Box' \tag{1.21}$$

— then according to an observer in S', the equation reads

$$\left(\Box' + m^2\right)\phi(\vec{x}', t') = 0 \tag{1.22}$$

which has the same form as in the original frame S. This covariance of the equation is required by special relativity, since otherwise the form of the equation could be used in order to determine how fast one is moving.

We first look for plane wave solutions of the Klein–Gordon equation, having the form

$$\phi(\vec{x}, t) = \exp\left(i\vec{p} \cdot \vec{x} - iEt\right) = \exp(-ip_\mu x^\mu) \ . \tag{1.23}$$

Since

$$\begin{aligned}
0 &= \left(\frac{\partial^2}{\partial t^2} - \vec{\nabla}^2 + m^2\right) \exp\left(i\vec{p} \cdot \vec{x} - iEt\right) \\
&= \left(-E^2 + \vec{p}^2 + m^2\right) \exp\left(i\vec{p} \cdot \vec{x} - iEt\right)
\end{aligned} \tag{1.24}$$

we see that $\phi(\vec{x}, t)$ does indeed satisfy the Klein–Gordon equation provided

$$E^2 = \vec{p}^2 + m^2 \tag{1.25}$$

which was our starting point. However, the wavefunction $\phi(\vec{x}, t)$ depends not upon E^2, but rather upon E, which has the values

$$E = \pm\sqrt{\vec{p}^2 + m^2} \ . \tag{1.26}$$

The energy can be positive or *negative*, which was somewhat disconcerting to the researchers first studying this equation. An additional problem was the inability to construct a conserved probability density. For the Schrödinger equation one has a probability density

$$\rho = \psi^* \psi \tag{1.27}$$

and probability current density

$$\vec{j} = \frac{i}{2m} \left(\psi^* \vec{\nabla} \psi - \left(\vec{\nabla} \psi^*\right) \psi\right) \tag{1.28}$$

with the property that

$$\frac{\partial}{\partial t}\rho + \vec{\nabla} \cdot \vec{j} = 0 \ . \tag{1.29}$$

This insures that

$$\frac{d}{dt} \int_V \rho \, d^3r = -\int_S \vec{j} \cdot d\vec{S} \tag{1.30}$$

and says that any probability which flows out of the volume V must pass through the surface. Thus probability is *locally* conserved, which is also required in a relativistic theory. (Simultaneous appearance and disappearance of probability at two spacelike separated points in one frame would not be simultaneous in another and would thus lead to trouble.)

It is easy to construct a conserved current density for the relativistic case via

$$j^\mu = (\rho, \vec{j}) = \frac{i}{2m} \left(\phi^* \partial^\mu \phi - (\partial^\mu \phi^*) \phi\right) \ . \tag{1.31}$$

We observe that

$$\partial_\mu j^\mu = \frac{\partial \rho}{\partial t} + \vec{\nabla} \cdot \vec{j} = \frac{i}{2m} \left((\partial^\mu \phi^*) \partial_\mu \phi + \phi^* \Box \phi - (\Box \phi^*) \phi - (\partial_\mu \phi^*) \partial^\mu \phi \right)$$
$$= \frac{i}{2m} \left(-m^2 \phi^* \phi + m^2 \phi^* \phi \right) = 0 \tag{1.32}$$

and since, $\partial_\mu j^\mu$ is a scalar, this local conservation holds in all frames. The problem is that if we calculate j^μ for the plane wave solution $\phi(\vec{x}, t)$ we find

$$j^\mu = \left(\frac{E}{m}, \frac{\vec{p}}{m} \right) = \frac{p^\mu}{m} , \tag{1.33}$$

whereby

$$\rho = \frac{E}{m} \tag{1.34}$$

can be either positive or negative, depending upon the sign of the energy. Such a negative "probability" density is obviously unsatisfactory.

Another peculiar feature of the relativistic wave equation is that it is *second* order in time, as opposed to the Schrödinger equation which is first order. This difference is an important one. For the Schrödinger equation, this means that, given the state vector $|\psi(0)\rangle$ at time $t = 0$, one can determine the state at all future times, via

$$|\psi(t)\rangle = e^{-i\hat{H}t} |\psi(0)\rangle . \tag{1.35}$$

On the other hand, for the Klein–Gordon equation one requires *two* initial conditions — both the wavefunction $\phi(\vec{x}, 0)$ *and* its time derivative $\dot{\phi}(\vec{x}, 0)$.

These problems are resolved by the realization that a properly relativistic wave equation involves of necessity *both* particle *and* antiparticle degrees of freedom. If we identify the positive energy solution $\phi_{E>0}(\vec{x}, t)$ with the particle, then the corresponding antiparticle solution is constructed from the negative energy solution via

$$\phi_{\text{antiparticle}}(\vec{x}, t) = \phi^*_{E<0}(\vec{x}, t) . \tag{1.36}$$

Then, for example, in the case of a plane wave

$$\phi_{\text{antiparticle}}(\vec{x}, t) = \left(e^{i\vec{p}\cdot\vec{x} + i|E|t} \right)^* = e^{-i\vec{p}\cdot\vec{x} - i|E|t} \tag{1.37}$$

which corresponds to positive energy time development.

This identification is made secure by writing the Klein–Gordon equation in the presence of a vector potential A^μ. Via the aforementioned minimal substitution

$$i\nabla^\mu \to i\nabla^\mu - eA^\mu \tag{1.38}$$

where e is the particle charge, the Klein–Gordon equation becomes, for a particle (positive energy) solution

$$\left((\nabla_\mu + ieA_\mu)(\nabla^\mu + ieA^\mu) + m^2 \right) \phi_{E>0}(\vec{x}, t) = 0 . \tag{1.39}$$

Taking the complex conjugate equation for a negative energy solution, we find

$$\begin{aligned}&\left((\nabla_\mu + ieA_\mu)^* \left(\nabla^\mu + ieA^\mu\right)^* + m^2\right) \phi^*_{E<0}(x,t) \\ &= \left((\nabla_\mu - ieA_\mu)\left(\nabla^\mu - ieA^\mu\right) + m^2\right) \phi_{\text{antiparticle}}(\vec{x},t) = 0\end{aligned} \quad (1.40)$$

which is the Klein–Gordon equation for a particle of opposite charge. The existence of both particle *and* antiparticle degrees of freedom in the wavefunction is the reason that one needs *two* boundary conditions at time $t = 0$ in order to predict the future behavior.

The antiparticle degrees of freedom also resolve the problem of the negative "probability" density. Multiplying by the particle charge e, we identify

$$j^\mu = \frac{ie}{2m}\left(\phi^* \partial^\mu \phi - \left(\partial^\mu \phi^*\right)\phi\right) \quad (1.41)$$

as the electromagnetic current density. Then ρ is the *charge* density which is positive (negative) for positive (negative) energy, *i.e.*, particle (antiparticle) solutions, as expected.

Returning to the plane wave solutions, we note that they are clearly Lorentz invariant since $p_\mu x^\mu$ is a Lorentz scalar. Thus an observer in the particle rest frame ($E = m$, $\vec{p} = 0$) writes the solution as e^{-imt}, while an observer in a frame moving with velocity $\vec{v} = -v\hat{k}$ with respect to the rest frame sees

$$E' = \frac{m}{\sqrt{1-\vec{v}^2}} \quad , \quad \vec{p}' = \frac{mv\hat{k}}{\sqrt{1-\vec{v}^2}} \quad (1.42)$$

and writes the wave function as

$$e^{i\vec{p}'\cdot\vec{x}' - iE't'} \quad (1.43)$$

which is identical since $p_\mu x^\mu$ has the same value in both frames.

The form of the current density is as expected since if we consider a region of space d^3x in the rest frame containing charge

$$dq = \rho(\vec{x},t)d^3x \quad (1.44)$$

then in the primed frame the corresponding volume is

$$d^3x' = \sqrt{1-\vec{v}^2}\, d^3x \quad (1.45)$$

since the dimension of the volume in the direction of the boost is reduced by Lorentz contraction. The same amount of charge must be contained in this contracted region so

$$dq = \rho'(\vec{x}',t')d^3x' \quad . \quad (1.46)$$

We find then

$$\frac{\rho'(\vec{x}',t')}{\rho(x,t)} = \frac{1}{\sqrt{1-\vec{v}^2}} = \frac{E'}{m} \quad . \quad (1.47)$$

VI.1 DERIVATION AND COVARIANCE

Also, we expect the spatial current density to be given by

$$\vec{j}'(\vec{x}',t') = \rho(\vec{x}',t')\vec{v} = \rho(\vec{x}',t')\frac{\vec{p}'}{E'}$$
$$= \frac{\vec{p}'}{m}\rho(\vec{x},t) \ . \tag{1.48}$$

These anticipated behaviors are obviously satisfied by the form

$$j^\mu = e\frac{p^\mu}{m} \tag{1.49}$$

arising from the plane wave solution.

Two-Component Form

In order to deal with a more general class of solutions it is useful to write the Klein–Gordon equation in two-component form

$$\chi_1 = \frac{1}{2}\left[\phi + \frac{i}{m}\left(\partial^0 + ieA^0\right)\phi\right]$$
$$\chi_2 = \frac{1}{2}\left[\phi - \frac{i}{m}\left(\partial^0 + ieA^0\right)\phi\right] \ . \tag{1.50}$$

In terms of this notation the charge density becomes

$$\rho = \frac{ie}{2m}\left[\phi^*\left(\partial^0 + ieA^0\right)\phi - \left(\left(\partial^0 - ieA^0\right)\phi^*\right)\phi\right]$$
$$= \frac{ie}{2m}\left[-im\left(\chi_1 - \chi_2\right)\left(\chi_1 + \chi_2\right)^* - im\left(\chi_1 - \chi_2\right)^*\left(\chi_1 + \chi_2\right)\right] \tag{1.51}$$
$$= e\left(|\chi_1|^2 - |\chi_2|^2\right) \ .$$

The components χ_1, χ_2 obey the coupled equations

$$\left(i\partial^0 - eA^0\right)\chi_1 = \left[\frac{1}{2m}\left(-i\vec{\nabla} - e\vec{A}\right)^2 + \frac{m}{2}\right](\chi_1 + \chi_2) + \frac{m}{2}(\chi_1 - \chi_2)$$
$$= \frac{1}{2m}\left(-i\vec{\nabla} - e\vec{A}\right)^2(\chi_1 + \chi_2) + m\chi_1 \tag{1.52}$$

$$\left(i\partial^0 - eA^0\right)\chi_2 = -\left[\frac{1}{2m}\left(-i\vec{\nabla} - e\vec{A}\right)^2 + \frac{m}{2}\right](\chi_1 + \chi_2) + \frac{m}{2}(\chi_1 - \chi_2)$$
$$= -\frac{1}{2m}\left(-i\vec{\nabla} - e\vec{A}\right)^2(\chi_1 + \chi_2) - m\chi_2 \tag{1.53}$$

These results are displayed most conveniently by defining a two-component "spinor"

$$\chi = \begin{pmatrix} \chi_1 \\ \chi_2 \end{pmatrix} \tag{1.54}$$

in terms of which the Klein–Gordon equation has the form

$$(i\partial^0 - eA^0)\chi = \left[\frac{1}{2m}\left(-i\vec{\nabla} - e\vec{A}\right)^2 (\tau_3 + i\tau_2) + m\tau_3\right]\chi \qquad (1.55)$$

where

$$\tau_1 = \begin{pmatrix} 0 & 1 \\ 1 & 0 \end{pmatrix}, \tau_2 = \begin{pmatrix} 0 & -i \\ i & 0 \end{pmatrix}, \tau_3 = \begin{pmatrix} 1 & 0 \\ 0 & -1 \end{pmatrix} \qquad (1.56)$$

are the Pauli matrices. The charge density may be written as

$$\rho = e\chi^\dagger \tau_3 \chi \qquad (1.57)$$

and the normalization condition becomes

$$\langle \chi | \chi \rangle = \int d^3x\, \chi^\dagger \tau_3 \chi = \pm 1, \qquad (1.58)$$

the sign being determined by whether we start with particles (+) or antiparticles (−).

The Klein–Gordon Hamiltonian

$$H = \frac{1}{2m}\left(-i\vec{\nabla} - e\vec{A}\right)^2 (\tau_3 + i\tau_2) + m\tau_3 + eA^0 \qquad (1.59)$$

does not appear to be Hermitian since

$$(\tau_3 + i\tau_2)^\dagger = \tau_3 - i\tau_2 \neq \tau_3 + i\tau_2 \,. \qquad (1.60)$$

However, since the norm is defined using τ_3 — cf. Eq. 1.58 — we have

$$\langle \chi' | H | \chi \rangle = \int d^3x\, \chi^{\dagger\prime}(x) \tau_3 H \chi(x) \qquad (1.61)$$

and

$$\begin{aligned}\langle \chi' | H | \chi \rangle^* &= \left(\int d^3x\, \chi'^\dagger(x)\tau_3 H\chi(x)\right)^* \\ &= \int d^3x\, \chi^\dagger(x) H^\dagger \tau_3 \chi'(x) = \int d^3x\, \chi^\dagger(x) \tau_3 \left(\tau_3 H^\dagger \tau_3\right) \chi'(x) \\ &= \langle \chi | \tau_3 H^\dagger \tau_3 | \chi' \rangle \,.\end{aligned} \qquad (1.62)$$

The Hamiltonian is thus Hermitian provided

$$\tau_3 H^\dagger \tau_3 = H \,, \qquad (1.63)$$

and since

$$\begin{aligned}\tau_3 (\tau_3 + i\tau_2)^\dagger \tau_3 &= \tau_3 (\tau_3 - i\tau_2) \tau_3 \\ &= \tau_3 + i\tau_2\end{aligned} \qquad (1.64)$$

this condition is satisfied.

Now let's take a look at the free particle solutions ($A^0 = 0$, $\vec{A} = 0$) in terms of this notation. A positive energy solution with momentum $\vec{p}$ and normalized to unit density is given by

$$\phi(x) = \sqrt{\frac{m}{E}} e^{i\vec{p}\cdot\vec{x} - iEt} \equiv \sqrt{\frac{m}{E}} e^{-ip\cdot x} . \tag{1.65}$$

This can be written in two component form as

$$\phi(x) = \chi^{(+)}(\vec{p}) \, e^{-ip\cdot x} \tag{1.66}$$

where

$$\chi^{(+)}(\vec{p}) = \frac{1}{2\sqrt{mE}} \begin{pmatrix} m + E \\ m - E \end{pmatrix} . \tag{1.67}$$

A corresponding negative energy solution

$$\phi(x) = \sqrt{\frac{m}{|E|}} e^{-i\vec{p}\cdot\vec{x} + i|E|t} \tag{1.68}$$

has the two component form

$$\chi^{(-)}(\vec{p}) = \frac{1}{2\sqrt{m|E|}} \begin{pmatrix} m - |E| \\ m + |E| \end{pmatrix} . \tag{1.69}$$

Of course, $\chi^{(+)}(\vec{p})$ is orthogonal to $\chi^{(-)}(\vec{p})$

$$\left\langle \chi^{(+)}(\vec{p}) | \chi^{(-)}(\vec{p}) \right\rangle = \chi^{(+)\dagger}(\vec{p}) \tau_3 \chi^{(-)}(\vec{p}) = 0 \tag{1.70}$$

and, by completeness, any wavepacket can be expanded in terms of a linear combination of positive and negative energy solutions. That is, we can write

$$\begin{aligned}\phi(\vec{x}, t) &= \int \frac{d^3p}{(2\pi)^3} e^{i\vec{p}\cdot\vec{x}} \left(a_{\vec{p}}^{(+)}(t) \chi^{(+)}(\vec{p}) + a_{-\vec{p}}^{(-)}(t) \chi^{(-)}(-\vec{p}) \right) \\ &= \int \frac{d^3p}{(2\pi)^3} \left(a_{\vec{p}}^{(+)}(t) \chi^{(+)}(\vec{p}) e^{i\vec{p}\cdot\vec{x}} + a_{\vec{p}}^{(-)}(t) \chi^{(-)}(\vec{p}) e^{-i\vec{p}\cdot\vec{x}} \right) .\end{aligned} \tag{1.71}$$

Clearly then $a_{\vec{p}}^{(+)}(t)$ is the amplitude to be in the positive charge, positive energy state $\chi^{(+)}(\vec{p})$ while $a_{\vec{p}}^{(-)}(t)$ is the amplitude to be in negative charge, negative energy state $\chi^{(-)}(\vec{p})$. These amplitudes are given by

$$\begin{aligned} a_{\vec{p}}^{(+)}(t) &= \int d^3x \, e^{-i\vec{p}\cdot\vec{x}} \chi^{(+)\dagger}(\vec{p}) \tau_3 \phi(\vec{x}, t) \\ a_{\vec{p}}^{(-)}(t) &= -\int d^3x \, e^{i\vec{p}\cdot\vec{x}} \chi^{(-)\dagger}(\vec{p}) \tau_3 \phi(\vec{x}, t) .\end{aligned} \tag{1.72}$$

If the wavefunction is normalized to ± 1 we have

$$\pm 1 = \langle \phi | \phi \rangle = \int d^3x\, \phi^\dagger(\vec{x},t)\tau_3\phi(\vec{x},t)$$
$$= \int \frac{d^3p}{(2\pi)^3}\left(\left|a_{\vec{p}}^{(+)}(t)\right|^2 - \left|a_{\vec{p}}^{(-)}(t)\right|^2\right) \tag{1.73}$$

while the energy and momentum expectation values are given by

$$E = \langle \phi | H_0 | \phi \rangle = \int d^3x\, \phi^\dagger(\vec{x},t)\tau_3 H_0\phi(\vec{x},t)$$
$$= \int \frac{d^3p}{(2\pi)^3}\sqrt{m^2+\vec{p}^2}\left(\left|a_{\vec{p}}^{(+)}(t)\right|^2 + \left|a_{\vec{p}}^{(-)}(t)\right|^2\right) \tag{1.74}$$

$$\vec{P} = \langle \phi | -i\vec{\nabla} | \phi \rangle = \int d^3x\, \phi^\dagger(\vec{x},t)\tau_3(-i\vec{\nabla}_x)\phi(\vec{x},t)$$
$$= \int \frac{d^3p}{(2\pi)^3}\vec{p}\left(\left|a_{\vec{p}}^{(+)}(t)\right|^2 + \left|a_{\vec{p}}^{(-)}(t)\right|^2\right). \tag{1.75}$$

PROBLEM VI.1.1

The Free Klein-Gordon Particle in a Magnetic Field

Consider a free charged Klein-Gordon particle of mass m and charge e immersed in a uniform magnetic field B in the z direction. Using the gauge

$$\vec{A} = \frac{1}{2}(\vec{B} \times \vec{r})$$

show that motion is quantized with energy

$$E_n = \sqrt{m^2 + p_z^2 + eB(2n+1)} \quad n = 0,1,2,...$$

PROBLEM VI.1.2

Pair Production by a Time Varying Electric Field

A rapidly varying electric field can lead to the creation of particle-antiparticle pairs. Calculate to lowest order in α the probability per unit volume per unit time of producing such pairs in the presence of an external electric field

$$\vec{E}(t) = \hat{e}_x a\cos\omega t$$

and show that

$$\text{Prob.} = VT\frac{\alpha a^2}{24}\left(1 - \frac{4m^2}{\omega^2}\right)^{\frac{3}{2}}\theta(\omega - 2m)$$

Suggestion: Use as an interaction potential the usual form

$$H_{\text{int}} = e\int d^3x\, j_\mu A^\mu$$

where
$$j_\mu = \frac{i}{2m}(\phi^* \partial_\mu \phi - \partial \phi^* \phi)$$
$$\vec{A}(t) = -\hat{e}_x \frac{a}{\omega} \sin \omega t$$

Utilize normalized plane wave solutions of the Klein-Gordon equation

$$\phi(x) = \sqrt{\frac{m}{E}} \exp(i\vec{p} \cdot \vec{x} - iEt) \quad \text{with} \quad E = \sqrt{\vec{p}^2 + m^2}$$

and simple first order perturbation theory

$$\text{Amp} = -i \int_{-\frac{T}{2}}^{\frac{T}{2}} \langle f | H_{\text{int}}(t) | 0 \rangle \, dt$$

VI.2 KLEIN'S PARADOX AND ZITTERBEWEGUNG

So far everything seems quite reasonable. However, there exist at least a few unexpected results within this formalism.

Zitterbewegung

Consider the construction of a wavepacket containing *only* positive energy components.

$$\phi(\vec{x}, t) = \int \frac{d^3p}{(2\pi)^3} e^{i\vec{p} \cdot \vec{x}} a_{\vec{p}}^{(+)}(t) \chi^{(+)}(\vec{p}) \tag{2.1}$$

with

$$a_{\vec{p}}^{(+)}(t) = e^{-iE_p t} f\left((\vec{p} - \vec{p}_0)^2\right) . \tag{2.2}$$

We choose some function f which is peaked about the origin, say

$$f\left((\vec{p} - \vec{p}_0)^2\right) \sim N \exp -\frac{(\vec{p} - \vec{p}_0)^2}{\Delta^2} . \tag{2.3}$$

and consider the expectation value of the position operator $\vec{x}$

$$\vec{X}(t) = \langle \phi | \vec{x} | \phi \rangle = \int d^3x \, \phi^\dagger(\vec{x}, t) \tau_3 \vec{x} \phi(\vec{x}, t)$$
$$= \int d^3x \int \frac{d^3p}{(2\pi)^3} \int \frac{d^3p'}{(2\pi)^3} \vec{x} \, e^{i(\vec{p}' - \vec{p}) \cdot \vec{x}} a_{\vec{p}}^{(+)*}(t) a_{\vec{p}'}^{(+)}(t) \tag{2.4}$$
$$\times \chi^{(+)\dagger}(\vec{p}) \tau_3 \chi^{(+)}(\vec{p}') .$$

If we write $\vec{x} \, e^{i\vec{p}' \cdot \vec{x}} = -i\vec{\nabla}_{\vec{p}'} e^{i\vec{p}' \cdot \vec{x}}$ and integrate by parts

$$\vec{X}(t) = i \int \frac{d^3p}{(2\pi)^3} a_{\vec{p}}^{(+)*}(t) \vec{\nabla}_{\vec{p}} a_{\vec{p}}^{(+)}(t)$$
$$+ i \int \frac{d^3p}{(2\pi)^3} \left|a_{\vec{p}}^{(+)}(t)\right|^2 \chi^{(+)\dagger}(\vec{p}) \tau_3 \nabla_{\vec{p}} \chi^{(+)}(\vec{p}) . \tag{2.5}$$

However,

$$\vec{\nabla}_{\vec{p}}\chi^{(+)}(\vec{p}) = \frac{1}{2\sqrt{mE_p}}\frac{\vec{p}}{E_p}\begin{pmatrix}1\\-1\end{pmatrix} - \frac{1}{4\sqrt{mE_p^3}}\frac{\vec{p}}{E_p}\begin{pmatrix}m+E_p\\m-E_p\end{pmatrix}$$

$$= -\frac{\vec{p}}{2E_p^2}\chi^{(-)}(\vec{p})$$
(2.6)

so that the term $\chi^{(+)\dagger}(\vec{p})\tau_3\vec{\nabla}_{\vec{p}}\chi^{(+)}(\vec{p})$ vanishes by orthogonality. Then

$$\vec{X}(t) = i\int\frac{d^3p}{(2\pi)^3}a_{\vec{p}}^{(+)*}(t)\vec{\nabla}_{\vec{p}}a_{\vec{p}}^{(+)}(t)$$

$$= \int\frac{d^3p}{(2\pi)^3}\left(\frac{\vec{p}}{E_p}tf^2\left((\vec{p}-\vec{p}_0)^2\right) + i2(\vec{p}-\vec{p}_0)f\left((\vec{p}-\vec{p}_0)^2\right)f'\left((\vec{p}-\vec{p}_0)^2\right)\right)$$

$$\cong \frac{\vec{p}_0}{E_{p_0}}t \ .$$
(2.7)

Thus the wave packet begins from the origin at $t=0$ and moves with uniform velocity $\vec{v}_0 = \frac{\vec{p}_0}{E_{p_0}}$, as expected. Now consider the width of the packet. For simplicity pick $\vec{p}=0$. Then

$$\left\langle(\vec{x}-\bar{\vec{x}})^2\right\rangle = \langle\phi|\vec{x}^2|\phi\rangle = -\int\frac{d^3p}{(2\pi)^3}\left[a_{\vec{p}}^{(+)*}(t)\vec{\nabla}_{\vec{p}}^2a_{\vec{p}}^{(+)}(t) + \frac{\vec{p}^2}{4E_p^4}\left|a_{\vec{p}}^{(+)}(t)\right|^2\right]$$ (2.8)

which has two components. By dimensional analysis the first piece gives the expected result

$$-\int\frac{d^3p}{(2\pi)^3}a_{\vec{p}}^{(+)*}(t)\vec{\nabla}_{\vec{p}}^2a_{\vec{p}}^{(+)}(t) \sim \frac{1}{\Delta^2}$$ (2.9)

where Δ is determined by the functional form of $f(\vec{p}^2)$. The second piece is more interesting, yielding

$$-\int\frac{d^3p}{(2\pi)^3}\frac{\vec{p}^2}{4E_p^4}\left|a_{\vec{p}}^{(+)}(t)\right|^2 \sim \frac{1}{m^2}K\left(\frac{m}{\Delta}\right) \ .$$ (2.10)

where K is a function depending on the specific form of $a_{\vec{p}}^{(+)}(t)$. We see that even if Δ is very large so that the first component of the width is very small, there is still a contribution $\delta x \sim \frac{1}{m}$ of order the Compton wavelength of the particle. This represents a minimum width of the wavepacket and cannot be made smaller so long as only positive energy components are present.

We may explore this point further by attempting to construct a wavepacket localized at the origin

$$\phi(\vec{x},t=0) = \delta^3(x)\begin{pmatrix}\rho\\\sigma\end{pmatrix}$$ (2.11)

where ρ, σ are arbitrary numbers. We find then

$$a_{\vec{p}}^{(+)}(t) = e^{-iE_p t} \frac{1}{2\sqrt{mE_p}} [(E_p + m)\rho + (E_p - m)\sigma]$$
$$a_{\vec{p}}^{(-)*}(t) = -e^{iE_p t} \frac{1}{2\sqrt{mE_p}} [(E_p - m)\rho + (E_p + m)\sigma] \quad , \quad (2.12)$$

so that $a_{\vec{p}}^{(-)}(t) \neq 0$. That is, if one wishes to localize a wavepacket within a distance smaller than a Compton wavelength negative energy components are *required*.

Such negative energy components arise naturally in another way as well. Suppose we start with a positive-energy-only wavepacket and apply the position operator $\vec{x}$. Then, as shown earlier

$$\vec{x}\phi(\vec{x},t) = \vec{x} \int \frac{d^3p}{(2\pi)^3} e^{i\vec{p}\cdot\vec{x}} a_{\vec{p}}^{(+)}(t) \chi^{(+)}(\vec{p})$$
$$= i \int \frac{d^3p}{(2\pi)^3} e^{i\vec{p}\cdot\vec{x}} \left[\chi^{(+)}(\vec{p}) \vec{\nabla}_{\vec{p}} a_{\vec{p}}^{(+)}(t) - \chi^{(-)}(\vec{p}) \frac{\vec{p}}{2E_p^2} a_{\vec{p}}^{(+)}(t) \right] \quad . \quad (2.13)$$

so that the position operator introduces a negative energy piece into the wavefunction even if there was none present originally. Equivalently, multiplication by the potential energy $eA^0(\vec{x})$ introduces such negative energy states, so that whenever a wavepacket interacts with a potential we should not be surprised to find negative energy states appearing.

This mixture of positive and negative energy components in the wavepacket has an interesting consequence if we evaluate the expectation value of the position operator

$$\langle\phi(t)|\vec{x}|\phi(t)\rangle = \int \frac{d^3p}{(2\pi)^3} \frac{\vec{p}}{E_p} t \left(\left|a_{\vec{p}}^{(+)}(t)\right|^2 + \left|a_{\vec{p}}^{(-)}(t)\right|^2 \right)$$
$$- \text{Re} \int \frac{d^3p}{(2\pi)^3} \frac{\vec{p}}{E_p^2} a_{\vec{p}}^{(+)*}(t) a_{\vec{p}}^{(-)}(t) \quad . \quad (2.14)$$

The first component represents just the expected uniform velocity motion of the packet. However, the second piece is more interesting. Since

$$a_{\vec{p}}^{(+)*}(t) a_{\vec{p}}^{(-)}(t) \sim e^{2i|E_p|t} \quad (2.15)$$

this term represents a rapid—$\omega \geq 2m$—wiggling of the position of the particle about its central location due to the interference of positive and negative energy components. This rapid movement — called zitterbewegung or jitter motion— is the price one pays for localization with $\delta x \lesssim \frac{1}{m}$, or for interaction with a potential. In the latter case, since positive and negative energies correspond to positive and negative charges, the particle and antiparticle components of the wavepacket travel in opposite directions. Thus the interference damps out after a time $\Delta t \sim \frac{1}{m}$ once interaction with the potential has ceased. An exception is when the potential is very strong — $V - E > m$. This problem is called Klein's paradox for reasons which will become apparent.

Klein's Paradox

Imagine a Klein–Gordon particle of mass m, charge e, and energy $E = \sqrt{p^2 + m^2}$ incident from the left upon a potential step

$$V(x) = V_0 \theta(x) = eA^0(x) \tag{2.16}$$

located at the origin, as shown in Figure VI.1. Since the potential

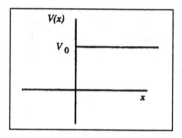

Fig. VI.1: Step potential used for the discussion of the Klein paradox.

is constant, the solutions can be represented in terms of plane waves. We look for stationary state solutions

$$\phi(x) = \begin{cases} a\,e^{ikx} + b\,e^{-ikx} & x < 0 \\ c\,e^{iqx} & x > 0 \end{cases} \qquad \begin{array}{l} k = \sqrt{E^2 - m^2} \\ q = \sqrt{(E - V_0)^2 - m^2} \end{array} \tag{2.17}$$

These are seen to be solutions of the Klein–Gordon equation in the regions $x > 0$, $x < 0$, respectively. Now, as usual, match the wavefunction and its first spatial derivative at the origin, yielding

$$\begin{aligned} \phi(0^+) &= c = a + b = \phi(0^-) \\ \phi'(0^+) &= iqc = ik(a - b) = \phi'(0^-) \end{aligned} \tag{2.18}$$

whose solution is

$$\frac{c}{a} = \frac{2}{1 + \frac{q}{k}}, \qquad \frac{b}{a} = \frac{1 - \frac{q}{k}}{1 + \frac{q}{k}}. \tag{2.19}$$

The transmission and reflection coefficients are calculated as

$$\begin{aligned} T &= \frac{q}{k}\left|\frac{c}{a}\right|^2 = 4\frac{q}{k}\frac{1}{\left|1 + \frac{q}{k}\right|^2} \\ R &= \left|\frac{b}{a}\right|^2 = \left|\frac{1 - \frac{q}{k}}{1 + \frac{q}{k}}\right|^2. \end{aligned} \tag{2.20}$$

If the kinetic energy $E - m$ is above the height of the barrier — $E - m > V_0$ — then k, q are both real and positive, yielding

$$T = 4\frac{qk}{(k+q)^2}, \qquad R = \left(\frac{k-q}{k+q}\right)^2, \qquad R + T = 1. \tag{2.21}$$

Thus the incident beam is partly reflected and partly transmitted, as expected from our experience in the corresponding non-relativistic problem. On the other hand, if the incident kinetic energy is less than the height of the barrier, but $|V_0 - E| < m$ we see that q is imaginary. Then

$$R = 1 \ , \qquad T = 0 \ , \qquad R + T = 1 \qquad (2.22)$$

which again agrees with the non-relativistic analog.

Suppose, however, that there exists a *very* strong potential — $V_0 > E + m$. In this case q becomes real again but *negative*. Then

$$R = \left(\frac{k + |q|}{k - |q|}\right)^2 > 1 \ , \qquad T = -\frac{4k|q|}{(k - |q|)^2} < 0 \qquad (2.23)$$

but

$$R + T = 1 \ . \qquad (2.24)$$

Probability is still conserved, but only at the cost of a *negative* transmission coefficient and a reflection coefficient which exceeds unity. This is the paradoxical result which confronted Klein and others.

In light of our present knowledge there exists no paradox. In the case that

$$V_0 - E > m \qquad (2.25)$$

the potential is sufficiently strong to create particle-antiparticle pairs. The antiparticles are *attracted* by the potential and create a negatively charged current moving to the right. This is the origin of the negative transmission coefficient. The particles, on the other hand, are reflected from the barrier and combine with the incident particle beam (which is completely reflected) leading to a positively charged current, moving to the left and with magnitude greater than that of the incident beam. Thus $R > 1$, as found.

Another way of thinking of this is in terms of what happens to the energy spectrum of the Klein–Gordon equation when a potential $V > 0$ is turned on adiabatically [Sa 67]. When $V = 0$ this spectrum ranges from $m < E < \infty$ and $-\infty < E < -m$. Now consider a positive energy solution as shown in Figure VI.2. As V is increased from zero, this energy level first finds itself in the forbidden region where solutions are strongly damped. However, when $E < V - m$ we are again in a region of oscillatory solutions. From Figure VI.2 it is clear that even though $E > 0$ this is essentially an antiparticle solution. The tunneling from region I to region III should be considered a transition from a particle state (when $V = 0$) to an antiparticle state (when $V > E + m$) as described above.

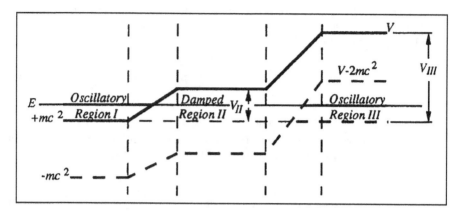

Fig. VI.2 Klein–Gordon energy levels in the presence of a potential.

There is no mystery then. Any problems which arise come from attempting to apply a simple single-particle wavefunction picture to what is obviously a many-body situation. The correct way in which to handle all the subtlety of this problem is via the formalism of quantum field theory. Nevertheless, the elementary wavefunction paradigm allows a reasonable sketch of the physics involved.

VI.3 THE COULOMB SOLUTION: MESONIC ATOMS

The Coulomb bound state problem can also straightforwardly analyzed. Before looking at the exact solution, however, it is useful to make the connection with the Schrödinger formalism by deriving an effective Hamiltonian for the situation that the Klein–Gordon particle is non-relativistic.

Effective Schrödinger Equation

Consider the two component formalism — χ_1, χ_2 — and look for a stationary state solution with

$$i\partial^0 \chi_j \approx (m + W)\chi_j \qquad j = 1, 2 \tag{3.1}$$

where $W \ll m$. Then

$$(m + W - e\phi)\chi_1 = \left(m + \frac{1}{2m}(\vec{p} - e\vec{A})^2\right)\chi_1 + \frac{1}{2m}(\vec{p} - e\vec{A})^2\chi_2$$

$$(m + W - e\phi)\chi_2 = -\left(m + \frac{1}{2m}(\vec{p} - e\vec{A})^2\right)\chi_2 - \frac{1}{2m}(\vec{p} - e\vec{A})^2\chi_1 \ . \tag{3.2}$$

Solving the second equation for χ_2 we find

$$\chi_2 \approx -\frac{1}{2m + W - e\phi} \frac{1}{2m}(\vec{p} - e\vec{A})^2 \chi_1$$

$$\approx -\frac{1}{4m^2}\left(1 - \frac{W - e\phi}{2m}\right)(\vec{p} - e\vec{A})^2 \chi_1 \ . \tag{3.3}$$

Substitution into the top equation yields

$$(W - e\phi)\chi_1 \cong \frac{1}{2m}(\vec{p} - e\vec{A})^2\chi_1 - \frac{1}{8m^3}(\vec{p} - e\vec{A})^2\left(1 - \frac{W - e\phi}{2m}\right)(\vec{p} - e\vec{A})^2\chi_1 \quad (3.4)$$

or

$$W\left(1 - \frac{1}{16m^4}(\vec{p} - e\vec{A})^4\right)\chi_1$$

$$= \left(\frac{(\vec{p} - e\vec{A})^2}{2m} + e\phi - \frac{(\vec{p} - e\vec{A})^4}{8m^3} - (\vec{p} - e\vec{A})^2\frac{e\phi}{16m^4}(\vec{p} - e\vec{A})^2\right)\chi_1 \quad . \quad (3.5)$$

However, χ_1 is not normalized to unity. Rather

$$1 = \int d^3x \left(\chi_1^\dagger \chi_1 - \chi_2^\dagger \chi_2\right) \approx \int d^3x\, \chi_1^\dagger \left(1 - \frac{(\vec{p} - e\vec{A})^4}{16m^4}\right)\chi_1$$

$$\approx \int d^3x\, \chi_1'^\dagger \chi_1' \quad (3.6)$$

where we have defined

$$\chi_1' = \left(1 - \frac{(\vec{p} - e\vec{A})^4}{32m^4}\right)\chi_1 \quad . \quad (3.7)$$

Thus multiply Eq. 3.5 by the factor $\left(1 + \frac{(\vec{p} - e\vec{A})^4}{32m^4}\right)$, yielding

$$W\chi_1' = \left(\frac{(\vec{p} - e\vec{A})^2}{2m} - \frac{(\vec{p} - e\vec{A})^4}{8m^3} - \frac{1}{16m^4}(\vec{p} - e\vec{A})^2 e\phi (\vec{p} - e\vec{A})^2 \right.$$

$$\left. + \left(1 + \frac{(\vec{p} - e\vec{A})^4}{32m^4}\right)e\phi\left(1 + \frac{(\vec{p} - e\vec{A})^4}{32m^4}\right)\right)\chi_1' + \ldots$$

$$= \left(\frac{(\vec{p} - e\vec{A})^2}{2m} - \frac{(\vec{p} - e\vec{A})^4}{8m^3} + e\phi + \frac{1}{32m^4}\left[(\vec{p} - e\vec{A})^2, \left[(\vec{p} - e\vec{A})^2, e\phi\right]\right]\right)\chi_1' \quad . \quad (3.8)$$

The effective Schrödinger Hamiltonian is then

$$H = \frac{1}{2m}(-i\vec{\nabla} - e\vec{A})^2 - \frac{1}{8m^3}(-i\vec{\nabla} - e\vec{A})^4 + e\phi$$

$$+ \frac{1}{32m^4}\left[(-i\vec{\nabla} - e\vec{A})^2, \left[(-i\vec{\nabla} - eA)^2, e\phi\right]\right] \quad . \quad (3.9)$$

and we can identify its various components as follows:

a) The terms

$$\frac{(\vec{p} - e\vec{A})^2}{2m} + e\phi \quad (3.10)$$

represent the usual non-relativistic energy.

b) Relativistically the kinetic energy is

$$T = \sqrt{m^2 + (\vec{p} - e\vec{A})^2} - m \tag{3.11}$$

whose non-relativistic approximation is

$$T = \frac{(\vec{p} - e\vec{A})^2}{2m} - \frac{(\vec{p} - e\vec{A})^4}{8m^3} + \ldots \tag{3.12}$$

Thus the piece $-\frac{1}{8m^3}(\vec{p} - e\vec{A})^4$ represents a relativistic $\mathcal{O}\left(\frac{v^2}{c^2}\right)$ correction to the usual kinetic energy.

c) The origin of the double commutator term is more subtle and is associated with the zitterbewegung motion discussed earlier. By completeness we can expand a bound state solution in terms of plane wave solutions. This expansion will involve, in general, a combination of positive and negative energy solutions and will thus lead to zitterbewegung motion of magnitude $(\delta x)^2 \sim \frac{1}{m^2}$ about the usual trajectory. This leads, in general, to a shift in the energy of magnitude

$$\begin{aligned} V(x + \delta x) - V(x) &\approx \nabla_i V(x) \langle \delta x_i \rangle + \frac{1}{2!} \nabla_i \nabla_j V(x) \langle \delta x_i \delta x_j \rangle + \ldots \\ &\approx \frac{1}{6m^2} \vec{\nabla}^2 V(x) + \ldots \end{aligned} \tag{3.13}$$

In the case of spin-1/2 we shall see that a term of precisely this form — the Darwin term — appears in the effective Hamiltonian. For the case of spinless particles the zitterbewegung term arises, however, in $\mathcal{O}\left(\frac{v^4}{c^4}\right)$.

Mesonic Atoms

A spin-zero "atom" actually exists in nature when a π^- or K^- meson is captured by a nucleus. This object is called a "pionic" or "kaonic" atom. We can calculate the energy levels of such a system approximately by treating the $\vec{p}^{\,4}$ term as a perturbation. In lowest order we have

$$H_0 = \frac{\vec{p}^{\,2}}{2m} - \frac{Z\alpha}{r} \tag{3.14}$$

which is simply hydrogen-like atomic Hamiltonian, yielding eigenvalues

$$E_n = -\frac{Z^2 \alpha^2}{2n^2} m \tag{3.15}$$

and eigenfunctions $\psi_n(\vec{r})$ with

$$H_0 \psi_n(\vec{r}) = \left(\frac{\vec{p}^{\,2}}{2m} - \frac{Z\alpha}{r}\right) \psi_n(\vec{r}) = E_n \psi_n(\vec{r}). \tag{3.16}$$

Then
$$\frac{\vec{p}^4}{8m^3}\psi_n(r) = \frac{1}{2m}\left(E_n + \frac{Z\alpha}{r}\right)^2 \psi_n(r)$$
$$= \frac{1}{2m}\left(m^2\frac{Z^4\alpha^4}{4n^4} - m\frac{Z^3\alpha^3}{n^2 r} + \frac{Z^2\alpha^2}{r^2}\right)\psi_n(r) \ . \tag{3.17}$$

Since
$$\left\langle\frac{1}{r}\right\rangle = m\frac{Z\alpha}{n^2} \ , \quad \left\langle\frac{1}{r^2}\right\rangle = m^2\frac{Z^2\alpha^2}{n^3\left(\ell+\frac{1}{2}\right)} \tag{3.18}$$

we find
$$\left\langle\frac{\vec{p}^4}{8m^3}\right\rangle = \frac{1}{2m}\left(m^2\frac{Z^4\alpha^4}{4n^4} - m^2\frac{Z^4\alpha^4}{n^4} + m^2\frac{Z^4\alpha^4}{n^3\left(\ell+\frac{1}{2}\right)}\right)$$
$$= \frac{mZ^4\alpha^4}{2n^3}\left(\frac{1}{\ell+\frac{1}{2}} - \frac{3}{4n}\right) \ . \tag{3.19}$$

Also, we note that
$$\left[\vec{p}^2, \phi(\vec{r})\right] = -Ze\delta^3(r) \ . \tag{3.20}$$

Then
$$\left\langle\psi_n\left|\left[\vec{p}^2, \left[\vec{p}^2, \phi\right]\right]\right|\psi_n\right\rangle = Ze\int d^3r\,\delta^3(r)\left(\left(\vec{\nabla}^2\psi_n^*(r)\right)\psi_n(r) - \psi_n^*(r)\vec{\nabla}^2\psi_n(r)\right) = 0 \tag{3.21}$$
so the Darwin term does not contribute.

The energy levels become
$$E_{n\ell} = -m\frac{Z^2\alpha^2}{2n^2}\left[1 + \frac{Z^2\alpha^2}{n^2}\left(\frac{n}{\ell+\frac{1}{2}} - \frac{3}{4}\right) + \ldots\right] \ . \tag{3.22}$$

We observe that the $\vec{p}^4$ term acts as a fine structure term, removing the ℓ degeneracy of the hydrogen atom and lowering the energy for states of smaller angular momentum.

We can also solve this system exactly. Using
$$e\phi(r) = -\frac{Z\alpha}{r} \tag{3.23}$$

the Klein–Gordon equation becomes
$$\left(\left(E + \frac{Z\alpha}{r}\right)^2 + \vec{\nabla}^2 - m^2\right)\psi(\vec{r}) = 0 \ . \tag{3.24}$$

If we look for solutions having a definite angular momentum ℓ
$$\psi(\vec{r}) = \psi_\ell(r)Y_\ell^m(\hat{r}) \ . \tag{3.25}$$

we require

$$\left(\left(E+\frac{Z\alpha}{r}\right)^2 + \frac{1}{r}\frac{\partial^2}{\partial r^2}r - \frac{\ell(\ell+1)}{r^2} - m^2\right)\psi_\ell(r)$$
$$= \left(E^2 - m^2 + 2\frac{Z\alpha E}{r} + \frac{1}{r}\frac{\partial^2}{\partial r^2}r - \frac{\ell(\ell+1) - Z^2\alpha^2}{r^2}\right)\psi_\ell(r) = 0 \ . \quad (3.26)$$

Then making the identification

$$\ell(\ell+1) - Z^2\alpha^2 = \ell'(\ell'+1)$$
$$E^2 - m^2 = k^2 = 2m'E' \quad (3.27)$$
$$E = m'$$

Eq. 3.26 becomes

$$\left(\frac{1}{r}\frac{\partial^2}{\partial r^2}r - \frac{\ell'(\ell'+1)}{r^2} + \frac{2m'Z\alpha}{r} + k^2\right)\psi(r) = 0 \quad (3.28)$$

which is identical to the differential equation which arises in the usual Schrödinger equation solution of the hydrogen atom, except that in the present case ℓ' is *not* an integer. It is necessary then to analytically continue the hydrogen solutions in order to apply them here. We have

$$E'_s = -\frac{Z^2\alpha^2}{2s^2}m' \quad (3.29)$$

and

$$E = \frac{m}{\left(1 + \frac{Z^2\alpha^2}{s^2}\right)^{1/2}} \ . \quad (3.30)$$

However, s is not an integer but is defined rather in terms of ℓ' †—

$$s = n + \ell' - \ell \ , \quad (3.31)$$

with ℓ' determined by

$$\ell'(\ell'+1) + \frac{1}{4} = \left(\ell' + \frac{1}{2}\right)^2 = \ell(\ell+1) - Z^2\alpha^2 + \frac{1}{4}$$

i.e.,

$$\ell' = -\frac{1}{2} \pm \sqrt{\left(\ell + \frac{1}{2}\right)^2 - Z^2\alpha^2} \ . \quad (3.32)$$

---

† The positive sign in front of $\ell' - \ell$ is determined by the hydrogen atom condition that $n - \ell$ is a non-negative integer.

Although mathematically either sign is allowed, only the positive sign allows a normalizable solution as $\alpha \to 0$. Then

$$s = n - \frac{1}{2} - \ell + \sqrt{\left(\ell + \frac{1}{2}\right)^2 - Z^2\alpha^2} \tag{3.33}$$

so that

$$E = m\left[1 + \frac{Z^2\alpha^2}{\left(n - \ell - \frac{1}{2} + \sqrt{\left(\ell + \frac{1}{2}\right)^2 - Z^2\alpha^2}\right)^2}\right]^{-1/2} \tag{3.34}$$

Noting that

$$s \approx n - \frac{1}{2}\frac{Z^2\alpha^2}{\ell + \frac{1}{2}} \tag{3.35}$$

we find

$$\begin{aligned}E &\approx m\left(1 - \frac{1}{2}\frac{Z^2\alpha^2}{s^2} + \frac{3}{8}\frac{Z^4\alpha^4}{s^4} + \ldots\right) \\ &\approx m\left(1 - \frac{1}{2}\frac{Z^2\alpha^2}{n^2}\left(1 + \frac{Z^2\alpha^2}{n^2}\left(\frac{n}{\ell + \frac{1}{2}} - \frac{3}{4}\right)\right)\ldots\right) \ .\end{aligned} \tag{3.36}$$

which is in complete agreement with Eq. 3.22 obtained perturbatively.

In comparing this prediction to experimental data on pionic atoms, various corrections are required:

i) the reduced mass

$$\mu = \frac{mM}{m + M} \tag{3.37}$$

must be utilized in place of the pion mass m;

ii) account must be taken of the fact that the central nucleus is not a point charge but has a radius $R \sim 1.2 \times A^{\frac{1}{3}}$ fm;

iii) correction must be made for the so-called vacuum polarization, wherein a virtual photon, responsible for the Coulomb potential between pion and nucleus, transforms temporarily into an electron–positron pair;

iv) finally, since the radius of the lowest Bohr orbit

$$r_\pi = \frac{1}{Z\alpha m_\pi} \tag{3.38}$$

is $m_\pi/m_e \sim 300$ times smaller than the corresponding electron Bohr radius, the pion wavefunction has significant overlap with the central nucleus, requiring a correction factor for the strong pion-nuclear interaction.

When these modifications are made, agreement is excellent over a wide range of nuclei.

PROBLEM VI.3.1

Relativistic Zeeman Effect

Suppose a pionic atom is placed in a uniform magnetic field, described by the vector potential

$$\vec{A} = \frac{1}{2}(\vec{B} \times \vec{r}) \ .$$

i) Neglecting the quadratic term (justify this) show that this problem can be exactly solved to yield the energy levels

$$E = E_{B=0}(1 - 2\omega_L \frac{m}{m_\pi})^{\frac{1}{2}}$$

where $\omega_L = eB/2m_\pi$ is the Larmor frequency and m is the eigenvalue of L along the direction of the magnetic field.

ii) Evaluate the nonrelativistic limit of the Klein-Gordon equation in this case and show that the effective Hamiltonian is

$$H = H_{B=0} - \frac{e}{2m_\pi}\vec{L}\cdot\vec{B}(1 - \frac{\vec{p}^2}{2m_\pi^2} + ...)$$

Thus the usual Bohr magneton $e/2m_\pi$ is reduced by relativistic effects to

$$\frac{e}{2m_\pi}(1 - \frac{\vec{p}^2}{2m_\pi^2})$$

iii) Calculate the energy shift induced by the magnetic field using perturbation theory and show that this result agrees with the exact answer to first order in $\vec{B}$.

PROBLEM VI.3.2

High Z Mesonic Atoms

We have seen that the energy levels of a mesonic atom are given by

$$E = m\left[1 + \frac{Z^2\alpha^2}{(n - \ell - \frac{1}{2} + \sqrt{(\ell + \frac{1}{2})^2 - Z^2\alpha^2})^2}\right]^{-\frac{1}{2}} .$$

i) Show that the ground state energy for any mesonic atom heavier that Z=69 is complex. Explain what this complex energy means.
ii) Mesonic atoms have been well studied at places like Los Alamos and it has been found that the ground states of atoms as heavy as lead (Z=82) or Uranium (Z=92) are quite stable. How do you reconcile this fact with the result obtained above? Be as quantitative as you can.

CHAPTER VII

THE DIRAC EQUATION

VII.1 DERIVATION AND COVARIANCE

Historically the Klein–Gordon equation was written down before the Dirac equation. However, it was abandoned for a period of time due to the problems with negative energy states and the inability to construct a positive definite probability density. Dirac then developed his formalism and demonstrated the connection between negative energy states and antiparticles, at which point the Klein–Gordon equation was resurrected. It is in this spirit then that we interrupt our discussion of the Klein–Gordon equation to present the Dirac equation.

Intuitive Derivation

Consider the Schrödinger equation which describes a spin-1/2 particle in the presence of an electromagnetic field. The wavefunction $\psi(\vec{x}, t)$ is a two-component object

$$\psi(\vec{x}, t) = \begin{pmatrix} \psi_1(\vec{x}, t) \\ \psi_2(\vec{x}, t) \end{pmatrix} \tag{1.1}$$

describing a particle with spin

$$\vec{S} = \left\langle \psi \left| \frac{1}{2}\vec{\sigma} \right| \psi \right\rangle . \tag{1.2}$$

In addition to the pieces of the Hamiltonian

$$H \sim \frac{\left(\vec{p} - e\vec{A}\right)^2}{2m} + e\phi \tag{1.3}$$

expected from the spinless case, we must also append a term which accounts for the interaction of the magnetic moment with a magnetic field. Recalling that for an electron the moment is given by

$$\vec{\mu} = \frac{g_e e}{2m} \vec{S} \tag{1.4}$$

with gyromagnetic ratio $g_e = 2$ Bohr magnetons, this leads to an additional term in the Hamiltonian

$$H' = -\vec{\mu} \cdot \vec{B} = -\frac{e}{m} \vec{S} \cdot \vec{B}$$
$$= -\frac{e}{2m} \vec{\sigma} \cdot \vec{B} \tag{1.5}$$

so that the Schrödinger equation becomes

$$\left(\frac{1}{2m} \left(\vec{p} - e\vec{A}\right)^2 + e\phi - \frac{e}{2m} \vec{\sigma} \cdot \vec{\nabla} \times \vec{A} \right) \psi(\vec{x}, t) = i\frac{\partial}{\partial t} \psi(\vec{x}, t) . \tag{1.6}$$

Using the identity
$$\sigma_i \sigma_j = \delta_{ij} + i\epsilon_{ijk}\sigma_k \tag{1.7}$$
we note that
$$\vec{\sigma} \cdot \left(\vec{p} - e\vec{A}\right) \vec{\sigma} \cdot \left(\vec{p} - e\vec{A}\right) = \left(\vec{p} - e\vec{A}\right) \cdot \left(\vec{p} - e\vec{A}\right) + i\vec{\sigma} \cdot \left(\vec{p} - e\vec{A}\right) \times \left(\vec{p} - e\vec{A}\right)$$
$$= \left(\vec{p} - e\vec{A}\right)^2 - e\vec{\sigma} \cdot \vec{\nabla} \times \vec{A} \tag{1.8}$$

where $\vec{p}$ denotes the operator $-i\vec{\nabla}$. Thus we may write the Schrödinger equation in the suggestive form
$$\left(\frac{1}{2m}\vec{\sigma} \cdot \left(\vec{p} - e\vec{A}\right) \vec{\sigma} \cdot \left(\vec{p} - e\vec{A}\right) + e\phi\right)\psi = i\frac{\partial \psi}{\partial t} . \tag{1.9}$$

Defining the four-vector
$$\pi_\mu = i\nabla_\mu - eA_\mu \tag{1.10}$$
we can write the previously discussed wave equations as

Non-Relativistic	Spin 0	$\frac{1}{2m}\vec{\pi} \cdot \vec{\pi}\psi = \pi_0 \psi$
Relativistic	Spin 0	$\left(\pi_0^2 - \vec{\pi} \cdot \vec{\pi} - m^2\right)\psi = 0$
Non-Relativistic	Spin $\frac{1}{2}$	$\frac{1}{2m}\vec{\sigma} \cdot \vec{\pi}\, \vec{\sigma} \cdot \vec{\pi}\psi = \pi_0 \psi$

and from this tabulation we might well guess that the relativistic version of the spin-1/2 equation would take the form
$$\left(\pi_0^2 - \vec{\sigma} \cdot \vec{\pi}\, \vec{\sigma} \cdot \vec{\pi} - m^2\right)\psi = 0 . \tag{1.11}$$

In fact this is almost but not quite right. Rather the correct form of the Dirac equation is given by [Fe 62]
$$\left((\pi_0 - \vec{\sigma} \cdot \vec{\pi})(\pi_0 + \vec{\sigma} \cdot \vec{\pi}) - m^2\right)\psi = 0 \tag{1.12}$$

where here ψ is a two-component spinor. [Note that if $\pi_0, \vec{\pi}$ were simply numbers and not operators we would have
$$(\pi_0 - \vec{\sigma} \cdot \vec{\pi})(\pi_0 + \vec{\sigma} \cdot \vec{\pi}) = \pi_0^2 - \vec{\sigma} \cdot \vec{\pi}\, \vec{\sigma} \cdot \vec{\pi}$$
$$= \pi_0^2 - \vec{\pi}^2 . \tag{1.13}$$

However, this is not in general the case.]

Eq. 1.12, while correct, is *not* the conventional form of the Dirac equation. Instead Dirac chose to write his equation in Hamiltonian form — *i.e.*, first order in time. As we saw in the case of the Klein–Gordon equation, this requires a doubling of the number of components. We must deal with a *four*-component spinor —

VII.1 DERIVATION AND COVARIANCE

two components for particles, two for antiparticles. We define then a pair of *two-component* objects ρ, χ such that

$$\begin{aligned}(\pi_0 + \vec{\sigma}\cdot\vec{\pi})\rho &= m\chi \\ (\pi_0 - \vec{\sigma}\cdot\vec{\pi})\chi &= m\rho\end{aligned} \quad (1.14)$$

Eq. 1.14 is an equivalent version of Eq. 1.12, but is still not the standard form of the Dirac equation. Instead the conventional version is found via use of the linear combinations

$$\psi_a = \chi + \rho \;, \qquad \psi_b = \chi - \rho \quad (1.15)$$

which satisfy

$$\begin{aligned}\pi_0 \psi_a - \vec{\sigma}\cdot\vec{\pi}\psi_b &= m\psi_a \\ \vec{\sigma}\cdot\vec{\pi}\psi_a - \pi_0 \psi_b &= m\psi_b\end{aligned} \quad (1.16)$$

Eq. 1.16 can be represented most succinctly by employing a *four*-component object

$$\psi \equiv \begin{pmatrix}\psi_a \\ \psi_b\end{pmatrix} \quad (1.17)$$

and the 4×4 matrices

$$\gamma^0 = \begin{pmatrix}1 & | & 0 \\ - & | & - \\ 0 & | & -1\end{pmatrix} \quad (1.18a)$$

$$\vec{\gamma} = \begin{pmatrix}0 & | & \vec{\sigma} \\ - & | & - \\ -\vec{\sigma} & | & 0\end{pmatrix} \quad (1.18b)$$

in terms of which we can write

$$\left(\gamma^0 \pi_0 - \vec{\gamma}\cdot\vec{\pi}\right)\psi = m\psi \;. \quad (1.19)$$

[Check:

$$\begin{aligned}\left(\gamma^0 \pi_0 - \vec{\gamma}\cdot\vec{\pi}\right)\psi &= \begin{pmatrix}\pi_0 & -\vec{\sigma}\cdot\vec{\pi} \\ \vec{\sigma}\cdot\vec{\pi} & -\pi_0\end{pmatrix}\begin{pmatrix}\psi_a \\ \psi_b\end{pmatrix} \\ &= m\begin{pmatrix}\psi_a \\ \psi_b\end{pmatrix} = m\psi\end{aligned} \quad (1.20)$$

which agrees with the coupled equations in Eq. 1.16.]
This is the conventional form of the Dirac equation and is usually written as

$$(\gamma^\mu \pi_\mu - m)\psi = 0 \;. \quad (1.21)$$

Often in the literature one finds this result expressed via a shorthand notation due to Feynman wherein an arbitrary four vector A^μ contracted with the Dirac matrices γ^μ is denoted by using a slash through the four-vector

$$A_\mu \gamma^\mu \equiv \slashed{A} \ . \tag{1.22}$$

Then the Dirac equation assumes the simple form

$$(\slashed{\pi} - m)\psi = (i\slashed{\nabla} - e\slashed{A} - m)\psi = 0 \ . \tag{1.23}$$

It is useful at this point to identify certain properties of the γ^μ matrices which we shall later exploit. We note

$$\gamma^{0\dagger} = \begin{pmatrix} 1 & 0 \\ 0 & -1 \end{pmatrix} = \gamma^0 \ , \qquad \vec{\gamma}^\dagger = \begin{pmatrix} 0 & -\vec{\sigma} \\ \vec{\sigma} & 0 \end{pmatrix} = -\vec{\gamma} \ . \tag{1.24}$$

Also

$$(\gamma^0)^2 = \begin{pmatrix} 1^2 & 0 \\ 0 & (-1)^2 \end{pmatrix} = 1 \ , \qquad (\gamma^i)^2 = \begin{pmatrix} -\sigma_i^2 & 0 \\ 0 & -\sigma_i^2 \end{pmatrix} = -1 \ . \tag{1.25}$$

Since any two different γ's anticommute

$$\gamma^0 \gamma^i + \gamma^i \gamma^0 = \begin{pmatrix} 0 & \sigma_i \\ \sigma_i & 0 \end{pmatrix} + \begin{pmatrix} 0 & -\sigma_i \\ -\sigma_i & 0 \end{pmatrix} = 0$$

$$\gamma^i \gamma^j + \gamma^j \gamma^i = \begin{pmatrix} -\sigma_i \sigma_j - \sigma_j \sigma_i & 0 \\ 0 & -\sigma_i \sigma_j - \sigma_j \sigma_i \end{pmatrix} = 0 \text{ if } i \neq j \tag{1.26}$$

we can represent Eqs. 1.25, 1.26 in terms of the relation

$$\gamma^\mu \gamma^\nu + \gamma^\nu \gamma^\mu = 2\eta^{\mu\nu} \tag{1.27}$$

where

$$\eta^{\mu\nu} = \eta_{\mu\nu} = \begin{pmatrix} 1 & 0 & 0 & 0 \\ 0 & -1 & 0 & 0 \\ 0 & 0 & -1 & 0 \\ 0 & 0 & 0 & -1 \end{pmatrix} \tag{1.28}$$

is the metric tensor.

Hamiltonian Form

It is important to note that Dirac's original presentation of the relativistic equation was somewhat different than given above and was written in Hamiltonian form

$$i\frac{\partial \psi}{\partial t} = H_D \psi \ . \tag{1.29}$$

We can reproduce this version by noting that since $(\gamma^0)^2 = 1$

$$\gamma^0 \slashed{\pi} = \gamma^0 \left(\gamma^0 \pi_0 - \vec{\gamma} \cdot \vec{\pi}\right) = \pi_0 - \gamma^0 \vec{\gamma} \cdot \vec{\pi} \ . \tag{1.30}$$

Thus Eq. 1.23 becomes

$$\left(\pi_0 - \gamma^0 \vec{\gamma} \cdot \vec{\pi} - \gamma^0 m\right)\psi = 0 \ . \tag{1.31}$$

Dirac's notation was to define

$$\beta \equiv \gamma^0 \qquad \vec{\alpha} \equiv \gamma^0 \vec{\gamma} \qquad (1.32)$$

whereby

$$i\frac{\partial}{\partial t}\psi = \left(\vec{\alpha} \cdot \left(\vec{p} - e\vec{A}\right) + e\phi + \beta m\right)\psi \qquad (1.33)$$
$$\equiv H_D \psi \ .$$

Here

$$\beta = \begin{pmatrix} 1 & 0 \\ 0 & -1 \end{pmatrix} \text{ and } \vec{\alpha} = \begin{pmatrix} 1 & 0 \\ 0 & -1 \end{pmatrix}\begin{pmatrix} 0 & \vec{\sigma} \\ -\vec{\sigma} & 0 \end{pmatrix} = \begin{pmatrix} 0 & \vec{\sigma} \\ \vec{\sigma} & 0 \end{pmatrix} \qquad (1.34)$$

so that $\beta^\dagger = \beta$ and $\vec{\alpha}^\dagger = \vec{\alpha}$— Dirac's Hamiltonian H_D is explicitly Hermitian.

Covariance

The crucial issue, of course, is the covariance of the equation — does it have an identical form in all Lorentz frames? That is, if in one frame the Dirac equation is written

$$\left(\gamma_\mu\left(i\nabla^\mu - eA^\mu(x)\right) - m\right)\psi(x) = 0 \qquad (1.35)$$

does it in some other frame read

$$\left(\gamma_\mu\left(i\nabla^{\mu'} - eA^{\mu'}(x')\right) - m\right)\psi'(x') = 0 \qquad (1.36)$$

where

$$x'^\mu = a^\mu{}_\nu x^\nu \qquad (1.37)$$

is the point into which x transforms and, since A^μ, ∇^μ are also four vectors

$$A'^\mu = a^\mu{}_\nu A^\nu \ , \qquad \nabla'^\mu = a^\mu{}_\nu \nabla^\nu \ . \qquad (1.38)$$

Although the Dirac matrices γ^μ are written with Greek indices, they are *not* four vectors. Rather, they have the same value in *every* frame. On the other hand, the Dirac spinor ψ does change under a Lorentz transformation

$$\psi(x) \longrightarrow \psi'(x') = S(a)\psi(x) \qquad (1.39)$$

where S represents an as yet underdetermined matrix function of $a^\mu{}_\nu$. One should not be surprised that the spinor undergoes such a change. Indeed even in the case of a non-relativistic two-component spinor, there exists such an effect. For example, under a rotation by angle $\delta\phi$ about an axis specified by $\hat{n}$, we have

$$\psi(x) = \begin{pmatrix} \psi_1(x) \\ \psi_2(x) \end{pmatrix} \longrightarrow \psi'(x') = \exp\left(i\frac{\delta\phi}{2}\vec{\sigma}\cdot\hat{n}\right)\begin{pmatrix} \psi_1(x) \\ \psi_2(x) \end{pmatrix}$$
$$= \exp\left[i\delta\phi\hat{n}\cdot\left(\vec{L} + \frac{\vec{\sigma}}{2}\right)\right]\begin{pmatrix} \psi_1(x') \\ \psi_2(x') \end{pmatrix} \qquad (1.40)$$

where $\vec{L} = \vec{r} \times \vec{p}$ is the orbital angular momentum operator.

Similarly in the relativistic case we seek an operator $S(a)$ such that $\psi'(x') = S(a)\psi(x)$ and

$$(\gamma_\mu i \nabla'^\mu - m) S(a)\psi(x) = (\gamma_\mu i a^\mu_{\ \nu} \nabla^\nu - m) S(a)\psi(x) = 0 \tag{1.41}$$

If we multiply on the left by $S^{-1}(a)$, yielding

$$\left(S^{-1}(a)\gamma_\mu S(a) a^\mu_{\ \nu} i \nabla^\nu - m\right) \psi(x) = 0 \ . \tag{1.42}$$

then in order to reproduce the original form of the Dirac equation, it is required that

$$S^{-1}(a)\gamma_\mu S(a) a^\mu_{\ \nu} = \gamma_\nu \ . \tag{1.43}$$

Noting that the covariant component of a four-vector must transform as

$$x'_\nu = x_\lambda (a^{-1})^\lambda_{\ \nu} \tag{1.44}$$

in order that

$$x'_\nu x'^\nu = x_\lambda (a^{-1})^\lambda_{\ \nu} \times a^\nu_{\ \epsilon} x^\epsilon = x_\lambda \delta^\lambda_{\ \epsilon} x^\epsilon = x_\lambda x^\lambda \ , \tag{1.45}$$

we see that Eq. 1.43 can be written in the alternate form

$$\gamma_\nu (a^{-1})^\nu_{\ \lambda} = S^{-1}(a)\gamma_\mu S(a) \, a^\mu_{\ \nu} (a^{-1})^\nu_{\ \lambda} = S^{-1}(a)\gamma_\lambda S(a) \tag{1.46}$$

or in terms of its contravariant version

$$S^{-1}(a)\gamma^\lambda S(a) = a^\lambda_{\ \nu} \gamma^\nu \ . \tag{1.47}$$

Rather than present a detailed derivation, we shall merely quote the appropriate forms for $S(a)$:

i) Rotations:

Consider a rotation by angle ϕ about the z-axis, with

$$\begin{aligned} x'^1 &= x^1 \cos\phi + x^2 \sin\phi \\ x'^2 &= -x^1 \sin\phi + x^2 \cos\phi \\ x'^3 &= x^3 \\ x'^0 &= x^0 \ . \end{aligned} \tag{1.48}$$

Then

$$\begin{aligned} S &= \exp\left(-\frac{\phi}{2}\gamma^1\gamma^2\right) = \cos\frac{\phi}{2} - \gamma^1\gamma^2 \sin\frac{\phi}{2} \\ S^{-1} &= \exp\left(\frac{\phi}{2}\gamma^1\gamma^2\right) = \cos\frac{\phi}{2} + \gamma^1\gamma^2 \sin\frac{\phi}{2} \end{aligned} \tag{1.49}$$

Check:
$$S^{-1}\gamma^0 S = \gamma^0 S^{-1} S = \gamma^0 \qquad \text{since } [\gamma^1\gamma^2, \gamma^0] = 0$$
$$S^{-1}\gamma^3 S = \gamma^3 S^{-1} S = \gamma^3 \qquad \text{since } [\gamma^1\gamma^2, \gamma^3] = 0 \ .$$
(1.50)

On the other hand, since $\{\gamma^1\gamma^2, \gamma^i\} = 0 \quad i = 1, 2$

$$S^{-1}\gamma^1 S = \gamma^1(S)^2 = \gamma^1\left(\cos^2\frac{\phi}{2} - \sin^2\frac{\phi}{2} - 2\sin\frac{\phi}{2}\cos\frac{\phi}{2}\gamma^1\gamma^2\right)$$
$$= \cos\phi\gamma^1 + \sin\phi\gamma^2 \qquad (1.51a)$$

$$S^{-1}\gamma^2 S = \gamma^2(S)^2 = \gamma^2\left(\cos^2\frac{\phi}{2} - \sin^2\frac{\phi}{2} - 2\sin\frac{\phi}{2}\cos\frac{\phi}{2}\gamma^1\gamma^2\right)$$
$$= \cos\phi\gamma^2 - \sin\phi\gamma^1 \ . \qquad (1.51b)$$

Note also that
$$(\gamma^1\gamma^2)^\dagger = \gamma^{2\dagger}\gamma^{1\dagger} = -\gamma^1\gamma^2 \qquad (1.52)$$

so that
$$S^\dagger = S^{-1} \ . \qquad (1.53)$$

Since under rotations $d^3x' = d^3x$ and
$$\psi'^\dagger\psi' = \psi^\dagger S^{-1} S\psi = \psi^\dagger\psi \qquad (1.54)$$

we see that the normalization $\int d^3x\, \psi^\dagger\psi$ is preserved.

ii) Lorentz Boost:

Consider a Lorentz transformation with velocity v along the z-axis, with

$$\begin{aligned}
x'^0 &= \cosh\theta x^0 - \sinh\theta x^3 \\
x'^3 &= \cosh\theta x^3 - \sinh\theta x^0 \\
x'^2 &= x^2 \\
x'^1 &= x^1
\end{aligned} \qquad (1.55)$$

where we have defined
$$\cosh\theta = \frac{1}{\sqrt{1-v^2}} \quad \text{and} \quad \sinh\theta = \frac{v}{\sqrt{1-v^2}} \ . \qquad (1.56)$$

[Note
$$\cosh^2\theta - \sinh^2\theta = \frac{1}{1-v^2} - \frac{v^2}{1-v^2} = 1 \ .] \qquad (1.57)$$

Then
$$\begin{aligned}
S &= \exp\frac{\theta}{2}\gamma^3\gamma^0 = \cosh\frac{\theta}{2} + \gamma^3\gamma^0\sinh\frac{\theta}{2} \\
S^{-1} &= \exp-\frac{\theta}{2}\gamma^3\gamma^0 = \cosh\frac{\theta}{2} - \gamma^3\gamma^0\sinh\frac{\theta}{2} \ .
\end{aligned} \qquad (1.58)$$

Check: Since $[\gamma^3\gamma^0, \gamma^1] = 0$ $i = 1, 2$

$$S^{-1}\gamma^1 S = \gamma^1 S^{-1} S = \gamma^1$$
$$S^{-1}\gamma^2 S = \gamma^2 S^{-1} S = \gamma^2 \ . \tag{1.59}$$

Since $\{\gamma^3\gamma^0, \gamma^i\} = 0$ $i = 0, 3$

$$S^{-1}\gamma^0 S = \gamma^0 (S)^2 = \gamma^0 \left(\cosh^2 \frac{\theta}{2} + \sinh^2 \frac{\theta}{2} + 2\cosh \frac{\theta}{2} \sinh \frac{\theta}{2} \gamma^3\gamma^0 \right)$$
$$= \gamma^0 \cosh\theta - \gamma^3 \sinh\theta \tag{1.60a}$$

$$S^{-1}\gamma^3 S = \gamma^3 (S)^2 = \gamma^3 \left(\cosh^2 \frac{\theta}{2} + \sinh^2 \frac{\theta}{2} + 2\cosh \frac{\theta}{2} \sinh \frac{\theta}{2} \gamma^3\gamma^0 \right)$$
$$= \gamma^3 \cosh\theta - \gamma^0 \sinh\theta \tag{1.60b}$$

Note also that
$$\left(\gamma^3\gamma^0\right)^\dagger = \gamma^{0\dagger}\gamma^{3\dagger} = -\gamma^0\gamma^3 = \gamma^3\gamma^0 \tag{1.61}$$

so that
$$S^\dagger = S \ . \tag{1.62}$$

Then
$$\psi'^\dagger \psi' = \psi^\dagger S^\dagger S \psi = \psi^\dagger (S)^2 \psi \neq \psi^\dagger \psi \ . \tag{1.63}$$

so that the normalization is *not* preserved. However, we should *expect* a change to occur since because of Lorentz contraction $d^3x' \neq d^3x$. Rather we should have

$$\psi'^\dagger \psi' d^3x' = \psi^\dagger \psi \, d^3x \tag{1.64}$$

which does not require unitarity of S.

We have shown then via i) and ii) that for arbitrary rotations and boosts it is possible to construct an operator S such that the Dirac equation *is* covariant. For later use we note that covariance can also be verified under a rather different kind of transformation:

iii) Spatial Inversion: (*i.e.*, Parity Transformation)
with
$$\begin{aligned} x'^0 &= x^0 \\ \vec{x}' &= -\vec{x} \end{aligned} \tag{1.65}$$

Then
$$S = S^{-1} = S^\dagger = \gamma^0 \ . \tag{1.66}$$

Check:
$$\begin{aligned} S^{-1}\gamma^0 S &= \gamma^0 S^{-1} S = \gamma^0 \\ S^{-1}\vec{\gamma} S &= -\vec{\gamma} S^{-1} S = -\vec{\gamma} \end{aligned} \ . \tag{1.67}$$

Of course, $\psi'^\dagger \psi' = \psi^\dagger S^\dagger S \psi = \psi^\dagger \psi$ so that the normalization is preserved.

Conserved Current Density

There exists one additional requirement on S — that we be able to construct a properly conserved probability current density—$j^\mu = (\rho, \vec{j})$ —satisfying

$$\nabla^\mu j_\mu = \frac{\partial \rho}{\partial t} + \vec{\nabla} \cdot \vec{j} = 0 \ . \tag{1.68}$$

in all frames. For the probability density we expect

$$\rho = \psi^\dagger \psi \tag{1.69}$$

which is properly non-negative. However, does there exist a corresponding $\vec{j}$? In order to answer this question we note that

$$\psi^\dagger \left(i\frac{\partial \psi}{\partial t} + \left(i\vec{\alpha} \cdot \vec{\nabla} - \beta m \right) \psi \right) = 0$$
$$\left(-i\frac{\partial \psi^\dagger}{\partial t} + \psi^\dagger \left(-i\vec{\alpha} \cdot \overleftarrow{\nabla} - \beta m \right) \right) \psi = 0 \ . \tag{1.70}$$

Subtracting, we find

$$i\left(\psi^\dagger \frac{\partial \psi}{\partial t} + \frac{\partial \psi^\dagger}{\partial t} \psi \right) + i\psi^\dagger \left(\vec{\alpha} \cdot \vec{\nabla} + \vec{\alpha} \cdot \overleftarrow{\nabla} \right) \psi = 0$$
$$= i\left(\frac{\partial}{\partial t} \psi^\dagger \psi + \vec{\nabla} \cdot \psi^\dagger \vec{\alpha} \psi \right) \ , \tag{1.71}$$

whereby we identify

$$\vec{j} = \psi^\dagger \vec{\alpha} \psi \ . \tag{1.72}$$

In order that the conservation equation

$$\nabla_\mu j^\mu = 0 \tag{1.73}$$

be Lorentz invariant and thus valid in an arbitrary frame, it is necessary that the four-current density

$$j^\mu = \left(\psi^\dagger \psi, \psi^\dagger \vec{\alpha} \psi \right) \tag{1.74}$$

transform as a four-vector. It is conventional to express the current density not in terms of $\psi^\dagger$ but rather in terms of

$$\bar{\psi} \equiv \psi^\dagger \gamma^0 \ . \tag{1.75}$$

Then

$$j^\mu = \bar{\psi} \gamma^\mu \psi \ . \tag{1.76}$$

which in a new frame becomes

$$j'^\mu = \psi^\dagger S^\dagger \gamma^0 \gamma^\mu S \psi = \psi^\dagger (\gamma^0)^2 S^\dagger \gamma^0 \gamma^\mu S \psi$$
$$= \bar{\psi} \gamma^0 S^\dagger \gamma^0 \gamma^\mu S \psi \ . \tag{1.77}$$

Since we already know that
$$S^{-1}\gamma^\mu S = \gamma'^\mu \tag{1.78}$$
the current density $\bar{\psi}\gamma^\mu\psi$ will be a four-vector provided
$$\gamma^0 S^\dagger \gamma^0 = S^{-1} . \tag{1.79}$$

This requirement is easily verified:

i) Rotations: $[\gamma^0, \gamma^i\gamma^j] = 0 \quad i,j = 1,2,3$
   Then
$$\gamma^0 S^\dagger \gamma^0 = S^\dagger = S^{-1} . \tag{1.80}$$

ii) Boosts: $\{\gamma^0, \gamma^0\gamma^i\} = 0 \quad i = 1,2,3$
   Then
$$\gamma^0 S^\dagger \gamma^0 = \gamma^0 S \gamma^0 = S^{-1} . \tag{1.81}$$

iii) Spatial Inversion: Then
$$\gamma^0 S^\dagger \gamma^0 = S^{-1} . \tag{1.82}$$

We have in general
$$j^\mu = \bar{\psi}\gamma^\mu\psi \longrightarrow \bar{\psi}\gamma'^\mu\psi = a^\mu{}_\nu j^\nu = j'^\mu, \tag{1.83}$$
so that, as required,
$$0 = \nabla^\mu j_\mu \longrightarrow \nabla^{\mu'} j'_\mu = 0 \tag{1.84}$$
i.e., current conservation obtains in all frames. Eq. 1.84 guarentees that the normalization is preserved in time. Thus
$$\frac{d}{dt}\int d^3x\, \rho = -\int_{\text{Vol}} d^3x\, \nabla \cdot \vec{j} = -\int_{\text{Surf}} \vec{j}\cdot d\vec{S} \tag{1.85}$$
where we have used Gauss' theorem. For a localized wavefunction the current density $\vec{j}$ vanishes on a sufficiently large surface and we find
$$\frac{d}{dt}\int d^3x\, \rho = 0 \qquad \text{q.e.d.} \tag{1.86}$$

Since
$$\rho = \psi^\dagger \psi \tag{1.87}$$
is non-negative we shall for the moment be able to interpret ψ in terms of a single particle wavefunction with ρ as the probability density. Later on we shall find that at least some of the same problems which plagued the Klein–Gordon equation occur in the Dirac case. The ultimate solution to these difficulties is quantization of the Dirac field. For now, however, we proceed with the single particle wavefunction interpretation.

VII.2 BILINEAR FORMS

We have seen that although the γ^μ are constant matrices which have the same value in *all* frames, the bilinear quantity
$$j^\mu = \bar{\psi}\gamma^\mu\psi \tag{2.1}$$
transforms as a four-vector—under a Lorentz transformation
$$x^\mu \to x'^\mu = a^\mu_{\;\nu} x^\nu \tag{2.2}$$
we find
$$j^\mu \longrightarrow j'^\mu = a^\mu_{\;\nu} j^\nu \tag{2.3}$$

Completeness

Since $\bar{\psi}$, ψ are simply four-component row, column vectors, a bilinear
$$\bar{\psi}\mathcal{O}\psi \tag{2.4}$$
can be always decomposed into a combination of $4 \times 4 = 16$ linearly independent matrices. Defining
$$\gamma_5 = -i\gamma^0\gamma^1\gamma^2\gamma^3 = -i\begin{pmatrix} 0 & \sigma_1 \\ \sigma_1 & 0 \end{pmatrix}\begin{pmatrix} -i\sigma_1 & 0 \\ 0 & -i\sigma_1 \end{pmatrix} = \begin{pmatrix} 0 & -1 \\ -1 & 0 \end{pmatrix} \tag{2.5}$$
and the antisymmetric tensor
$$\sigma^{\mu\nu} = \frac{i}{2}(\gamma^\mu\gamma^\nu - \gamma^\nu\gamma^\mu) \tag{2.6}$$
with
$$\sigma^{00} = \sigma^{ii} = 0$$
$$\sigma^{0i} = i\begin{pmatrix} 0 & \sigma_i \\ \sigma_i & 0 \end{pmatrix} \quad \sigma^{ij} = \epsilon_{ijk}\begin{pmatrix} \sigma_k & 0 \\ 0 & \sigma_k \end{pmatrix} \tag{2.7}$$
we may choose these sixteen matrices to be
$$1 \quad \gamma^\mu \quad \sigma^{\mu\nu} \quad \gamma^\mu\gamma_5 \quad \gamma_5 \tag{2.8}$$
There exist four γ^μ's, four $\gamma^\mu\gamma_5$'s, one unit operator and one γ_5 matrix. The 4×4 matrices $\sigma^{\mu\nu}$ are antisymmetric in μ, ν. Since an antisymmetric 4×4 matrix must have the structure
$$A^{\mu\nu} \sim \begin{pmatrix} 0 & c_1 & c_2 & c_3 \\ -c_1 & 0 & c_4 & c_5 \\ -c_2 & -c_4 & 0 & c_6 \\ -c_3 & -c_5 & -c_6 & 0 \end{pmatrix} \tag{2.9}$$
we see that there exist only six *independent* elements. Thus the total number of linearly independent matrices is found to be

$$\begin{array}{ccccc} 1 & \gamma^\mu & \sigma^{\mu\nu} & \gamma^\mu\gamma_5 & \gamma_5 \\ 1 \;+ & 4 \;+ & 6 \;+ & 4 \;+ & 1 \;= 16 \end{array}$$

as required.

Transformation Properties

All Dirac matrices are simply constants and have the same value in all Lorentz frames. However, when contracted with $\bar{\psi}, \psi$ different bilinears have their own distinct transformation properties and we study each in turn:

1: $\quad \bar{\psi}\psi \longrightarrow \left(\psi^\dagger S^\dagger\right) \gamma^0 S\psi = \psi^\dagger \gamma^0 (\gamma^0 S^\dagger \gamma^0) S\psi = \bar{\psi} S^{-1} S\psi = \bar{\psi}\psi$. (2.10)

Thus $\bar{\psi}\psi$ is a Lorentz scalar, transforming into itself under boosts, rotations and spatial inversions.

Before looking at γ_5 we note that since the matrix is a product of the four Dirac matrices, it must correspondingly anticommute with any of these—

$$\{\gamma_5, \gamma^\mu\} = 0 \qquad (2.11)$$

—so that

$\gamma_5: \quad \bar{\psi}\gamma_5\psi \longrightarrow \psi^\dagger S^\dagger \gamma^0 \gamma_5 S\psi = \psi^\dagger \gamma^0 \left(\gamma^0 S^\dagger \gamma^0\right) \gamma_5 S\psi = \bar{\psi} S^{-1} \gamma_5 S\psi$. (2.12)

Since γ_5 anticommutes with any single γ^μ, it must commute with a product

$$[\gamma_5, \gamma^\mu \gamma^\nu] = 0 \ . \qquad (2.13)$$

Thus for rotations and/or boosts

$$\bar{\psi}\gamma_5\psi \xrightarrow[\substack{\text{Rot}\\\text{Boost}}]{} \bar{\psi}\gamma_5 S^{-1} S\psi = \bar{\psi}\gamma_5\psi \qquad (2.14)$$

but under spatial inversion

$$\bar{\psi}\gamma_5\psi \xrightarrow[P]{} -\bar{\psi}\gamma_5 S^{-1} S\psi = -\bar{\psi}\gamma_5\psi \ . \qquad (2.15)$$

The bilinear $\bar{\psi}\gamma_5\psi$ then transforms as a pseudoscalar.

Classical physics examples of scalar and pseudoscalar quantities are charge (scalar) and "magnetic charge" (pseudoscalar) — if it exists! In order to see this, note that an ordinary four-vector such as $x^\mu = (t, \vec{x})$ has the behavior under a spatial inversion

$$x^\mu = (t, \vec{x}) \xrightarrow[P]{} (t, -\vec{x}) = x_\mu \qquad (2.16)$$

Since A^μ itself is a four-vector then

$$A^\mu = (\phi, \vec{A}) \xrightarrow[P]{} (\phi, -\vec{A}) = A_\mu \ . \qquad (2.17)$$

This means, however, that

$$\begin{aligned} \vec{E} &= -\vec{\nabla}\phi - \frac{\partial \vec{A}}{\partial t} \xrightarrow[P]{} -(-\vec{\nabla})\phi - \frac{\partial}{\partial t}(-\vec{A}) = -\vec{E} \\ \vec{B} &= \vec{\nabla} \times \vec{A} \xrightarrow[P]{} (-\vec{\nabla}) \times (-\vec{A}) = \vec{B} \end{aligned} \qquad (2.18)$$

so that $\vec{E}$ and $\vec{B}$ behave oppositely under spatial inversion. Consider the electric field from a point charge. Since

$$\vec{E} = \frac{q}{4\pi r^2}\hat{r} \xrightarrow[P]{} \frac{q_P}{4\pi r^2}(-\hat{r}) = -\vec{E} \qquad (2.19)$$

we see that q_P (the electric charge in the inverted frame) must be the same as q—q is a scalar. However, if a magnetic monopole were to exist, with magnetic charge g

$$\vec{B} = \frac{g}{r^2}\hat{r} \xrightarrow[P]{} \frac{g_P}{r^2}(-\hat{r}) = +\vec{B} \qquad (2.20)$$

which requires that

$$g = -g_P \qquad (2.21)$$

i.e., magnetic charge is a pseudoscalar quantity.

Now consider the transformation properties of $\bar{\psi}\gamma^\mu\psi$, $\bar{\psi}\gamma^\mu\gamma_5\psi$.

$$\gamma^\mu: \quad \bar{\psi}\gamma^\mu\psi \xrightarrow[\text{boost}]{\text{rot}} \psi^\dagger S^\dagger \gamma^0 \gamma^\mu S\psi = \psi^\dagger \gamma^0 \left(\gamma^0 S^\dagger \gamma^0\right)\gamma^\mu S\psi$$
$$= \bar{\psi} S^{-1}\gamma^\mu S\psi = a^\mu{}_\nu \bar{\psi}\gamma^\nu\psi = \bar{\psi}\gamma'^\mu\psi \qquad (2.22)$$

so that $\bar{\psi}\gamma^\mu\psi$ transforms as a Lorentz four-vector. Also, under a spatial inversion we note that

$$\bar{\psi}\gamma^\mu\psi \xrightarrow[P]{} \bar{\psi}\gamma^0\gamma^\mu\gamma^0\psi = \bar{\psi}\gamma_\mu\psi \ . \qquad (2.23)$$

Thus $\bar{\psi}\gamma^\mu\psi$ transforms under parity in the same manner as $x^\mu = (t,\vec{x})$ and is a four-vector or polar vector.

If, however, we consider $\bar{\psi}\gamma^\mu\gamma_5\psi$

$$\gamma^\mu\gamma_5: \quad \bar{\psi}\gamma^\mu\gamma_5\psi \xrightarrow[\text{boost}]{\text{rot}} \psi^\dagger S^\dagger \gamma^0 \gamma^\mu \gamma_5 S\psi = \psi^\dagger\gamma^0\left(\gamma^0 S^\dagger\gamma^0\right)\gamma^\mu\gamma_5 S\psi$$
$$= \bar{\psi}S^{-1}\gamma^\mu S\gamma_5\psi = \bar{\psi}\gamma'^\mu\gamma_5\psi \qquad (2.24)$$

but

$$\bar{\psi}\gamma^\mu\gamma_5\psi \xrightarrow[P]{} \bar{\psi}\gamma^0\gamma^\mu\gamma_5\gamma^0\psi = -\bar{\psi}\gamma_\mu\gamma_5\psi \ . \qquad (2.25)$$

Thus under ordinary Lorentz transformations (rotations/boosts) this quantity transforms like a four-vector. However, under spatial inversion an extra minus sign arises. Such a quantity is termed a "pseudo-vector" or axial vector.

There exist many examples of polar vectors in classical physics, such as

$$\vec{r} \xrightarrow[P]{} -\vec{r}$$
$$\vec{v} = \frac{d\vec{r}}{dt} \xrightarrow[P]{} -\frac{d\vec{r}}{dt} = -\vec{v} \qquad (2.26)$$
$$\vec{a} = \frac{d^2\vec{r}}{dt^2} \xrightarrow[P]{} -\frac{d^2\vec{r}}{dt^2} = -\vec{a} \ .$$

Perhaps the most familiar example of an axial vector is angular momentum

$$\vec{L} = \vec{r} \times m\vec{v} \xrightarrow{P} (-\vec{r}) \times (-m\vec{v}) = +\vec{L} \ . \tag{2.27}$$

Similarly one requires that the spin be an axial vector

$$\vec{S} \xrightarrow{P} +\vec{S} \tag{2.28}$$

in order that the total angular momentum

$$\vec{J} = \vec{L} + \vec{S} \tag{2.29}$$

have the property of transforming into itself under spatial inversion.

Finally, for the bilinear $\bar{\psi}\sigma^{\mu\nu}\psi$

$$\sigma_{\mu\nu}: \quad \bar{\psi}\sigma^{\mu\nu}\psi = \bar{\psi}\frac{i}{2}(\gamma^{\mu}\gamma^{\nu} - \gamma^{\nu}\gamma^{\mu})\psi \xrightarrow[\text{boost}]{\text{rot}} \psi^{\dagger}S^{\dagger}\gamma^{0}\frac{i}{2}(\gamma^{\mu}\gamma^{\nu} - \gamma^{\nu}\gamma^{\mu})S\psi$$

$$= \psi^{\dagger}\gamma^{0}\gamma^{0}S^{\dagger}\gamma^{0}\frac{i}{2}(\gamma^{\mu}\gamma^{\nu} - \gamma^{\nu}\gamma^{\mu})S\psi = \bar{\psi}S^{-1}\frac{i}{2}(\gamma^{\mu}\gamma^{\nu} - \gamma^{\nu}\gamma^{\mu})S\psi$$

$$= \bar{\psi}\frac{i}{2}(S^{-1}\gamma^{\mu}S\,S^{-1}\gamma^{\nu}S - S^{-1}\gamma^{\nu}S\,S^{-1}\gamma^{\mu}S)\psi$$

$$= \bar{\psi}\frac{i}{2}a^{\mu}{}_{\lambda}a^{\nu}{}_{\sigma}(\gamma^{\lambda}\gamma^{\sigma} - \gamma^{\sigma}\gamma^{\lambda})\psi = \bar{\psi}\sigma'^{\mu\nu}\psi \ . \tag{2.30}$$

Under spatial inversion

$$\bar{\psi}\sigma^{\mu\nu}\psi \xrightarrow{P} \bar{\psi}\gamma^{0}\sigma^{\mu\nu}\gamma^{0}\psi = \bar{\psi}\sigma_{\mu\nu}\psi \ . \tag{2.31}$$

so that $\bar{\psi}\sigma^{\mu\nu}\psi$ transforms as a an antisymmetric second rank Lorentz tensor, an example of which in the sector of classical physics is the electromagnetic field tensor $F^{\mu\nu}$.

We have already seen the utility of one of these bilinear forms — the four-vector $\bar{\psi}\gamma^{\mu}\psi$ has been identified as the conserved probability current density. Other uses will arise in subsequent discussions.

VII.3 NONRELATIVISTIC REDUCTION

It is helpful to construct the effective Schrödinger equation which applies in the non-relativistic limit of the Dirac equation, as the resulting form must contain various familiar structures.

Effective Schrödinger Equation

We seek a (positive energy) stationary state solution

$$\psi(\vec{x},t) = \psi(\vec{x})\exp(-iEt) \tag{3.1}$$

with
$$E = m + W \quad \text{and} \quad W \ll m \ . \tag{3.2}$$
Then for the two-component coupled equations involving ψ_a, ψ_b we find
$$\begin{aligned}(m + W - e\phi)\psi_a - \vec{\sigma}\cdot\vec{\pi}\psi_b &= m\psi_a \\ \vec{\sigma}\cdot\vec{\pi}\psi_a - (m + W - e\phi)\psi_b &= m\psi_b \ . \end{aligned} \tag{3.3}$$
We may solve the second of these equations for ψ_b in terms of ψ_a
$$\psi_b = \frac{1}{2m + W - e\phi}\vec{\sigma}\cdot\vec{\pi}\psi_a \ . \tag{3.4}$$
If $W, e\phi$ are both much smaller than m, then since $\pi \sim mv$ we have
$$\psi_b \sim v\psi_a \ll \psi_a \quad \text{for} \quad \frac{v}{c} \ll 1 \ . \tag{3.5}$$
For this reason ψ_a (ψ_b) is often called the large (small) component of the Dirac equation. Substitution of Eq. 3.4 into Eq. 3.3a gives a relation for ψ_a alone
$$\begin{aligned}\left(\vec{\sigma}\cdot\vec{\pi}\frac{1}{2m + W - e\phi}\vec{\sigma}\cdot\vec{\pi} + e\phi\right)\psi_a &= W\psi_a \\ \approx \left(\frac{1}{2m}\vec{\sigma}\cdot\vec{\pi}\vec{\sigma}\cdot\vec{\pi} + \frac{1}{(2m)^2}\vec{\sigma}\cdot\vec{\pi}(e\phi - W)\vec{\sigma}\cdot\vec{\pi} + e\phi\right)\psi_a \ . \end{aligned} \tag{3.6}$$
This is not quite the effective Schrödinger equation since ψ_a is not normalized to unity. Instead we have
$$\begin{aligned}1 &= \int d^3x\, \psi^\dagger(\vec{x})\psi(\vec{x}) = \int d^3x \left(\psi_a^\dagger\psi_a + \psi_b^\dagger\psi_b\right) \\ &\approx \int d^3x\, \psi_a^\dagger \left(1 + \frac{\vec{\sigma}\cdot\vec{\pi}}{2m}\frac{\vec{\sigma}\cdot\vec{\pi}}{2m}\right)\psi_a \approx \int d^3x\, \chi^\dagger\chi \end{aligned} \tag{3.7}$$
where we have defined
$$\chi \equiv \left(1 + \frac{(\vec{\sigma}\cdot\vec{\pi})^2}{8m^2}\right)\psi_a \ . \tag{3.8}$$
Then if we add
$$\frac{1}{8m^2}\left[(\vec{\sigma}\cdot\vec{\pi})^2(W - e\phi) + (W - e\phi)(\vec{\sigma}\cdot\vec{\pi})^2\right]\psi_a \tag{3.9}$$
to each side of Eq. 3.6 we find
$$\begin{aligned}&\left(\frac{1}{2m}(\vec{\sigma}\cdot\vec{\pi})^2 + \frac{1}{8m^2}\left((\vec{\sigma}\cdot\vec{\pi})^2(W - e\phi)\right.\right. \\ &\left.\left. - 2\vec{\sigma}\cdot\vec{\pi}(W - e\phi)\vec{\sigma}\cdot\vec{\pi} + (W - e\phi)(\vec{\sigma}\cdot\vec{\pi})^2\right)\right)\psi_a \\ &\cong \left(1 + \frac{(\vec{\sigma}\cdot\vec{\pi})^2}{8m^2}\right)(W - e\phi)\left(1 + \frac{(\vec{\sigma}\cdot\vec{\pi})^2}{8m^2}\right)\psi_a\end{aligned} \tag{3.10}$$

or in terms of the wavefunction χ

$$\left(1 + \frac{(\vec{\sigma} \cdot \vec{\pi})^2}{8m^2}\right)(W - e\phi)\chi \cong \frac{1}{2m}(\vec{\sigma} \cdot \vec{\pi})^2 \left(1 - \frac{(\vec{\sigma} \cdot \vec{\pi})^2}{8m^2}\right)\chi$$
$$+ \frac{1}{8m^2}\left((\vec{\sigma} \cdot \vec{\pi})^2(W - e\phi) - 2\vec{\sigma} \cdot \vec{\pi}(W - e\phi)\vec{\sigma} \cdot \vec{\pi} + (W - e\phi)(\vec{\sigma} \cdot \vec{\pi})^2\right)\chi \ . \tag{3.11}$$

We can rewrite Eq. 3.11 by use of the identity

$$\hat{A}^2\hat{B} - 2\hat{A}\hat{B}\hat{A} + \hat{B}\hat{A}^2 = \hat{A}(\hat{A}\hat{B} - \hat{B}\hat{A}) - (\hat{A}\hat{B} - \hat{B}\hat{A})\hat{A}$$
$$= \left[\hat{A}, [\hat{A}, \hat{B}]\right] \tag{3.12}$$

as

$$(W - e\phi)\chi \cong \frac{1}{2m}(\vec{\sigma} \cdot \vec{\pi})^2 \left(1 - \frac{(\vec{\sigma} \cdot \vec{\pi})^2}{4m^2}\right)\chi$$
$$+ \frac{1}{8m^2}\left[(\vec{\sigma} \cdot \vec{\pi}), [(\vec{\sigma} \cdot \vec{\pi}), W - e\phi]\right]\chi \ . \tag{3.13}$$

Since, assuming $\vec{A}(x)$ to be independent of t,

$$[\vec{\sigma} \cdot \vec{\pi}, [\vec{\sigma} \cdot \vec{\pi}, W - e\phi]] = \left[\vec{\sigma} \cdot \vec{\pi}, ie\vec{\sigma} \cdot \vec{\nabla}\phi\right] = -ie\left[\vec{\sigma} \cdot \vec{\pi}, \vec{\sigma} \cdot \vec{E}\right]$$
$$= e\left(-\vec{\nabla} \cdot \vec{E} + 2\vec{\sigma} \cdot \vec{\pi} \times \vec{E}\right) \tag{3.14}$$

we have, finally

$$W\chi \cong \left(\frac{1}{2m}(\vec{p} - e\vec{A})^2 - \frac{1}{8m^3}(\vec{p} - e\vec{A})^4\right.$$
$$\left. - \frac{e}{2(m+W)}\vec{\sigma} \cdot \vec{B} + e\phi - \frac{e}{8m^2}\left(\vec{\nabla} \cdot \vec{E} - 2\vec{\sigma} \cdot \vec{\pi} \times \vec{E}\right)\right)\chi \tag{3.15}$$

which is the effective Schrödinger equation we seek.

Interpretation

Interpretation of the various terms can now be undertaken. Thus in the spinless case, we evaluated the relativistic Hamiltonian to be used in the presence of an electromagnetic field described by the vector potential $A_\mu = (\phi, \vec{A})$, yielding

$$H = \sqrt{m^2 + (\vec{p} - e\vec{A})^2} + e\phi$$
$$\approx m + \frac{1}{2m}(\vec{p} - e\vec{A})^2 - \frac{1}{8m^3}(p - e\vec{A})^4 + e\phi + \ldots \tag{3.16}$$

in the non-relativistic limit. Likewise in Eq. 3.15 we recognize

$$\frac{1}{2m}(\vec{p} - e\vec{A})^2 + e\phi \tag{3.17}$$

as the usual Schrödinger Hamiltonian and

$$-\frac{1}{8m^3}\left(\vec{p}-e\vec{A}\right)^4 \tag{3.18}$$

as a relativistic correction to the kinetic energy. Similarly we identify the term

$$-\frac{e}{2(m+W)}\vec{\sigma}\cdot\vec{B} \approx -\frac{e}{2m}\vec{\sigma}\cdot\vec{B} \tag{3.19}$$

as the energy of interaction of the magnetic moment of a spin-1/2 particle (with gyromagnetic ratio g=2) with an external magnetic field.[†] As discussed above, this arises automatically if we use

$$\frac{1}{2m}\vec{\sigma}\cdot\vec{\pi}\,\vec{\sigma}\cdot\vec{\pi} = \frac{1}{2m}(\vec{p}-e\vec{A})^2 - \frac{e}{2m}\vec{\sigma}\cdot\vec{B} \tag{3.20}$$

for the kinetic energy term in the Hamiltonian rather than the simple spinless form

$$\frac{1}{2m}\vec{\pi}\cdot\vec{\pi} = \frac{1}{2m}\left(\vec{p}-e\vec{A}\right)^2 \;. \tag{3.21}$$

We can also identify the remaining terms in a straight-forward fashion. For example, for an electron bound in a hydrogen atom the operator

$$-\frac{e}{4m^2}\vec{\sigma}\cdot\vec{E}\times\vec{p} \tag{3.22}$$

is simply the usual spin-orbit term since for $\phi = \phi(r)$

$$-\frac{e}{4m^2}\vec{\sigma}\cdot\vec{E}\times\vec{p} = \frac{e}{4m^2}\frac{1}{r}\frac{d\phi}{dr}\vec{\sigma}\cdot\vec{r}\times\vec{p} = \frac{e}{4m^2}\frac{1}{r}\frac{d\phi}{dr}\vec{\sigma}\cdot\vec{L} \;. \tag{3.23}$$

The "classical" derivation of this form involves looking at the problem of the atom from the perspective of the electron rest frame, in which case the central nucleus is seen to be orbiting at the distance of a Bohr radius. In this frame, however, there exists a *magnetic* field due to the Lorentz transformation of the electromagnetic fields involved

$$\vec{B} = -\vec{v}\times\vec{E} \;. \tag{3.24}$$

The energy associated with the interaction of the electron magnetic moment with this induced field is

$$H = -\vec{\mu}\cdot\vec{B} = \frac{e}{2m}\vec{\sigma}\cdot\vec{v}\times\vec{E} = \frac{e}{2m^2}\frac{1}{r}\frac{d\phi}{dr}\vec{\sigma}\cdot\vec{L} \;, \tag{3.25}$$

which differs from Eq. 3.23 by a factor of two due to the "Thomas precession."

The point is [Ja 80] that in an inertial frame the energy

$$U = -\vec{\mu}\cdot\vec{B} \tag{3.26}$$

[†] Note that the relativistic Bohr magneton is $\frac{e}{2E}$.

associated with the interaction of a magnetic moment $\vec{\mu}$ with a magnetic field $\vec{B}$ corresponds to an equation of motion

$$\frac{d\vec{S}}{dt} = \vec{S} \times \frac{e\vec{B}}{m} \qquad (3.27)$$

where the have used the relation

$$\vec{\mu} = \frac{e}{m}\vec{S} \qquad (3.28)$$

between the spin and magnetic moment. However, in a non-inertial frame of reference one has

$$\frac{d\vec{S}}{dt} = \left.\frac{d\vec{S}}{dt}\right|_{\text{non-rotating}} - \vec{\omega} \times \vec{S} \qquad (3.29)$$

where $\vec{\omega}$ is the angular velocity of the rotating frame, and the corresponding interaction energy is

$$U' = -\vec{S} \cdot \vec{\omega} \ . \qquad (3.30)$$

In order to find the angular velocity $\vec{\omega}$ corresponding to our case, consider the trajectory of an accelerating electron as shown in Figure VII.1.

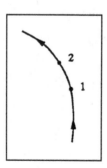

Fig. VII.1: *Trajectory of an accelerating electron.*

Suppose that at time t the electron is at position #1 with velocity $\vec{v}$ while at time $t+\delta t$ the electron is located at position #2 with velocity $\vec{v}+\delta\vec{v}$. Let the laboratory frame be denoted by W, the rest frames of the electron at positions #1 and #2 by W_1 and W_2, respectively. Then one can reach W_1 from W by a simple Lorentz transformation with velocity $\vec{v}$

$$W \xrightarrow[\vec{v}]{} W_1 \qquad (3.31)$$

while one can reach W_2 from W by a Lorentz transformation with velocity $\vec{v} + \delta\vec{v}$

$$W \xrightarrow[\vec{v}+\delta\vec{v}]{} W_2 \ . \qquad (3.32)$$

However, the transformation between the frames W_1 and W_2 is in general a combination of a boost plus a rotation. To first order in $\delta\vec{v}$ we find

$$t_2 = t_1 - \vec{x}_1 \cdot \frac{1}{\sqrt{1-v^2}} \left(\delta\vec{v} + \left(\frac{1}{\sqrt{1-v^2}} - 1 \right) \hat{v}\hat{v} \cdot \delta\vec{v} \right)$$

$$\vec{x}_2 = \vec{x}_1 - t_1 \frac{1}{\sqrt{1-v^2}} \left(\delta\vec{v} + \left(\frac{1}{\sqrt{1-v^2}} - 1 \right) \hat{v}\hat{v} \cdot \delta\vec{v} \right) \quad (3.33)$$

$$+ \left(\frac{1}{\sqrt{1-v^2}} - 1 \right) \vec{x}_1 \times (\vec{v} \times \delta\vec{v}) \frac{1}{v^2}$$

which corresponds to a Lorentz transformation with velocity

$$\Delta\vec{v} = \frac{1}{\sqrt{1-v^2}} \left(\delta\vec{v} + \left(\frac{1}{\sqrt{1-v^2}} - 1 \right) \hat{v}\hat{v} \cdot \delta\vec{v} \right) \quad (3.34)$$

accompanied by a rotation through angle

$$\delta\vec{\theta} = \left(\frac{1}{\sqrt{1-v^2}} - 1 \right) \vec{v} \times \delta\vec{v} \frac{1}{v^2} . \quad (3.35)$$

The corresponding angular velocity is

$$\vec{\omega} = \frac{\delta\vec{\theta}}{\delta t} = \left(\frac{1}{\sqrt{1-v^2}} - 1 \right) \vec{v} \times \vec{a} \frac{1}{v^2} \approx \frac{1}{2} \vec{v} \times \vec{a} , \quad (3.36)$$

leading to an additional interaction energy

$$U' = -\vec{\omega} \cdot \vec{S} = -\frac{1}{2} \vec{v} \times \vec{a} \cdot \vec{S}$$

$$= -\frac{1}{2} \vec{v} \times -\frac{e}{r} \frac{d\phi}{dr} \vec{r} \frac{1}{m} \cdot \vec{S} \quad (3.37)$$

$$= -\frac{e}{4m^2} \frac{1}{r} \frac{d\phi}{dr} \vec{L} \cdot \vec{\sigma}.$$

Adding Eqs. 3.25 and 3.37 we find

$$U_{\text{tot}} = \frac{e}{4m^2} \frac{1}{r} \frac{d\phi}{dr} \vec{L} \cdot \vec{\sigma} \quad (3.38)$$

as found in reduction of the Dirac equation.

The origin of the final piece of the effective interaction, the Darwin term

$$-\frac{e}{8m^2} \vec{\nabla} \cdot \vec{E}, \quad (3.39)$$

has already been noted in our discussion of the Klein–Gordon equation, where we showed that the zitterbewegung motion associated with the interference between positive and negative energy components leads to a shift of the potential energy in the amount

$$\Delta U \simeq \frac{1}{2!} \delta r_i \, \delta r_j \frac{\partial^2}{\partial r_i \partial r_j} e\phi(\vec{r}) \sim \frac{e}{2 \cdot 3m^2} \vec{\nabla}^2 \phi$$

$$= -\frac{e}{6m^2} \vec{\nabla} \cdot \vec{E} . \quad (3.40)$$

Except for the factor of 1/6 rather than 1/8 this is clearly the additional term under discussion. Because of this identification the Darwin or zitterbewegung term has no classical analogy.

Hydrogen Atom Energy Levels: Perturbative Approach

Now examine the effects of these perturbations on the energy levels of the hydrogen atom. For the zitterbewegung term we find that only S-waves are affected

$$\begin{aligned}(\Delta E)^{n\ell}_{\text{zitt}} &= -\frac{e}{8m^2}\left\langle \vec{\nabla}\cdot\vec{E}\right\rangle_{n\ell} = \frac{e^2}{8m^2}\left\langle \delta^3(r)\right\rangle_{n\ell} \\ &= \frac{e^2}{8m^2}\int d^3r\,\delta^3(r)\,\psi^*_{n\ell}(\vec{r})\psi_{n\ell}(\vec{r}) \\ &= \frac{\pi\alpha}{2m^2}\delta_{\ell 0}|\psi_{n0}(0)|^2 = \frac{\pi\alpha}{2m^2}\delta_{\ell 0}\left(\frac{1}{\pi n^3 a_0^3}\right) \\ &= m\frac{\alpha^4}{2n^3}\delta_{\ell 0}\ .\end{aligned} \qquad (3.41)$$

while for the spin-orbit term S-waves are not altered in energy, but other angular momentum states are shifted. Recalling that for a given value of orbital angular momentum ℓ, the total angular momentum

$$\vec{J} = \vec{L} + \vec{S} \qquad (3.42)$$

can have the value $\ell + 1/2$ or $\ell - 1/2$, we find

$$\begin{aligned}(\Delta E)^{n\ell}_{so} &= \frac{\alpha}{4m^2}\left\langle r^{-3}\right\rangle_{n\ell} 2\vec{S}\cdot\vec{L} \\ &= \frac{\alpha}{4m^2}\left\langle r^{-3}\right\rangle_{n\ell}(J^2 - L^2 - S^2) = \frac{\alpha}{4m^2}\left\langle r^{-3}\right\rangle_{n\ell}\begin{cases}\ell & j=\ell+\tfrac{1}{2} \\ -(\ell+1) & j=\ell-\tfrac{1}{2}\end{cases} \\ &= \frac{\alpha}{4m^2}\left(\frac{1}{n^3 a_0^3\,\ell(\ell+1)(\ell+\tfrac{1}{2})}\right)\begin{cases}\ell & j=\ell+\tfrac{1}{2} \\ -(\ell+1) & j=\ell-\tfrac{1}{2}\end{cases} \\ &= m\frac{\alpha^4}{2n^3(2\ell+1)}\begin{cases}(\ell+1)^{-1} & j=\ell+\tfrac{1}{2} \\ -\ell^{-1} & j=\ell-\tfrac{1}{2}\end{cases}\ .\end{aligned} \qquad (3.43)$$

Finally, we note that for the relativistic kinetic energy term

$$\begin{aligned}(\Delta E)^{n\ell}_{p^4} &= -\frac{1}{2m}\left\langle \left(\frac{p^2}{2m}\right)^2\right\rangle_{n\ell} \\ &= -\frac{1}{2m}\left\langle (E_n - V(r))^2\right\rangle_{n\ell} \\ &= -\frac{1}{2m}\left\langle E_n^2 - 2E_n V(r) + V^2(r)\right\rangle_{n\ell}\ .\end{aligned} \qquad (3.44)$$

According to the virial theorem

$$\langle V(r)\rangle_{n\ell} = 2E_n \qquad (3.45)$$

and by direct calculation

$$\langle V^2(r)\rangle_{n\ell} = \alpha^2 \langle r^{-2}\rangle_{n\ell} = \alpha^2 \frac{1}{n^3 a_0^2(\ell+1/2)} \qquad (3.46)$$
$$= m^2 \frac{\alpha^4}{n^3(\ell+1/2)}.$$

Thus

$$(\Delta E)^{n\ell}_{p^4} = -m\frac{\alpha^4}{2n^3}\left(\frac{1}{\ell+1/2} - \frac{3}{4n}\right). \qquad (3.47)$$

Our final result then, writing $E = m + W$ is

$$W = -m\frac{\alpha^2}{2n^2}\left(1 + \frac{\alpha^2}{n}\left(\frac{1}{j+1/2} - \frac{3}{4n}\right) + \cdots\right). \qquad (3.48)$$

We observe that the energy depends only upon j — it is independent of ℓ. For example, the $2S_{1/2}$ and $2P_{1/2}$ states are degenerate, as used earlier in our discussions of the Lamb shift. Also, we see that the shift has a similar form to that found in the case of the Klein–Gordon atom, but with

$$\begin{matrix} K-G \\ \ell + \tfrac{1}{2} \end{matrix} \quad \text{replaced by} \quad \begin{matrix} \text{Dirac} \\ j + \tfrac{1}{2} \end{matrix}. \qquad (3.49)$$

This may seem like a small difference, but it is easily measurable and agreement with experiment for the hydrogen atom *requires* the electron to be a spin-1/2 particle.

PROBLEM VII.3.1

The Runge-Lenz Vector and the Hydrogen Atom

One of the remarkable features of motion in a Coulomb (or gravitational) field is that there is no precession of the classical orbits. By use of the so-called Runge-Lenz vector it is possible both to understand this result and to derive the energy levels of a hydrogen atom without use of any differential equations.

i) Verify this result by showing classically that

$$\frac{d\vec{R}}{dt} = 0 \quad \text{and} \quad \vec{L}\cdot\vec{R} = 0$$

where

$$\vec{R} = \frac{1}{m}\vec{p}\times\vec{L} - \frac{\alpha}{r}\vec{r}$$

is the Runge-Lenz vector. Show that for gravitational motion $\vec{R}$ points along perihelion so that no precession takes place in the planetary orbits for an exact $1/r$ potential.

ii) If we define quantum mechanically

$$\vec{R} = \frac{1}{2m}(\vec{p}\times\vec{L} - \vec{L}\times\vec{p}) - \frac{\alpha}{r}\vec{r}$$

show that
$$[H, \vec{L}] = [H, \vec{R}] = 0 \qquad \vec{R} \cdot \vec{L} = \vec{L} \cdot \vec{R} = 0$$

iii) Verify that
$$R^2 = \alpha^2 + \frac{2}{m} H(L^2 + 1)$$
so that the Hamiltonian can be written in terms of two constants of the motion.

iv) Define $\vec{K} = \sqrt{\frac{-m}{2H}} \vec{R}$ and
$$\vec{M} = \frac{1}{2}(\vec{L} + \vec{K}) \qquad \vec{N} = \frac{1}{2}(\vec{L} - \vec{K}).$$

Show that
$$[M_i, M_j] = i\epsilon_{ijk} M_k$$
$$[N_i, N_j] = i\epsilon_{ijk} N_k$$
$$[M_i, N_j] = 0$$

Thus $\vec{M}$ and $\vec{N}$ obey commutation relations for angular momenta and commute with each other and also with the Hamiltonian
$$H = -\frac{m\alpha^2}{2(2M^2 + 2N^2 + 1)}.$$

We can then find simultaneous eigenstates of H, M^2, N^2, M_z, N_z with
$$H|E, m, n, m_z, n_z\rangle = E|E, m, n, m_z, n_z\rangle$$
$$M^2|E, m, n, m_z, n_z\rangle = m(m+1)|E, m, n, m_z, n_z\rangle$$
$$N^2|E, m, n, m_z, n_z\rangle = n(n+1)|E, m, n, m_z, n_z\rangle$$
$$M_z|e, m, n, m_z, n_z\rangle = m_z|e, m, n, m_z, n_z\rangle$$
$$N_z|E, m, n, m_z, n_z\rangle = n_z|E, m, n, m_z, n_z\rangle.$$

v) Show that $\vec{R} \cdot \vec{L} = 0$ implies $\vec{K} \cdot \vec{L} = 0$ and hence that $M^2 = N^2$.

vi) Show that this in turn implies
$$E = -\frac{m\alpha^2}{2k^2}$$
where k=1,2,3,... with degeneracy factor $2n^2$, as required.

PROBLEM VII.3.2

The Anomalous Magnetic Moment

The Dirac equation describing the interaction of a proton or neutron with an applied external electromagnetic field has an additional term

$$\left(i\not\nabla - Q_i \not A + \frac{\kappa_i |e|}{4m}\sigma_{\mu\nu}F^{\mu\nu} - m\right)\psi(x) = 0$$

involving the so-called anomalous magnetic moment. (For the proton, of course, $Q_i = |e|$ and for the neutron $Q_i = 0$.)

i) Verify that the choice

$$\kappa_p = 1.79 \ , \qquad \kappa_n = -1.91$$

corresponds to the observed magnetic moments of these particles, and

ii) show that the additional interaction disturbs neither the Lorentz covariance of the equation nor the hermiticity of the Hamiltonian.

PROBLEM VII.3.3

The Aharonov-Casher Effect

In Section III.3 we discussed the Aharonov-Bohm effect which shows that in quantum mechanics the behavior of particles can be altered by the presence of a non-zero vector potential even though the magnetic field vanishes in all regions of space accessible to these particles. More recently Aharonov and Casher [AhC 84] pointed out another interesting quantum mechanical process whereby the behavior of *magnetic* dipoles is altered by the presence of an *electric* field. In this problem we explore this effect.

i) The Dirac equation which describes an electron in the presence of an external vector potential A_μ is

$$(i\not\nabla - e\not A - m)\psi(x) = 0$$

Writing the energy as $E = m + W$ and making a nonrelativistic reduction as done in the text, show that the effective Schrödinger equation which results is

$$\frac{\vec\sigma\cdot(\vec p - e\vec A)\vec\sigma\cdot(\vec p - e\vec A)}{2m}\psi(x) = (W - e\phi)\psi(x)$$

ii) For the standard version of the A-B effect one uses an infinite solenoid positioned along the z-axis so that in cylindrical coordinates

$$\vec A(x) = \frac{BR^2}{2r}\hat e_\phi \qquad r > R$$

where R is the solenoidal radius. Show that for this geometry the effective Hamiltonian becomes

$$H_{\text{eff}} = \frac{1}{2m}\sum_{i=1}^{2}(p_i - eA_i)^2$$

iii) As shown in problem VII.3.2, the Dirac equation describing the interaction of a neutral spin 1/2 particle with an external electromagnetic field is given by

$$(i\,\not\nabla + \frac{\kappa|e|}{4m}\sigma_{\mu\nu}F^{\mu\nu} - m)\psi(x) = 0$$

where κ is the magnetic moment. Now consider a beam of neutrons polarized along the z-axis interacting with a line charge with charge per unit length λ aligned along the z-direction. Perform a nonrelativistic reduction of the above relativistic equation and demonstrate that the effective Schrödinger equation is

$$\frac{\vec{\sigma}\cdot(\vec{p}-i\kappa'\vec{E})\vec{\sigma}\cdot(\vec{p}+i\kappa'\vec{E})}{2m}\psi(x) = W\psi(x)$$

where $\kappa' = \kappa|e|/2m$ and

$$\vec{E} = \frac{\lambda}{2\pi r}\hat{r}$$

is the electric field generated by the line charge.

iv) Show that this Hamiltonian is equivalent to the form

$$H_{\text{eff}} = \frac{(\vec{p}-\vec{E}\times\vec{\mu})^2}{2m} - \frac{\kappa'^2 E^2}{2m}$$

for the geometry at hand, where $\vec{\mu} = \kappa'\vec{\sigma}$. Replacing $H_{\text{eff}} \to H_{\text{eff}}' \equiv \chi_i^\dagger H_{\text{eff}}\chi_i$ verify that

$$H_{\text{eff}}' = \frac{1}{2m}\sum_{i=1}^{2}[p_i - (\vec{E}\times\kappa'\hat{e}_z)_i]^2 \quad,$$

which is completely equivalent to the A-B Hamiltonian provided we make the replacement $\kappa' E \to eA$.

We observe then that the Hamiltonian is the same and hence there must exist an effect on the magnetic dipoles for the A-C situation in complete analogy to that on electric charges for the A-B geometry. Recently this prediction was confirmed experimentally [Ci 89].

VII.4 COULOMB SOLUTION

Although we have derived the $2S_{1/2} - 2P_{1/2}$ degeneracy within the context of first order perturbation theory, the result is more general and is valid to *all orders* in the fine structure constant α, as we shall demonstrate.

Hydrogen Atom Energy Levels: Relativistic Approach

We begin with the Dirac equation in its usual representation

$$(\gamma_\mu \pi^\mu - m)\psi(x) = 0 \quad \text{where} \quad \pi_\mu = i\nabla_\mu - eA_\mu \tag{4.1}$$

and introduce the projection operators

$$P_1, P_2 = \frac{1}{2}(1 - \gamma_5), \frac{1}{2}(1 + \gamma_5) \ . \tag{4.2}$$

[Note that since $\gamma_5^2 = 1$ we have $P_1^2 = P_2^2 = 1$, $P_1 P_2 = P_2 P_1 = 0$.] If we define

$$\psi_1 \equiv P_1 \psi \ , \qquad \psi_2 \equiv P_2 \psi \tag{4.3}$$

then since

$$P_2 \gamma_\mu = \gamma_\mu P_1 \tag{4.4}$$

we find

$$\begin{aligned} P_2 \gamma_\mu \pi^\mu \psi &= \gamma_\mu \pi^\mu P_1 \psi = \gamma_\mu \pi^\mu \psi_1 \\ &= m P_2 \psi = m \psi_2 \end{aligned} \tag{4.5}$$

or

$$\psi_2 = \frac{1}{m} \gamma_\mu \pi^\mu \psi_1 \ . \tag{4.6}$$

Also, since $P_1 + P_2 = 1$ we find

$$\psi = \psi_1 + \psi_2 = \left(1 + \frac{1}{m}\gamma_\mu \pi^\mu\right)\psi_1 \tag{4.7}$$

so that knowledge of ψ_1 is equivalent to knowledge of ψ itself. From Eq. 4.7 we see that ψ_1 obeys the equation

$$m(\gamma_\mu \pi^\mu - m)\psi = (\gamma_\mu \pi^\mu - m)(\gamma_\mu \pi^\mu + m)\psi_1 = 0 \ . \tag{4.8}$$

Also since

$$P_1 = \frac{1}{2}\begin{pmatrix} 1 & 1 \\ 1 & 1 \end{pmatrix} \tag{4.9}$$

a stationary state solution ψ_1 having definite orbital angular momentum ℓ must be of the form

$$\psi_1 = \begin{pmatrix} \chi_\ell(\vec{r}) \\ \chi_\ell(\vec{r}) \end{pmatrix} e^{-iEt} \tag{4.10}$$

where the χ_ℓ's represent *two*-component spinors which satisfy the equation

$$\left(\frac{1}{r}\frac{\partial^2}{\partial r^2}r + 2\alpha\frac{E}{r} - \frac{\ell(\ell+1) - \alpha^2 - i\alpha\vec{\sigma}\cdot\hat{r}}{r^2} + E^2 - m^2\right)\chi_\ell(\vec{r}) = 0 \ . \tag{4.11}$$

Except for the term involving the Pauli spin matrix $\vec{\sigma}$, this differential equation is identical to that studied in the case of the Klein-Gordon atom

$$\left(\frac{1}{r}\frac{\partial^2}{\partial r^2}r - \frac{\ell'(\ell'+1)}{r^2} + \frac{2m'\alpha}{r} + k^2\right)\psi(r) = 0 \tag{4.12}$$

provided we make the substitution

$$\ell(\ell+1) - \alpha^2 = \ell'(\ell'+1)$$
$$E^2 - m^2 = k^2 = 2m'E' \tag{4.13}$$
$$E = m'$$

Eq. 4.11 then becomes

$$\left(\frac{1}{r}\frac{\partial^2}{\partial r^2}r - \frac{\ell'(\ell'+1)}{r^2} + \frac{2m'\alpha}{r} + k^2 + i\alpha\frac{1}{r^2}\vec{\sigma}\cdot\hat{r}\right)\chi_\ell(\vec{r}) = 0 \tag{4.14}$$

and can be diagonalized in terms of functions $\phi_{jm}^\pm$ which are eigenstates of J^2, L^2, S^2, J_z:

$$\phi_{jm}^\pm = \sum_{p,q} C_{\ell\frac{1}{2};j=\ell\pm\frac{1}{2}}^{p\,q;m} Y_\ell^p(\theta,\phi)\chi_{\frac{1}{2}}^q \ . \tag{4.15}$$

Explicitly, we find

$$\phi_{j,m}^+ = \begin{pmatrix} ((j+m)/2j)^{1/2} \ Y_{j-1/2}^{m-1/2}(\theta,\phi) \\ ((j-m)/2j)^{1/2} \ Y_{j-1/2}^{m+1/2}(\theta,\phi) \end{pmatrix}$$

$$\phi_{j,m}^- = \begin{pmatrix} ((j+1-m)/2(j+1))^{1/2} \ Y_{j+1/2}^{m-1/2}(\theta,\phi) \\ ((j+1+m)/2(j+1))^{1/2} \ Y_{j+1/2}^{m+1/2}(\theta,\phi) \end{pmatrix} \tag{4.16}$$

Observe that $\phi_{j,m}^\pm$ have opposite parities—if $\hat{\Pi}$ is the spatial inversion operator

$$\hat{\Pi}\phi_{j,m}^\pm = (-1)^{j\mp 1/2}\phi_{j,m}^\pm \tag{4.17}$$

—and that they are connected via the operator $\vec{\sigma}\cdot\hat{r}$

$$\vec{\sigma}\cdot\hat{r}\,\phi_{j,m}^\pm = \phi_{j,m}^\mp \ . \tag{4.18}$$

This result is clear since
  i) $\vec{\sigma}\cdot\hat{r}$ is odd under parity;

ii) $(\vec{\sigma} \cdot \hat{r})^2 = 1$;

iii) $\left[\vec{J}, \vec{\sigma} \cdot \hat{r}\right] = 0$;

so that eigenstates of $\vec{J}$ are also eigenstates of $\vec{\sigma} \cdot \hat{r}$. [For the latter result note that for J_z:

$$L_z = -i\frac{\partial}{\partial \phi} : \quad [L_z, \sigma_x \sin\theta \cos\phi + \sigma_y \sin\theta \sin\phi + \sigma_z \cos\theta]$$
$$= -i(-\sigma_x \sin\theta \sin\phi + \sigma_y \sin\theta \cos\phi) \quad (4.19)$$

$$S_z = \frac{\sigma_z}{2} : \quad [S_z, \sigma_x \sin\theta \cos\phi + \sigma_y \sin\theta \sin\phi + \sigma_z \cos\theta]$$
$$= i\sigma_y \sin\theta \cos\phi - i\sigma_x \sin\theta \sin\phi \ .$$

Thus (and other components may be proved similarly)

$$[J_z, \vec{\sigma} \cdot \hat{r}] = [L_z, \vec{\sigma} \cdot \hat{r}] + [S_z, \vec{\sigma} \cdot \hat{r}] = 0 \ . \quad (4.20)$$

Now choose the linear combinations

$$F^{\pm}_{j,m} \equiv \phi^{\pm}_{j,m} \pm \frac{i}{\alpha}\left(j + \frac{1}{2} - s\right)\phi^{\mp}_{j,m} \quad (4.21)$$

where

$$s = \left(\left(j + \frac{1}{2}\right)^2 - \alpha^2\right)^{1/2} \ . \quad (4.22)$$

Then we have

$$\left(L^2 - \alpha^2 - i\alpha\vec{\sigma} \cdot \hat{r}\right) F^{\pm}_{j,m} = \left(L^2 - \alpha^2 \pm \left(j + \frac{1}{2} - s\right)\right)\phi^{\pm}_{j,m}$$
$$+ \left(\pm\frac{i}{\alpha}\left(j + \frac{1}{2} - s\right)(L^2 - \alpha^2) - i\alpha\right)\phi^{\mp}_{j,m} \ . \quad (4.23)$$

But

$$\left(L^2 - \alpha^2 \pm \left(j + \frac{1}{2} - s\right)\right)\phi^{\pm}_{j,m} = u_{\pm}(u_{\pm} + 1)\phi^{\pm}_{j,m} \quad (4.24)$$

with

$$u_{\pm} = s - \frac{1}{2} \mp \frac{1}{2} \ . \quad (4.25)$$

[Check:

$$\left(L^2 - \alpha^2 \pm \left(j + \frac{1}{2}\right)\right)\phi^{\pm}_{j,m} = \begin{cases} \left(\left(j - \frac{1}{2}\right)\left(j + \frac{1}{2}\right) - \alpha^2 + j + \frac{1}{2}\right)\phi^{+}_{j,m} \\ \left(\left(j + \frac{1}{2}\right)\left(j + \frac{3}{2}\right) - \alpha^2 - j - \frac{1}{2}\right)\phi^{+}_{j,m} \end{cases}$$
$$= \left(\left(j + \frac{1}{2}\right)^2 - \alpha^2\right)\phi^{\pm}_{j,m} = s^2\phi^{\pm}_{j,m} \quad (4.26)$$

$$s^2 - s = s(s-1) = u_+(u_+ + 1)$$
$$s^2 + s = s(s+1) = u_-(u_- + 1) \quad \text{q.e.d.}]$$

Also

$$\left(\pm \frac{i}{\alpha}\left(j + \frac{1}{2} - s\right)(L^2 - \alpha^2) - i\alpha\right) \phi_{j,m}^{\mp}$$
$$= \pm \frac{i}{\alpha}\left(j + \frac{1}{2} - s\right)\left[L^2 - \alpha^2 \mp \frac{\alpha^2}{j + \frac{1}{2} - s}\right] \phi_{j,m}^{\pm} \qquad (4.27)$$
$$= \pm \frac{i}{\alpha}\left(j + \frac{1}{2} - s\right)\left[L^2 - \alpha^2 \mp \left(j + \frac{1}{2} + s\right)\right] \phi_{j,m}^{\mp}$$
$$= \pm \frac{i}{\alpha}\left(j + \frac{1}{2} - s\right) u_\pm (u_\pm + 1) \phi_{j,m}^{\mp} \ .$$

Thus

$$\left(L^2 - \alpha^2 - i\alpha \vec{\sigma} \cdot \hat{r}\right) F_{j,m}^{\pm} = u_\pm(u_\pm + 1) F_{j,m}^{\pm} \qquad (4.28)$$

and Eq. 4.14 becomes

$$\left(\frac{1}{r}\frac{\partial^2}{\partial r^2}r + \frac{2m'\alpha}{r} + k^2 - \frac{u_\pm(u_\pm + 1)}{r^2}\right) F_{j,m}^{\pm} = 0 \qquad (4.29)$$

which is completely identical to the Klein–Gordon case provided we make the substitution

$$\ell'(\ell' + 1) \longrightarrow u_\pm(u_\pm + 1) \ . \qquad (4.30)$$

The associated energy levels are then given by (*cf.* Eq. VI.3.34)

$$E_{n,j} = m\left(1 + \frac{\alpha^2}{\left(n + s - (j + \frac{1}{2})\right)^2}\right)^{-1/2} \quad n = 1, 2, \ldots$$
$$\approx m\left(1 - \frac{1}{2}\frac{\alpha^2}{n^2}\left(1 + \frac{\alpha^2}{n^2}\left(\frac{n}{j + \frac{1}{2}} - \frac{3}{4}\right) + \ldots\right)\right) \ . \qquad (4.31)$$

which is identical to that found in the Klein-Gordon case, but with

$$\begin{array}{ccc} K - G & & \text{Dirac} \\ & \text{replaced by} & \\ \ell + \frac{1}{2} & & j + \frac{1}{2} \end{array} \qquad (4.32)$$

in agreement with experiment for the hydrogen atom. Also, in the Dirac case there exists a twofold degeneracy in all levels with the same value of j and n, so that *e.g.*, as claimed earlier, the $2P_{1/2}$ and $2S_{1/2}$ levels are degenerate to all orders in α.

VII.5 PLANE WAVE SOLUTIONS

It is particularly useful to examine plane wave solutions of the Dirac equation, corresponding to a freely moving particle, since it is with such wavefunctions that one can develop an intuitive feel for the physics.

Derivation

We begin by writing

$$\psi(x) = u(p)\, e^{-iEt+i\vec{p}\cdot\vec{x}}$$
$$= u(p)\, e^{-ip\cdot x} \tag{5.1}$$

where $u(p)$ is a four-component spinor. Since

$$i\nabla^\mu e^{-ip\cdot x} = p^\mu e^{-ip\cdot x} \tag{5.2}$$

the free particle Dirac equation

$$(i\gamma_\mu \nabla^\mu - m)\,\psi(x) = 0 \tag{5.3}$$

becomes an algebraic relation

$$(\gamma_\mu p^\mu - m)u(p) = (\not{p} - m)u(p) = 0 \tag{5.4}$$

which is equivalent to four linear homogeneous equations. In order for a solution to exist, we must require $\det(\not{p} - m)$ to vanish, and one could proceed to solve the system formally in this fashion. However, we shall utilize an alternative, more intuitive, approach, writing

$$u(p) = \begin{pmatrix} u_a \\ u_b \end{pmatrix} \tag{5.5}$$

where u_a, u_b are both two-component spinors. Then the validity of the Dirac equation requires

$$(\gamma^0 E - \vec{\gamma}\cdot\vec{p})\,u(p) = \begin{pmatrix} E & -\vec{\sigma}\cdot\vec{p} \\ \vec{\sigma}\cdot\vec{p} & -E \end{pmatrix} \begin{pmatrix} u_a \\ u_b \end{pmatrix} = m \begin{pmatrix} u_a \\ u_b \end{pmatrix} \tag{5.6}$$

which yields the pair of coupled equations

$$\begin{aligned} E u_a - \vec{\sigma}\cdot\vec{p}\, u_b &= m u_a \\ -E u_b + \vec{\sigma}\cdot\vec{p}\, u_a &= m u_b \end{aligned} \tag{5.7}$$

There exist two relations between the spinors u_a, u_b which must *both* be satisfied

$$u_a = \frac{\vec{\sigma}\cdot\vec{p}}{E-m} u_b, \qquad u_b = \frac{\vec{\sigma}\cdot\vec{p}}{E+m} u_a. \tag{5.8}$$

We thus require

$$u_b = \frac{\vec{\sigma}\cdot\vec{p}}{E+m} u_a = \frac{\vec{\sigma}\cdot\vec{p}\,\vec{\sigma}\cdot\vec{p}}{E^2 - m^2} u_b. \tag{5.9}$$

Using the identity

$$\vec{\sigma}\cdot\vec{p}\,\vec{\sigma}\cdot\vec{p} = \vec{p}^2 \tag{5.10}$$

we see that Eq. 5.9 is satisfied provided that

$$E^2 - m^2 = \vec{p}^2 \tag{5.11}$$

which is the desired relation between the relativistic energy E and momentum $\vec{p}$. Of course, for a given value of the momentum $\vec{p}$ there exist *two* solutions for the energy

$$E = \pm\sqrt{\vec{p}^2 + m^2} \tag{5.12}$$

as found in the case of the Klein–Gordon equation.

Consider first the positive energy solutions. We define

$$u_a = N\chi \tag{5.13}$$

where χ is a two-component spinor and N is a normalization constant. The lower component u_b is then

$$u_b = N\frac{\vec{\sigma}\cdot\vec{p}}{E+m}\chi \tag{5.14}$$

and our solution takes the form

$$u(p) = N\begin{pmatrix} \chi \\ \frac{\vec{\sigma}\cdot\vec{p}}{E+m}\chi \end{pmatrix} . \tag{5.15}$$

There exist two linearly independent spinors $u(p)$ corresponding to the two linearly independent values for χ — call these χ_1 and χ_2 (e.g. we could take $\chi_1 = \binom{1}{0}$ and $\chi_2 = \binom{0}{1}$) with $\chi_1^\dagger\chi_2 = \chi_2^\dagger\chi_1 = 0$ and $\chi_1^\dagger\chi_1 = \chi_2^\dagger\chi_2 = 1$).

In order to normalize the Dirac spinor, one's first thought might be to place a normalization condition upon $u^\dagger(p)u(p)$. However, this is a non-relativistic way of thinking and would not be Lorentz invariant. Defining

$$\bar{u} \equiv u^\dagger\gamma^0 \tag{5.16}$$

we have already shown that $\bar{u}u$ is a Lorentz scalar quantity. Thus we may set

$$\bar{u}(p)u(p) = 1 . \tag{5.17}$$

as our normalization condition and this will hold in all frames. Since

$$u(p) = N\begin{pmatrix} \chi \\ \frac{\vec{\sigma}\cdot\vec{p}}{E+m}\chi \end{pmatrix} , \quad \bar{u}(p) = N^*\left(\chi^\dagger, -\chi^\dagger\frac{\vec{\sigma}\cdot\vec{p}}{E+m}\right) \tag{5.18}$$

Eq. 5.17 becomes

$$\begin{aligned} 1 &= |N|^2\left(1 - \frac{\vec{\sigma}\cdot\vec{p}\,\vec{\sigma}\cdot\vec{p}}{(E+m)^2}\right) = |N|^2\left(1 - \frac{\vec{p}^2}{(E+m)^2}\right) \\ &= |N|^2\left(1 - \frac{E^2 - m^2}{(E+m)^2}\right) = |N|^2\left(1 - \frac{E-m}{E+m}\right) = |N|^2\frac{2m}{E+m} . \end{aligned} \tag{5.19}$$

Thus we choose
$$N = \sqrt{\frac{E+m}{2m}} \qquad (5.20)$$
so
$$u(p) = \sqrt{\frac{E+m}{2m}} \begin{pmatrix} \chi \\ \frac{\vec{\sigma}\cdot\vec{p}}{E+m}\chi \end{pmatrix} . \qquad (5.21)$$

We have then *two* linearly independent positive energy solutions $u_1(p), u_2(p)$ with
$$\bar{u}_1(p)u_1(p) = \bar{u}_2(p)u_2(p) = 1 \qquad (5.22a)$$
and
$$\bar{u}_1(\vec{p})u_2(\vec{p}) = \bar{u}_2(\vec{p})u_1(\vec{p}) = 0 . \qquad (5.22b)$$

Sometimes one picks the corresponding Pauli spinors χ_1, χ_2 by the requirement that
$$\vec{\sigma}\cdot\hat{p}\chi_1 = \chi_1 , \qquad \vec{\sigma}\cdot\hat{p}\chi_2 = -\chi_2 . \qquad (5.23)$$

Then the Dirac spinor constructed using χ_1 is said to be in a positive "helicity" state, while that constructed from χ_2 is said to have negative "helicity." Positive, negative helicity corresponds to the spin being parallel, antiparallel to the direction of momentum.

Now consider the two remaining linearly independent negative energy solutions. We shall construct these spinors for the case that $\vec{p}$ is the negative of what it was above. That is, we define
$$p'^\mu = (-E, -\vec{p}) = -p^\mu , \qquad (5.24)$$
and assume a solution of the form
$$\psi'(x) = v(p)\, e^{-ip'\cdot x} = v(p)\, e^{ip\cdot x} . \qquad (5.25)$$

As before, the Dirac equation becomes an algebraic equation
$$(\gamma_\mu p^\mu + m)v(p) = 0 . \qquad (5.26)$$

Writing
$$v = \begin{pmatrix} v_a \\ v_b \end{pmatrix} \quad \text{we have} \quad \begin{matrix} -Ev_a + \vec{\sigma}\cdot\vec{p}\, v_b = mv_a \\ -\vec{\sigma}\cdot\vec{p}\, v_a + Ev_b = mv_b \end{matrix} \qquad (5.27)$$

which yields
$$v_a = \frac{\vec{\sigma}\cdot\vec{p}}{E+m}v_b , \qquad v_b = \frac{\vec{\sigma}\cdot\vec{p}}{E-m}v_a . \qquad (5.28)$$

Picking
$$v_b = N\chi' . \qquad (5.29)$$
we find
$$v_a = N\frac{\vec{\sigma}\cdot\vec{p}}{E+m}\chi' \qquad (5.30)$$

and
$$v(p) = N \begin{pmatrix} \frac{\vec{\sigma}\cdot\vec{p}}{E+m}\chi' \\ \chi' \end{pmatrix} . \tag{5.31}$$

Thus
$$\bar{v}(p)v(p) = |N|^2 \left(\chi'^\dagger \frac{\vec{\sigma}\cdot\vec{p}}{E+m}, -\chi'^\dagger\right) \begin{pmatrix} \frac{\vec{\sigma}\cdot\vec{p}}{E+m}\chi' \\ \chi' \end{pmatrix}$$
$$= -|N|^2 \chi'^\dagger \left(-\frac{\vec{\sigma}\cdot\vec{p}\,\vec{\sigma}\cdot\vec{p}}{(E+m)^2} + 1\right)\chi' = -|N|^2 \frac{2m}{E+m} \tag{5.32}$$

so that if we normalize to
$$\bar{v}(p)v(p) = -1 \tag{5.33}$$

we have, as before
$$N = \sqrt{\frac{E+m}{2m}} . \tag{5.34}$$

Note that the positive and negative energy solutions are orthogonal in that
$$\bar{u}(p)v(p) = |N|^2 \left(\chi^\dagger, -\chi^\dagger \frac{\vec{\sigma}\cdot\vec{p}}{E+m}\right) \begin{pmatrix} \frac{\vec{\sigma}\cdot\vec{p}}{E+m}\chi' \\ \chi' \end{pmatrix} = 0 . \tag{5.35}$$

We can summarize these results concisely as
$$\begin{aligned} \slashed{p}u(p) &= mu(p) & \slashed{p}v(p) &= -mv(p) \\ \bar{u}(p)u(p) &= 1 & \bar{v}(p)v(p) &= -1 \end{aligned} \tag{5.36}$$

with the orthogonality condition
$$\bar{v}(p)u(p) = \bar{u}(p)v(p) = 0 . \tag{5.37}$$

One can also write the Dirac equation in its conjugate form. Thus
$$\gamma_\mu p^\mu u(p) = mu(p) \quad \text{implies} \quad u^\dagger(p)\gamma_\mu^\dagger p^\mu = mu^\dagger(p) . \tag{5.38}$$

Since
$$\gamma_\mu^\dagger = \gamma^\mu \tag{5.39}$$

and
$$\gamma_\mu^\dagger \gamma^0 = \gamma^\mu \gamma^0 = \gamma^0 \gamma_\mu \tag{5.40}$$

we can write Eq. 5.38 as
$$\bar{u}(p)\slashed{p} = m\bar{u}(p) \quad \text{and also} \quad \bar{v}(p)\slashed{p} = -mv(p) . \tag{5.41}$$

Boosts

We can understand the form of these plane-wave spinors in an alternative fashion by using the boost operator derived earlier. Considering a positive energy electron at rest, we have the spinor

$$u(\vec{p}=0) = \begin{pmatrix} \chi \\ 0 \end{pmatrix}. \tag{5.42}$$

If we view this state from a frame moving with velocity

$$\vec{v} = -v\hat{k} \tag{5.43}$$

so that

$$t' = \frac{t+vz}{\sqrt{1-v^2}}, \quad z' = \frac{z+vt}{\sqrt{1-v^2}}, \quad y' = y, \quad x' = x \tag{5.44}$$

the corresponding Lorentz transformation matrix is

$$S(a) = \exp -\frac{\theta}{2}\gamma^3\gamma^0 \quad \text{with} \quad \cosh\theta = \frac{E}{m} = \frac{1}{\sqrt{1-v^2}}, \quad \sinh\theta = \frac{p}{m} = \frac{v}{\sqrt{1-v^2}} \tag{5.45}$$

and the Dirac spinor as viewed in this frame becomes

$$u(p) = S(a)u(\vec{p}=0) = \left(\cosh\frac{\theta}{2} - \gamma^3\gamma^0 \sinh\frac{\theta}{2}\right)\begin{pmatrix}\chi\\0\end{pmatrix} = \begin{pmatrix}\cosh\frac{\theta}{2}\chi\\\sinh\frac{\theta}{2}\sigma_3\chi\end{pmatrix}. \tag{5.46}$$

Since

$$\cosh^2\frac{\theta}{2} = \frac{1}{2}(1+\cosh\theta) = \frac{E+m}{2m}$$
$$\sinh^2\frac{\theta}{2} = \frac{1}{2}(\cosh\theta - 1) = \frac{E-m}{2m} \tag{5.47}$$

we have

$$u(p) = \sqrt{\frac{E+m}{2m}}\begin{pmatrix}\chi\\\sqrt{\frac{E-m}{E+m}}\vec{\sigma}\cdot\hat{p}\chi\end{pmatrix} = \sqrt{\frac{E+m}{2m}}\begin{pmatrix}\chi\\\frac{\vec{\sigma}\cdot\vec{p}}{E+m}\chi\end{pmatrix} \tag{5.48}$$

in agreement with Eq. 5.21. Similarly for a negative energy spinor, we find

$$v(\vec{p}) = S(a)v(\vec{p}=0) = S(a)\begin{pmatrix}0\\\chi\end{pmatrix} = \sqrt{\frac{E+m}{2m}}\begin{pmatrix}\frac{\vec{\sigma}\cdot\vec{p}}{E+m}\chi\\\chi\end{pmatrix} \tag{5.49}$$

which agrees with Eq. 5.31.

Rotations

It is also useful to examine the free Dirac spinors from the point of view of rotations in order to get a feel for the physics involved. This is most easily displayed in the case of an *infinitesimal* rotation by angle $\delta\phi$ about say the z-axis

$$\vec{x}' = \vec{x} + \delta\vec{x} \qquad \delta\vec{x} = (-x_2\delta\phi, x_1\delta\phi, 0) \qquad (5.50)$$
$$= -\vec{x} \times \delta\vec{\phi}$$

For a spinless particle, *e.g.* the Klein–Gordon equation, it is well-known that the angular momentum operator $\vec{L}$ is the "generator" of rotations in that after rotation we have[†]

$$\phi'(x) = \exp\left(i\delta\vec{\phi}\cdot\vec{L}\right)\phi(x) \approx \left(1 + i\delta\vec{\phi}\cdot\vec{L}\right)\phi(x) . \qquad (5.51)$$

On the other hand for the Dirac case, we expect rotations to be generated by the *total* angular momentum $\vec{J} = \vec{L} + \frac{1}{2}\vec{\Sigma}$

$$\psi'(x) = \exp\left(i\delta\vec{\phi}\cdot\vec{J}\right)\psi(x)$$
$$\cong \left(1 + i\delta\vec{\phi}\cdot\left(\vec{L} + \frac{1}{2}\vec{\Sigma}\right)\right)\psi(x) , \qquad (5.52)$$

where

$$\frac{1}{2}\vec{\Sigma} = \begin{pmatrix} \frac{1}{2}\vec{\sigma} & 0 \\ 0 & \frac{1}{2}\vec{\sigma} \end{pmatrix} \qquad (5.53)$$

is the spin operator. We can verify this conjecture since we know the form of the Dirac solution under rotations

$$\psi'(x') = S(\delta\vec{\phi})\psi(x) \qquad (5.54)$$

where

$$S(\delta\vec{\phi}) = \exp\left(-\frac{\delta\phi}{2}\gamma^1\gamma^2\right) \cong 1 - \gamma^1\gamma^2\frac{\delta\phi}{2} . \qquad (5.55)$$

Note that

$$\gamma^1\gamma^2 = \begin{pmatrix} 0 & \sigma_1 \\ -\sigma_1 & 0 \end{pmatrix}\begin{pmatrix} 0 & \sigma_2 \\ -\sigma_2 & 0 \end{pmatrix} = \begin{pmatrix} -i\sigma_3 & 0 \\ 0 & -i\sigma_3 \end{pmatrix} . \qquad (5.56)$$

Then

$$\gamma^i\gamma^j \equiv -i\epsilon_{ijk}\Sigma_k \qquad (5.57)$$

[†] Usually Eq. 5.51 is written in the equivalent form

$$\phi'(x') = \phi(x)$$

where x is the point which rotates into x'.

and
$$\begin{aligned}\psi'(x') &= \left(1 - x_2\delta\phi\frac{\partial}{\partial x_1} + x_1\delta\phi\frac{\partial}{\partial x_2} + i\frac{1}{2}\Sigma_3\delta\phi\right)\psi(x') \\ &= \left(1 + i\delta\phi\left(x_1 p_2 - x_2 p_1 + \frac{1}{2}\Sigma_3\right)\right)\psi(x') \\ &= \left(1 + i\delta\vec{\phi}\cdot\left(\vec{L} + \frac{1}{2}\vec{\Sigma}\right)\right)\psi(x')\end{aligned} \qquad (5.58)$$

as expected.

We also verify that although

$$\begin{aligned}\left[H, \vec{L}\right] &= \left[\vec{\alpha}\cdot\left(\vec{p} - e\vec{A}\right) + e\phi + \beta m, \vec{r}\times\vec{p}\right] \\ &= -i\vec{\alpha}\times\vec{p} + e\left[\phi, \vec{L}\right] - e[\vec{\alpha}\cdot\vec{A}, \vec{L}] \neq 0\end{aligned} \qquad (5.59)$$

and

$$\begin{aligned}\left[H, \frac{1}{2}\vec{\Sigma}\right] &= \left[\vec{\alpha}\cdot\left(\vec{p} - e\vec{A}\right) + e\phi + \beta m, \frac{1}{2}\vec{\Sigma}\right] \\ &= i\vec{\alpha}\times\left(\vec{p} - e\vec{A}\right) \neq 0\end{aligned} \qquad (5.60)$$

if we consider a particle moving under the influence of a spherically symmetric potential $\phi(r)$ so that

$$\vec{A} = 0 , \qquad \left[\vec{L}, \phi(r)\right] = 0 \qquad (5.61)$$

we find

$$\left[H, \vec{J}\right] = \left[H, \vec{L} + \frac{1}{2}\vec{\Sigma}\right] = 0 . \qquad (5.62)$$

—the total angular momentum is a constant of the motion, as expected.

Helicity

It is particularly interesting to examine the time development of the helicity $\vec{\Sigma}\cdot\left(\vec{p} - e\vec{A}\right)$, where $\vec{p} - e\vec{A}$ is the so-called "mechanical" momentum $m\dot{\vec{x}}$ (as opposed to the canonical momentum $\vec{p}$). Since

$$\begin{aligned}\left[H, \vec{p} - e\vec{A}\right] &= \left[\vec{\alpha}\cdot\left(\vec{p} - e\vec{A}\right) + e\phi + \beta m, \vec{p} - e\vec{A}\right] \\ &= ie\vec{\nabla}\phi - ie\vec{\alpha}\times(\vec{\nabla}\times\vec{A})\end{aligned} \qquad (5.63)$$

we have

$$\begin{aligned}\left[H, \vec{\Sigma}\cdot(\vec{p} - e\vec{A})\right] &= \left[H, \vec{\Sigma}\right]\cdot(\vec{p} - e\vec{A}) + \vec{\Sigma}\cdot\left[H, \vec{p} - e\vec{A}\right] \\ &= ie\vec{\Sigma}\cdot\vec{\nabla}\phi .\end{aligned} \qquad (5.64)$$

For a free particle — $\phi = 0$, $\vec{A} = 0$ — we find

$$\left[H, \vec{\Sigma}\cdot\vec{p}\right] = 0 \qquad (5.65)$$

so that helicity is a constant of the motion. However, this result is more general. Imagine an electron moving in a region where $\vec{E} = 0$ (take $\vec{\nabla}\phi = \partial \vec{A}/\partial t = 0$) but $\vec{B} \neq 0$. It follows then from Eq. 5.64 that

$$\left[H, \vec{\Sigma} \cdot (\vec{p} - e\vec{A})\right] = 0 \tag{5.66}$$

which means that the helicity will be also be unchanged in this circumstance. In particular imagine a longitudinally polarized electron moving in a uniform magnetic field $\vec{B}$. Then the trajectory of the electron will be a circle as shown in Figure VII.2, and the spin vector will exactly track the mechanical momentum.

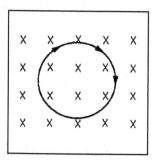

Fig. VII.2: *The electron trajectory in the presence of a uniform magnetic field is a circle of radius mv/eB.*

The reason for this is clear. According to classical physics the rotation frequency of the electron — the Larmor frequency ω_L — is found to be

$$\omega_L = \frac{v}{r} = \frac{eB}{m} \qquad (i.e.,\ \frac{mv^2}{r} = evB) \ . \tag{5.67}$$

On the other hand, the spin precession frequency ω_S is determined from the torque equation

$$\frac{d\vec{S}}{dt} = \vec{S} \times \frac{g_e e \vec{B}}{2m} \tag{5.68}$$

which yields

$$\omega_S = \frac{g_e e B}{2m} \ . \tag{5.69}$$

We see that the Larmor and spin precession frequencies are identical (for $g_e=2$), and this is the origin of the constancy of the helicity.

Current Density

Finally, consider the current density

$$j^\mu = \bar{\psi}\gamma^\mu\psi \ . \tag{5.70}$$

VII.5 PLANE WAVE SOLUTIONS

Between identical plane wave states—$u(p)\, e^{-ip\cdot x}$—j^μ assumes the form

$$j^\mu = \bar{u}(p)\gamma^\mu u(p) = \frac{1}{2}\bar{u}(p)\left(\frac{1}{m}\slashed{p}\gamma^\mu + \gamma^\mu \slashed{p}\frac{1}{m}\right)u(p)$$

$$= \text{(using the identity } \gamma^\mu\gamma^\nu + \gamma^\nu\gamma^\mu = 2\eta^{\mu\nu})\quad \bar{u}(p)\frac{p^\mu}{m}u(p) \qquad (5.71)$$

$$= \frac{p^\mu}{m}\ .$$

This proportionality to p^μ/m is easy to understand. Since the probability density is $j^0(x)$ the probability of finding this particle in a volume d^3x in the rest frame is

$$j^0(x)d^3x = \frac{m}{m}d^3x = d^3x\ . \qquad (5.72)$$

As viewed from a frame moving with velocity

$$\vec{v} = -v\hat{k} \qquad (5.73)$$

the probability density becomes

$$j'^0(x') = \frac{E}{m} = \gamma \quad \text{where} \quad \gamma = \frac{1}{\sqrt{1-v^2}}\ . \qquad (5.74)$$

However, because of the Lorentz contraction, the region of space being examined is correspondingly smaller

$$d^3x' = \frac{1}{\gamma}d^3x\ . \qquad (5.75)$$

so that the probability of being found within this volume is unchanged, as required.

$$j'^0(x')d^3x' = \frac{E}{m}d^3x' = \gamma \times \frac{1}{\gamma}d^3x = d^3x = j^0(x)d^3x\ . \qquad (5.76)$$

The corresponding three-vector component of the current density is also found as expected via

$$\vec{j}'(x') = j'^0(x')\vec{v} = \frac{E}{m}\cdot\frac{\vec{p}}{E} = \frac{\vec{p}}{m}\ . \qquad (5.77)$$

Considering the so-called transition current density, taken between different solutions of the Dirac equation ψ_i, ψ_f, we have

$$j^\mu(x) = \bar{\psi}_f(x)\gamma^\mu\psi_i(x) = \frac{1}{2m}\left(\bar{\psi}_f(x)\gamma^\mu i\gamma^\nu\partial_\nu\psi_i(x) - (i\partial_\nu\bar{\psi}_f(x))\gamma^\nu\gamma^\mu\psi_i(x)\right)\ . \qquad (5.78)$$

Now write

$$i\gamma^\mu\gamma^\nu = \frac{i}{2}\{\gamma^\mu,\gamma^\nu\} + \frac{i}{2}[\gamma^\mu,\gamma^\nu] = i\eta^{\mu\nu} + \sigma^{\mu\nu}\ . \qquad (5.79)$$

so that

$$j^\mu(x) = \frac{i}{2m}\left(\bar{\psi}_f(x)\partial^\mu\psi_i(x) - \partial^\mu\bar{\psi}_f(x)\psi_i(x)\right)$$
$$+ \frac{1}{2m}\left(\bar{\psi}_f(x)\sigma^{\mu\nu}\partial_\nu\psi_i(x) + \partial_\nu\bar{\psi}_f(x)\sigma^{\mu\nu}\psi_i(x)\right) \equiv j^{(1)\mu} + j^{(2)\mu}\ . \qquad (5.80)$$

We observe that there exists a more or less natural separation of the current density into two components:

i) For the first piece — $j^{(1)\mu}$ — since for positive energy solutions the lower component of the Dirac spinor is $\mathcal{O}(v/c)$ compared to the upper component, we have

$$\bar{\psi}\psi \equiv \begin{pmatrix} \psi^{\dagger}_{\text{upper}} & \psi^{\dagger}_{\text{lower}} \end{pmatrix} \begin{pmatrix} 1 & 0 \\ 0 & -1 \end{pmatrix} \begin{pmatrix} \psi_{\text{upper}} \\ \psi_{\text{lower}} \end{pmatrix}$$
$$= \psi^{\dagger}_{\text{upper}}\psi_{\text{upper}}\left(1 + \mathcal{O}\left(\frac{v^2}{c^2}\right)\right) \approx \psi^{\dagger}\psi \ . \tag{5.81}$$

We can write this contribution to the transition density as

$$j^{(1)\mu} \simeq \frac{i}{2m}\left(\psi^{\dagger}_f \partial_\mu \psi_i - \partial_\mu \psi^{\dagger}_f \psi_i\right) \tag{5.82}$$

which is identical to the usual Schrödinger form.

ii) For the second piece — $j^{(2)\mu}$ — we note that its contribution to a Lagrange density would be of the form

$$\mathcal{L}_{\text{int}} = -ej^\mu A_\mu \sim -\frac{e}{2m}A_\mu \partial_\nu(\bar{\psi}_f \sigma^{\mu\nu}\psi_i) \ . \tag{5.83}$$

Writing Eq. 5.83 as

$$\mathcal{L}_{\text{int}} = -\frac{e}{2m}\left(\partial_\nu\left(A_\mu \bar{\psi}_f \sigma^{\mu\nu}\psi_i\right) - \partial_\nu A_\mu \bar{\psi}_f \sigma^{\mu\nu}\psi_i\right) \ . \tag{5.84}$$

we see that the first piece may be discarded, as it is a total derivative and contributes only a constant to the Lagrangian. For the second term we may use the antisymmetry of $\sigma^{\mu\nu}$ to write

$$\mathcal{L}_{\text{int}} = \frac{e}{2m}\frac{1}{2}(\partial_\nu A_\mu - \partial_\mu A_\nu)\bar{\psi}_f \sigma^{\mu\nu}\psi_i$$
$$= -\frac{e}{4m}F_{\mu\nu}\bar{\psi}_f \sigma^{\mu\nu}\psi_i \ . \tag{5.85}$$

Looking at the non-relativistic limit, we have

$$\bar{\psi}_f \sigma^{ij}\psi_i \sim \epsilon_{ijk}\psi^{\dagger}_{\text{upper}}\sigma_k \psi_{\text{upper}}$$
$$\bar{\psi}_f \sigma^{0i}\psi_i \sim \mathcal{O}\left(\frac{v}{c}\right)\psi^{\dagger}_{\text{upper}}\sigma_i \psi_{\text{upper}} \ . \tag{5.86}$$

Hence, keeping only the piece involving σ_{ij} and noting that

$$F_{ij} = -\epsilon_{ijk}B_k \tag{5.87}$$

we find

$$\mathcal{L}_{\text{int}} = \frac{e}{4m}\epsilon_{ijk}B_k \epsilon_{ij\ell}\psi^{\dagger}_{\text{upper}}\sigma_\ell \psi_{\text{upper}}$$
$$= \frac{e}{2m}\vec{B}\cdot\psi^{\dagger}_{\text{upper}}\vec{\sigma}\psi_{\text{upper}} \ . \tag{5.88}$$

The corresponding Hamiltonian density is

$$\mathcal{H}_{\text{int}} = -\mathcal{L}_{\text{int}} = -\frac{e}{2m}\vec{B}\cdot\psi^{\dagger}_{\text{upper}}\vec{\sigma}\psi_{\text{upper}} \tag{5.89}$$

which is the usual energy of interaction of the electron magnetic moment

$$\vec{\mu} = \frac{e}{2m}\psi^\dagger \vec{\sigma}\psi \tag{5.90}$$

with an external magnetic field. For these reasons $j_\mu^{(1)}$ is called the "convection" current density while $j_\mu^{(2)}$ is referred to as the "magnetization" current density.

Experimental values of the gyromagnetic ratio for real spin-1/2 particles are found to be

$$\begin{aligned}
\text{electron} \quad & g_{\text{exp}} = 2\left(1 + \frac{\alpha}{2\pi} + \ldots\right) \\
\text{proton} \quad & g_{\text{exp}} = 2\left(1 + 1.79\right) \\
\text{neutron} \quad & g_{\text{exp}} = 2\left(0 - 1.91\right) \ .
\end{aligned} \tag{5.91}$$

Does this mean that these are *not* Dirac particles? Not really. In the case of the proton and neutron, a microscopic view of these systems reveals that they are far from being simple pointlike spin-1/2 structures. Rather they are composed of three pointlike particles called quarks and the bound state wavefunction, which extends over distances of the order of 10^{-13} cm, clearly does not represent a pointlike structure. At a second level particles like the proton/neutron can fragment virtually into a nucleon-meson system with which the photon can interact, as shown in Figure VII.3.

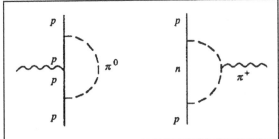

Fig. VII.3: Mesonic corrections to the nucleon-photon interaction.

Again we should expect substantial deviations from the result g=2 expected for a pointlike particle. In the case of the electron, there exists no quark substructure to deal with (as far as we know the electron really is a point particle). However, the electron can fragment into an $e - \gamma$ system as shown in Figure VII.4

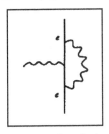

Fig. VII.4: Radiative corrections to the electron-photon interaction.

which yields a modification of the *g*-factor, but only at $\mathcal{O}(\alpha)$, as will be shown in the next chapter.

PROBLEM VII.5.1

Electric Dipole Moment of the Electron [BjD 64]

Suppose that the electron had a static electric dipole moment d analogous to its magnetic moment.

i) Show that this could be accomodated by modifying the Dirac equation to become
$$(i\,\slashed{\nabla} - e\,\slashed{A} - i\frac{ed}{4m}\sigma_{\mu\nu}\gamma_5 F^{\mu\nu} - m)\psi(x) = 0 \ .$$

ii) Demonstrate that this equation is covariant but *not* invariant under a parity transformation.

iii) Show that this interaction would lead to a mixing between the $2S_{\frac{1}{2}}$ and $2P_{\frac{1}{2}}$ levels of the hydrogen atom and from the observed agreement between calculated and measured values of the Lamb shift at the level of a 0.05 MHz obtain an upper bound on the electric dipole moment of the electron.

Note: The relevant matrix element vanishes if the nonrelativistic wavefunctions are used—you will need the appropriate relativistic solutions.

PROBLEM VII.5.2

Chirality and the Dirac Equation

The operators
$$P_L = \frac{1}{2}(1 + \gamma_5) \qquad P_R = \frac{1}{2}(1 - \gamma_5)$$
are projection operators which are said to identify states of definite chirality (handedness).

i) Show that P_L, P_R are legitimate projection operators in that
$$P_L^2 = P_L \qquad P_R^2 = P_R \qquad P_L P_R = P_R P_L = 0 \ .$$

ii) Demonstrate that in the limit of high energy—$E/m \gg 1$—or equivalently in the massless limit that the Dirac spinors for positive helicity (right-handed) and negative helicity (left-handed) states of momentum $\vec{p}$ are given by
$$u_\pm(\vec{p}) = \sqrt{\frac{1}{2}} \begin{pmatrix} \chi_{\pm\hat{p}} \\ \pm\chi_{\pm\hat{p}} \end{pmatrix}$$
where $\chi_{\pm\hat{p}}$ are spinors such that
$$\vec{\sigma}\cdot\hat{p}\,\chi_{\pm\hat{p}} = \pm\chi_{\pm\hat{p}} \ .$$

Note: In order to conveniently deal with massless particles, it is important to use the normalization $u(p)^\dagger u(p) = 1$. The appropriate Dirac spinors can then be found by multiplying the usual forms by the factor $\sqrt{\frac{m}{E}}$. Demonstrate this.

iii) Show that

$$P_L u_-(p) = u_-(p) \qquad P_R u_+(p) = u_+(p) \qquad P_L u_+(p) = P_R u_-(p) = 0$$

so that the chirality operator is equivalent to the helicity operator in this limit.

PROBLEM VII.5.3
Electron in a Magnetic Field

Consider an electron immersed in a uniform magnetic field

$$\vec{B} = B_0 \hat{k} \ .$$

i) Obtain the most general four-component positive energy eigenfunctions and demonstrate that the energy eigenvalues are given by

$$E = \sqrt{m^2 + p_3^2 + 2neB_0} \qquad n = 0, 1, 2, \ldots$$

ii) Compare your answer with what is expected nonrelativistically.

VII.6 NEGATIVE ENERGY SOLUTIONS AND ANTIPARTICLES

It is important at this point to address the question of the meaning of the negative energy solutions. We know that quantum mechanics is based upon the prescription

$$p^\mu \longrightarrow i\partial^\mu \quad \text{not} \quad p^\mu \longrightarrow -i\partial^\mu \ . \tag{6.1}$$

But physical observables are real numbers and so cannot depend on this choice of $+i$ vs. $-i$. Does this freedom correspond to any freedom in the physical world? One answer is yes — it represents the particle/antiparticle duality seen in nature. We have already observed this duality in the case of the Klein–Gordon equation, wherein a particle solution corresponds to the positive energy Klein–Gordon wavefunction

$$\phi_{\text{part}}(x) = \phi_{E>0}(x) \tag{6.2}$$

while the antiparticle solution corresponds to the complex conjugate of the negative energy solution

$$\phi_{\text{antipart}}(x) = \phi^*_{E<0}(x) \ . \tag{6.3}$$

Antiparticles and the Dirac Equation

Similar results obtain for the Dirac equation. If we begin with a positive energy solution

$$\psi(x) = e^{-iEt}\psi(\vec{x}) \qquad E > 0 \tag{6.4}$$

which satisfies the equation

$$(i\not{\nabla} - e\not{A} - m)\psi(x) = 0 \tag{6.5}$$

then we may identify this as the wavefunction of a particle with charge e.
On the other hand, if we consider the negative energy solution

$$\psi(x) = e^{-iEt}\psi(\vec{x}) \qquad E = -W < 0 \qquad (6.6)$$

and take the complex conjugate

$$\psi^*(x) = e^{-iWt}\psi^*(\vec{x}) \qquad (6.7)$$

we see that this wavefunction obeys the differential equation

$$\left(-i\nabla^\mu \gamma_\mu^* - eA^\mu \gamma_\mu^* - m\right)\psi^*(x) = 0 \ . \qquad (6.8)$$

In order to reproduce the Dirac equation, we need a transformation under which

$$\gamma_\mu^* \to -\gamma_\mu \ . \qquad (6.9)$$

That is, we seek the "charge conjugation" operator $C\gamma^0$ which satisfies

$$\left(C\gamma^0\right)\gamma_\mu^*\left(C\gamma^0\right)^{-1} = -\gamma_\mu \ . \qquad (6.10)$$

Since

$$\gamma_0^* = \gamma^0 \ , \quad \gamma_1^* = \gamma_1 \ , \quad \gamma_2^* = -\gamma_2 \ , \quad \gamma_3^* = \gamma_3 \qquad (6.11)$$

we observe that the choice

$$C\gamma^0 = i\gamma_2 \ , \quad \left(C\gamma^0\right)^{-1} = i\gamma_2 \qquad (6.12)$$

will suffice. Under this operation the Dirac equation becomes

$$C\gamma^0\left(-i\nabla^\mu \gamma_\mu^* - eA^\mu \gamma_\mu^* - m\right)\psi^*(x) = \left(i\nabla^\mu \gamma_\mu + eA^\mu \gamma_\mu - m\right)C\gamma^0\psi^*(x) = 0 \ . \qquad (6.13)$$

If we define the antiparticle solution to be

$$\psi_{\text{antipart}}(x) = C\gamma^0\psi^*(x) \qquad (6.14)$$

we see that $\psi_{\text{antipart}}(x)$ obeys the Dirac equation for a particle of charge $-e$ and carries the correct (positive energy) time evolution.
[Note: Since

$$\psi^\dagger = \tilde{\psi}^* \qquad (6.15)$$

where ~ indicates the transpose we have

$$\bar{\psi} = \tilde{\gamma}^0 \tilde{\psi}^\dagger = \gamma^0 \psi^* \ . \qquad (6.16)$$

Thus the antiparticle solution is sometimes written as

$$\psi_{\text{antiparticle}}(x) = C\tilde{\bar{\psi}} \ .] \qquad (6.17)$$

An interesting feature has to do with the intrinsic parities of these particle/antiparticle solutions. For a particle (positive energy) solution $\psi(\vec{x},t)$ we have (cf. sect. VII.2)

$$\hat{\Pi}\psi(\vec{x},t) = \gamma^0 \psi(-\vec{x},t) = \gamma^0 \begin{pmatrix} \psi_a(-\vec{x},t) \\ \psi_b(-\vec{x},t) \end{pmatrix}$$
$$= \begin{pmatrix} \psi_a(-\vec{x},t) \\ -\psi_b(-\vec{x},t) \end{pmatrix} , \quad (6.18)$$

where $\hat{\Pi}$ is the spatial inversion or parity operator. Thus upper and lower components of the wavefunction then have differing behaviors under a parity transform. However, this is to be expected, since we have

$$\psi_b = \frac{\left(\vec{\sigma} \cdot (-i\vec{\nabla} - e\vec{A})\right)}{E+m} \psi_a . \quad (6.19)$$

and the extra minus sign arises from the behavior of $\vec{\nabla}, \vec{A}$ under parity

$$\vec{\nabla}, \vec{A} \xrightarrow{P} -\vec{\nabla}, -\vec{A} . \quad (6.20)$$

However, if we look at the analogous negative energy or antiparticle solution, we find

$$\hat{\Pi}\psi_{\text{antipart}}(\vec{x},t) = \gamma^0 C \gamma^0 \psi^*(-\vec{x},t) = -C\gamma^0 \begin{pmatrix} \psi_a^*(-\vec{x},t) \\ -\psi_b^*(-\vec{x},t) \end{pmatrix} . \quad (6.21)$$

Thus the intrinsic parity of Dirac particles and antiparticles are opposite! (This is to be contrasted to the Klein–Gordon case wherein one finds identical parities for particle/antiparticle solutions.) This feature is verified experimentally in study of positronium decay or in the observation that ground (S-wave) states of quark-antiquark bound systems — i.e., π, K, η mesons — are determined to be *pseudo*scalar rather than scalar quantities.

Dirac Sea

When Dirac found these negative energy solutions, he did not in the beginning understand their significance. (Recall that the Dirac equation was written down in 1928, but Anderson did not find the positron until 1932.) At first Dirac was worried that, since nature always prefers to lower the energy, positive energy electrons would radiate photons (of energy $\gtrsim 2m$) and fall into negative energy states. However, once in a negative energy state the electron could reduce its energy even further by radiating additional photons, thus lowering its energy indefinitely. Since this does not happen — experimentally the electron is stable— Dirac was faced with a dilemma. He solved this crisis by postulating that *all* negative energy states were already filled. Then, according to the Pauli exclusion principle, positive energy states are unable to make transitions to negative energy levels. Nevertheless, it is possible for an energetic ($\omega \gtrsim 2m$) photon to cause a negative energy electron to make a transition to a positive energy state leaving a "hole" in the set of of negative energy states.

The vacuum (*i.e.*, lowest energy state) in this picture consists of a "Dirac sea" of filled negative energy states, and the net charge of a given state must be defined with respect to this vacuum. Thus a hole state behaves as if it had charge

$$Q_{\text{hole}} = (Q_{\text{vacuum}} - (e)) - Q_{\text{vacuum}} = -e \qquad (6.22)$$

i.e., the *negative* of the electron charge! (Of course, Q_{vacuum} is infinite but we have seen such infinite renormalizations before.) Similarly if the momentum of this (negative energy) state is $\vec{p}$, the hole, upon renormalization, will behave as if it has momentum

$$\vec{P}_{\text{hole}} = \left(\vec{P}_{\text{vacuum}} - \vec{p}\right) - \vec{P}_{\text{vacuum}} = -\vec{p} \quad . \qquad (6.23)$$

(In this case we expect $\vec{P}_{\text{vacuum}} = 0$ since for each negative energy state with momentum $\vec{p}$ there is another with momentum $-\vec{p}$.) Finally, for the energy and spin we have

$$E_{\text{hole}} = (E_{\text{vacuum}} - (-E)) - E_{\text{vacuum}} = +E$$
$$\frac{1}{2}\vec{\Sigma}_{\text{hole}} = \left(\frac{1}{2}\vec{\Sigma}_{\text{vacuum}} - \frac{1}{2}\vec{\Sigma}\right) - \frac{1}{2}\vec{\Sigma}_{\text{vacuum}} = -\frac{1}{2}\vec{\Sigma} \quad . \qquad (6.24)$$

so that the hole state behaves as a positive energy, positive charge state of momentum $-\vec{p}$ and spin $-\frac{1}{2}\vec{\Sigma}$. We recognize this state as a positron, whose existence was predicted by Dirac *prior* to its discovery. (Actually, Dirac first identified this antiparticle solution with the proton, but soon realized that its mass must be identical to that of the electron.)

We see then that the process by which an energetic photon ejects a negative energy electron from the Dirac sea, knocking it into a positive energy level and leaving behind a negative energy hole is to be interpreted as the process of pair creation

$$\gamma \longrightarrow e^+ e^- \quad . \qquad (6.25)$$

Since only a very energetic ($\omega \gtrsim 2m$) photon can bring about such a transition, one might be tempted to think that antiparticle states should not play an important role in low energy quantum mechanics. However, this is *not* correct. Consider the scattering of photons by a free positive energy electron. (This is the relativistic analog of the Thomson scattering process discussed previously.) Writing the Dirac equation in Hamiltonian form

$$i\frac{\partial}{\partial t}\psi = (\vec{\alpha}\cdot\vec{p} + \beta m + e\gamma^0 \slashed{A})\psi$$
$$\equiv (H_0 + V)\psi \qquad (6.26)$$

we recognize the interaction potential as

$$V = e\gamma^0 \slashed{A} \quad . \qquad (6.27)$$

and can apply canonical time-dependent perturbation theory. Two obvious diagrams which arise in second order are shown below

$$\left\langle \vec{p}_2; \vec{k}_2, \hat{\epsilon}_2 \left| \hat{V} \frac{1}{E_{\vec{p}_1} + \omega_1 - \hat{H}_0} \hat{V} \right| \vec{p}_1; \vec{k}_1, \hat{\epsilon}_1 \right\rangle = \frac{e^2}{\sqrt{2\omega_1 2\omega_2}}$$

$$\times \frac{\left\langle \vec{p}_2 \left| \vec{\alpha} \cdot \hat{\epsilon}_2^* e^{-i\vec{k}_2 \cdot \vec{x}} \right| \vec{p}_1 + \vec{k}_1 \right\rangle \left\langle \vec{p}_1 + \vec{k}_1 \left| \vec{\alpha} \cdot \hat{\epsilon}_1 e^{i\vec{k}_1 \cdot \vec{x}} \right| \vec{p}_1 \right\rangle}{E_{\vec{p}_1} + \omega_1 - E_{\vec{p}_1 + \vec{k}_1}}$$

$$\times \frac{\left\langle \vec{p}_2 \left| \vec{\alpha} \cdot \hat{\epsilon}_1 e^{i\vec{k}_1 \cdot \vec{x}} \right| \vec{p}_1 - \vec{k}_2 \right\rangle \left\langle \vec{p}_1 - \vec{k}_2 \left| \vec{\alpha} \cdot \hat{\epsilon}_2^* e^{-i\vec{k}_2 \cdot \vec{x}} \right| \vec{p}_1 \right\rangle}{E_{\vec{p}_1} - \omega_2 - E_{\vec{p}_1 - \vec{k}_2}}$$

(6.28)

which correspond to similar diagrams discussed in the analogous non-relativistic case. However, as Eq. 6.27 contains no term in A^2, there exists no analog of the seagull diagrams for the relativistic situation. Also since in the non-relativistic limit

$$u(p) \longrightarrow \begin{pmatrix} \chi \\ 0 \end{pmatrix} \qquad (6.29)$$

while

$$\vec{\alpha} = \begin{pmatrix} 0 & \vec{\sigma} \\ \vec{\sigma} & 0 \end{pmatrix} \qquad (6.30)$$

couples upper and lower components so that

$$u^\dagger(p')\vec{\alpha}u(p) \approx 0, \qquad (6.31)$$

there exists essentially *no* contribution to the scattering from the above pole diagrams.

The paradox is resolved if we include the contribution from the diagrams involving negative energy solutions as shown below

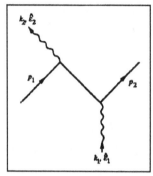

$$\times - \frac{\langle \vec{p}_2 | \vec{\alpha} \cdot \hat{\epsilon}_2^* e^{-i\vec{k}_2 \cdot \vec{x}} | \overline{\vec{p}_2 + \vec{k}_2} \rangle \langle \overline{\vec{p}_2 + \vec{k}_2} | \vec{\alpha} \cdot \hat{\epsilon}_1 e^{i\vec{k}_1 \cdot \vec{x}} | \vec{p}_1 \rangle}{-E_{\vec{p}_2} - \omega_2 - |E_{\vec{p}_2 + \vec{k}_2}|}$$

$$\times - \frac{\langle \vec{p}_2 | \vec{\alpha} \cdot \hat{\epsilon}_1 e^{i\vec{k}_1 \cdot \vec{x}} | \overline{\vec{p}_2 - \vec{k}_1} \rangle \langle \overline{\vec{p}_2 - \vec{k}_1} | \vec{\alpha} \cdot \hat{\epsilon}_2^* e^{-i\vec{k}_2 \cdot \vec{x}} | \vec{p}_1 \rangle}{-E_{\vec{p}_2} + \omega_1 - |E_{\vec{p}_2 - \vec{k}_1}|}$$

(6.32)

Then if $\omega \ll m$ and working in the non-relativistic limit where

$$v^\dagger(p') \vec{\alpha} u(p) \approx \chi'^\dagger \vec{\sigma} \chi \qquad (6.33)$$

the sum of these negative energy diagrams yields

$$\text{Amp} \approx \frac{e^2}{\sqrt{2\omega_1 2\omega_2}} \frac{1}{2m} \sum_{i=1}^{2} \left(\chi_2^\dagger \vec{\sigma} \cdot \hat{\epsilon}_2^* \chi_i \chi_i^\dagger \vec{\sigma} \cdot \hat{\epsilon}_1 \chi_1 + \chi_2^\dagger \vec{\sigma} \cdot \hat{\epsilon}_1 \chi_i \chi_i^\dagger \vec{\sigma} \cdot \hat{\epsilon}_2^* \chi_1 \right)$$

$$= \frac{e^2}{\sqrt{2\omega_1 2\omega_2}} \frac{1}{2m} \chi_2^\dagger \left(\vec{\sigma} \cdot \hat{\epsilon}_2^* \vec{\sigma} \cdot \hat{\epsilon}_1 + \vec{\sigma} \cdot \hat{\epsilon}_1 \vec{\sigma} \cdot \hat{\epsilon}_2^* \right) \chi_1 \qquad (6.34)$$

$$= \frac{e^2}{\sqrt{2\omega_1 2\omega_2}} \frac{1}{2m} 2 \hat{\epsilon}_2^* \cdot \hat{\epsilon}_1 \chi_2^\dagger \chi_1$$

which is identical to the seagull contribution found in the analogous non-relativistic case. We observe then that the inclusion of these negative energy states is absolutely crucial. Without these pieces the Dirac picture of Compton scattering would not reduce to the simple Thomson process.

Zitterbewegung

We have already discussed the presence of zitterbewegung associated with the origin of the $\vec{\nabla} \cdot \vec{E}$ term in the effective non-relativistic Hamiltonian. We can study this phenomenon in more detail by examining the velocity operator $\vec{\alpha}$. That the Dirac matrix $\vec{\alpha}$ is related to the relativistic velocity is clear from the relation

$$\psi^\dagger(x) \vec{\alpha} \psi(x) = |N|^2 \bar{u}(\vec{p}) \vec{\gamma} u(\vec{p}) = |N|^2 \frac{\vec{p}}{m} \qquad (6.35)$$

valid for the plane wave solution

$$\psi(x) = Nu(p)e^{-ip\cdot x} \ . \tag{6.36}$$

The constant N is determined by the normalization condition

$$1 = \int d^3x\, \rho(\vec{x}, t) = |N|^2 \frac{E}{m} \int d^3x = |N|^2 \frac{E}{m} \tag{6.37}$$

using unit volume $\left(\int d^3x = 1\right)$. Thus

$$|N|^2 = \frac{m}{E} \tag{6.38}$$

and

$$\psi^\dagger(x)\vec{\alpha}\psi(x) = \frac{\vec{p}}{E} = \vec{v} \ . \tag{6.39}$$

[Note: This is also suggested by the relation

$$\begin{aligned}\left[H, \vec{p} - e\vec{A}\right] &= ie\vec{\nabla}\phi - ie\vec{\alpha} \times \vec{B} \\ &= -ie\left(\vec{E} + \vec{\alpha} \times \vec{B}\right) \quad \text{if} \quad \frac{\partial \vec{A}}{\partial t} = 0\end{aligned} \tag{6.40}$$

since

$$\frac{d}{dt}\mathcal{O} = i[H, \mathcal{O}] \ .] \tag{6.41}$$

For a free particle, using the Heisenberg representation, we have

$$\frac{d}{dt}\vec{x} = i[H, \vec{x}] = \vec{\alpha} \tag{6.42}$$

and

$$\begin{aligned}\frac{d}{dt}\vec{\alpha} &= i[H, \vec{\alpha}] = i(-2m\vec{\alpha}\beta + 2i\vec{\Sigma} \times \vec{p}) \\ &= i\left(-2\vec{\alpha}H + 2\vec{p}\right)\end{aligned} \tag{6.43}$$

We see that

$$\vec{\alpha}(t) = \vec{p}H^{-1} + \left(\vec{\alpha}(0) - \vec{p}H^{-1}\right)e^{-2iHt} \ . \tag{6.44}$$

[Check:

$$\begin{aligned}\frac{d\vec{\alpha}}{dt} &= \left(\vec{\alpha}(0) - \vec{p}H^{-1}\right) \times (-2iH)\, e^{-2iHt} \\ &= -2i\left(\vec{\alpha}(t) - \vec{p}H^{-1}\right)H \\ &= i\left(-2\vec{\alpha}H + 2\vec{p}\right) \quad . \ ]\end{aligned} \tag{6.45}$$

and thereby

$$\vec{x}(t) = \vec{x}(0) + \vec{p}H^{-1}t + \frac{i}{2}\left(\vec{\alpha}(0) - \vec{p}H^{-1}\right)H^{-1}\left(e^{-2iHt} - 1\right) \ . \tag{6.46}$$

[Check:
$$\frac{d\vec{x}}{dt} = \vec{p}H^{-1} + \left(\vec{\alpha}(0) - \vec{p}H^{-1}\right)e^{-2iHt} = \vec{\alpha} \; . \;] \qquad (6.47)$$

Thus, for a wavepacket

$$\vec{x}(t) = \vec{x}(0) + \left\langle \frac{\vec{p}}{E} \right\rangle t + \text{terms oscillating} \sim e^{-2imt} \; . \qquad (6.48)$$

This last term represents the zitterbewegung.

We can see this more directly by explicitly constructing a wavepacket

$$\psi(\vec{x},t) = \int \frac{d^3p}{(2\pi)^3} \sum_s \left(c(\vec{p},s)\sqrt{\frac{m}{E}}e^{-ip\cdot x}u(p,s) + d(\vec{p},s)\sqrt{\frac{m}{E}}e^{ip\cdot x}v(\vec{p},s) \right) \; . \qquad (6.49)$$

Then

$$\langle \vec{\alpha} \rangle = \int d^3x \, \psi^\dagger(\vec{x},t)\vec{\alpha}\psi(\vec{x},t)$$

$$= \int \frac{d^3p}{(2\pi)^3} \sum_s \left(|c(\vec{p},s)|^2 \frac{\vec{p}}{E} + |d(\vec{p},s)|^2 \frac{\vec{p}}{E} \right.$$

$$+ \sum_{s'} \left(d^*(\vec{p},s')c(\vec{p},s)\frac{m}{E}v^\dagger(\vec{p},s')\vec{\alpha}u(\vec{p},s)e^{-i2Et} \right.$$

$$\left.\left. + c^*(\vec{p},s)d(\vec{p},s')\frac{m}{E}u^\dagger(\vec{p},s)\vec{\alpha}v(\vec{p},s')e^{i2Et} \right) \right) \; ,$$

and since

$$\frac{d}{dt}\langle \vec{x} \rangle = \frac{d}{dt}\int d^3x \, \psi^\dagger(\vec{x},t)\vec{x}\psi(\vec{x},t) = \int d^3x \, \psi^\dagger(\vec{x},t)\vec{\alpha}\psi(\vec{x},t) = \langle \vec{\alpha} \rangle \qquad (6.50)$$

we find

$$\langle \vec{x}(t) \rangle = \langle \vec{x}(0) \rangle + \int \frac{d^3p}{(2\pi)^3} \sum_s \left(\frac{\vec{p}t}{E}\left(|c(\vec{p},s)|^2 - |d(\vec{p},s)|^2 \right) \right.$$

$$+ \sum_{s'} \left(d^*(\vec{p},s')c(\vec{p},s)\frac{im}{2E^2}\left(e^{-i2Et} - 1\right)v^\dagger(\vec{p},s')\vec{\alpha}u(p,s) \right. \qquad (6.51)$$

$$\left.\left. - c^*(\vec{p},s)d(\vec{p},s')\frac{im}{2E^2}\left(e^{i2Et} - 1\right)u^\dagger(\vec{p},s)\vec{\alpha}v(\vec{p},s') \right) \right) \; .$$

Again we observe the uniform motion of the packet accompanied by violent oscillation at frequencies $\omega \sim 2m$. As in the Klein–Gordon analysis, we see that this zitterbewegung motion is absent for wavepackets constructed from only positive (or only negative) energy solutions.

PROBLEM VII.6.1
External Fields and Negative Energy Transitions

Consider a positive energy spin 1/2 particle at rest. Suppose that at t=0 we apply an external (classical) vector potential

$$\vec{A} = -\hat{e}_x \frac{a}{\omega} \sin \omega t$$

which corresponds to an electric field of the form

$$\vec{E} = \hat{e}_x a \cos \omega t \ .$$

Show that for $t > 0$ there exists a finite probability of finding the particle in a negative energy state if such negative energy states are assumed to be originally empty. In particular, work out quantitatively the two cases: $\omega \ll 2m$ and $\omega \approx 2m$ and comment.

VII.7 PERTURBATION THEORY: INTRODUCTION

We wish to construct a relativistic perturbation theory in analogy to what we have already formulated for the Schrödinger equation. Actually, one could straightforwardly employ the previous formalism provided that we utilize the Hamiltonian form of the Dirac equation, as shown in our earlier discussion of Thomson scattering. While this procedure is rigorous and correct, it is also awkward and cumbersome because of the proliferation of negative energy diagrams. Instead we shall develop the formalism, due to Feynman, which treats positive and negative energy states simultaneously.

Intuitive Arguments

In order to motivate this approach, we consider a particle moving in a potential $V(t)$. We might be interested, for example, in a scattering process which is described to lowest order by the diagram below

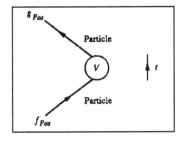

$$\text{Amp}^{(1)} = -i \int_{-\infty}^{\infty} dt \ \langle g_{\text{Pos}}(t) | V(t) | f_{\text{Pos}}(t) \rangle$$

(7.1)

where $|f_{\text{Pos}}(t)\rangle$ represents a (positive energy) state $|f_{\text{Pos}}\rangle$ in the remote past ($t = -\infty$) which has propagated freely until time t, while $\langle g_{\text{Pos}}(t)|$ represents a (positive energy) state which when propagated freely from time t will evolve into the state $\langle g_{\text{Pos}}|$ in the remote future ($t = +\infty$). If f, g are definite (positive) energy states, we write

$$|f_{\text{Pos}}(t)\rangle = |f_{\text{Pos}}\rangle e^{-iE_f t} \ , \qquad \langle g_{\text{Pos}}(t)| = e^{iE_g t} \langle g_{\text{Pos}}| \ .$$

(7.2)

Expanding $V(t)$ in terms of its Fourier components

$$V(t) = \int_{-\infty}^{\infty} \frac{d\omega}{2\pi} e^{-i\omega t} \mathcal{V}(\omega) \tag{7.3}$$

we find

$$\begin{aligned}
\text{Amp}^{(1)} &= -i \int_{-\infty}^{\infty} \frac{d\omega}{2\pi} \int_{-\infty}^{\infty} dt\, e^{i(E_g - E_f - \omega)t} \langle g_{\text{Pos}} | \mathcal{V}(\omega) | f_{\text{Pos}} \rangle \\
&= -i \int_{-\infty}^{\infty} \frac{d\omega}{2\pi} 2\pi \delta(E_g - E_f - \omega) \langle g_{\text{Pos}} | \mathcal{V}(\omega) | f_{\text{Pos}} \rangle \\
&= -i \langle g_{\text{Pos}} | \mathcal{V}(E_g - E_f) | f_{\text{Pos}} \rangle \\
&= -i \int d^3x\, g^*_{\text{Pos}}(x) \mathcal{V}(E_g - E_f, x) f_{\text{Pos}}(x) \ .
\end{aligned} \tag{7.4}$$

Suppose we define f = entry state and g = exit state, as indicated by the arrows in the diagram above which imply whether the state involved sits after (exit) or before (entry) the potential. Then it is of interest to ask whether we can assign any physical meaning if one or both of these energies is negative. For example, suppose that the entry state has negative energy

$$E_{\text{entry}} = -W_{\text{entry}} \ . \tag{7.5}$$

Then we find

$$\text{Amp}^{(1)} = -i \int_{-\infty}^{\infty} \frac{d\omega}{2\pi} \int_{-\infty}^{\infty} dt\, e^{i(E_g + W_f - \omega)t} \int d^3x\, g^*_{\text{Pos}}(x) \mathcal{V}(\omega, x) f_{\text{Neg}}(x) \ . \tag{7.6}$$

We see that E_g, W_f are treated on an equal footing, which suggests the speculation that this matrix element corresponds to pair production. The matrix element becomes

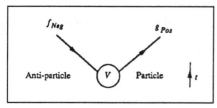

$$\begin{aligned}
\text{Amp}^{(1)} &= -i \int d^3x\, g^*_{\text{Pos}}(x) \\
&\quad \times \mathcal{V}(E_g + W_f, x) f_{\text{Neg}}(x) \\
&= -i \int d^3x\, g^*_{\text{Pos}}(x) \\
&\quad \times \mathcal{V}(E_g + W_f, x) \left(f^*_{\text{Neg}}(x)\right)^*
\end{aligned} \tag{7.7}$$

where f^*_{Neg} is the antiparticle wavefunction describing an antiparticle of charge $-e$ and energy $W_{\text{entry}} = -E_{\text{entry}}$.

Similarly, if we consider a diagram wherein the exit state has a negative energy $W_{\text{exit}} = -E_{\text{exit}}$ the transition amplitude becomes

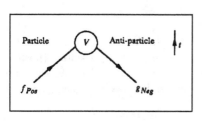

$$\begin{aligned}
\text{Amp}^{(1)} &= \\
&- i \int_{-\infty}^{\infty} \frac{d\omega}{2\pi} \int_{-\infty}^{\infty} dt\, e^{-i(W_g + E_f + \omega)t} \\
&\times \int d^3x\, g^*_{\text{Neg}}(x) \mathcal{V}(\omega, x) f_{\text{Pos}}(x) \\
&= -i \int d^3x\, g^*_{\text{Neg}}(x) \mathcal{V}(-E_f - W_g, x) f_{\text{Pos}}(x)
\end{aligned} \tag{7.8}$$

where we recognize g^*_{Neg} as the antiparticle wavefunction describing an antiparticle of energy $W_g = -E_g$. This amplitude then represents annihilation of a particle-antiparticle pair.

Finally, in the case that both E_f and E_g are negative, with

$$W_f = -E_f \quad \text{and} \quad W_g = -E_g \tag{7.9}$$

we have

$$\text{Amp}^{(1)} = -i \int_{-\infty}^{\infty} \frac{d\omega}{2\pi} \int_{-\infty}^{\infty} dt\, e^{i(W_f - W_g - \omega)t} \int d^3x\, g^*_{\text{Neg}}(x) \mathcal{V}(\omega, x) f_{\text{Neg}}(x) \;. \tag{7.10}$$

We speculate that this amplitude describes the scattering of the antiparticle from the potential V, with

$$\begin{aligned}\text{Amp}^{(1)} &= -i \int d^3x\, g^*_{\text{Neg}}(x) \mathcal{V}(W_f - W_g, x) f_{\text{Neg}}(x) \\ &= -i \int d^3x\, g^*_{\text{Neg}}(x) \mathcal{V}(W_f - W_g, x) \left(f^*_{\text{Neg}}(x)\right)^* \;.\end{aligned}$$

(7.11)

We recognize g^*_{Neg}, f^*_{Neg} as the antiparticle solutions describing antiparticles of energy W_g, W_f, respectively.

	Particle	antiparticle
Entry	Positive Past	Negative Future
Exit	Positive Future	Negative Past

Table VII.1: *Intuitive connection between entry/exit and past/future states for particles and antiparticles.*

The above arguments suggest the validity of the chart which appears in Table VII.1. This is merely a speculation, however, and our next task is to construct an actual theory on this basis.

VII.8 DIRAC PROPAGATOR

Based upon our discussion in the previous section it is suggestive to attempt to invent a propagator theory for the Dirac equation, which includes consideration of negative energy states. Suppose we are analyzing the amplitude to make a transition from a positive energy state $f_{\text{Pos}}(1)$ at time t_1 to a positive energy state $g_{\text{Pos}}(2)$ at time t_2, a diagrammatic representation of which is given in Figure VII.5.

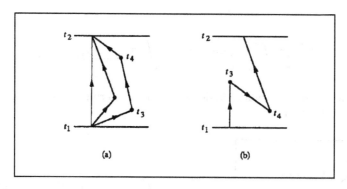

Fig. VII.5: Diagrammatic representation of the relativistic perturbation expansion.

We encounter no difficulties until second order in the potential. In Figure 5a if $t_4 > t_3$ we cannot permit negative energy states to play any role in the propagation from 3 to 4. This propagator is the same as that from 4 to 2 (with $t_2 > t_4$) which cannot contain negative energies since we have g_{Pos} at point 2. However, there is an additional way in which the particle can propagate from 1 to 2 by means of a second order process—Figure 5b, wherein at vertex 3 there is a pair annihilation, while at vertex 4 there is a pair creation. Thus for $t_4 < t_3$ we need to describe things in terms of a propagator involving only negative energies.

Fourier Transform Approach

First we must construct the free particle propagator, which should be a Green's function for the Dirac equation

$$(i \not{\nabla}_2 - m) S_F^{(0)}(2,1) = \delta(2,1) \ . \tag{8.1}$$

where $\delta(2,1) \equiv \delta^4(2-1)$. In order to solve Eq. 8.1 we examine its Fourier transform and express the solution in terms of a Fourier integral.

$$S_F^{(0)}(2,1) = \int \frac{dp_0 \, d^3p}{(2\pi)^4} e^{i\vec{p}\cdot(\vec{r}_2-\vec{r}_1) - ip_0(t_2-t_1)} S_F^{(0)}(p_0, \vec{p}) \tag{8.2}$$

Introducing the shorthand $dp_0 \, d^3p \equiv d^4p$ with $p^\mu = (p_0, \vec{p})$ we have

$$S_F^{(0)}(2,1) = \int \frac{d^4p}{(2\pi)^4} e^{-ip\cdot(x_2-x_1)} S_F^{(0)}(p) \tag{8.3}$$

and
$$(i\,\slashed{\nabla}_2 - m)\,S_F^{(0)}(2,1) = \int \frac{d^4p}{(2\pi)^4} e^{-ip\cdot(x_2-x_1)} (\slashed{p}-m)\,S_F^{(0)}(p)$$
$$= \int \frac{d^4p}{(2\pi)^4} e^{-ip\cdot(x_2-x_1)} = \delta(2,1) \ . \tag{8.4}$$

We require then
$$(\slashed{p}-m)\,S_F^{(0)}(p) = 1 \quad \text{or} \quad S_F^{(0)}(p) = \frac{1}{\slashed{p}-m} \ . \tag{8.5}$$

Note that this is a matrix equation since $\slashed{p}$ is a 4×4 matrix. However,
$$(\slashed{p}+m)(\slashed{p}-m) = \slashed{p}\,\slashed{p} - m^2 \tag{8.6}$$

and
$$\slashed{p}\,\slashed{p} = \frac{1}{2}\left(p_\mu \gamma^\mu p_\nu \gamma^\nu + p_\nu \gamma^\nu p_\mu \gamma^\mu\right)$$
$$= p_\mu p_\nu \frac{1}{2}\{\gamma^\mu, \gamma^\nu\} = p_\mu p_\nu \eta^{\mu\nu} = p^2 \tag{8.7}$$

so that
$$(\slashed{p}+m)(\slashed{p}-m) = p^2 - m^2 \ . \tag{8.8}$$

Then
$$S_F^{(0)}(p) = \frac{1}{\slashed{p}-m} = (\slashed{p}+m)\frac{1}{\slashed{p}+m}\frac{1}{\slashed{p}-m} = \frac{\slashed{p}+m}{p^2-m^2} \tag{8.9}$$

and
$$S_F^{(0)}(2,1) = \int \frac{d^4p}{(2\pi)^4} \frac{\slashed{p}+m}{p^2-m^2} e^{-ip\cdot(x_2-x_1)}$$
$$= (i\,\slashed{\nabla}_2 + m) \int \frac{d^4p}{(2\pi)^4} \frac{e^{-ip\cdot(x_2-x_1)}}{p^2-m^2} \ . \tag{8.10}$$

This representation is advantageous since it allows us to evaluate the matrix properties *after* the integration has been done. The integral which remains is just the propagator for a spin zero particle, which satisfies
$$(\Box_2 + m^2)\,D_F(2,1) = -\delta(2-1)$$
$$\text{i.e.,} \quad D_F(2,1) = \int \frac{d^4p}{(2\pi)^4} \frac{e^{-ip\cdot(x_2-x_2)}}{p^2-m^2} \ . \tag{8.11}$$

and will be studied in detail in the next chapter (*cf.* Sect. VIII.2). Now write
$$\int \frac{d^3p}{(2\pi)^3} \int_{-\infty}^{\infty} \frac{dp_0}{2\pi} \frac{e^{-ip_0(t_2-t_1)} e^{i\vec{p}\cdot(\vec{r}_2-\vec{r}_1)}}{p_0^2 - \vec{p}^2 - m^2}$$
$$= \int \frac{d^3p}{(2\pi)^3} e^{i\vec{p}\cdot(\vec{r}_2-\vec{r}_1)} \int_{-\infty}^{\infty} \frac{dp_0}{2\pi} \frac{e^{-ip_0(t_2-t_1)}}{(p_0-E_p)(p_0+E_p)} \quad \text{with} \quad E_p = \sqrt{\vec{p}^2+m^2} \tag{8.12}$$

and replace
$$p_0 - E_p \to p_0 - E_p + i\epsilon$$
$$p_0 + E_p \to p_0 + E_p - i\epsilon \ . \tag{8.13}$$

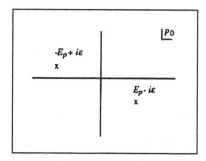

Fig. VII.6: Singularity structure of the Dirac propagator.

Then, if $t_2 > t_1$ we must close the contour in the lower half plane and we pick up the pole at $p_0 = +E_p - i\epsilon$ while if $t_2 < t_1$ we must close the contour in the upper half plane and we pick up the pole at $p_0 = -E_p + i\epsilon$ (cf. Figure VII.6). This is precisely the behavior we desire and is equivalent to giving the mass a small negative imaginary part

$$m \to m - i\epsilon \tag{8.14}$$

since then

$$\sqrt{\vec{p}^2 + m^2} \to \sqrt{\vec{p}^2 + m^2 - i\epsilon} \approx \sqrt{\vec{p}^2 + (m - i\epsilon)^2} \;. \tag{8.15}$$

Finally, we have

$$iS_F^{(0)}(2,1) = \int \frac{d^4p}{(2\pi)^4} e^{-ip\cdot(x_2-x_1)} \frac{i(\not{p}+m)}{p^2 - m^2 + i\epsilon}$$
$$= \begin{cases} t_2 > t_1 & \int \frac{d^3p}{(2\pi)^3} e^{i\vec{p}\cdot(\vec{r}_1-\vec{r}_1)-iE_p(t_2-t_1)} \frac{1}{2E_p}(\not{p}+m) \\ t_2 < t_1 & -\int \frac{d^3p}{(2\pi)^3} e^{-i\vec{p}\cdot(\vec{r}_2-\vec{r}_1)+iE_p(t_2-t_1)} \frac{1}{2E_p}(\not{p}-m) \end{cases} \tag{8.16}$$

where we have changed variables from $\vec{p}$ to $-\vec{p}$ in obtaining the last line. To see that this propagator satisfies our original prescription, we note that

$$e^{iE_p(t_2-t_1)} = \left(e^{-iE_pt_2}\right)^* \left(e^{-iE_pt_1}\right) \;. \tag{8.17}$$

Then $e^{iE_pt_2} = (e^{-iE_pt_2})^*$ corresponds to the creation of a positron at time t_2 (i.e., positron exit at time t_2, negative energy electron entry), while $e^{-iE_pt_1}$ corresponds to the annihilation of a positron at time t_1 (i.e., positron entry at time t_1, negative energy electron exit).

Intuitive Approach

We can obtain additional insight into Eq. 8.16 by obtaining the propagator on less formal grounds and comparing with the previous result. We require that the propagator should serve to propagate positive energy solutions from past to future and negative energy solutions from future to past. Let $H^{(0)}$ represent the free Dirac Hamiltonian and

$$\psi_{\text{Pos}}(\vec{x}_1, t_1) = \text{linear superposition of positive energy solutions}$$
$$= \sum_{\text{Pos } E_n} a_n \phi_n(\vec{x}_1) \qquad (8.18)$$
$$H^{(0)} \phi_n(\vec{x}) = E_n \phi_n(\vec{x}) \ .$$

The Dirac Hamiltonian has a complete orthonormal set of eigenstates consisting of negative as well as positive energies: $\{\phi_n(x)\}$. Treating the index n as if it were discrete, from orthogonality we find

$$a_n = \int \phi_{n_{\text{Pos}}}^\dagger(\vec{x}_1) \psi_{\text{Pos}}(\vec{x}_1, t_1) d^3 x_1 \equiv \int \overline{\phi}_{n_{\text{Pos}}}(\vec{x}_1) \gamma^0 \psi_{\text{Pos}}(\vec{x}_1, t_1) d^3 x_1 \ . \qquad (8.19)$$

At time $t_2 > t_1$ the wave function has evolved into

$$\psi_{\text{Pos}}(\vec{x}_2, t_2) = \sum_{\text{Pos } E_n} a_n \phi_n(\vec{x}_2) \, e^{-iE_n(t_2-t_1)}$$
$$= \sum_{\text{Pos } E_n} \int d^3 x_1 \phi_n(\vec{x}_2) e^{-iE_n(t_2-t_1)} \overline{\phi}_n(\vec{x}_1) \gamma^0 \psi_{\text{Pos}}(\vec{x}_1, t_1) \qquad (8.20)$$
$$\equiv i \int S_F^{(0)}(2,1) \gamma^0 \psi_{\text{Pos}}(\vec{x}_1, t_1) d^3 x_1$$

so that

$$i S_F^{(0)}(2,1) = \sum_{\text{Pos } E_n} \phi_n(\vec{x}_2) \overline{\phi}_n(\vec{x}_1) \, e^{-iE_n(t_2-t_1)} \qquad t_2 > t_1 \ . \qquad (8.21)$$

If

$$\phi_n(1) \equiv \phi_n(\vec{x}_1) \, e^{-iE_n t_1}$$
$$\phi_n(2) \equiv \phi_n(\vec{x}_2) \, e^{-iE_n t_2} \qquad (8.22)$$

we can write

$$i S_F^{(0)}(2,1) = \sum_{\text{Pos } E_n} \phi_n(2) \overline{\phi}_n(1) \qquad t_2 > t_1 \qquad (8.23)$$

Now define a second quantity

$$i S_0(2,1) = \sum_{\text{all } E_n} \phi_n(2) \overline{\phi}_n(1) \qquad t_2 > t_1 \ . \qquad (8.24)$$

Then clearly for any wave function ψ

$$\psi(\vec{x}_2, t_2) = \int d^3 x_1 \, i S_0(2,1) \gamma^0 \psi(\vec{x}_1, t_1) \qquad t_2 > t_1 \qquad (8.25)$$

so

$$\lim_{t_2 \to t_1^+} i S_0(2,1) \gamma^0 = \delta^3(\vec{x}_2 - \vec{x}_1) \ . \qquad (8.26)$$

If we require $S_0(2,1) \equiv 0$ for $t_2 < t_1$ so that

$$(i \not{\nabla}_2 - m) S_0(2,1) = 0 \qquad t_2 \neq t_1 \ . \tag{8.27}$$

then $S_0(2,1)$ obeys the differential equation

$$(i \not{\nabla}_2 - m) S_0(2,1) = \delta(2,1) \tag{8.28}$$

since

$$\lim_{\epsilon \to 0} \int_{t_1-\epsilon}^{t_1+\epsilon} dt_2 \, (i \not{\nabla}_2 - m) S_0(2,1) = i\gamma^0 \left(\lim_{t_2 \to t_1+\epsilon} - \lim_{t_2 \to t_1-\epsilon} \right) S_0(2,1)$$
$$= \gamma^0 \left(\gamma^0 \delta^3 (\vec{x}_2 - \vec{x}_1) - 0 \right) \tag{8.29}$$
$$= \delta^3 (\vec{x}_2 - \vec{x}_1)$$

and also

$$\lim_{\epsilon \to 0} \int_{t_1-\epsilon}^{t_1+\epsilon} dt_2 \delta(2,1) = \delta^3 (\vec{x}_2 - \vec{x}_1) \ . \tag{8.30}$$

As $S_F^{(0)}(2,1)$ also satisfies Eq. 8.28, both $S_F^{(0)}(2,1)$ and $S_0(2,1)$ are Green's functions for the Dirac equation. Thus

$$i S_F^{(0)}(2,1) - i S_0(2,1) = - \sum_{\text{Neg } E_n} \phi_n(2) \bar{\phi}_n(1) \qquad t_2 > t_1 \tag{8.31}$$

must be a solution of the *homogeneous* Dirac equation and must have the same form for $t_2 < t_1$ as for $t_2 > t_1$. Finally, since $S_0(2,1) \equiv 0$ for $t_2 < t_1$ we have

$$i S_F^{(0)}(2,1) = - \sum_{\text{Neg } E_n} \phi_n(2) \bar{\phi}_n(1) \qquad t_2 < t_1 \ . \tag{8.32}$$

If we have a negative energy wavepacket

$$\psi_{\text{Neg}}(\vec{x}_1, t_1) = \sum_{\text{Neg } E_n} b_n \phi_n(\vec{x}_1)$$

with

$$b_n = \int d^3x_1 \, \bar{\phi}_n(\vec{x}_1) \gamma_0 \psi_{\text{Neg}}(\vec{x}_1, t_1) \ . \tag{8.33}$$

then for $t_2 < t_1$

$$\psi_{\text{Neg}}(t_2, \vec{x}_2) = \sum_{\text{Neg } E_n} \int d^3x_1 \, \phi_n(\vec{x}_2) e^{-i E_n (t_2 - t_1)} \bar{\phi}_n(\vec{x}_1) \gamma^0 \psi_{\text{Neg}}(\vec{x}_1, t_1)$$
$$= \int d^3x_1 \sum_{\text{Neg } E_n} \phi_n(2) \bar{\phi}_n(1) \gamma^0 \psi_{\text{Neg}}(\vec{x}_1, t_1) \tag{8.34}$$
$$= - \int d^3x_1 \, i S_F^{(0)}(2,1) \gamma^0 \psi_{\text{Neg}}(\vec{x}_1, t_1) \ .$$

VII.8 DIRAC PROPAGATOR

Propagation of Entry States   Propagation of Exit States

$$f_{\text{Pos}}(2) = \int i S_F^{(0)}(2,1) \gamma^0 f_{\text{Pos}}(1) d^3x_1 \qquad \bar{g}_{\text{Pos}}(1) = \int \bar{g}_{\text{Pos}}(2) \gamma^0 i S_F^{(0)}(2,1) d^3x_2$$

$$f_{\text{Neg}}(2) = -\int i S_F^{(0)}(2,1) \gamma^0 f_{\text{Neg}}(1) d^3x_1 \qquad \bar{g}_{\text{Neg}}(1) = -\int \bar{g}_{\text{Neg}}(2) \gamma^0 i S_F^{(0)}(2,1) d^3x_2$$

Table VII.2: *Propagation of positive and negative Dirac plane wave states.*

Note the characteristic minus sign, which is associated with the negative energy solutions. We have then the rules
Associated with the index n we choose normalized plane wave states

$$\text{Positive } E_n \quad \phi^{(r)}_{p_{\text{Pos}}}(x) = \sqrt{\frac{m}{E_p}} u^{(r)}(p) e^{-ip \cdot x} \quad r = 1,2$$

$$\text{Negative } E_n \quad \phi^{(r)}_{p_{\text{Neg}}}(x) = \sqrt{\frac{m}{E_p}} v^{(r)}(p) e^{+ip \cdot x} \quad r = 1,2$$
(8.35)

where $u^{(r)}(p)$ and $v^{(r)}(p)$ are spinors which are solutions to the Dirac equation for positive and negative energies, respectively, *i.e.*,

$$\begin{aligned}
\not{p} u^{(r)}(p) &= m u^{(r)}(p) & \not{p} v^{(r)}(p) &= -m v^{(r)}(p) \\
\bar{u}^{(r)}(p) \not{p} &= \bar{u}^{(r)}(p) m & \bar{v}^{(r)}(p) \not{p} &= -\bar{v}^{(r)}(p) m \\
u^{(r)\dagger}(p) u^{(r')}(p) &= \frac{E_p}{m} \delta_{rr'} & v^{(r)\dagger}(p) v^{(r')}(p) &= \delta_{rr'} \frac{E_p}{m} \\
\bar{u}^{(r)}(p) u^{(r')}(p) &= \delta_{rr'} & \bar{v}^{(r)}(p) v^{(r')}(p) &= -\delta_{rr'} \; .
\end{aligned}$$
(8.36)

Then

$$\begin{aligned}
t_2 > t_1 \quad i S_F^{(0)}(2,1) &= \sum_{\text{Pos } E_n} \phi_n(2) \bar{\phi}_n(1) \\
&= \int \frac{d^3 p}{(2\pi)^3} \frac{m}{E_p} \sum_{r=1}^{2} u^{(r)}(p) \bar{u}^{(r)}(p) e^{-ip \cdot (x_2 - x_1)} \\
t_2 < t_1 \quad i S_F^{(0)}(2,1) &= -\sum_{\text{Neg } E_n} \phi_n(2) \bar{\phi}_n(1) \\
&= -\int \frac{d^3 p}{(2\pi)^3} \frac{m}{E_p} \sum_{r=1}^{2} v^{(r)}(p) \bar{v}^{(r)}(p) e^{ip \cdot (x_2 - x_1)}
\end{aligned}$$
(8.37)

We can express an arbitrary four component spinor w as a linear combination of $u^{(r)}, v^{(r)}$

$$w = \sum_{r=1}^{2}(a_r u^{(r)}(p) + b_r v^{(r)}(p)) \tag{8.38}$$

where $a_r = \bar{u}^{(r)}(p)w$ and $b_r = -\bar{v}^{(r)}(p)w$. Then

$$w = \sum_{r=1}^{2}\left(u^{(r)}(p)\bar{u}^{(r)}(p) - v^{(r)}(p)\bar{v}^{(r)}(p)\right)w \tag{8.39}$$

so that (since w is arbitrary)

$$1 = \sum_{r=1}^{2}\left(u^{(r)}(p)\bar{u}^{(r)}(p) - v^{(r)}(p)\bar{v}^{(r)}(p)\right) \tag{8.40}$$

and

$$\frac{\not{p}+m}{2m} = \frac{\not{p}+m}{2m}\sum_{r=1}^{2}\left(u^{(r)}(p)\bar{u}^{(r)}(p) - v^{(r)}(p)\bar{v}^{(r)}(p)\right) = \sum_{r=1}^{2}u^{(r)}(p)\bar{u}^{(r)}(p)$$

$$\frac{\not{p}-m}{2m} = \frac{\not{p}-m}{2m}\sum_{r=1}^{2}\left(u^{(r)}(p)\bar{u}^{(r)}(p) - v^{(r)}(p)\bar{v}^{(r)}(p)\right) = \sum_{r=1}^{2}v^{(r)}(p)\bar{v}^{(r)}(p) \ .$$
$$\tag{8.41}$$

Substitution into Eq. 8.37 yields

$$iS_F^{(0)}(2,1) = \begin{cases} \int \frac{d^3p}{(2\pi)^3}\frac{1}{2E_p}e^{-ip\cdot(x_2-x_1)}(\not{p}+m) & t_2 > t_1 \\ = -\int \frac{d^3p}{(2\pi)^3}\frac{1}{2E_p}e^{ip\cdot(x_2-x_1)}(\not{p}-m) & t_2 < t_1 \end{cases}$$
$$= \int \frac{d^4p}{(2\pi)^4}e^{-ip\cdot(x_2-x_1)}i\frac{\not{p}+m}{p^2-m^2+i\epsilon} \tag{8.42}$$

as found in Eq. 8.16 by more formal methods. Armed now with the form of the Dirac propagator we can proceed with our program to develop a perturbation theory for the Dirac equation.

VII.9 COVARIANT PERTURBATION THEORY

Propagation in an External Potential

Consider the motion of an electron in an external potential $A^\mu(\vec{x}, t)$. Let the potential vanish in the remote past and in the remote future. The propagator in the presence of the potential is $S_F^A(2, 1)$, where

$$(i\not{\nabla}_2 - m - e\not{A}(2))S_F^A(2,1) = \delta(2,1) \ . \tag{9.1}$$

VII.9 COVARIANT PERTURBATION THEORY

We can find an integral equation for $S_F^A(2,1)$ as follows. We know how to solve the problem of finding a function ϕ obeying the inhomogeneous equation

$$(i\not\nabla_2 - m)\phi(2) = Q(2) \tag{9.2}$$

—*viz.*, from the definition of the free propagator we must have

$$\phi(2) = \int S_F^{(0)}(2,3)Q(3)d\tau_3 + \text{solution of the homogeneous equation} \tag{9.3}$$

where we have introduced the notation $d\tau_3$ as a shorthand for d^4x_3. Now apply this same method to Eq. 9.1 for $S_F^A(2,1)$. Writing

$$(i\not\nabla_2 - m)S_F^A(2,1) = \{\delta(2,1) - ie\not{A}(2)iS_F^A(2,1)\}$$

we find
$$S_F^A(2,1) = \int S_F^{(0)}(2,3)\{\delta(3,1) - ie\not{A}(3)iS_F^A(3,1)\}d\tau_3$$
$$+ \text{solution of the homogeneous equation} \tag{9.4}$$
$$= S_F^{(0)}(2,1) - ie\int S_F^{(0)}(2,3)\not{A}(3)iS_F^A(3,1)d\tau_3$$
$$+ \text{solution of the homogeneous equation.}$$

Since physics should require that S_F^A reduce to $S_F^{(0)}$ as $A^\mu \to 0$, the homogeneous solution must vanish and we find

$$S_F^A(2,1) = S_F^{(0)}(2,1) - ie\int S_F^{(0)}(2,3)\not{A}(3)iS_F^A(3,1)d\tau_3 \tag{9.5}$$

which can be solved by iteration

$$S_F^A(2,1) = S_F^{(0)}(2,1) - ie\int S_F^{(0)}(2,3)\not{A}(3)iS_F^{(0)}(3,1)d\tau_3$$
$$+ (-ie)^2\int S_F^{(0)}(2,3)\not{A}(3)iS_F^{(0)}(3,4)\not{A}(4)iS_F^{(0)}(4,1)d\tau_3\,d\tau_4 + \ldots \quad. \tag{9.6}$$

We could now use this expansion for $S_F^A(2,1)$ directly to obtain a perturbation series for transition amplitudes in scattering problems. Alternatively, the desired result is reached if we consider the scattering integral equation for the Dirac wave function itself

$$(i\not\nabla - m)\psi = e\not{A}\psi\ . \tag{9.7}$$

If $f(x)$ is the entry state wave function (solution of the free particle equation) the same procedure as used above gives

$$\psi(2) = f(2) - ie\int iS_F^{(0)}(2,1)\not{A}(1)\psi(1)d\tau_1\ . \tag{9.8}$$

Suppose the entry state to be of positive energy and let the potential be non-vanishing over the interval $-\frac{T}{2}, \frac{T}{2}$. We may consider two choices for exit state $\bar{g}$

i) positive energy $\bar{g}_{\text{Pos}}$ in the remote future

ii) negative energy $\bar{g}_{\text{Neg}}$ in the remote past.

Case (i) gives the amplitude for scattering of an electron from state f_{Pos} (in the past) to state g_{Pos} (in the future). Case (ii) gives the amplitude for annihilation of an electron from state f_{Pos} (in the past) and a positron described by the effectively *positive* energy wave function $\bar{g}_{\text{Neg}}$ (in the past)

Case (i): Electron Scattering

Assume f_{Pos} and g_{Pos} are orthogonal— $\left[\int \bar{g}_{\text{Pos}}(2)\gamma^0 f_{\text{Pos}}(2)d^3x_2 = 0\right]$ and take t_2 to be in the remote future. The scattering amplitude is

$$\text{Amp} = \int \bar{g}_{\text{Pos}}(2)\gamma^0 \psi(2) d^3 x_2$$
$$= -ie \iint \bar{g}_{\text{Pos}}(2) i S_F^{(0)}(2,1)\, \slashed{A}\,(1)\psi(1) d^3 x_2 d\tau_1 \quad . \tag{9.9}$$

This expression is exact, with $\psi(1)$ beomg the exact wave function. From Table VII.2 we may perform the integration on $d^3 x_2$ to find

$$\int \bar{g}_{\text{Pos}}(2)\gamma^0 i S_F^{(0)}(2,1) d^3 x_2 = \bar{g}_{\text{Pos}}(1) \tag{9.10}$$

so that

$$\text{Amp} = -ie \int d\tau_1\, \bar{g}_{\text{Pos}}(1)\, \slashed{A}\,(1)\psi(1) \quad . \tag{9.11}$$

We see that this result is Lorentz invariant. It is also clear that we can generate a perturbation series for the transition amplitude by successive iteration of Eq. 9.8:

$$\text{Amp} = \sum_{n=0}^{\infty} \text{Amp}^{(n)}$$
$$\text{Amp}^{(0)} = \int d^3 x\, \bar{g}_{\text{Pos}} \gamma^0 f_{\text{Pos}} = 0 \quad \text{assuming} \quad \vec{p}_i \neq \vec{p}_f$$
$$\text{Amp}^{(1)} = -ie \int d\tau_1\, \bar{g}_{\text{Pos}}(1)\, \slashed{A}\,(1) f_{\text{Pos}}(1) \tag{9.12}$$
$$\text{Amp}^{(2)} = (-ie)^2 \int d\tau_1\, d\tau_2\, \bar{g}_{\text{Pos}}(1)\, \slashed{A}\,(1) i S_F^{(0)}(1,2)\, \slashed{A}\,(2) f_{\text{Pos}}(2)$$

etc.

With each term in perturbation theory we may associate a Feynman diagram, in familiar fashion. But it is important to realize that the properties of the free propagator $S_F^{(0)}(2,1)$ allow contributions from $t_2 > t_1$ *as well as* from $t_2 < t_1$ (*cf.* Figure VII.7). The theory was designed to take this into account, and hence $\text{Amp}^{(2)}$ contains both.

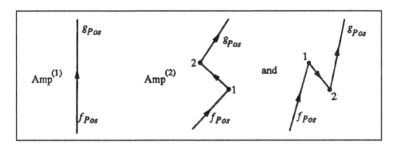

Fig. VII.7: Terms in the relativistic perturbation expansion for electron scattering.

Recall the convention: Arrows on the world line in a Feynman diagram keep track of entry and exit at each vertex. An arrow forward in time implies positive E_n, while an arrow backward in time signifies negative E_n.

Case (ii):   Annihilation of a Pair

In this case the exit state is $\bar{g}_{\text{Neg}}$ in the remote past and

$$\begin{aligned}
\text{Amp} &= \int \bar{g}_{\text{Neg}}(2)\gamma^0\psi(2)d^3x_2 \\
&= -ie \iint d^3x_2\,d\tau_1\, \bar{g}_{\text{Neg}}(2)\gamma^0 iS_F^{(0)}(2,1)\,\slashed{A}(1)\psi(1) \qquad (9.13) \\
&= +ie \int d\tau_1\, \bar{g}_{\text{Neg}}(1)\,\slashed{A}(1)\psi(1)\ .
\end{aligned}$$

The resulting perturbation series is

$$\begin{aligned}
\text{Amp} = &-(-ie)\int d\tau_1\, \bar{g}_{\text{Neg}}(1)\,\slashed{A}(1)f_{\text{Pos}}(1) \\
&-(-ie)^2 \int d\tau_1\,d\tau_2\, \bar{g}_{\text{Neg}}(1)\,\slashed{A}(1) iS_F^{(0)}(1,2)\,\slashed{A}(2)f_{\text{Pos}}(2) + \ldots
\end{aligned} \qquad (9.14)$$

Diagrams

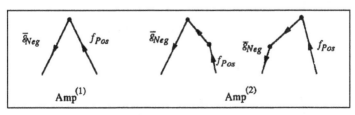

Fig. VII.8: Terms in the relativistic perturbation expansion for pair-annihilation.

Now suppose that the entry state is of negative energy, f_{Neg}, corresponding to an effectively positive energy antiparticle state in the remote future. Again we may consider two choices for the exit state $\bar{g}$.

iii) positive energy $\bar{g}_{\text{Pos}}$ in the remote future
iv) negative energy $\bar{g}_{\text{Neg}}$ in the remote past.

Case (iii) describes the scattering of a positron from a state (in the past) with effectively positive energy wavefunction $\bar{g}_{\text{Neg}}$ to a positron (in the future) in effectively positive energy state f_{Neg}. Case (iv) gives the amplitude for creation of an electron in positive energy state $\bar{g}_{\text{Pos}}$ (in the future) and a positron described by an effectively positive energy wavefunction f_{Neg} (in the future).

Case (iii): Positron Scattering:

The exit state $\bar{g}_{\text{Neg}}$ is in the remote past and the transition amplitude becomes

$$\text{Amp} = -(-ie) \int \bar{g}_{\text{Neg}}(1) \, \slashed{A}(1) \psi(1) d\tau_1$$
$$= -(-ie) \int d\tau_1 \, \bar{g}_{\text{Neg}}(1) \, \slashed{A}(1) f_{\text{Neg}}(1) \qquad (9.15)$$
$$- (-ie)^2 \int d\tau_1 \, d\tau_2 \, \bar{g}_{\text{Neg}}(1) \, \slashed{A}(1) i S_F^{(0)}(1,2) \, \slashed{A}(2) f_{\text{Neg}}(2) + \ldots$$

Diagrams

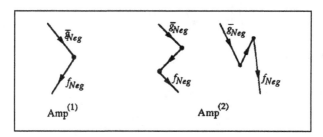

Fig. VII.9: *Terms in the relativistic perturbation expansion for positron scattering.*

Case (iv): Pair Creation:

The exit state $\bar{g}_{\text{Pos}}$ is in the remote future and the transition amplitude is

$$\text{Amp} = -ie \int d\tau_1 \, \bar{g}_{\text{Pos}}(1) \, \slashed{A}(1) \psi(1)$$
$$= -ie \int d\tau_1 \, \bar{g}_{\text{Pos}}(1) \, \slashed{A}(1) f_{\text{Neg}}(1) \qquad (9.16)$$
$$+ (-ie)^2 \int d\tau_1 \, d\tau_2 \, \bar{g}_{\text{Pos}}(1) \, \slashed{A}(1) i S_F^{(0)}(1,2) \, \slashed{A}(2) f_{\text{Neg}}(2) + \ldots$$

Diagrams

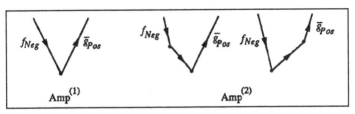

Fig. VII.10: *Terms in the relativistic perturbation expansion for pair creation.*

Having written the various transition amplitudes we note that we could have derived identical results from the perturbative expression Eq. 9.6 for the propagator in the presence of an external electromagnetic field. However, the wavefunction approach is somewhat more intuitive.

The simple expression for the Fourier transform of the propagator suggests the utility of momentum space for scattering problems. Thus instead of dealing with a vector potential $A^\mu(x)$ in coordinate space, consider

$$A^\mu(q) = \int d^4x \, e^{-iq\cdot x} A^\mu(x) \ . \tag{9.17}$$

For example, the Coulomb potential

$$A^\mu(x) = \left(-\frac{Ze}{4\pi|\vec{x}|}, \vec{0}\right) \tag{9.18}$$

becomes

$$\begin{aligned} A^0(q) &= -\int d^4x \, e^{-iq\cdot x} \frac{Ze}{4\pi|\vec{x}|} \\ &= -2\pi\delta(q_0) \int d^3x \, e^{i\vec{q}\cdot\vec{x}} \frac{Ze}{4\pi|\vec{x}|} \\ &= -2\pi\delta(q_0) \frac{Ze}{\vec{q}^2} \end{aligned} \tag{9.19}$$

$$\vec{A}(q) = 0 \ .$$

Now consider electron scattering with

$$\begin{aligned} \text{Entry state} \quad & f_{\text{Pos}} = u_i(p_i) \, e^{-ip_i \cdot x} \\ \text{Exit state} \quad & g_{\text{Pos}} = u_f(p_f) \, e^{-ip_f \cdot x} \ . \end{aligned} \tag{9.20}$$

We can write the transition matrix element M as a perturbation series

$$M = M^{(0)} + M^{(1)} + M^{(2)} + \cdots \tag{9.21}$$

where

$$M^{(0)} = \int d^3x \, \bar{g}_{\text{Pos}}(x) \gamma^0 f_{\text{Pos}}(x) = 0 \tag{9.22}$$

assuming $p_f \neq p_i$ and

$$
\begin{aligned}
M^{(1)} &= -ie \int d\tau_1 \, \bar{g}_{\text{Pos}}(1) \, \slashed{A}(1) f_{\text{Pos}}(1) \\
&= -ie \, \bar{u}_f(p_f) \int d^4x \, e^{ip_f \cdot x} \, \slashed{A}(x) \, e^{-ip_i \cdot x} u_i(p_i) \\
&= -ie \, \bar{u}_f(p_f) \, \slashed{A}(p_i - p_f) u_i(p_i) \; .
\end{aligned}
\quad (9.23)
$$

That is, the first order matrix element involves the Fourier component of the potential with momentum transfer $q = p_i - p_f$.

For the second order term we find

$$
\begin{aligned}
M^{(2)} &= (-ie)^2 \int d\tau_1 \int d\tau_2 \, \bar{g}_{\text{Pos}}(2) \, \slashed{A}(2) i S_F^{(0)}(2,1) \, \slashed{A}(1) f_{\text{Pos}}(1) \\
&= (-ie)^2 \bar{u}_f(p_f) \int d^4x_2 \int d^4x_1 \, e^{ip_f \cdot x_2} \, \slashed{A}(x_2) \\
&\quad \times \int \frac{d^4p}{(2\pi)^4} e^{-ip \cdot (x_2 - x_1)} \frac{i}{\slashed{p} - m + i\epsilon} \, \slashed{A}(x_1) e^{-ip_i \cdot x_1} u_i(p_i) \\
&= (-ie)^2 \bar{u}_f(p_f) \int \frac{d^4p}{(2\pi)^4} \, \slashed{A}(p - p_f) \frac{i}{\slashed{p} - m + i\epsilon} \, \slashed{A}(p_i - p) u_i(p_i) \; .
\end{aligned}
\quad (9.24)
$$

The structure of the general term should now be apparent and is most readily visualized in terms of *relativistic* Feynman diagrams with particle lines labelled by their respective four-momenta—there is no distinction between particle and antiparticle propagators here since the Feynman prescription does both simultaneously. For the potential at each vertex it is useful to draw a wavy line, labelled by momentum transfer q. Examination of the previous analytic form of the transition matrix element shows that with this method of labeling we may regard four-momentum as being conserved at each vertex. For example, for the lowest order amplitude, we find

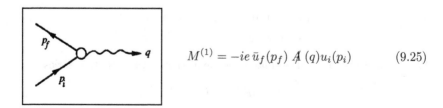

$$M^{(1)} = -ie \, \bar{u}_f(p_f) \, \slashed{A}(q) u_i(p_i) \quad (9.25)$$

where four momentum conservation requires: $p_i = q + p_f$.

Likewise for the next order amplitudes we find the diagrams

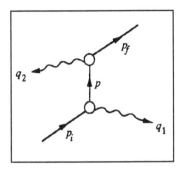

$$M^{(2)} = (-ie)^2 \bar{u}_f(p_f) \int \frac{d^4p}{(2\pi)^4} \slashed{A}(q_2) \frac{i}{\slashed{p} - m + i\epsilon}$$
$$\times \slashed{A}(q_1) u_i(p_i)$$

(9.26)

where, by four momentum conservation:

$$p = q_2 + p_f, \qquad p_i = p + q_1$$

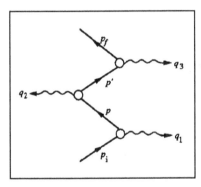

$$M^{(3)} = (-ie)^3 \bar{u}_f(p_f) \int \frac{d^4p\, d^4p'}{(2\pi)^8} \slashed{A}(q_3)$$
$$\times \frac{i}{\slashed{p}' - m + i\epsilon} \slashed{A}(q_2) \frac{i}{\slashed{p} - m + i\epsilon} \slashed{A}(q_1) u_i(p_i)$$

(9.27)

where due to four momentum conservation:

$$p' = q_3 + p_f, \qquad p = q_2 + p', \qquad p_i = q_1 + p \ .$$

The general structure for an arbitrary diagram can be summarized in terms of five simple rules:

1) an electron of four-momentum p propagating in an intermediate state contributes a factor $i/\slashed{p} - m + i\epsilon$;
2) a potential carrying four-momentum q contributes $-ie\,\slashed{A}(q)$;
3) four-momentum is conserved at each vertex; and
4) all intermediate momenta are summed over $\int d^4p/(2\pi)^4$.

For future use it is important to note one additional rule

5) a diagram containing a closed fermion loop must include an extra minus sign. The reason behind this rule will be explained in the next chapter and is associated with Fermi statistics.

PROBLEM VII.9.1

Pair Production by a Time Varying Electric Field: Fermions

As we saw in the last chapter, a rapidly varying electric field can lead to the creation of particle-antiparticle pairs. Calculate to lowest order the probability per unit volume per unit time of producing fermion pairs in the presence of an external electric field

$$\vec{E}(t) = \hat{e}_x a \cos \omega t$$

and show that

$$\text{Prob} = \text{VT} \frac{e^2 a^2}{24\pi} (1 - \frac{4m^2}{\omega^2})^{\frac{1}{2}} (1 + \frac{2m^2}{\omega^2}) \ .$$

Suggestion: Utilize normalized plane wave solutions of the Dirac equation

$$\psi(x) = \sqrt{\frac{m}{E}} u(p) \exp(i\vec{p} \cdot \vec{x} - iEt) \quad \text{with} \quad E = \sqrt{\vec{p}^2 + m^2}$$

and simple first order perturbation theory

$$\text{Amp} = -i \int_{-\frac{T}{2}}^{\frac{T}{2}} dt \, \langle f \, |H_{\text{int}}(t)| \, i \rangle$$

with $H_{\text{int}} = e \int d^3x j_\mu A^\mu$ as in the Klein-Gordon case.

VII.10 ELECTROMAGNETIC INTERACTIONS

Having developed this treatment of relativistic perturbation theory, we can now test it in applications to various realistic situations.

Relativistic Potential Scattering

We first try out this formalism for the case of an electron scattering from a Coulomb potential. We use

$$f_{\text{Pos}}(x) = \sqrt{\frac{m}{E_i}} u_i(p_i) \, e^{-ip_i \cdot x} \qquad g_{\text{Pos}}(x) = \sqrt{\frac{m}{E_f}} u_f(p_f) \, e^{-ip_f \cdot x}$$

$$\int f_{\text{Pos}}^\dagger(x) f_{\text{Pos}}(x) d^3x = 1 \qquad \int g_{\text{Pos}}^\dagger(x) g_{\text{Pos}}(x) d^3x = 1 \tag{10.1}$$

as the asymptotic entry, exit wavefunctions, assumed to be plane waves, and

$$A^\mu(q) = \left(-2\pi\delta(q_0)\frac{Ze}{|\vec{q}|^2}, \vec{0}\right) \tag{10.2}$$

for the vector potential. [We assume the electron to scatter from a massive target of charge $+Z|e|$.] The transition amplitude, in lowest order perturbation theory, is given by

$$\text{Amp}^{(1)} = -ie \frac{m}{\sqrt{E_i E_f}} \bar{u}_f(p_f) \, \slashed{A}\, (p_i - p_f) u_i(p_i)$$

$$= i4\pi Z\alpha \frac{m}{\sqrt{E_i E_f}} 2\pi\delta(E_f - E_i) \frac{1}{(\vec{p}_f - \vec{p}_i)^2} \bar{u}_f(p_f) \gamma^0 u_i(p_i) \ . \tag{10.3}$$

VII.10 ELECTROMAGNETIC INTERACTIONS

and the transition probability summed over a range of final states is

$$\sum_f P_{fi} = (4\pi Z\alpha)^2 \frac{m}{E_i} \int \frac{d^3 p_f}{(2\pi)^3} \frac{m}{E_f} (2\pi \delta(E_f - E_i))^2 \frac{1}{(\vec{p}_f - \vec{p}_i)^4} \left| \bar{u}_f(p_f) \gamma^0 u_i(p_i) \right|^2 \quad (10.4)$$

Treating the square of the delta function in the usual fashion, we find

$$\sum_f P_{fi} = T(4\pi Z\alpha)^2 \frac{m}{E_i} \int \frac{d^3 p_f}{(2\pi)^3} \frac{m}{E_f} 2\pi \delta(E_f - E_i) \times \frac{1}{(\vec{p}_f - \vec{p}_i)^4} \left| \bar{u}_f(p_f) \gamma^0 u_i(p_i) \right|^2 \quad (10.5)$$

where T is the total time of interaction. The cross section is

$$d\sigma = \frac{\frac{1}{T} \sum_f P_{fi}}{\text{Incident flux}} \quad (10.6)$$

where the incident flux is given by the component of the incident probability current density in the direction $\hat{v}_i$

$$\text{Incident flux} = \hat{v}_i \cdot \bar{f}_{\text{Pos}} \vec{\gamma} f_{\text{Pos}} = \hat{v}_i \cdot \frac{m}{E_i} \bar{u}_i(p_i) \vec{\gamma} u_i(p_i) = \hat{v}_i \cdot \frac{m}{E_i} \frac{\vec{p}_i}{m} = \frac{p_i}{E_i} , \quad (10.7)$$

(*i.e.*, the incident velocity) and assumes the form

$$d\sigma = \frac{E_i}{p_i} (4\pi Z\alpha)^2 \frac{m^2}{E_i^2} \int d\Omega_f p_f^2 \, dp_f \frac{1}{(2\pi)^2} \delta(E_f - E_i) \frac{1}{(\vec{p}_f - \vec{p}_i)^4} \left| \bar{u}_f(p_f) \gamma^0 u_i(p_i) \right|^2$$

$$= (4\pi Z\alpha)^2 \frac{m^2}{E_i p_i} d\Omega_f \frac{1}{(2\pi)^2} \frac{p_f^2}{dE_f/dp_f} \frac{1}{(\vec{p}_f - \vec{p}_i)^4} \left| \bar{u}_f(p_f) \gamma^0 u_i(p_i) \right|^2 . \quad (10.8)$$

Since

$$\frac{dE_f}{dp_f} = \frac{p_f}{E_f} = \frac{p_i}{E_i} \quad (10.9)$$

we find

$$\frac{d\sigma}{d\Omega_f} = \frac{4Z^2 \alpha^2 m^2}{(\vec{p}_f - \vec{p}_i)^4} \left| \bar{u}_f(p_f) \gamma^0 u_i(p_i) \right|^2 . \quad (10.10)$$

For the Dirac matrix element, we have

$$\bar{u}_f(p_f) \gamma^0 u_i(p_i) = u_f^\dagger(p_f) u_i(p_i)$$

$$= \sqrt{\frac{E_f + m}{2m}} \sqrt{\frac{E_i + m}{2m}} \begin{pmatrix} \chi_f^\dagger & \chi_f^\dagger \frac{\vec{\sigma} \cdot \vec{p}_f}{E_f + m} \end{pmatrix} \begin{pmatrix} \chi_i \\ \frac{\vec{\sigma} \cdot \vec{p}_i}{E_i + m} \chi_i \end{pmatrix} \quad (10.11)$$

$$= \frac{E_i + m}{2m} \chi_f^\dagger \left(1 + \frac{\vec{\sigma} \cdot \vec{p}_f \, \vec{\sigma} \cdot \vec{p}_i}{(E_i + m)^2} \right) \chi_i .$$

Generally one is interested in the cross section for the case of an unpolarized incident beam and without detection of the final polarization state. Then we must sum over final state spinors χ_f and average over initial state spinors χ_i. In this way, using standard trace methods, we find[†]

$$\begin{aligned}
\frac{1}{2}\sum_{s_f}\sum_{s_i}|\bar{u}_f(p_f)\gamma^0 u_i(p_i)|^2 &= \frac{1}{2}\left(\frac{E_i+m}{2m}\right)^2 \text{Tr}\left(1+\frac{\vec{\sigma}\cdot\vec{p}_f\,\vec{\sigma}\cdot\vec{p}_i}{(E_i+m)^2}\right) \\
&\quad \times \left(1+\frac{\vec{\sigma}\cdot\vec{p}_i\,\vec{\sigma}\cdot\vec{p}_f}{(E_i+m)^2}\right) \\
&= \frac{1}{2}\left(\frac{E_i+m}{2m}\right)^2 \text{Tr}\left(1+\frac{\vec{\sigma}\cdot\vec{p}_f\,\vec{\sigma}\cdot\vec{p}_i+\vec{\sigma}\cdot\vec{p}_i\,\vec{\sigma}\cdot\vec{p}_f}{(E_i+m)^2}+\frac{\vec{\sigma}\cdot\vec{p}_f\,\vec{\sigma}\cdot\vec{p}_i\,\vec{\sigma}\cdot\vec{p}_i\,\vec{\sigma}\cdot\vec{p}_f}{(E_i+m)^4}\right) \\
&= \left(\frac{E_i+m}{2m}\right)^2 \frac{1}{2}\text{Tr}\left(1+\frac{2\vec{p}_f\cdot\vec{p}_i}{(E_i+m)^2}+\frac{p_i^2 p_f^2}{(E_i+m)^4}\right) \\
&= \left(\frac{E_i+m}{2m}\right)^2 \left[1+\frac{2p_i^2\cos\theta}{(E_i+m)^2}+\frac{p_i^4}{(E_i+m)^4}\right]
\end{aligned}$$
(10.12)

where θ is the scattering angle. Writing the quantity in brackets as

$$\begin{aligned}
[\ ] &= \left(1+\frac{p_i^2}{(E_i+m)^2}\right)^2 - 2\frac{p_i^2}{(E_i+m)^2}(1-\cos\theta) \\
&= \frac{1}{(E_i+m)^2}\left(4E_i^2 - 4p_i^2\sin^2\frac{\theta}{2}\right)
\end{aligned}$$
(10.13)

the scattering cross section is found to be

$$\frac{d\sigma}{d\Omega_f} = \frac{Z^2\alpha^2}{4E_i^2\beta^4\sin^4\frac{\theta}{2}}\left(1-\beta^2\sin^2\frac{\theta}{2}\right)$$
(10.14)

where

$$\beta = \frac{p_i}{E_i}$$
(10.15)

is the electron velocity and we have used the relation

$$\vec{q}^2 = p_i^2 + p_f^2 - 2\vec{p}_f\cdot\vec{p}_i = 2p_i^2(1-\cos\theta) = 4p_i^2\sin^2\frac{\theta}{2} \ .$$
(10.16)

[†] Recall that for an initial state spinor χ_i with polarization $\vec{s}_i$ and a final state spinor χ_f with polarization $\vec{s}^f$ [Me 70]

$$|\chi_f\mathcal{O}\chi_i|^2 = \frac{1}{4}\text{Tr}(1+\vec{\sigma}\cdot\vec{s}_f)\mathcal{O}(1+\vec{\sigma}\cdot\vec{s}_i)\mathcal{O}^\dagger \ ,$$

so that for an unpolarized incident beam with the outgoing polarization undetected we have

$$\frac{1}{2}\sum_{s_i}\sum_{s_f}|\chi_f\mathcal{O}\chi_i|^2 = \frac{1}{2}\text{Tr}\mathcal{O}\mathcal{O}^\dagger \ .$$

In the non-relativistic limit — $\beta \ll 1$ — Eq. 10.14 reduces to

$$\frac{d\sigma}{d\Omega_f} = \left| \frac{Z\alpha m}{2p_i^2 \sin^2 \frac{\theta}{2}} \right|^2 \tag{10.17}$$

which agrees with the non-relativistic Born approximation result for the scattering amplitude

$$f_{p_i}(\theta) = \frac{m}{2\pi} \int d^3 r \, e^{i\vec{q}\cdot\vec{r}} \frac{Z\alpha}{r} = \frac{2mZ\alpha}{\vec{q}^{\,2}} = \frac{mZ\alpha}{2p_i^2 \sin^2 \frac{\theta}{2}} \ . \tag{10.18}$$

The meaning of this relativistic form is explored in Problem VII.10.1.

Dirac Trace Techniques

Before proceeding, it is useful to demonstrate a formal procedure by which one can perform spin averages on the absolute square of a Dirac matrix element. Above in evaluating the cross section for electron scattering from a Coulomb potential, we reduced the problem to one involving Pauli spinors and employed traces over the appropriate 2×2 matrices in this space. However, it is much easier in general and more powerful to stay with Dirac notation and to take traces in 4×4 matrix space.

Consider the generic Dirac matrix element

$$\bar{u}_f(p_f) \Gamma u_i(p_i) \tag{10.19}$$

where Γ is some combination of Dirac matrices. We note that

$$\begin{aligned}|\bar{u}_f(p_f) \Gamma u_i(p_i)|^2 &= \bar{u}_f(p_f) \Gamma u_i(p_i) \, (\bar{u}_f(p_f) \Gamma u_i(p_i))^* \\ &= \bar{u}_f(p_f) \Gamma u_i(p_i) u_i^\dagger(p_i) \Gamma^\dagger \gamma^0 u_f(p_f) \\ &= \bar{u}_f(p_f) \Gamma u_i(p_i) \bar{u}_i(p_i) \bar{\Gamma} u_f(p_f)\end{aligned} \tag{10.20}$$

where we have defined

$$\bar{\Gamma} = \gamma^0 \Gamma^\dagger \gamma^0 \ . \tag{10.21}$$

For example

$$\bar{\gamma}_\mu = \gamma^0 \gamma_\mu^\dagger \gamma^0 = \gamma^0 \gamma^\mu \gamma^0 = \gamma_\mu \tag{10.22}$$

and

$$\begin{aligned}\overline{\not{d}\not{b}\not{c}\ldots} &= \gamma_0 \ldots c^\mu \gamma_\mu^\dagger b^\nu \gamma_\nu^\dagger a^\lambda \gamma_\lambda^\dagger \gamma^0 \\ &= \ldots c^\mu \bar{\gamma}_\mu b^\nu \bar{\gamma}_\nu a^\lambda \bar{\gamma}_\lambda = \ldots \not{c}\not{b}\not{d}\end{aligned} \tag{10.23}$$

etc. Recalling that

$$\begin{aligned}\sum_{s=1}^{2} u_i^s(p_i) \bar{u}_i^s(p_i) &= \frac{1}{2m}(\not{p}_i + m) \\ \sum_{r=1}^{2} u_f^r(p_f) \bar{u}_f^r(p_f) &= \frac{1}{2m}(\not{p}_f + m)\end{aligned} \tag{10.24}$$

we find†

$$\frac{1}{2}\sum_{r=1}^{2}\sum_{s=1}^{2}|\bar{u}_f^r(p_f)\Gamma u_i^s(p_i)|^2$$

$$=\frac{1}{2}\sum_{r=1}^{2}\sum_{s=1}^{2}\sum_{\alpha\beta\gamma\delta}\bar{u}_f^r(p_f)_\beta \Gamma_{\beta\alpha} u_i^s(p_i)_\alpha \bar{u}_i^s(p_i)_\delta \bar{\Gamma}_{\delta\gamma} u_f^r(p_f)_\gamma \qquad (10.25)$$

$$=\frac{1}{2}\sum_{\alpha\beta\gamma\delta}\Gamma_{\beta\alpha}\left(\frac{\not{p}_i+m}{2m}\right)_{\alpha\delta}\bar{\Gamma}_{\delta\gamma}\left(\frac{\not{p}_f+m}{2m}\right)_{\gamma\beta}$$

$$=\frac{1}{8m^2}\operatorname{Tr}\Gamma(\not{p}_i+m)\bar{\Gamma}(\not{p}_f+m).$$

Typical traces which arise can be evaluated using a few simple theorems:

i) $\operatorname{Tr}\mathbf{1} = 4$.

ii) $\operatorname{Tr}\not{a}\not{b} = 4a\cdot b$.
Proof:

$$\operatorname{Tr}\not{a}\not{b} = \frac{1}{2}\operatorname{Tr}(\not{a}\not{b}+\not{b}\not{a}) = \frac{1}{2}a^\mu b^\nu \operatorname{Tr}(\gamma_\mu\gamma_\nu+\gamma_\nu\gamma_\mu)$$
$$= a^\mu b^\nu \eta_{\mu\nu}\operatorname{Tr}\mathbf{1} = 4a\cdot b \qquad \text{q.e.d.} \qquad (10.26)$$

iii) $\operatorname{Tr}\not{a}\not{b}\not{c}\not{d} = 4(a\cdot b\,c\cdot d - a\cdot c\,b\cdot d + a\cdot d\,b\cdot c)$
Proof:

$$\operatorname{Tr}\not{a}\not{b}\not{c}\not{d} = \operatorname{Tr}(2a\cdot b - \not{b}\not{a})\not{c}\not{d}$$
$$= 2a\cdot b\times 4c\cdot d - \operatorname{Tr}\not{b}(2a\cdot c - \not{c}\not{a})\not{d}$$
$$= 2a\cdot b\times 4c\cdot d - 2a\cdot c\times 4b\cdot d + \operatorname{Tr}\not{b}\not{c}(2a\cdot d - \not{d}\not{a})$$
$$= 2a\cdot b\times 4c\cdot d - 2a\cdot c\times 4b\cdot d + 2a\cdot d\times 4b\cdot c - \operatorname{Tr}\not{a}\not{b}\not{c}\not{d} \quad \text{q.e.d.}$$
$$(10.27)$$

iv) The trace of the product of an odd number of γ matrices vanishes, i.e., $\operatorname{Tr}\not{a} = \operatorname{Tr}\not{a}\not{b}\not{c} = \ldots = 0$
Proof: Since $\gamma_5^2 = 1$

$$\operatorname{Tr}\not{a}\not{b}\not{c}\ldots = \operatorname{Tr}\gamma_5^2 \not{a}\not{b}\not{c}\ldots$$
$$= -\operatorname{Tr}\gamma_5 \not{a}\not{b}\not{c}\ldots\gamma_5 = -\operatorname{Tr}\gamma_5^2 \not{a}\not{b}\not{c}\ldots \qquad (10.28)$$
$$= -\operatorname{Tr}\not{a}\not{b}\not{c}\ldots \qquad \text{q.e.d.}$$

---

† For completeness we note that since

$$\not{s}_i\gamma_5 u^{s_i}(p_i) = u^{s_i}(p_i)$$

where s_i^μ is the four-vector defined as $(0,\vec{s}_i)$ in the rest frame with $\vec{\sigma}\cdot\vec{s}_i\chi_i = \chi_i$, the projection operator for an electron with polarization vector s^μ is

$$\frac{1}{2m}(\not{p}_i+m)\frac{1}{2}(1+\not{s}_i\gamma_5)\ .$$

where we have anticommuted γ_5 through the odd number of $\not{a}, \not{b}, \not{c}$, etc.

Applying these methods to the problem of Coulomb potential scattering, we find

$$\frac{1}{2}\sum_{s_i}\sum_{s_f}|\bar{u}_f(p_f)\gamma^0 u_i(p_i)|^2 = \frac{1}{8m^2}\operatorname{Tr}\gamma^0(\not{p}_i+m)\gamma^0(\not{p}_f+m)$$

$$= \frac{1}{8m^2}\left(\operatorname{Tr}\gamma^0\not{p}_i\gamma^0\not{p}_f + m^2\operatorname{Tr}\gamma_0^2\right) = \frac{1}{8m^2}\left(8E_iE_f - 4p_i\cdot p_f + 4m^2\right) \quad (10.29)$$

$$= \frac{1}{8m^2}4\left(E_i^2 + m^2 + p_i^2\cos\theta\right) = \frac{1}{m^2}\left(E_i^2 - p_i^2\sin^2\frac{\theta}{2}\right)$$

which agrees precisely with Eq. 10.13 obtained by reducing Dirac to Pauli spinors. Using the Dirac trace theorems derived above one can also straightforwardly deal with other, more complex, Dirac matrix elements, as we shall see.

Relativistic Two-Body Scattering

Of course, electron scattering from a fixed potential makes no sense relativistically. In order to deal with a more realistic situation consider the scattering of an electron from an "ideal" proton —i.e., a pointlike particle having gyromagnetic ratio $g_p = 2$ which obeys the Dirac equation. We shall treat the scattering to lowest order in perturbation theory —$\mathcal{O}(\alpha)$ — and imagine a particle a with charge e_a, mass m scattered by an effective vector potential $A_\mu^{(b)}$ produced by the motion of a particle b with charge e_b, mass M. [For definiteness we take particle a as the electron and particle b as the "proton."] We have then to lowest order in the perturbative expansion

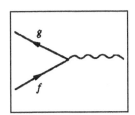

$$\operatorname{Amp}^{(1)} = -ie_a\int d^4x\,\bar{g}_a(x)\,\not{A}^b(x)f_a(x)\ . \quad (10.30)$$

where $A_\mu^b(x)$ is the vector potential produced at the location of particle a due to the motion of particle b. We see that the transition amplitude has the structure of an interaction

$$\mathcal{L}(x) = -e_a j_\mu^{(a)}(x)A_{(b)}^\mu(x) \quad (10.31)$$

between an transition current density

$$e_a j_\mu^{(a)}(x) = e_a\bar{g}_a(x)\gamma_\mu f_a(x) \quad (10.32)$$

and an external vector potential $A_{(b)}^\mu(x)$ so that Eq. 10.30 can be written in the suggestive form

$$\operatorname{Amp}^{(1)} = -ie_a\int d^4x\,j_\mu^{(a)}(x)A_{(b)}^\mu(x)\ . \quad (10.33)$$

The question is what to use for the vector potential $A_\mu^{(b)}(x)$ produced by particle b. By analogy with Eq. 10.32 the transition current density associated with particle b must be

$$e_b j_\mu^{(b)}(y) = e_b \bar{g}_b(y) \gamma_\mu f_b(y) \tag{10.34}$$

where $\bar{g}_b$, f_b are the appropriate exit, entry states for the scattering in question. The vector potential $A_\mu^{(b)}(x)$ produced by this current density is determined by the Maxwell equation

$$\Box_x A_\mu^{(b)}(x) = e_b j_\mu^{(b)}(x) \tag{10.35}$$

provided that $A_\mu^{(b)}(x)$ satisfies the Lorentz condition $\partial^\mu A_\mu^{(b)}(x) = 0$. We can solve Eq. 10.35 by use of Green's function methods. We define $D_{\mu\nu}(x,y)$ in terms of its Fourier transform

$$D_{\mu\nu}(x,y) = \int \frac{d^4q}{(2\pi)^4} e^{-iq\cdot(x-y)} D_{\mu\nu}(q) \tag{10.36}$$

and demand that the differential equation

$$\Box_x D_{\mu\nu}(x,y) = \eta_{\mu\nu} \delta^4(x-y) \tag{10.37}$$

be satisfied. We require then

$$\begin{aligned}\Box_x D_{\mu\nu}(x,y) &= -\int \frac{d^4q}{(2\pi)^4} e^{-iq\cdot(x-y)} q^2 D_{\mu\nu}(q) \\ &= \eta_{\mu\nu} \int \frac{d^4q}{(2\pi)^4} e^{-iq\cdot(x-y)}\end{aligned} \tag{10.38}$$

or

$$D_{\mu\nu}(q) = -\frac{\eta_{\mu\nu}}{q^2 + i\epsilon} \tag{10.39}$$

where we have inserted the $+i\epsilon$ prescription, as before, in order to assure the proper integration contour. The Green's function thus defined is the propagator for the quanta of the Maxwell field — (virtual) photons. That the photons are virtual is clear since for real photons, the energy-momentum vector q_μ satisfies $q^2 = 0$, whence the propagator becomes infinite. Likewise for the Dirac equation, the propagator becomes infinite when the quanta are on their "mass shell" — $p^2 - m^2 = 0$.

In terms of the Green's function we can evaluate the vector potential

$$A_\mu^{(b)}(x) = -ie_b \int d^4y\, iD_{\mu\nu}(x-y) j_b^\nu(y) \ . \tag{10.40}$$

(We could also, of course, append a solution of the homogeneous equation here. However, the solution which we wish for $A_\mu^{(b)}(x)$ is that which vanishes in the absence of a transition density $j_\mu^b(x)$.) The field $A_\mu^b(x)$ then is clearly a solution of Eq. 10.35 and it is straightforward to verify that the Lorentz condition is satisfied

$$[\text{Check}: \quad \partial^\mu A_\mu^{(b)}(x) = -ie_b \int d^4y\, \partial_x^\mu iD_{\mu\nu}(x-y)j_{(b)}^\nu(y)$$

$$= +ie_b \int d^4y\, \partial_y^\mu iD_{\mu\nu}(x-y)j_{(b)}^\nu(y)$$

$$= (\text{integrating by parts}) - ie_b \int d^4y\, iD_{\mu\nu}(x-y)\partial_y^\mu j_{(b)}^\nu(y) \,. \quad (10.41)$$

However, defining $D_{\mu\nu}(x-y) = \eta_{\mu\nu}D(x-y)$ we have

$$\partial^\mu A_\mu^{(b)}(x) = -ie_b \int d^4y\, iD(x-y)\partial^\mu j_\mu^{(b)}(y)$$

$$= -e_b \int d^4y\, iD(x-y)\left(\bar{g}_b(y)i\overrightarrow{\nabla}f_b(y) + \bar{g}_b(y)i\overleftarrow{\nabla}f_b(y)\right) \quad (10.42)$$

$$= -e_b \int d^4y\, iD(x-y)\left(m\bar{g}_b(y)f_b(y) - m\bar{g}_b(y)f_b(y)\right)$$

$$= 0.]$$

We now return to the transition amplitude, Eq. 10.33, which becomes

$$\text{Amp}^{(1)} = (-ie_a)(-ie_b) \int d^4x\, d^4y\, j_{(a)}^\mu(x)iD_{\mu\nu}(x-y)j_{(b)}^\nu(y)$$

$$= (-i)^2 e_a e_b \int d^4x\, d^4y\, \bar{g}_a(x)\gamma^\mu f_a(x)iD_{\mu\nu}(x-y)\bar{g}_b(y)\gamma^\nu f_b(y)\,. \quad (10.43)$$

It is reassuring to observe the symmetry of this amplitude in particles a, b in spite of the apparent asymmetry in its derivation. The interpretation of Eq. 10.43 is that the interaction of two charged particles is due to the exchange of virtual photons between them, as illustrated in Figure VII.11

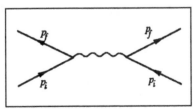

Fig. VII.11: *One-photon exchange interaction between electrons.*

For the case at hand, we are treating the electromagnetic interaction in lowest order so only a single photon is involved, and the two particle transition amplitude is given by

$$\text{Amp}^{(1)} = (-ie_a)(-ie_b)\frac{m}{\sqrt{\epsilon_i\epsilon_f}}\frac{M}{\sqrt{E_iE_f}}\int d^4x \int d^4y\, e^{i(p_f-p_i)\cdot x}\bar{u}_f(p_f)\gamma^\mu u_i(p_i)$$

$$\times \int \frac{d^4q}{(2\pi)^4}e^{-iq\cdot(x-y)}\frac{-i\eta_{\mu\nu}}{q^2+i\epsilon}e^{i(P_f-P_i)\cdot y}\bar{u}_f(P_f)\gamma^\nu u_i(P_i)\,. \quad (10.44)$$

The integrations over x and y can be performed first, yielding delta functions $(2\pi)^4\delta^4(p_f - p_i - q)$ and $(2\pi)^4\delta^4(P_f - P_i + q)$, respectively. The presence of these delta functions allows the q integration to take place, yielding

$$\text{Amp}^{(1)} = (-ie_a)(-ie_b)\frac{m}{\sqrt{\epsilon_i \epsilon_f}}\frac{M}{\sqrt{E_i E_f}}(2\pi)^4\delta^4(p_i + P_i - p_f - P_f)M_{fi} \quad (10.45)$$

where

$$M_{fi} = \bar{u}_f(p_f)\gamma^\mu u_i(p_i)\frac{-i\eta_{\mu\nu}}{(p_i - p_f)^2 + i\epsilon}\bar{u}_f(P_f)\gamma^\nu u_i(P_i) \quad (10.46)$$

is a Lorentz invariant quantity.

We calculate the cross section in the usual way

$$d\sigma = \frac{\frac{1}{T}\sum_f P_{fi}}{\text{Incident Flux}} \quad (10.47)$$

where

$$\text{Incident Flux} = \left|\vec{v}_i - \vec{V}_i\right| \quad (10.48)$$

is the relative velocity and

$$\sum_f P_{fi} = \int \left|\text{Amp}^{(1)}\right|^2 \frac{d^3 p_f}{(2\pi)^3}\frac{d^3 P_f}{(2\pi)^3}$$

$$= \int \frac{d^3 p_f d^3 P_f}{(2\pi)^6}\left((2\pi)^4\delta^4(p_i + P_i - p_f - P_f)\right)^2 \frac{m^2 M^2}{\epsilon_i \epsilon_f E_i E_f}e_a^2 e_b^2 |M_{fi}|^2 \, . \quad (10.49)$$

We deal with the square of the delta function via

$$((2\pi)^4\delta^4(p_i + P_i - p_f - P_f))^2 = (2\pi)^4\delta^4(p_i + P_i - p_f - P_f)$$

$$\times \int_{-T/2}^{T/2} dt \, e^{it(E_i + \epsilon_i - E_f - \epsilon_f)} \int_V d^3 r \, e^{-i\vec{r}\cdot(\vec{p}_i + \vec{P}_i - \vec{p}_f - \vec{P}_f)}$$

$$= (2\pi)^4\delta^4(p_i + P_i - p_f - P_f)\int_{-T/2}^{T/2} dt \int_V d^3 r = T(2\pi)^4\delta^4(p_i + P_i - p_f - P_f) \quad (10.50)$$

where we have assumed unit volume

$$\int_V d^3 r = \text{Volume} = 1 \quad (10.51)$$

and T is the total interaction time. We find then

$$d\sigma = \frac{1}{\left|\vec{v}_i - \vec{V}_i\right|}e_a^2 e_b^2 \frac{mM}{\epsilon_i E_i}\int \frac{d^3 p_f d^3 P_f}{(2\pi)^6}(2\pi)^4\delta^4(p_i + P_i - p_f - P_f)\frac{mM}{\epsilon_f E_f}|M_{fi}|^2 \, . \quad (10.52)$$

At this stage the problem is formally solved and we could content ourselves with the above expression. However, it is of interest to evaluate things explicitly in order to see when this result reduces (and when it does not) to the previously evaluated potential scattering cross section. Assuming that both beam and target are unpolarized and that final spin is not detected, we find, using trace techniques

$$\frac{1}{4}\sum_{\substack{s_i\\s_f}}\sum_{\substack{S_i\\S_f}}|M_{fi}|^2 = \frac{1}{4(p_i-p_f)^4}\frac{1}{(2m)^2}\frac{1}{(2M)^2}\mathrm{Tr}\,(\slashed{p}_i+m)\gamma_\mu(\slashed{p}_f+m)\gamma_\nu$$

$$\times \mathrm{Tr}\,(\slashed{P}_i+M)\gamma^\mu(\slashed{P}_f+M)\gamma^\nu$$

$$=\frac{1}{4(p_i-p_f)^4\,16m^2M^2}4\left(p_{i\mu}p_{f\nu}+p_{i\nu}p_{f\mu}-\eta_{\mu\nu}(p_i\cdot p_f-m^2)\right)$$

$$\times 4\left(P_i^\mu P_f^\nu+P_f^\mu P_i^\nu-\eta^{\mu\nu}(P_i\cdot P_f-M^2)\right)$$

$$=\frac{1}{4(p_i-p_f)^4 m^2 M^2}2\Big[P_f\cdot p_f\,P_i\cdot p_i+P_f\cdot p_i\,P_i\cdot p_f+P_f\cdot P_i(m^2-p_i\cdot p_f)$$

$$+p_f\cdot p_i(M^2-P_i\cdot P_f)+2(m^2-p_i\cdot p_f)(M^2-P_i\cdot P_f)\Big]$$

$$=\frac{1}{2(p_i-p_f)^4 m^2 M^2}\Big[P_f\cdot p_f\,P_i\cdot p_i+P_f\cdot p_i\,P_i\cdot p_f$$

$$-M^2 p_f\cdot p_i-m^2 P_f\cdot P_i+2m^2M^2\Big]$$

(10.53)

and our general result becomes

$$d\sigma = \frac{e_a^2 e_b^2}{2\epsilon_i E_i|\vec{v}_i-\vec{V}_i|}\int\frac{d^3p_f\,d^3P_f}{(2\pi)^3\epsilon_f(2\pi)^3 E_f}(2\pi)^4\delta^4(p_i+P_i-p_f-P_f)$$

$$\times\frac{1}{(p_i-p_f)^4}\left[P_f\cdot p_f\,P_i\cdot p_i+P_f\cdot p_i\,P_i\cdot p_f-M^2 p_f\cdot p_i-m^2 P_f\cdot P_i+2m^2M^2\right].$$

(10.54)

We shall evaluate the cross section in the laboratory frame, wherein the target proton is at rest and in the limit where the incident electron energy ϵ_i is much less than the proton rest mass M—$\epsilon_i/M \ll 1$. We find then

$$d\sigma = \frac{(4\pi\alpha)^2}{2M\epsilon_i v_i}\frac{1}{(2\pi)^2}\int\frac{p_f^2\,dp_f\,d\Omega_f}{\epsilon_f M}\delta(\epsilon_i+M-\epsilon_f-M)$$

$$\times\frac{M^2}{(\vec{p}_i-\vec{p}_f)^4}\left(\epsilon_f\epsilon_i+\vec{p}_f\cdot\vec{p}_i+m^2\right)$$

$$= d\Omega_f\,2\alpha^2\frac{(\epsilon_i^2+p_i^2\cos\theta+m^2)}{(\vec{p}_f-\vec{p}_i)^4}$$

$$= d\Omega_f\,\frac{\alpha^2\epsilon_i^2}{4p_i^4\sin^4\frac{\theta}{2}}\left(1-\beta^2\sin^2\frac{\theta}{2}\right)$$

(10.55)

which agrees precisely with the potential scattering result Eq. 10.14.

On the other hand, still in the laboratory frame but in the ultrarelativistic limit — $\epsilon_i \gg m$ — we have

$$d\sigma \cong \frac{e_a^2 e_b^2}{2M\epsilon_i} \frac{1}{(2\pi)^2} \int p_f^2 dp_f \, d\Omega_f \frac{1}{\epsilon_f E_f} \delta\left(\epsilon_i + M - \epsilon_f - \sqrt{M^2 + (\vec{p}_i - \vec{p}_f)^2}\right)$$

$$\times \frac{1}{(p_i - p_f)^4} \left[P_i \cdot p_f(P_i + p_i - p_f) \cdot p_i + P_i \cdot p_i(P_i + p_i - p_f) \cdot p_f - M^2 p_f \cdot p_i\right]$$

$$\simeq d\Omega_f \frac{(2\alpha)^2}{2M\epsilon_i} \frac{\epsilon_f}{E_f} \frac{1}{1 + \frac{p_f - p_i \cos\theta}{E_f}} \frac{1}{(p_i - p_f)^4}$$

$$\times \left[2M^2 \epsilon_f \epsilon_i + \epsilon_f \epsilon_i (1 - \cos\theta)(\epsilon_i - \epsilon_f - M)M\right] \quad .$$
(10.56)

We can simplify this expression by noting that

$$E_f + p_f - p_i \cos\theta \approx M + \epsilon_i(1 - \cos\theta) = M + 2\epsilon_i \sin^2 \frac{\theta}{2}$$
(10.57)

$$q^2 = (p_f - p_i)^2 = 2m^2 - 2\epsilon_f \epsilon_i (1 - \cos\theta) \approx -4\epsilon_f \epsilon_i \sin^2 \frac{\theta}{2} \quad .$$

Also

$$P_f^2 = M^2 = (p_i + P_i - p_f)^2 = q^2 + 2P_i \cdot (p_i - p_f) + M^2$$
so that $\quad q^2 = -2M(\epsilon_i - \epsilon_f) \quad .$
(10.58)

Then we can write

$$\left[2M^2 \epsilon_f \epsilon_i + \epsilon_f \epsilon_i (1 - \cos\theta)(\epsilon_i - \epsilon_f - M)M\right]$$

$$= \epsilon_f \epsilon_i \left[2M^2 + 2\sin^2 \frac{\theta}{2}\left(-M^2 - \frac{1}{2}q^2\right)\right]$$
(10.59)

$$= 2M^2 \epsilon_f \epsilon_i \left(\cos^2 \frac{\theta}{2} - \frac{q^2}{2M^2} \sin^2 \frac{\theta}{2}\right)$$

which yields

$$\frac{d\sigma}{d\Omega_f} = \frac{2\alpha^2}{M} \frac{\epsilon_f}{\epsilon_i} \frac{1}{M + 2\epsilon_i \sin^2 \frac{\theta}{2}} \frac{2\epsilon_f \epsilon_i M^2 \left(\cos^2 \frac{\theta}{2} - \frac{q^2}{2M^2} \sin^2 \frac{\theta}{2}\right)}{16\epsilon_f^2 \epsilon_i^2 \sin^4 \frac{\theta}{2}}$$

$$= \frac{\alpha^2}{4\epsilon_i^2} \frac{\cos^2 \frac{\theta}{2} - \frac{q^2}{2M^2} \sin^2 \frac{\theta}{2}}{\sin^4 \frac{\theta}{2} \left(1 + \frac{2\epsilon_i}{M} \sin^2 \frac{\theta}{2}\right)} \quad .$$
(10.60)

Now look at the cross section in the limit $\epsilon_i \ll M$, but of course $\epsilon_i \gg m$, in which case

$$\frac{d\sigma}{d\Omega} = \frac{\alpha^2}{4\epsilon_i^2} \frac{\left(1 - \sin^2 \frac{\theta}{2}\right)}{\sin^4 \frac{\theta}{2}}$$
(10.61)

which agrees with the ultrarelativistic limit of the relativistic potential scattering cross section Eq. 10.14

$$\frac{d\sigma}{d\Omega} = \frac{\alpha^2}{4\epsilon_i^2} \frac{1 - \beta^2 \sin^2 \frac{\theta}{2}}{\beta^4 \sin^4 \frac{\theta}{2}} \xrightarrow[\beta \to 1]{} \frac{\alpha^2}{4\epsilon_i^2} \frac{1 - \sin^2 \frac{\theta}{2}}{\sin^4 \frac{\theta}{2}} \quad .$$
(10.62)

However, Eq. 10.60 is correct not only in the limit $\epsilon_i \ll M$, where the proton recoil is negligible and the potential scattering cross section emerges, but also in the ultrarelativsitic limit $\epsilon_i \gg M$, where proton recoil is very important.

Low Energy Limit

It is interesting that in the "exact" calculation of $e - p$ scattering described above we obtain the potential scattering result for energies $\epsilon_i \ll M$. It must be then that the exact one-photon exchange term

$$\text{Amp} = (2\pi)\delta^4(p_i + P_i - p_f - P_f)4\pi\alpha \frac{mM}{\sqrt{\epsilon_i\epsilon_f E_i E_f}}$$

$$\times \bar{u}_f(p_f)\gamma_\mu u_i(p_i)\frac{-i}{q^2 + i\epsilon}\bar{u}_f(P_f)\gamma^\mu u_i(P_i) \quad \text{with} \quad q = p_i - p_f = P_f - P_i \tag{10.63}$$

contains somehow the Coulomb interaction at low energies plus "something else" at higher energies. Let's see if we can understand the physics content of the low energy photon exchange interaction:

By current conservation we have

$$q^\mu \bar{u}_f(p_f)\gamma_\mu u_i(p_i) = \bar{u}_f(p_f)(\slashed{p}_i - \slashed{p}_f)u_i(p_i) = 0 \tag{10.64}$$

and

$$q^\mu \bar{u}_f(P_f)\gamma_\mu u_i(P_i) = \bar{u}_f(P_f)(\slashed{P}_f - \slashed{P}_i)u_i(P_i) = 0 \ . \tag{10.65}$$

Then if we define

$$\slashed{q} = q_0 \gamma^0 - |\vec{q}|\gamma^{\hat{q}} \tag{10.66}$$

we can write

$$\bar{u}_f(p_f)\gamma^{\hat{q}} u_i(p_i) = \frac{q_0}{|\vec{q}|}\bar{u}_f(p_f)\gamma^0 u_i(p_i) \tag{10.67}$$

and hence

$$4\pi\alpha \bar{u}_f(p_f)\gamma_\mu u_i(p_i)\frac{1}{q^2}\bar{u}_f(P_f)\gamma^\mu u_i(P_i)$$

$$= \frac{4\pi\alpha}{q^2}\Big[\bar{u}_f(p_f)\gamma^0 u_i(p_i)\bar{u}_f(P_f)\gamma^0 u_i(P_i) - \bar{u}_f(p_f)\gamma^{\hat{q}} u_i(p_i)\bar{u}_f(P_f)\gamma^{\hat{q}} u_i(P_i)$$

$$- \sum_{\substack{\text{2 transverse} \\ \text{directions}}} \bar{u}_f(p_f)\gamma^{\text{tr}} u_i(p_i)\bar{u}_f(P_f)\gamma^{\text{tr}} u_i(P_i)\Big]$$

$$= \frac{4\pi\alpha}{q^2}\Big[\Big(1 - \frac{q_0^2}{\vec{q}^2}\Big)\bar{u}_f(p_f)\gamma^0 u_i(p_i)\bar{u}_f(P_f)\gamma^0 u_i(P_i)$$

$$- \sum_{\substack{\text{2 transverse} \\ \text{directions}}} \bar{u}_f(p_f)\gamma^{\text{tr}} u_i(p_i)\bar{u}_f(P_f)\gamma^{\text{tr}} u_i(P_i)\Big]$$

$$= -\frac{4\pi\alpha}{|\vec{q}|^2}\bar{u}_f(p_f)\gamma^0 u_i(p_i)\bar{u}_f(P_f)\gamma^0 u_i(P_i)$$

$$- \frac{4\pi\alpha}{q^2}\sum_{\substack{\text{2 transverse} \\ \text{directions}}} \bar{u}_f(p_f)\gamma^{\text{tr}} u_i(p_i)\bar{u}_f(P_f)\gamma^{\text{tr}} u_i(P_i) \ . \tag{10.68}$$

If $\epsilon_i \ll M$ we can write (here ˜ denotes that we are dealing with the proton)
$$\bar{u}_f(P_f)\gamma^0 u_i(P_i) \approx \tilde{\chi}_f^\dagger \tilde{\chi}_i = \delta_{S_f,S_i} \ . \tag{10.69}$$

since the proton will be nearly at rest. Then the first term of Eq. 10.68 becomes

$$-\frac{4\pi\alpha}{|\vec{q}|^2}\bar{u}_f(p_f)\gamma^0 u_i(p_i) \tag{10.70}$$

in the elastic scattering ($\tilde{\chi}_f = \tilde{\chi}_i$) limit. This is simply the Coulomb potential scattering amplitude. We can make the connection with the coordinate space picture by noting that since the lowest order transition amplitude[†]

$$T_{fi}(\vec{q}) = \left\langle f|\hat{V}|i\right\rangle = \int d^3r\, e^{-i\vec{p}_f\cdot\vec{r}} V(\vec{r})\, e^{i\vec{p}_i\cdot\vec{r}}$$
$$= \int d^3r\, e^{i\vec{q}\cdot\vec{r}} V(\vec{r}) \tag{10.71}$$

is the Fourier transform of the potential, we can determine the potential by taking the inverse Fourier transform of the transition amplitude

$$V(\vec{r}) = \int \frac{d^3q}{(2\pi)^3} e^{-i\vec{q}\cdot\vec{r}} T_{fi}(\vec{q}) \ . \tag{10.72}$$

Using the non-relativistic reduction

$$\frac{m}{\sqrt{\epsilon_f\epsilon_i}}\bar{u}_f(p_f)\gamma^0 u_i(p_i) \approx \chi_f^\dagger \left(1 + \frac{1}{4m^2}\vec{p}_f\cdot\vec{p}_i + \frac{i}{4m^2}\vec{\sigma}\cdot(\vec{p}_f\times\vec{p}_i)\right)\chi_i$$
$$\times \left(1 - \frac{p_i^2 + p_f^2}{8m^2} + \cdots\right), \tag{10.73}$$

the leading piece yields

$$V_0(r) = -\int \frac{d^3q}{(2\pi)^3} e^{-i\vec{q}\cdot\vec{r}} \frac{4\pi\alpha}{\vec{q}^2}$$
$$= -\alpha\frac{2}{\pi}\int_0^\infty dq\, j_0(qr)$$
$$= -\frac{\alpha}{r}\frac{2}{\pi}\int_0^\infty dx\, \frac{\sin x}{x} = -\frac{\alpha}{r} \tag{10.74}$$

[†] Here the transition amplitude is defined as usual via
$$S_{fi} = \delta_{fi} - i(2\pi)^4\delta^4(p_i + P_i - p_f - P_f)T_{fi}(\vec{q}) \quad \text{with} \quad \vec{q} = \vec{p}_i - \vec{p}_f \ .$$

which is the simple Coulomb interaction. However, there is more. For a spherically symmetric potential $V_0(r)$ one has

$$\begin{aligned}
\int d^3 r\, e^{-i\vec{p}_f \cdot \vec{r}} \frac{1}{r} \frac{dV_0}{dr} \vec{\sigma} \cdot \left(\vec{r} \times (-i\vec{\nabla}) \right) e^{i\vec{p}_i \cdot \vec{r}} \\
= -\vec{\sigma} \cdot \vec{p}_i \times \int d^3 r\, \vec{r}\, e^{i\vec{q}\cdot\vec{r}} \frac{1}{r} \frac{dV_0}{dr} \\
= i\vec{\sigma} \cdot \vec{p}_i \times \vec{q} \int d^3 r\, e^{i\vec{q}\cdot\vec{r}} V_0(r) \\
= -i\vec{\sigma} \cdot \vec{p}_i \times \vec{p}_f \int d^3 r\, e^{i\vec{q}\cdot\vec{r}} V_0(r)\ .
\end{aligned} \tag{10.75}$$

Thus the spin-dependent term in the photon exchange interaction becomes

$$\frac{1}{4m^2} \frac{1}{r} \frac{dV_0}{dr} \vec{\sigma} \cdot \vec{L} \tag{10.76}$$

which is the usual spin-orbit interaction.

Finally, writing the remaining portion of the amplitude as

$$T_{fi} = \ldots + \frac{4\pi\alpha}{\vec{q}^2} \cdot \frac{\vec{q}^2}{8m^2} \tag{10.77}$$

we find a contribution to the potential of the form

$$\frac{4\pi\alpha}{8m^2} \delta^3(r)\ . \tag{10.78}$$

The meaning of this term can be found by noting that

$$\vec{\nabla}^2 \frac{1}{r} = -4\pi \delta^3(r)\ . \tag{10.79}$$

Thus we can represent this piece of the interaction as

$$-e \frac{1}{8m^2} \vec{\nabla} \cdot \vec{E} \tag{10.80}$$

which we recognize as the "Darwin term."

Having identified each component above with a familiar piece of the effective non-relativistic Hamiltonian derived earlier,[†] we turn now to the transverse contribution of the photon-exchange interaction. Since in the non-relativistic limit, $\epsilon_i \ll m$,

---

[†] Of course, we have reproduced only the electromagnetic components of the effective potential. The additional kinetic terms arise from adding the contribution of the free Dirac Hamiltonian.

$$\begin{aligned}
H_{\text{free}} &= \bar{u}(p) \left(-i\vec{\alpha} \cdot \vec{\nabla} + \beta m \right) u(p) \\
&= p_0 \bar{u}(p) u(p) = p_0 \\
&\cong m + \frac{p^2}{2m} - \frac{p^4}{8m^3} + \ldots
\end{aligned}$$

VII THE DIRAC EQUATION

$$q_0 = \epsilon_i - \epsilon_f \approx m + \frac{p_i^2}{2m} - m - \frac{p_f^2}{2m} = \frac{p_i + p_f}{2m}(p_i - p_f) \quad (10.81)$$
$$\approx v(p_i - p_f) \sim v|\vec{q}| \ .$$

we have
$$q^2 = q_0^2 - \vec{q}^2 \approx -\vec{q}^2\left(1 + \mathcal{O}(\frac{v^2}{c^2})\right) \approx -\vec{q}^2 \ . \quad (10.82)$$

Also since
$$\bar{u}_f(p_f)\gamma^{\text{tr}} u_i(p_i) \approx \left(\chi_f^\dagger - \chi_f^\dagger \frac{\vec{\sigma}\cdot\vec{p}_f}{2m}\right)\begin{pmatrix} 0 & \sigma_{\text{tr}} \\ -\sigma_{\text{tr}} & 0 \end{pmatrix}\begin{pmatrix} \chi_i \\ \frac{\vec{\sigma}\cdot\vec{p}_i}{2m}\chi_i \end{pmatrix}$$
$$= \chi_f^\dagger \frac{\sigma_{\text{tr}}\vec{\sigma}\cdot\vec{p}_i + \vec{\sigma}\cdot\vec{p}_f \sigma_{\text{tr}}}{2m}\chi_i \quad (10.83)$$
$$= \chi_f^\dagger \left(\frac{(\vec{p}_i + \vec{p}_f)_{\text{tr}}}{2m} - i\frac{1}{2m}\vec{\sigma}\times(\vec{p}_i - \vec{p}_f)_{\text{tr}}\right)\chi_i$$

we see that the transverse component of the interaction becomes

$$\frac{4\pi\alpha}{\vec{q}^2}\sum_{\substack{2\,\text{transverse}\\ \text{directions}}} \chi_f^\dagger\left(\frac{\vec{p}_i+\vec{p}_f}{2m} - \frac{i}{2m}\vec{\sigma}\times(\vec{p}_i-\vec{p}_f)\right)_{\text{tr}}\chi_i$$
$$\cdot \tilde{\chi}_f^\dagger\left(\frac{\vec{P}_f+\vec{P}_i}{2M} - \frac{i}{2M}\vec{\sigma}\times(\vec{P}_i-\vec{P}_f)\right)_{\text{tr}}\tilde{\chi}_i \ . \quad (10.84)$$

Inasmuch as
$$\left(\frac{\vec{p}_i+\vec{p}_f}{2m}\right)_{\text{tr}} \quad (10.85)$$

represents the piece of the convective current density transverse to $\vec{q}$ while

$$-\frac{i}{2m}(\vec{\sigma}\times\vec{q})_{\text{tr}} \quad (10.86)$$

represents the component of the current density due to the magnetic dipole moment, we see that this transverse interaction correction to the Coulomb potential consists of three pieces:

i) a current-current interaction;
ii) a current-dipole interaction; and
iii) a dipole-dipole interaction.

Such contributions are expected from classical physics considerations and are often called the Breit-Fermi interaction. At least one of these terms is a very familiar one. We note that the dipole-dipole interaction component can be written as the potential

VII.10 ELECTROMAGNETIC INTERACTIONS

$$V_{\text{dipole-dipole}}(\vec{r}) = \int \frac{d^3q}{(2\pi)^3} \frac{4\pi\alpha}{\vec{q}^2} \left(\vec{\sigma}^{(e)} \times \vec{q}\right) \cdot \left(\vec{\sigma}^{(p)} \times \vec{q}\right) e^{-i\vec{q}\cdot\vec{r}} \frac{1}{2m \cdot 2M}$$

$$= -\vec{\sigma}^{(e)} \times \vec{\nabla} \cdot \vec{\sigma}^{(p)} \times \vec{\nabla} \frac{1}{2m \cdot 2M} 4\pi\alpha \int \frac{d^3q}{(2\pi)^3} e^{-i\vec{q}\cdot\vec{r}} \frac{1}{\vec{q}^2} \quad (10.87)$$

$$= -\frac{e}{2m}\vec{\sigma}^{(e)} \cdot \vec{\nabla} \times \left(\frac{e}{2M}\vec{\sigma}^{(p)} \times \vec{\nabla} \frac{1}{4\pi r}\right) \ .$$

We recognize

$$\vec{B}(\vec{r}) = \vec{\nabla} \times \left(\frac{e}{2M}\vec{\sigma}^{(p)} \times \vec{\nabla} \frac{1}{4\pi r}\right) \quad (10.88)$$

as the magnetic field produced by the proton dipole moment (assumed, since our proton is "ideal," to have $g_p = 2$). Thus the dipole-dipole interaction can be written as

$$U_{\text{dipole-dipole}} = -\vec{\mu}^{(e)} \cdot \vec{B} \quad (10.89)$$

and represents the interaction of the electron magnetic moment with the magnetic field produced by the proton moment. This is the hyperfine interaction. The term hyperfine is employed here because, while the splitting is the same order in α as found in the case of fine structure, there is an additional suppression in the amount m/M because of the smallness of the proton magneton. In the hydrogen atom the ground state is split by the hyperfine potential into a component with $F = 1$ and one with $F = 0$ where

$$\vec{F} = \vec{S}_e + \vec{S}_p \quad (10.90)$$

is the total atomic spin. The value of the resulting energy difference

$$\nu = \frac{\omega}{2\pi} = (1.420405751767 \pm 0.000000000001) \times 10^9 \text{ Hz} \quad (10.91)$$

is one of the most precisely measured constants in nature and is responsible for the well-known 21 cm radiation observed by radioastronomers.

A realistic interaction between proton and electron requires two additional modifications. The first is that experimentally the g-factor of the proton is not $g_p = 2$ as used above but is rather $g_p^{\text{exp}} = 5.58$, which should be utilized whenever the proton moment is required. This change in the g-factor is associated with the feature that the proton is *not* a point particle but rather has a complex structure in terms of quarks, mesons, *etc.*, as discussed in section VII.5.

The second modification is also associated with this non-pointlike nature. Describing the proton by a charge distribution $\rho(\vec{r})$, the Coulomb potential is not simply $\frac{-\alpha}{r}$ but rather is given by

$$V(\vec{r}) = \int d^3w \frac{-\alpha}{|\vec{r} - \vec{w}|} \rho(\vec{w}) \quad (10.92)$$

where we assume the normalization

$$\int d^3w \rho(\vec{w}) = 1 \ . \quad (10.93)$$

Because of this finite-size effect the zeroeth order e-p scattering amplitude reads

$$T_{fi}(\vec{q}) = \int d^3r\, e^{i\vec{q}\cdot\vec{r}} V(\vec{r}) = \int d^3r\, e^{i\vec{q}\cdot\vec{r}} \int d^3w\, \frac{-\alpha}{|\vec{r}-\vec{w}|} \rho(\vec{w}) \ . \qquad (10.94)$$

Changing variables to $\vec{w}$ and $\vec{s} = \vec{r} - \vec{w}$ Eq. 10.94 becomes

$$T_{fi} = \int d^3w\, e^{i\vec{q}\cdot\vec{w}} \rho(\vec{w}) \int d^3s\, \frac{-\alpha}{s} e^{i\vec{q}\cdot\vec{s}}$$

$$= F(\vec{q}) \times \frac{-4\pi\alpha}{\vec{q}^2} = F(\vec{q}) \times T_{fi}^{\text{point-charge}}(\vec{q}) \ . \qquad (10.95)$$

Thus the point-charge scattering amplitude is modified by the function $F(\vec{q})$, where $F(\vec{q}) = \int d^3r\, e^{i\vec{q}\cdot\vec{r}} \rho(\vec{r})$ is the "form factor" of the proton (*cf.* Problem II.3.1). If $\rho(\vec{r})$ is spherically symmetric

$$\rho(\vec{r}) = \rho(r) \qquad (10.96)$$

then we can expand

$$\begin{aligned} F(\vec{q}) &= \int d^3r\, e^{i\vec{q}\cdot\vec{r}} \rho(r) \\ &= \int d^3r \left(1 + i\vec{q}\cdot\vec{r} - \frac{1}{2}(\vec{q}\cdot\vec{r})^2 + \ldots\right) \rho(r) \qquad (10.97) \\ &= 1 - \frac{1}{6}\vec{q}^2 \langle r^2 \rangle + \ldots \end{aligned}$$

and the resulting cross section can be written as

$$\frac{d\sigma}{d\Omega} = \left(\frac{d\sigma}{d\Omega}\right)^{\text{point}} \times |F(\vec{q})|^2 = \left(\frac{d\sigma}{d\Omega}\right)^{\text{point}} \times \left(1 - \frac{1}{3}\vec{q}^2 \langle r^2 \rangle + \ldots\right) \ , \qquad (10.98)$$

which provides then a measure of the size of the object. Here

$$\vec{q}^2 = 4p_i^2 \sin^2 \frac{\theta}{2} \qquad (10.99)$$

and thus by measurement of the cross section at various angles and incident momenta one can evaluate the proton form factor $F(\vec{q})$. The associated charge density can then be found by taking the inverse Fourier transform

$$\rho(\vec{r}) = \int \frac{d^3q}{(2\pi)^3} e^{-i\vec{q}\cdot\vec{r}} F(\vec{q}) \ . \qquad (10.100)$$

We see then that the picture of the electromagnetic interaction via photon exchange includes automatically all expected features of the $e-p$ interaction — both relativistic and non-relativistic — and offers a concise and convenient representation of these effects.

Relativistic Compton Scattering

As a final example of the use of covariant perturbation theory we return to the problem of photon-electron scattering which was previously discussed using non-relativistic perturbation techniques. As mentioned above, there exist only two "pole" diagrams which contribute to this process in lowest order, as shown in Figure VII.12.

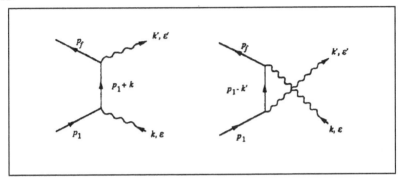

Fig. VII.12: Pole diagrams for electron-photon scattering.

and the corresponding transition amplitude is found to be

$$\text{Amp} = -(2\pi)^4 \delta^4(p_i + k - p_f - k') 4\pi\alpha \frac{m}{\sqrt{E_i E_f}} \frac{1}{\sqrt{2\omega\, 2\omega'}}$$
$$\times \bar{u}(p_f) \left(\not{\epsilon}' \frac{i}{\not{p}_i + \not{k} - m + i\epsilon} \not{\epsilon} + \not{\epsilon} \frac{i}{\not{p}_i - \not{k}' - m + i\epsilon} \not{\epsilon}' \right) u(p_i) \, . \tag{10.101}$$

(Note that the diagrams a and b are related by the substitution $\epsilon, k \leftrightarrow \epsilon', -k'$.) This form can be simplified considerably by working in the "laboratory" or target rest frame so that $p_i = (m, \vec{0})$ and by use of a gauge wherein the polarization vectors $\epsilon'_\mu, \epsilon_\mu$ are purely transverse

$$\epsilon'_\mu = (0, \hat{\epsilon}') \quad \text{with} \quad \vec{k}' \cdot \hat{\epsilon}' = 0$$
$$\epsilon_\mu = (0, \hat{\epsilon}) \quad \text{with} \quad \vec{k} \cdot \hat{\epsilon} = 0 \tag{10.102}$$

so that $p_i \cdot \epsilon = p_i \cdot \epsilon' = 0$. Since

$$(\not{p}_i + \not{k} + m)\not{\epsilon} u(p_i) = (2 p_i \cdot \epsilon - \not{\epsilon} \not{p}_i + \not{k} \not{\epsilon} + m \not{\epsilon}) u(p_i)$$
$$= \not{k} \not{\epsilon} u(p_i) = -\not{\epsilon} \not{k} u(p_i)$$
$$(\not{p}_i - \not{k}' + m)\not{\epsilon}' u(p_i) = (2 p_i \cdot \epsilon' - \not{\epsilon}' \not{p}_i - \not{k}' \not{\epsilon}' + m \not{\epsilon}') u(p_i) \tag{10.103}$$
$$= -\not{k}' \not{\epsilon}' u(p_i) = \not{\epsilon}' \not{k}' u(p_i)$$

we find that the transition amplitude simplifies to the form

$$\text{Amp} = i(2\pi)^4 \delta^4(p_i + k - p_i - k') 4\pi\alpha \sqrt{\frac{m}{E_f}} \frac{1}{\sqrt{2\omega\, 2\omega'}}$$
$$\times \bar{u}(p_f) \left(\frac{\not{\epsilon}' \not{\epsilon} \not{k}}{2 p_i \cdot k} + \frac{\not{\epsilon} \not{\epsilon}' \not{k}'}{2 p_i \cdot k'} \right) u(p_i) \tag{10.104}$$

Quantizing in a box of unit volume, the scattering cross section is given by

$$d\sigma = \frac{1}{T}\frac{1}{\text{Flux}}\int \frac{d^3 p_f}{(2\pi)^3}\frac{d^3 k'}{(2\pi)^3}\frac{1}{2}\sum_{s_i,s_f}|\text{Amp}|^2 \tag{10.105}$$

where we assume the electron target to be unpolarized and that the polarization of the final state electron is not detected. Since the incident beam consists of photons, with unit velocity, we have

$$\text{Flux} = v_i = 1 \;, \tag{10.106}$$

and treating the square of the delta function as in Eq. 10.50 we find

$$d\sigma = \frac{1}{2\omega}\int \frac{d^3 p_f}{(2\pi)^3}\frac{m}{E_f}\frac{d^3 k'}{(2\pi)^3}\frac{1}{2\omega'}(2\pi)^4 \delta^4(p_i + k - p_f - k')$$
$$\times (4\pi\alpha)^2 \frac{1}{2}\text{Tr}\frac{\slashed{p}_f + m}{2m}\left(\frac{\slashed{\epsilon}'\slashed{\epsilon}\slashed{k}}{2p_i\cdot k} + \frac{\slashed{\epsilon}\slashed{\epsilon}'\slashed{k}'}{2p_i\cdot k'}\right)\frac{\slashed{p}_i + m}{2m}\left(\frac{\slashed{k}\slashed{\epsilon}\slashed{\epsilon}'}{2p_i\cdot k} + \frac{\slashed{k}'\slashed{\epsilon}'\slashed{\epsilon}}{2p_i\cdot k'}\right) \tag{10.107}$$

where trace techniques have been employed in order to carry out the electron spin sums. Evaluation of the traces is straightforward but tedious:

$$K_1 \equiv \text{Tr}\frac{(\slashed{p}_f + m)}{2m}\frac{\slashed{\epsilon}'\slashed{\epsilon}\slashed{k}}{2p_i\cdot k}\frac{(\slashed{p}_i + m)}{2m}\frac{\slashed{k}\slashed{\epsilon}\slashed{\epsilon}'}{2p_i\cdot k}$$
$$= (\text{using } \slashed{k}\slashed{k} = k^2 = 0)\frac{1}{(2m)^2}\frac{1}{(2p_i\cdot k)^2}\text{Tr}\,\slashed{p}_f\slashed{\epsilon}'\slashed{\epsilon}\slashed{k}\slashed{p}_i\slashed{k}\slashed{\epsilon}\slashed{\epsilon}' \tag{10.108}$$
$$= (\text{using } \slashed{p}_i\slashed{k} = 2p_i\cdot k - \slashed{k}\slashed{p}_i)\frac{1}{(2m)^2}\frac{1}{2p_i\cdot k}\text{Tr}\,\slashed{p}_f\slashed{\epsilon}'\slashed{\epsilon}\slashed{k}\slashed{\epsilon}\slashed{\epsilon}'$$
$$= (\text{using } \slashed{\epsilon}\slashed{k} = -\slashed{k}\slashed{\epsilon})\frac{1}{(2m)^2}\frac{1}{2p_i\cdot k}\text{Tr}\,\slashed{p}_f\slashed{\epsilon}'\slashed{k}\slashed{\epsilon}'$$

Since $\epsilon'\cdot p_f = \epsilon'\cdot(p_i + k - k') = \epsilon'\cdot k$ and $p_f\cdot k = p_i\cdot k'$

$$\text{Tr}\,\slashed{p}_f\slashed{\epsilon}'\slashed{k}\slashed{\epsilon}' = 8p_f\cdot\epsilon' k\cdot\epsilon' + 4p_f\cdot k = 8(k\cdot\epsilon')^2 + 4p_i\cdot k' \tag{10.109}$$

and

$$K_1 = \frac{8(k\cdot\epsilon')^2 + 4p_i\cdot k'}{8m^2 p_i\cdot k} \tag{10.110}$$

Similarly since

$$K_2 \equiv \text{Tr}\frac{\slashed{p}_f + m}{2m}\frac{\slashed{\epsilon}\slashed{\epsilon}'\slashed{k}'}{2p_i\cdot k'}\frac{\slashed{p}_i + m}{2m}\frac{\slashed{k}'\slashed{\epsilon}'\slashed{\epsilon}}{2p_i\cdot k'} \tag{10.111}$$

is related to K_1 by the replacement $\epsilon \leftrightarrow \epsilon'$, $-k \leftrightarrow k'$ we have

$$K_2 = \frac{-8(k'\cdot\epsilon)^2 + 4p_i\cdot k}{8m^2 p_i\cdot k'} \;. \tag{10.112}$$

Also

$$K_3 \equiv \text{Tr} \frac{\not{p}_f + m}{2m} \frac{\not{\epsilon}'\not{\epsilon}\not{k}}{2p_i \cdot k} \frac{\not{p}_i + m}{2m} \frac{\not{k}'\not{\epsilon}'\not{\epsilon}}{2p_i \cdot k'}$$

$$= \frac{1}{16m^2 p_i \cdot k\, p_i \cdot k'} \text{Tr}\left[(\not{p}_i + \not{k} - \not{k}')\not{\epsilon}'\not{\epsilon}\not{k}\not{p}_i\not{k}'\not{\epsilon}'\not{\epsilon} + m^2 \not{\epsilon}'\not{\epsilon}\not{k}\not{k}'\not{\epsilon}'\not{\epsilon}\right]$$

$$= \frac{1}{16m^2 p_i \cdot k\, p_i \cdot k'}\left[2p_i \cdot k\, \text{Tr}\, \not{\epsilon}'\not{\epsilon}\not{p}_i\not{k}'\not{\epsilon}'\not{\epsilon} - 2k \cdot \epsilon'\, \text{Tr}\, \not{k}\not{p}_i\not{k}'\not{\epsilon}' + 2k' \cdot \epsilon\, \text{Tr}\, \not{\epsilon}\not{k}\not{p}_i\not{k}'\right]$$

$$= \frac{1}{16m^2 p_i \cdot k\, p_i \cdot k'}[-8p_i \cdot k'(\epsilon' \cdot k)^2 + 8p_i \cdot k(\epsilon \cdot k')^2$$
$$+ 16p_i \cdot k p_i \cdot k'(\epsilon \cdot \epsilon')^2 - 8p_i \cdot k p_i \cdot k']$$

$$= \frac{1}{2m^2}\left[2(\epsilon \cdot \epsilon')^2 - 1 + (\epsilon \cdot k')^2 \frac{1}{p_i \cdot k'} - (\epsilon' \cdot k)^2 \frac{1}{p_i \cdot k}\right]$$
(10.113)

so that the trace becomes

$$\text{Tr}\ldots = K_1 + K_2 + 2K_3 = \frac{1}{2m^2}\left[\frac{p_i \cdot k}{p_i \cdot k'} + \frac{p_i \cdot k'}{p_i \cdot k} + 4(\epsilon \cdot \epsilon')^2 - 2\right]. \quad (10.114)$$

Performing the phase space integration, we find

$$\frac{d\sigma}{d\Omega_{k'}} = \frac{\alpha^2}{2}\frac{k'}{k}\frac{m}{E_f}\frac{1}{1+\frac{k'-k\cos\theta}{E_f}} \cdot \text{Tr}\ldots$$
$$= \frac{\alpha^2 m}{2}\frac{k'}{k}\frac{1}{m+k(1-\cos\theta)} \cdot \text{Tr}\ldots \quad (10.115)$$

Finally since

$$p_f^2 = m^2 = (p_i + k - k')^2 = m^2 + 2p_i \cdot (k - k') + (k - k')^2$$
$$= m^2 + 2m(k - k') - 2kk'(1 - \cos\theta) \quad (10.116)$$

we find

$$k' = \frac{mk}{m + k(1 - \cos\theta)} \quad (10.117)$$

which is the familiar expression for the photon energy shift in Compton scattering. The cross section then becomes

$$\frac{d\sigma}{d\Omega_{k'}} = \frac{\alpha^2}{4m^2}\left[\frac{k}{k'} + \frac{k'}{k} + 4(\epsilon \cdot \epsilon')^2 - 2\right]\frac{k'^2}{k^2} \quad (10.118)$$

which is called the Klein–Nishina formula and expresses the cross section for photon electron scattering at arbitrary energy. We can make contact with the nonrelativistic result by working in the non-relativistic limit wherein $k \ll m$, $k' \approx k$. Then

$$\frac{d\sigma}{d\Omega_{k'}} \xrightarrow[k\ll m]{} \frac{\alpha}{m^2}(\epsilon \cdot \epsilon')^2 \quad (10.119)$$

which is the low-energy Thomson cross section derived from the seagull diagram.

PROBLEM VII.10.1

Relativistic Coulomb Scattering and the Dirac Equation

Above we calculated the cross section for the scattering of a spin-1/2 particle from a Coulomb potential

$$\frac{d\sigma}{d\Omega} = \frac{Z^2\alpha^2}{4E_i^2\beta^4 \sin^4\frac{\theta}{2}} (1 - \beta^2 \sin^2\frac{\theta}{2}).$$

In the high energy limit this becomes

$$\frac{d\sigma}{d\Omega} \approx \frac{Z^2\alpha^2 \cos^2\frac{\theta}{2}}{4E_i^2 \sin^4\frac{\theta}{2}}$$

which, as we shall demonstrate in the next chapter, differs from the corresponding spinless cross section by the factor $\cos^2\frac{\theta}{2}$. It is possible to understand the physics of this result by a simple argument.

i) Show that the Coulomb scattering amplitude can be written in the form

$$\text{Amp} = A_0(p_i - p_f) \times \frac{-ie}{(p_f - p_i)^2} \left(\bar{u}_L(p_f)\gamma_0 u_L(p_i) + \bar{u}_R(p_f)\gamma_0 u_R(p_i) \right)$$

i.e., in terms of a current matrix element which conserves chirality.

ii) Show that the amplitude for a spin 1/2 particle with spin aligned along the direction θ, ϕ to pass through an analyzing (*e.g.* Stern-Gerlach) device aligned along the +z direction is

$$\langle \theta, \phi | 0, 0 \rangle = \cos\frac{\theta}{2} e^{-i\frac{\phi}{2}}$$

iii) Using these results and Problem VII.5.2 explain why the high energy Coulomb cross section for a spin 1/2 particle must have the form given above.

PROBLEM VII.10.2

Electron Positron Annihilation

The process $e^+e^- \to q\bar{q}$, where q represents a generic spin 1/2 particle of mass M and charge $Q_q e$, is easily treated by means of the same techniques used to derive the ep scattering amplitude in Eq. 10.54.

i) Show that the annihilation cross section for electron-positron to $q\bar{q}$ is given by

$$d\sigma = \frac{m^2}{\epsilon_1\epsilon_2} \frac{e^4 Q_q^2}{|\vec{v}_1 - \vec{v}_2|} \int \frac{d^3P_1 d^3P_2}{(2\pi)^6} (2\pi)^4 \delta^4(p_1 + p_2 - P_1 - P_2) \frac{M^2}{E_1 E_2} \frac{1}{4} \sum_{\text{spins}} |M_{\text{fi}}|^2$$

where

$$\frac{1}{4}\sum_{\text{spins}}|M_{\text{fi}}|^2 = \frac{1}{2(p_1+p_2)^4 m^2 M^2}[P_2 \cdot p_2 P_1 \cdot p_1 + P_2 \cdot p_1 P_1 \cdot p_2 + M^2 p_1 \cdot p_2 + m^2 P_1 \cdot P_2 + 2m^2 M^2]$$

is obtained from Eq. 10.53 by replacing

$$p_i \to p_1 \quad p_f \to -p_2 \quad P_i \to -P_1 \quad P_f \to P_2.$$

ii) Evaluate the cross section in the center of mass frame in the high energy limit—$E_{\text{CM}} \gg m, M$—and show that it reduces to the simple result

$$\frac{d\sigma}{d\Omega} = \frac{\alpha^2 Q_q^2}{4 E_{\text{CM}}^2}(1+\cos^2\theta) \ .$$

This form has been used to analyze the reaction wherein electron and positron annhilate to a quark-antiquark pair. The fact that the experimental cross section for the high energy $e^+e^- \to q\bar{q}$ process behaves as $1+\cos^2\theta$ gives support to the fact that quarks are spin 1/2 objects.

PROBLEM VII.10.3

The Rosenbluth Scattering Cross Section

In Eq. 10.60 we derived the laboratory frame cross section for scattering of a high energy—$\epsilon \gg m$—electron from an "ideal" (structureless) proton. However, the proton matrix element of the electromagnetic current was assumed to have the naive form given in Eq. 10.32. In reality one should use the current matrix element

$$\langle P_f | J_\mu^{\text{em}} | P_i \rangle = \bar{u}_f(P_f)\left[\gamma_\mu F_1(q^2) - i\frac{\kappa}{2M}\sigma_{\mu\nu}q^\nu F_2(q^2)\right]u_i(P_i)$$

where $q = P_i - P_f$ is the momentum transfer, $\kappa = 1.79$ is the anomalous magnetic moment of the proton, and $F_i(q^2)$ are form factors which account for the finite hadronic size.

i) Calculate the high energy laboratory cross section for electron-proton scattering using the full current matrix element and show that

$$\frac{d\sigma}{d\Omega} = \frac{\alpha^2 \cos^2\frac{\theta}{2}}{4\epsilon_i^2 \sin^4\frac{\theta}{2}} \frac{1}{[1+\frac{2\epsilon_i \sin^2\frac{\theta}{2}}{M}]}$$

$$\times \left\{|F_1(q^2)|^2 - \frac{q^2}{4M^2}\left[2|F_1(q^2) + \kappa F_2(q^2)|^2 \tan^2\frac{\theta}{2} + \kappa^2|F_2(q^2)|^2\right]\right\}$$

where

$$q^2 = -\frac{4\epsilon_i^2 \sin^2\frac{\theta}{2}}{1+\frac{2\epsilon_i \sin^2\frac{\theta}{2}}{M}} \ .$$

ii) Verify that this expression reduces to that given in Eq. 10.32 in the limit that $\kappa \to 0$ and $F_1(q^2) \to 1$.

This is the Rosenbluth cross section which has been used in order to map out nuclear and particle charge distributions via electron scattering.

PROBLEM VII.10.4

21 Centimeter Radiation

i) Use the hyperfine piece of the $e - p$ potential, Eq. 10.87, to calculate the splitting between the $F = |\vec{S}_1 + \vec{S}_2|$ singlet and triplet levels of the hydrogen atom ground state. Use the potential

$$\phi(\vec{r}) = \int d^3s \frac{\rho(\vec{s})}{4\pi |\vec{r} - \vec{s}|}$$

with $\rho(\vec{s}) = e\delta^3(\vec{s})$ and show that

$$\Delta E = \frac{4m_e^2 \alpha^4}{3m_N} g_p$$

where $g_p = 5.58$ is the proton g-factor.

ii) Calculate the size of this splitting and compare with the experimental value given in Eq. 10.91.

iii) Why is your answer different? Include the effects of the anomalous electron magnetic moment and of the reduced mass and see if this makes the answer closer.

CHAPTER VIII

ADVANCED TOPICS

VIII.1 RADIATIVE CORRECTIONS

Thus far our discussion of relativistic perturbation theory has involved only "tree" diagrams wherein internal momenta are completely determined by energy-momentum conservation at the vertices so that no momentum integration is required. In this fashion, for example, we obtained a remarkably detailed picture of the electron-proton interaction in terms of a simple one-photon exchange diagram. However, this piece of the $e-p$ interaction is $\mathcal{O}(\alpha)$. Higher order contributions needed for precision tests of quantum electrodynamics, require inclusion of loop diagrams. We shall show below how these are evaluated in the case of radiative corrections to the basic electron photon interaction.

Self Energy and the Propagator: Renormalization

In order to begin this calculation it is useful to examine the radiative correction to the electron propagator which, to lowest order in e, is given by the simple expression

$$iS_F^{(0)}(p) = \frac{i}{\not{p} - m + i\epsilon} \tag{1.1}$$

and is represented by a straight line in a Feynman diagram. On the other hand, due to the effects of the electromagnetic interactions, this simple form must be modified. To first order in e^2 we have, as shown graphically in Figure VIII.1

$$iS_F(p) = iS_F^{(0)}(p) + iS_F^{(0)}(p) \times -i\Sigma(p) \times iS_F^{(0)}(p) + \ldots \tag{1.2}$$

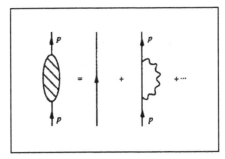

Fig. VIII.1: *The radiatively corrected Dirac propagator.*

where we have defined

$$\begin{aligned} -i\Sigma(p) &= (-ie)^2 \int \frac{d^4k}{(2\pi)^4} \frac{-i\eta^{\mu\nu}}{k^2 + i\epsilon} \gamma_\mu \frac{i}{\not{p} - \not{k} - m + i\epsilon} \gamma_\nu \\ &= (-ie)^2 \int \frac{d^4k}{(2\pi)^4} \frac{\eta^{\mu\nu}}{k^2 + i\epsilon} \frac{\gamma_\mu(\not{p} - \not{k} + m)\gamma_\nu}{k^2 - 2p \cdot k + p^2 - m^2 + i\epsilon} \end{aligned} \tag{1.3}$$

to represent the "bubble" diagram wherein the electron emits and reabsorbs a virtual photon. As we found in the corresponding non-relativistic analysis, multiple bubble insertions

$$iS_F(p) = iS_F^{(0)}(p) + iS_F^{(0)}(p) \times -i\Sigma(p) \times iS_F^{(0)}(p) \\ + iS_F^{(0)}(p) \times -i\Sigma(p) \times iS_F^{(0)}(p) \times -i\Sigma(p) \times iS_F^{(0)}(p) + \ldots \quad (1.4)$$

can be summed using the identity

$$\frac{1}{\hat{A} - \hat{B}} = \frac{1}{\hat{A}} + \frac{1}{\hat{A}}\hat{B}\frac{1}{\hat{A}} + \frac{1}{\hat{A}}\hat{B}\frac{1}{\hat{A}}\hat{B}\frac{1}{\hat{A}} + \ldots \quad (1.5)$$

to yield

$$iS_F(p) = \frac{i}{\not{p} - m - \Sigma(p)} + \ldots \quad (1.6)$$

where the ellipses represent the contribution from insertions of $\mathcal{O}(e^4)$ and higher.

From the fact that $\Sigma(p)$ must be a Lorentz as well as a Dirac scalar, it is clear that we must be able to represent

$$\Sigma(p) = A(p^2) + B(p^2)(\not{p} - m) \quad . \quad (1.7)$$

For electrons which are near the mass shell — $p^2 \approx m^2$ — we can write

$$\Sigma(p) = A(m^2) + \big(B(m^2) + 2mA'(m^2)\big)(\not{p} - m) + \ldots \quad (1.8)$$

where we have used

$$p^2 - m^2 = (\not{p} + m)(\not{p} - m) \approx 2m(\not{p} - m) \quad . \quad (1.9)$$

To order α then

$$iS_F(p) = \frac{i}{\not{p} - m - \Sigma(p)} \cong \frac{i}{\not{p} - m - A(m^2) - (\not{p} - m)\left(B(m^2) + 2mA'(m^2)\right)} \\ \approx \frac{i}{(1 - B(m^2) - 2mA'(m^2))(\not{p} - m - A(m^2))} \approx \frac{iZ_2}{\not{p} - m_{\text{phys}}} \quad (1.10)$$

where we have defined

$$m_{\text{phys}} = m + A(m^2) \equiv m + \delta m \\ Z_2 \cong 1 + B(m^2) + 2mA'(m^2) \equiv 1 + \tilde{B}(m^2) \quad . \quad (1.11)$$

Since the propagator has a pole at the physical mass of the particle, we see that $A(m^2)$ performs the role of a mass shift due to interaction with the electromagnetic field. The factor Z_2 is known as the wave function renormalization constant for the electron and represents the probability to find the "bare" electron state (*i.e.*, an electron unencumbered by electromagnetic effects) in the physical electron wavefunction. This effect is a familiar one from ordinary time independent perturbation theory, where we represent the normalized eigenstate of the full Hamiltonian H – $|\psi_n\rangle$ – in terms of "bare" eigenstates of the free Hamiltonian H_0 – $|\phi_n\rangle$ – as

$$|\psi_n\rangle = \sqrt{Z_n}|\phi_n\rangle + \sum_{m\neq n} \frac{\langle\phi_m|\hat{V}|\phi_n\rangle}{E_n^{(0)} - E_m^{(0)}}|\phi_m\rangle + \ldots \quad . \tag{1.12}$$

Obviously,

$$Z_n = 1 - \sum_{m\neq n} \frac{|\langle\phi_m|\hat{V}|\phi_n\rangle|^2}{(E_n^{(0)} - E_m^{(0)})^2} + \ldots \tag{1.13}$$

represents the "wave function renormalization" — the probability that the unperturbed eigenstate $|\phi_n\rangle$ is to be found in the corresponding full eigenstate $|\psi_n\rangle$ —

$$Z_n = |\langle\phi_n|\psi_n\rangle|^2 \quad . \tag{1.14}$$

So far, so good. However, we encounter problems when we attempt to actually calculate $A(p^2)$, $B(p^2)$ — they are divergent! This is easily seen from Eq. 1.3, which in the large k regime behaves as

$$\Sigma(p) \sim e^2 \int^\Lambda \frac{d^4k}{k^4} m \sim \alpha m \ln\left(\frac{\Lambda}{m}\right) \quad . \tag{1.15}$$

[Note: One might naively have expected a linear divergence. However,

$$\int \frac{d^4k}{k^4} k_\mu = 0 \tag{1.16}$$

because the integrand is an odd function of k.] Both the mass shift and wave function renormalization effects diverge logarithmically. This result should not be unexpected. Indeed the corresponding non-relativistic calculation discussed in Section IV.8 involves a *linear* divergence! The logarithmic relativistic form is much more manageable. Even if the cutoff Λ were as large as 1 TeV, the self-energy correction would represent only a small fraction — $\ll 10\%$ — of the electron mass. Still a divergence *is* a divergence and appears to be unsatisfactory.

In fact, as we shall see, such divergences do *not* enter into physically measurable predictions of the theory and cancel against one another when such predictions are expressed in terms of experimentally accessible quantities such as m_{phys} instead of mathematical artifacts such as m which appear in the zeroth order Lagrangian. Nevertheless, the use of some sort of cutoff or "regularization" procedure for divergent integrals is a useful intermediate step, since subtracting infinities from one another is a notoriously dangerous procedure. There exist two common regularization schemes in the literature. One, due to Pauli and Villars, introduces an ultraviolet (large momentum) cutoff Λ. The second is that of 't Hooft and involves integration in a space with other than four dimensions. We consider each in turn.

First we examine the cutoff (Pauli-Villars) procedure, which controls the ultraviolet behavior of QED by modifying the form of the photon propagator [Fe61]. Suppose we use

$$iD_F^{\mu\nu}(x,y) = \int \frac{d^4k}{(2\pi)^4} e^{-ik\cdot(x-y)} \frac{-i\eta^{\mu\nu}}{k^2 + i\epsilon} c(k^2) \tag{1.17}$$

where $c(k^2)$ is chosen so that $c(0) = 1$ and $c(k^2) \xrightarrow[k^2 \to \infty]{} 0$. Since $c(k^2)$ removes just the *high* frequency components, only the very short-distance behavior of the theory is modified and the length scale at which the change occurs can be chosen to be arbitrarily tiny. Still there exists a fundamental problem with this procedure in that the causal properties of the theory are altered. That is to say, if $c(k^2) = 1$ an electromagnetic signal sent from the origin can reach a point at distance r only at time t such that $t^2 - r^2 = 0$. However, if Λ^2 is a large number such that $c(k^2) \approx 0$ for $k^2 \gg \Lambda^2$ then in the modified theory signals obey

$$t^2 - r^2 \lesssim \frac{1}{\Lambda^2} \quad i.e., \quad t \lesssim \sqrt{r^2 + \frac{1}{\Lambda^2}} \tag{1.18}$$

which makes a negligible difference if Λ is sufficiently large, but nevertheless *does* have a nonzero effect. As a specific manifestation, suppose we pick

$$c(k^2) = \frac{-\Lambda^2}{k^2 - \Lambda^2 + i\epsilon}. \tag{1.19}$$

Then we can write

$$\frac{1}{k^2 + i\epsilon} c(k^2) = \frac{1}{k^2 + i\epsilon} \times \frac{-\Lambda^2}{k^2 - \Lambda^2 + i\epsilon} = \frac{1}{k^2 + i\epsilon} - \frac{1}{k^2 - \Lambda^2 + i\epsilon} \tag{1.20}$$

so that this modification is equivalent to introducing a neutral vector meson of mass Λ into the theory. The relative minus sign between the two propagators implies that the heavy meson is coupled with coupling constant $-e^2$ rather than $+e^2$ for the photon — i.e., the coupling constant is imaginary and the Hamiltonian is no longer Hermitian, leading to violation of unitarity. While this may seem a steep price to pay for the luxury of having regularized integrals, recall that this violation is only illusory. After all diagrams are added together the cutoff dependence disappears and at this point we can take the limit $\Lambda \to \infty$, restoring causality and unitarity to the theory.

In the alternative dimensional regularization procedure no such violations arise, and the photon propagator is always of the simple $(k^2 + i\eta)^{-1}$ form.[†] However, the calculation is performed in $d < 4$ dimensional spacetime and divergences of the form

$$\frac{1}{\epsilon} \quad \text{where} \quad \epsilon = 4 - d \tag{1.21}$$

[†] We temporarily use the symbol η to represent a positive infinitesimal, since the symbol ϵ is generally used to represent $4 - d$, as in Eq. 1.21.

occur instead of factors of log Λ which arise in the Pauli–Villars analysis. While the presence of such poles might seem strange, since certainly we live in a universe where d is identically four, there is again no real problem. These poles are merely a calculational artifact. When all diagrams are added together any ϵ^{-1} dependence disappears and the limit $\epsilon \to 0$ can be taken.

Now let's see how such regularization is actually carried out. We begin by simplifying the form of the self-energy $\Sigma(p)$, by use of the identify

$$\gamma_\mu \slashed{A} \gamma^\mu = \gamma_\mu \left(\{\slashed{A},\gamma^\mu\} - \gamma^\mu \slashed{A}\right) \\ = 2\gamma_\mu A^\mu - (\gamma_\mu \gamma^\mu)\, \slashed{A} \ . \tag{1.22}$$

Since $\gamma_\mu \gamma^\mu = d$, where d is the number of dimensions, we find

$$\gamma_\mu \slashed{A} \gamma^\mu = (2-d)\slashed{A} = (-2+\epsilon)\slashed{A} \ . \tag{1.23}$$

Then

$$\Sigma(p) = -ie^2 \mu^\epsilon \int \frac{d^d k}{(2\pi)^d} \frac{c(k^2)}{k^2 - \lambda^2 + i\eta} \frac{4m - 2\slashed{p} + 2\slashed{k} + \epsilon(\slashed{p} - \slashed{k} - m)}{(k-p)^2 - m^2 + i\eta} \tag{1.24}$$

where μ is an arbitrary constant having the dimension of mass, inserted in order that $\Sigma(p)$ retain the proper dimensionality if $d \neq 4$,

i) Pauli–Villars: $\quad \epsilon = 4 - d = 0 \ , \quad c(k^2) = -\dfrac{\Lambda^2}{k^2 - \Lambda^2 + i\eta}$

ii) Dimensional: $\quad c(k^2) = 1 \ , \quad \epsilon = 4 - d \neq 0$

and a small photon "mass" λ^2 has been included in order to regularize infrared ($k \to 0$) divergences which otherwise arise.

In Appendix VIII.1 we show how integrals of this form can be performed, using

$$\int \frac{d^d k}{(2\pi)^d} \frac{[1; k_\mu; k_\mu k_\nu]}{(k^2 - m_2^2)^\ell \left((k-p)^2 - m_1^2\right)^n} = (-1)^{n+\ell} \frac{i}{(4\pi)^{d/2}} \frac{\Gamma(\ell + n - \frac{d}{2})}{\Gamma(\ell)\Gamma(n)}$$

$$\times \int_0^1 dy\, y^{n-1}(1-y)^{\ell-1} \left[ym_1^2 + (1-y)m_2^2 - y(1-y)p^2\right]^{\frac{d}{2} - n - \ell}$$

$$\times \left[1; p_\mu y; p_\mu p_\nu y^2 - \frac{1}{2}\eta_{\mu\nu} \frac{1}{\ell + n - \frac{d}{2} - 1} \left(ym_1^2 + (1-y)m_2^2 - y(1-y)p^2\right)\right]. \tag{1.25}$$

For Pauli–Villars regularization we write

$$\frac{1}{k^2 - \lambda^2} \times \frac{-\Lambda^2}{k^2 - \Lambda^2} = \frac{1}{k^2 - \lambda^2} - \frac{1}{k^2 - \Lambda^2} = -\int_{\lambda^2}^{\Lambda^2} dL^2 \frac{1}{(k^2 - L^2)^2} \tag{1.26}$$

and find

$$\Sigma(p) = e^2 \frac{1}{(4\pi)^2} \frac{\Gamma(1)}{\Gamma(1)\Gamma(2)} \int_0^1 dy(1-y) \qquad (1.27)$$
$$\times \int_{\lambda^2}^{\Lambda^2} dL^2 \left(ym^2 + (1-y)L^2 - y(1-y)p^2\right)^{-1} (4m - 2\not{p}(1-y)) \ .$$

The L^2 integration is easily performed, yielding

$$\Sigma(p) = \frac{e^2}{(4\pi)^2} \int_0^1 dy \ln \frac{\Lambda^2(1-y)}{ym^2 - y(1-y)p^2 + (1-y)\lambda^2} (4m - 2\not{p}(1-y)) \qquad (1.28)$$

which, near the mass shell, becomes

$$\Sigma(p)\bigg|_{p^2 \approx m^2} = \frac{e^2}{(4\pi)^2} \int_0^1 dy \bigg[(4m - 2\not{p}(1-y)) \ln \frac{\Lambda^2(1-y)}{y^2 m^2 + (1-y)\lambda^2}$$
$$+ 2m(1+y) \frac{y(1-y)}{y^2 m^2 + (1-y)\lambda^2}(p^2 - m^2)\bigg] + \ldots \qquad (1.29)$$

Thus we read off (using $p^2 - m^2 = (\not{p} - m)(\not{p} + m) \simeq 2m(\not{p} - m)$)

$$A(m^2) = m\frac{\alpha}{4\pi} \int_0^1 dy\, 2(1+y) \ln \frac{\Lambda^2(1-y)}{m^2 y^2} = m\frac{\alpha}{4\pi}\left(3\ln \frac{\Lambda^2}{m^2} + \frac{3}{2}\right)$$

$$\tilde{B}(m^2) = -\frac{\alpha}{4\pi} \int_0^1 dy \left[2(1-y)\ln\frac{\Lambda^2(1-y)}{m^2 y^2} - 4m^2 \frac{y(1-y^2)}{y^2 m^2 + (1-y)\lambda^2}\right] \qquad (1.30)$$
$$= -\frac{\alpha}{4\pi}\left(\ln\frac{\Lambda^2}{m^2} - 2\ln\frac{m^2}{\lambda^2} + \frac{9}{2}\right)$$

and identify

$$\delta m = A(m^2) = m\frac{\alpha}{4\pi}\left(3\ln\frac{\Lambda^2}{m^2} + \frac{3}{2}\right) \qquad (1.31)$$

as the electromagnetic mass shift and

$$Z_2 = 1 + \tilde{B}(m^2)$$
$$= 1 - \frac{\alpha}{4\pi}\left(\ln\frac{\Lambda^2}{m^2} - 2\ln\frac{m^2}{\lambda^2} + \frac{9}{2}\right) \qquad (1.32)$$

as the wavefunction renormalization constant.

Similarly, we may evaluate the self-energy in the dimensional regularization scheme

$$\Sigma(p) = \frac{e^2}{(4\pi)^{d/2}} (\mu^2)^{2-\frac{d}{2}} \frac{\Gamma(2-\frac{d}{2})}{\Gamma(1)\Gamma(1)} \int_0^1 dy \left[ym^2 + (1-y)\lambda^2 - p^2 y(1-y)\right]^{\frac{d}{2}-2}$$
$$\times (4m - 2\not{p}(1-y) + \epsilon(\not{p}(1-y) - m)) \ . \qquad (1.33)$$

[Here, as mentioned above, μ is an arbitrary (but necessary) parameter having dimensions of mass inserted in order that the coupling e^2 remain dimensionless when we make the change from four to d dimensions. As with the Pauli–Villars cutoff Λ, physical results must be independent of μ, but *some* value of μ must be chosen in order to *define* the theory.]

Writing

$$\Gamma\left(2 - \frac{d}{2}\right) = \Gamma\left(\frac{\epsilon}{2}\right) = \frac{2}{\epsilon}\Gamma\left(1 + \frac{\epsilon}{2}\right) = \frac{2}{\epsilon} - \gamma + \mathcal{O}(\epsilon) \tag{1.34}$$

where $\gamma = -0.574216\ldots$ is Euler's constant, and using the identity

$$a^\epsilon = 1 + \epsilon \ln a + \mathcal{O}(\epsilon^2) \tag{1.35}$$

we determine

$$\Sigma(p) = \frac{e^2}{(4\pi)^2}\bigg[(3m - (\not{p} - m))\left(\frac{2}{\epsilon} - \gamma + \ln 4\pi + \ln \mu^2\right) + \not{p} - 2m$$
$$- \int_0^1 dy\, (2m(1+y) - 2(\not{p} - m)(1-y))\ln\left(m^2 y + (1-y)\lambda^2 - y(1-y)p^2\right)\bigg] \tag{1.36}$$

i.e.,

$$A(m^2) = m\frac{e^2}{(4\pi)^2}\left\{\frac{6}{\epsilon} - 1 - 3\gamma - 2\int_0^1 dy(1+y)\ln\frac{y^2 m^2}{4\pi\mu^2}\right\}$$
$$= m\frac{\alpha}{4\pi}\left\{\frac{6}{\epsilon} - 3\gamma - 3\ln\frac{m^2}{4\pi\mu^2} + 4\right\}$$

$$\tilde{B}(m^2) = \frac{e^2}{(4\pi)^2}\left\{-\frac{2}{\epsilon} + 1 + \gamma + 2\int_0^1 dy(1-y)\ln\frac{y^2 m^2}{4\pi\mu^2}\right. \tag{1.37}$$
$$\left. + 4m^2 \int_0^1 dy\,\frac{y(1-y^2)}{m^2 y^2 + (1-y)\lambda^2}\right\}$$
$$= -\frac{\alpha}{4\pi}\left\{\frac{2}{\epsilon} - \gamma - \ln\frac{m^2}{4\pi\mu^2} - 2\ln\frac{m^2}{\lambda^2} + 4\right\} \ .$$

We observe the connection

$$\ln\frac{\Lambda^2}{m^2} \longleftrightarrow \frac{2}{\epsilon} - \gamma - \ln\frac{m^2}{4\pi\mu^2} \tag{1.38}$$

between dimensional vs. Pauli–Villars regularized divergences.

Since both δm and Z_2 are divergent as $\epsilon \to 0$ or $\Lambda \to \infty$, neither can contribute to physical processes. The mass correction can be eliminated in the same fashion as in the non-relativistic case by rewriting the Dirac equation in terms of $m_{\text{phys}} = m + \delta m$

$$(i\not{\nabla} - m_{\text{phys}})\psi(x) = (e\not{A}(x) - \delta m)\psi(x) \ . \tag{1.39}$$

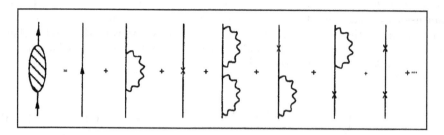

Fig. VIII.2: The radiatively corrected electron propagator, including effects of δm.

We observe that δm must be employed as a perturbation to the electron propagator to $\mathcal{O}(\alpha)$, as shown graphically in Figure VIII.2 where the cross indicates the perturbation introduced by δm.

Mathematically we have

$$iS_F(p) = iS_F^{(0)}(p) + iS_F^{(0)}(p) \times -i\Sigma(p) \times iS_F^{(0)}(p) + iS_F^{(0)}(p) \times i\delta m \times iS_F^{(0)}(p) + \ldots$$
$$= \frac{i}{\slashed{p} - m_{\text{phys}} - \Sigma(p) + \delta m} + \mathcal{O}(\alpha^2) \approx \frac{iZ_2}{\slashed{p} - m_{\text{phys}}}$$
(1.40)

so that the divergent mass shift correction has been eliminated. The wavefunction renormalization Z_2 remains, however, and in order to see how it disappears we must examine the interaction of the electron with an external photon.

To lowest order in e the electron photon vertex is given by

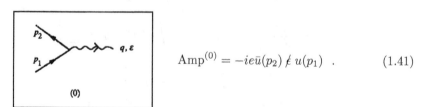

$$\text{Amp}^{(0)} = -ie\bar{u}(p_2) \slashed{\epsilon} u(p_1) \ .$$
(1.41)

However, in higher orders there exist modifications due to the diagrams shown in Figure VIII.3.

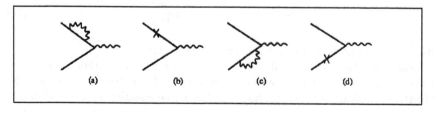

Fig. VIII.3: Radiative corrections to the electron-photon vertex.

Diagrams a through d have been discussed above, but involve a subtlety. Naively adding them to the lowest order result we seem to find

$$\text{Amp}(0 + a + b + c + d) \simeq -ie\bar{u}^{(0)}(p_2)$$
$$\times \left(\not{\epsilon} + (Z_2 - 1)(\not{p}_2 - m_{\text{phys}}) \frac{1}{\not{p}_2 - m_{\text{phys}}} \not{\epsilon} \right.$$
$$\left. + \not{\epsilon} \frac{1}{\not{p}_1 - m_{\text{phys}}} (\not{p}_1 - m_{\text{phys}})(Z_2 - 1) \right) u^{(0)}(p_1) \quad (1.42)$$
$$= -ie\bar{u}^{(0)}(p_2) \not{\epsilon} \, u^{(0)}(p_1) \, (1 + 2(Z_2 - 1)) \; .$$

However this is not quite correct—since the *external* electrons are already "dressed" by self energy effects, we should contract with *physical* spinors u, which are related to their "undressed" counterparts $u^{(0)}$ via Eq. 1.12

$$u(p) \sim \sqrt{Z_2} u^{(0)}(p) \; . \quad (1.43)$$

Then using

$$[1 + 2(Z_2 - 1)] \approx [1 + Z_2 - 1]^2 \approx Z_2[1 + Z_2 - 1] \quad (1.44)$$

we have

$$\text{Amp}(0 + a + b + c + d) \cong -ie\bar{u}^{(0)}(p_2) \not{\epsilon} \, u^{(0)}(p_1)[1 + 2(Z_2 - 1)]$$
$$\approx -ie Z_2 \bar{u}^{(0)}(p_2) \not{\epsilon} \, u^{(0)}(p_1) \, [1 + (Z_2 - 1)] \quad (1.45)$$
$$\approx -ie\bar{u}(p_2) \not{\epsilon} \, u(p_1)[1 + (Z_2 - 1)] \; .$$

Alternatively, we may say that since it is not necessary to "redress" the external spinors, the effect of the radiative correction in Figure VIII.3 with one external and one internal spinor is only *half* the value calculated in Eq. 1.42, wherein neither initial or final particles were contracted with an external spinor.

Vertex Correction

Next consider the vertex modification shown below, for which we define

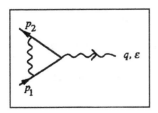

$$\text{Amp}(e) = -iee^{\lambda}\bar{u}(p_2)\Lambda_{\lambda}(p_2, p_1)u(p_1) \quad (1.46)$$

where

$$\Lambda_{\lambda}(p_2, p_1) = (-ie)^2 \int \frac{d^4k}{(2\pi)^4} \frac{-i\eta^{\mu\nu}}{k^2 + i\epsilon} \gamma_{\mu} \frac{i}{\not{p}_2 - \not{k} - m + i\epsilon} \gamma_{\lambda} \frac{i}{\not{p}_1 - \not{k} - m + i\epsilon} \gamma_{\nu} \; . \quad (1.47)$$

Before evaluating $\Lambda_\lambda(p_2, p_1)$ in full generality it is useful to calculate its value in the limit $q = p_1 - p_2 = 0$. Using again the identity

$$\frac{1}{\hat{A} - \hat{B}} = \frac{1}{\hat{A}} + \frac{1}{\hat{A}} \hat{B} \frac{1}{\hat{A}} + \ldots \tag{1.48}$$

we have

$$\frac{1}{\slashed{p} + \delta\slashed{p} - m} = \frac{1}{\slashed{p} - m} - \frac{1}{\slashed{p} - m} \delta\slashed{p} \frac{1}{\slashed{p} - m} + \ldots \tag{1.49}$$

so that

$$\frac{\partial}{\partial p^\lambda} \frac{1}{\slashed{p} - m} = -\frac{1}{\slashed{p} - m} \gamma_\lambda \frac{1}{\slashed{p} - m} . \tag{1.50}$$

We find then

$$-\frac{\partial}{\partial p_1^\lambda} \Sigma(p_1) \simeq (-ie)^2 \int \frac{d^4k}{(2\pi)^4} \frac{-i}{k^2 + i\epsilon} \times -i \frac{\partial}{\partial p_1^\lambda} \gamma_\mu \frac{i}{\slashed{p}_1 - \slashed{k} - m + i\epsilon} \gamma^\mu$$

$$= (-ie)^2 \int \frac{d^4k}{(2\pi)^4} \frac{-i}{k^2 + i\epsilon} \gamma_\mu \frac{i}{\slashed{p}_1 - \slashed{k} - m + i\epsilon} \gamma_\lambda \frac{i}{\slashed{p}_1 - \slashed{k} - m + i\epsilon} \gamma^\mu$$

$$= \Lambda_\lambda(p_1, p_1) \tag{1.51}$$

which relates the vertex correction at $q = 0$ to the self-energy. From the form

$$\Sigma(p_1) = A(m^2) + \tilde{B}(m^2)(\slashed{p}_1 - m) \tag{1.52}$$

we find

$$\Lambda_\lambda(p_1, p_1) = -\frac{\partial}{\partial p_1^\lambda} \left(A(m^2) + \tilde{B}(m^2)(\slashed{p}_1 - m) \right)$$

$$= -\tilde{B}(m^2)\gamma_\lambda = -(Z_2 - 1)\gamma_\lambda . \tag{1.53}$$

Then defining the general form of the vertex correction as

$$\Lambda_\lambda(p_2, p_1) \equiv \Lambda_\lambda(p_1, p_1) + \Lambda_\lambda^f(p_2, p_1) = -(Z_2 - 1)\gamma_\lambda + \Lambda_\lambda^f(p_2, p_1) \tag{1.54}$$

we find, adding all diagrams together

$$\text{Amp}^{\text{Tot}} = \text{Amp}(0 + a + b + c + d + e) = -ie\epsilon^\lambda \bar{u}(p_2)[\gamma_\lambda (Z_2 - (Z_2 - 1))$$
$$+ \Lambda_\lambda^f(p_2, p_1)] u(p_1) = -ie\epsilon^\lambda \bar{u}(p_2) \left(\gamma_\lambda + \Lambda_\lambda^f(p_2, p_1) \right) u(p_1) . \tag{1.55}$$

All ultraviolet divergences have cancelled and we are left with only the (ultraviolet) finite vertex modification produced by Λ_λ^f.

We may calculate the vertex correction explicitly by performing the integration in Eq. 1.47. In the Pauli–Villars scheme

$$\bar{u}(p_2) \Lambda_\lambda(p_2, p_1) u(p_1) = ie^2 \int_{\lambda^2}^{\Lambda^2} dL^2 \int \frac{d^4k}{(2\pi)^4}$$
$$\times \frac{1}{(k^2 - L^2 + i\epsilon)^2} \frac{1}{k^2 - 2p_1 \cdot k + i\epsilon} \frac{1}{k^2 - 2k \cdot p_2 + i\epsilon} \tag{1.56}$$
$$\times \bar{u}(p_2) \left(2p_{2\mu} - \gamma_\mu \slashed{k}\right) \gamma_\lambda \left(2p_1^\mu - \slashed{k} \gamma^\mu\right) u(p_1) .$$

The integration may be put into the form of Eq. 1.25 via the identity

$$\frac{1}{ab} = \int_0^1 dx\, [ax + b(1-x)]^{-2} \ . \tag{1.57}$$

Then, using $p_x \equiv p_1 x + p_2(1-x)$

$$\bar{u}(p_2)\Lambda_\lambda(p_2,p_1)u(p_1)$$
$$= ie^2 \int_0^1 dx \int_{\lambda^2}^{\Lambda^2} dL^2 \int \frac{d^4k}{(2\pi)^4}\, \frac{1}{(k^2 - L^2 + i\epsilon)^2}\, \frac{1}{(k^2 - 2k\cdot p_x + i\epsilon)^2}$$
$$\times \bar{u}(p_2)\, (4p_2\cdot p_1 \gamma_\lambda - 2(\not{p}_1 \not{k}\, \gamma_\lambda + \gamma_\lambda\, \not{k}\not{p}_2) - 2\not{k}\,\gamma_\lambda\,\not{k})\, u(p_1)$$
$$= -\frac{e^2}{(4\pi)^2} \int_0^1 dx \int_{\lambda^2}^{\Lambda^2} dL^2 \int_0^1 dy\, y(1-y)\bar{u}(p_2)\Big(-2\gamma_\lambda \frac{1}{y^2 p_x^2 + (1-y)L^2}$$
$$+ (4p_2\cdot p_1 \gamma_\lambda - 2y\,(\not{p}_1\not{p}_x\,\gamma_\lambda + \gamma_\lambda\,\not{p}_x\not{p}_2) - 2y^2\,\not{p}_x\,\gamma_\lambda\,\not{p}_x)\, \frac{1}{(y^2 p_x^2 + (1-y)L^2)^2}\Big)u(p_1). \tag{1.58}$$

Noting that $p_x^2 = m^2 - q^2 x(1-x)$, with $q = p_1 - p_2$, and performing the L^2 integration we find

$$\bar{u}(p_2)\Lambda_\lambda(p_2,p_1)u(p_1) = -\frac{e^2}{(4\pi)^2} \int_0^1 dx \int_0^1 dy\, y\bar{u}(p_2)\Big(-2\gamma_\lambda \ln\frac{\Lambda^2(1-y)}{y^2(m^2 - q^2 x(1-x))}$$
$$+ \gamma_\lambda\,(4m^2 - 2q^2 - 2y(4m^2 - q^2) - 2y^2\,(q^2 x(1-x) - m^2))$$
$$\times \frac{1}{y^2(m^2 - q^2 x(1-x)) + (1-y)\lambda^2}$$
$$+ (p_1 + p_2)_\lambda\, 4mxy(1-y) \frac{1}{y^2(m^2 - q^2 x(1-x)) + (1-y)\lambda^2}\Big)u(p_1) \ . \tag{1.59}$$

Finally, by use of the Gordon decomposition†

$$\bar{u}(p_2)\gamma_\mu u(p_1) = \frac{1}{2m}\bar{u}(p_2)\Big[(p_1 + p_2)_\mu - i\sigma_{\mu\nu}q^\nu\Big]u(p_1) \tag{1.60}$$

---

† This result is easily derived via

$$\bar{u}(p_2)\gamma_\mu u(p_1) = \frac{1}{2m}\bar{u}(p_2)\,(\not{p}_2\,\gamma_\mu + \gamma_\mu\,\not{p}_1)\, u(p_1)$$
$$= \frac{1}{2m}\bar{u}(p_2)\,(p_{2\mu} + i\sigma_{\mu\nu}p_2^\nu + p_{1\mu} - i\sigma_{\mu\nu}p_1^\nu) \ .$$
$$= \frac{1}{2m}\bar{u}(p_2)\,[(p_1 + p_2)_\mu - i\sigma_{\mu\nu}q^\nu]\, u(p_1)$$

Eq. 1.59 can be written as

$$\bar{u}(p_2)\Lambda_\lambda(p_2,p_1)u(p_1) = \frac{e^2}{(4\pi)^2}\left[\gamma_\lambda\left(\ln\frac{\Lambda^2}{m^2} - 2\ln\frac{m^2}{\lambda^2} + \frac{9}{2}\right.\right.$$
$$\left.\left. + \frac{4}{3}\frac{q^2}{m^2}\left(\ln\frac{m}{\lambda} - \frac{3}{8}\right)\right) - i\sigma_{\lambda\eta}q^\eta\frac{1}{m}\right]u(p_1) + \ldots \quad (1.61)$$

where we have expanded in powers of q^2/m^2. In the limit as $q \to 0$ we find

$$\bar{u}(p_1)\Lambda_\lambda(p_1,p_1)u(p_1) = -\bar{u}(p_2)\gamma_\lambda u(p_1)\tilde{B}(m^2) \quad (1.62)$$

as required by Eq. 1.54, and the radiatively corrected vertex becomes

$$\text{Amp}^{\text{Tot}} = -ie\epsilon^\lambda \bar{u}(p_2)\left(\gamma_\lambda + \Lambda_\lambda^f(p_2,p_1)\right)u(p_1)$$
$$= -ie\epsilon^\lambda \bar{u}(p_2)\left[\gamma_\lambda\left(1 + \frac{\alpha}{3\pi}\frac{q^2}{m^2}\left(\ln\frac{m}{\lambda} - \frac{3}{8}\right)\right) - i\frac{\alpha}{4\pi m}\sigma_{\lambda\eta}q^\eta\right]u(p_1) .$$
$$(1.63)$$

An identical result is found if dimensional regularization is employed.

We observe two modifications due to radiative corrections. One is the appearance of a *new* Dirac invariant — $i\sigma_{\lambda\eta}q^\eta$ — in the electromagnetic vertex. The second is that the lowest order structure γ_λ develops a dependence on q^2. The significance of the former result is particularly interesting. As discussed in the previous chapter this term has the form of a magnetic moment interaction — since $A^\lambda(x) = \epsilon^\lambda e^{iq\cdot x}$

$$-i\epsilon^\lambda \sigma_{\lambda\eta}q^\eta = -\frac{1}{2}\left(\partial^\eta A^\lambda - \partial^\lambda A^\eta\right)\sigma_{\lambda\eta} = \frac{1}{2}\sigma_{\lambda\eta}F^{\lambda\eta}$$
$$= -\vec{\sigma}\cdot\vec{B} \quad (1.64)$$

— and represents a radiative correction $\frac{\alpha}{2\pi}$ to the usual Dirac moment

$$\vec{\mu} = \frac{e}{2m}\vec{\sigma}\left(1 + \frac{\alpha}{2\pi} + \ldots\right) . \quad (1.65)$$

Here the ellipses represent higher order $(\mathcal{O}(\alpha^2) + \ldots)$ contributions to the magnetic moment, which have been calculated as

$$1 + \frac{\alpha}{2\pi} - 0.328\frac{\alpha^2}{\pi^2} + \ldots = 1.00115965218178 \pm 0.00000000000077 . \quad (1.66)$$

The electron magnetic moment is one of the best known quantities in nature and has been measured to be

$$\vec{\mu}^{exp} = \frac{e}{2m}\vec{\sigma}\cdot(1.00115965218073 \pm 0.00000000000028) \quad (1.67)$$

in excellent agreement with the theoretically calculated value. The $\mathcal{O}(\alpha)$ correction, first found by Schwinger, is well-verified!

As we shall see, the q^2-dependence of the charge form factor has also been well-tested. However, the expression given in Eq. 1.63 is certainly not satisfactory because of its dependence on the photon "mass" λ. If we set λ to its physical value — zero — the scattering amplitude diverges. In order to understand the resolution of this problem we need to study the problem of electron scattering accompanied by the emission of a low-energy photon — the so-called infrared problem.

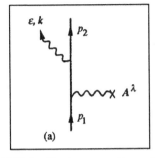 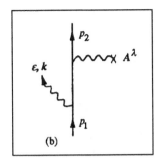

Fig. VIII.4: *Electromagnetic scattering accompanied by photon emission.*

The Infrared Catastrophe

Suppose that an electron scatters from an electromagnetic potential and emits a photon with $A^\rho(x) = \epsilon^\rho e^{-ik\cdot x}$. There exist two diagrams which contribute in lowest order, as indicated in Figure VIII.4, and the amplitude is given by

$$\text{Amp} = (-ie)^2 \bar{u}(p_2) \, \slashed{\epsilon}^* \, \frac{i}{\slashed{p}_2 + \slashed{k} - m + i\epsilon} \, \slashed{A}(p_2 + k - p_1) u(p_1) \\ + (-ie)^2 \bar{u}(p_2) \, \slashed{A}(p_2 + k - p_1) \frac{i}{\slashed{p}_1 - \slashed{k} - m + i\epsilon} \slashed{\epsilon}^* \, u(p_1) \; . \tag{1.68}$$

If the emitted photon is very soft — $k \ll m$ — we can approximate

$$\begin{aligned}
\text{Amp} &= -ie^2 \bar{u}(p_2) \left[\frac{\slashed{\epsilon}^* (\slashed{p}_2 + \slashed{k} + m) \slashed{A}(p_2 - p_1 + k)}{k^2 + 2p_2 \cdot k + i\epsilon} \right. \\
&\quad \left. + \frac{\slashed{A}(p_2 - p_1 + k)(\slashed{p}_1 - \slashed{k} + m) \slashed{\epsilon}^*}{k^2 - 2p_1 \cdot k + i\epsilon} \right] u(p_1) \\
&\approx -ie^2 \bar{u}(p_2) \, \slashed{A}(p_2 - p_1) \left(\frac{p_2 \cdot \epsilon^*}{p_2 \cdot k} - \frac{p_1 \cdot \epsilon^*}{p_1 \cdot k} \right) \\
&= e \left(\frac{p_2 \cdot \epsilon^*}{p_2 \cdot k} - \frac{p_1 \cdot \epsilon^*}{p_1 \cdot k} \right) \text{Amp}^{(0)}(p_2, p_1)
\end{aligned} \tag{1.69}$$

where $\text{Amp}^{(0)}(p_2, p_1) = -ie\bar{u}(p_2) \slashed{A}(p_2 - p_1) u(p_1)$ is the electron scattering amplitude *without* photon emission.

The scattering cross section with no emitted photon is given by

$$d\sigma^{(0)} = \frac{1}{\text{Flux}} \int \frac{d^3 p_2}{(2\pi)^3} \frac{m}{E_2} 2\pi \delta(E_2 - E_1) \times \frac{1}{2} \sum_{s_1, s_2} \left| \text{Amp}^{(0)}(p_2, p_1) \right|^2 \tag{1.70}$$

where

$$\text{Flux} = \frac{p_1}{E_1} \frac{E_1}{m} \tag{1.71}$$

is the incident flux. On the other hand

$$d\sigma = \frac{1}{\text{Flux}} \int \frac{d^3 p_2}{(2\pi)^3} \frac{m}{E_2} 2\pi\delta(E_2 + \omega_k - E_1) \times \frac{1}{2}\sum_{s_1,s_2} \left|\text{Amp}^{(0)}(p_2,p_1)\right|^2$$

$$\times e^2 \left|\frac{p_2 \cdot \epsilon}{p_2 \cdot k} - \frac{p_1 \cdot \epsilon}{p_1 \cdot k}\right|^2 \frac{d^3k}{(2\pi)^3} \frac{1}{2\omega_k} \qquad (1.72)$$

represents the cross-section for scattering of an electron with incoming momentum $\vec{p}_1$ into an outgoing electron with momentum $\vec{p}_2$ into solid angle $d\Omega_2$, accompanied by a photon of polarization ϵ and momentum $\vec{k}$. Thus

$$\frac{d\sigma}{d\Omega_2} = \left(\frac{d\sigma}{d\Omega_2}\right)^{(0)} \times \frac{1}{2} e^2 k^2 dk\, d\Omega_{\hat{k}} \left|\frac{p_2 \cdot \epsilon}{p_2 \cdot k} - \frac{p_1 \cdot \epsilon}{p_1 \cdot k}\right|^2 \frac{1}{(2\pi)^3} \qquad (1.73)$$

Note, however, the apparently troublesome feature that as $k \to 0$,

$$\frac{d\sigma}{d\Omega_2} \sim \frac{dk}{k} d\Omega_{\hat{k}} \ldots \qquad (1.74)$$

This phenomenon, whereby the cross-section for emission of very soft photons during the scattering process diverges logarithmically, is termed the "infrared catastrophy." This apparent infinity will not prove to be a problem due to the fact that it is impossible to prove that a scattering process occurs without emission of *any* accompanying photons. When the electron is scattered, the electromagnetic field must change from that associated with a charge e and momentum $\vec{p}_1$ to that of a charge e and momentum $\vec{p}_2$. This change in momentum represents an acceleration and is generally accompanied by the emission of radiation. Any apparatus which detects elastically scattered electrons also is sensitive to those scattered inelastically along with emission of a photon up to some detector resolution ΔE, so that for a consistent comparison with experiment one must *add* the elastic and "bremsstrahlung" cross-sections.

Before seeing how this is carried out, we evaluate the cross-section summed over photon helicity since such soft photons are not detected. We note that the bremsstrahlung amplitude is explicitly gauge invariant, *i.e.*, unmodified under the gauge change

$$\epsilon_\mu \to \epsilon_\mu + \lambda k_\mu \ . \qquad (1.75)$$

In order to exploit this invariance, we orient the coordinates so that $\hat{k} = \hat{e}_z$. Then since

$$k^\mu M_\mu = \omega_k (M_0 - M_z) = 0 \qquad (1.76)$$

we can write

$$\sum_{\text{pol}} \epsilon^{\mu*} M_\mu \epsilon^\nu M_\nu^* = |M_x|^2 + |M_y|^2 \qquad (1.77)$$

$$= |M_x|^2 + |M_y|^2 + |M_z|^2 - |M_0|^2 = -\eta^{\mu\nu} M_\mu^* M_\nu \ ,$$

so that the bremsstrahlung cross-section can be written as

$$\sum_{\text{pol}} \frac{d\sigma}{d\Omega_2} = -\left(\frac{d\sigma}{d\Omega}\right)^{(0)} \frac{e^2}{2(2\pi)^3} k dk d\Omega_{\hat{k}} \left(\frac{p_2}{p_2 \cdot k} - \frac{p_1}{p_1 \cdot k}\right)^2$$

$$= \left(\frac{d\sigma}{d\Omega}\right)^{(0)} \frac{\alpha}{4\pi^2} k dk d\Omega_{\hat{k}} \left(\frac{2p_2 \cdot p_1}{p_2 \cdot k p_1 \cdot k} - \frac{m^2}{(p_2 \cdot k)^2} - \frac{m^2}{(p_1 \cdot k)^2}\right). \tag{1.78}$$

The solid angle integration may be performed directly

$$\int \frac{d\Omega_{\hat{k}}}{4\pi} \frac{m^2}{(p_2 \cdot k)^2} = \frac{m^2}{k_0^2} \int \frac{d\Omega_{\hat{k}}}{4\pi} \frac{1}{\left(E_2 - \vec{p}_2 \cdot \hat{k}\right)^2}$$

$$= \frac{m^2}{2k_0^2} \int_{-1}^{1} d(\cos\theta) \frac{1}{(E_2 - p_2 \cos\theta)^2} = \frac{m^2}{k_0^2 (E_2^2 - p_2^2)} = \frac{1}{k_0^2}. \tag{1.79}$$

Similarly defining $p_x = p_1 x + p_2(1-x)$

$$\int \frac{d\Omega_{\hat{k}}}{4\pi} \frac{m^2}{k \cdot p_2 k \cdot p_1} = \int_0^1 dx \int \frac{d\Omega_{\hat{k}}}{4\pi} \frac{m^2}{(k \cdot p_x)^2} = \int_0^1 dx \frac{m^2}{k_0^2 p_x^2}$$

$$= \int_0^1 dx \frac{m^2}{k_0^2 (m^2 - q^2 x(1-x))} = \frac{1}{k_0^2}\left(1 + \frac{1}{6}\frac{q^2}{m^2} + \ldots\right). \tag{1.80}$$

We then have

$$\int d\Omega_{\hat{k}} \sum_{\text{pol}} \frac{d\sigma}{d\Omega_2} = \frac{\alpha}{\pi} \frac{dk}{k_0} \left(\frac{d\sigma}{d\Omega}\right)^{(0)} \left(\left(2 - \frac{q^2}{m^2}\right)\left(1 + \frac{1}{6}\frac{q^2}{m^2} + \ldots\right) - 2\right)$$

$$= -\frac{2\alpha}{3\pi} \frac{dk}{k_0} \left(\frac{d\sigma}{d\Omega}\right)^{(0)} \frac{q^2}{m^2} + \ldots \tag{1.81}$$

so that the cross-section to have scattered accompanied by emission of an unobserved photon of energy $k < \Delta E$ is given by

$$\int_\lambda^{\Delta E} \frac{dk}{k_0} \int d\Omega_{\hat{k}} \sum_{\text{pol}} \frac{d\sigma}{d\Omega_2} = -\frac{2\alpha}{3\pi} \ln\frac{\Delta E}{\lambda} \left(\frac{d\sigma}{d\Omega}\right)^{(0)} \frac{q^2}{m^2} + \ldots \tag{1.82}$$

We can now understand the resolution of the infrared problem. Any physical apparatus measuring elastic electron scattering will also be sensitive to soft photon emission up to its resolution ΔE, so that the measured cross-section is given by (for small q^2)

$$\frac{d\sigma}{d\Omega}_{\text{exptl}} = \frac{d\sigma}{d\Omega}(\text{elastic} + k < \Delta E) = \left(\frac{d\sigma}{d\Omega}\right)^{(0)} \left[1 + \frac{2\alpha}{3\pi}\frac{q^2}{m^2}\left(\ln\frac{m}{\lambda} - \frac{3}{8}\right) + \ldots\right]$$

$$- \left(\frac{d\sigma}{d\Omega}\right)^{(0)} \frac{2\alpha}{3\pi}\frac{q^2}{m^2} \ln\frac{\Delta E}{\lambda} + \ldots$$

$$= \left(\frac{d\sigma}{d\Omega}\right)^{(0)} \left[1 + \frac{2\alpha}{3\pi}\frac{q^2}{m^2}\left(\ln\frac{m}{\Delta E} - \frac{3}{8}\right) + \ldots\right] \tag{1.83}$$

where the ellipses represent $\mathcal{O}(\alpha^2)$ terms plus an $\mathcal{O}(\alpha)$ correction to the cross-section due to the anomalous moment. The dependence upon λ has disappeared and we can now take the limit $\lambda \to 0$. Bloch and Nordsieck [BlN 37] have shown that this procedure works to all orders in α.

A careful reader may well have noted an inconsistency in the calculation outlined above in that we have taken the lower limit of the photon energy integration as λ but we have *not* otherwise consistently assumed a finite photon mass. Indeed we should have summed over *three* — not two — photon polarizations and the propagators should have had the form

$$\frac{1}{\pm 2p \cdot k + \lambda^2 + i\epsilon} \quad . \tag{1.84}$$

When this is consistently carried out the bremsstrahlung cross-section becomes [Sa 67]

$$\frac{d\sigma}{d\Omega}(k < \Delta E) = -\frac{2\alpha}{3\pi}\frac{q^2}{m^2}\left(\ln\frac{\Delta E}{\lambda} + \ln 2 - \frac{5}{6}\right) \quad . \tag{1.85}$$

For small q^2 the full result then reads

$$\begin{aligned}\frac{d\sigma}{d\Omega}(\text{elastic} + k < \Delta E) &= \left(\frac{d\sigma}{d\Omega}\right)^{(0)}\left(1 + \frac{2\alpha}{3\pi}\frac{q^2}{m^2}\left(\ln\frac{m}{\lambda} - \frac{3}{8}\right)\right.\\ &\quad \left. - \frac{2\alpha}{3\pi}\frac{q^2}{m^2}\left(\ln\frac{2\Delta E}{\lambda} - \frac{5}{6}\right)\right)\\ &= \left(\frac{d\sigma}{d\Omega}\right)^{(0)}\left(1 + \frac{2\alpha}{3\pi}\frac{q^2}{m^2}\left(\ln\frac{m}{2\Delta E} + \frac{11}{24}\right) + \ldots\right)\end{aligned} \tag{1.86}$$

and is completely free of divergences. It would be tempting at this point to compare with experiment, but we have not yet included all radiative correction diagrams to $\mathcal{O}(\alpha)$.

Vacuum Polarization

There is still one term contributing to the $\mathcal{O}(\alpha)$ corrections to the electron-photon vertex which we have not considered. As shown in Figure VIII.5, this diagram accounts for the feature that the photon can dissociate into a virtual electron-positron pair.

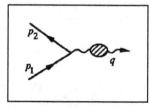

Fig. VIII.5: The vacuum polarization diagram.

VIII.1 RADIATIVE CORRECTIONS

This virtual dissociation process is called "vacuum polarization" and is best described in terms of its modification of the photon propagator, as indicated in Figure VIII.6.

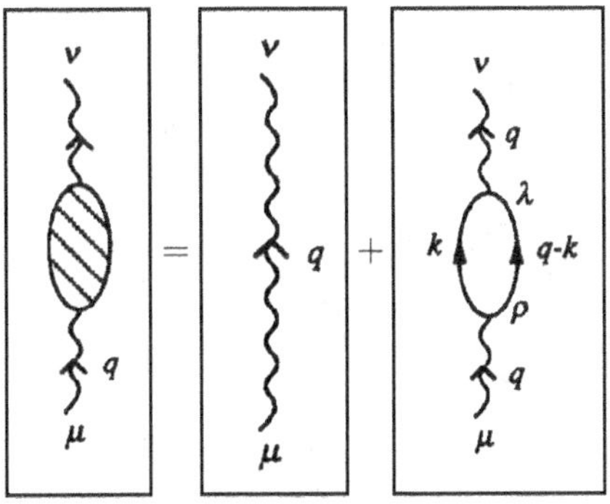

$$\text{(1.87)}$$

Fig. VIII.6: Modification of the photon propagator by vacuum polarization effects.

That is,

$$iD_F^{\mu\nu}(q) = -\frac{i\eta^{\mu\nu}}{q^2+i\epsilon} - \frac{i\eta^{\mu\rho}}{q^2+i\epsilon} \times (-)(-ie)^2 \int \frac{d^4k}{(2\pi)^4} (\gamma_\rho)_{ai} \left(\frac{i}{\slashed{k}-\slashed{q}-m+i\epsilon}\right)_{ij}$$
$$\times (\gamma_\lambda)_{j\ell} \left(\frac{i}{\slashed{k}-m+i\epsilon}\right)_{\ell a} \cdot -\frac{i\eta^{\lambda\nu}}{q^2+i\epsilon} + \cdots$$
$$= -\frac{i\eta^{\mu\nu}}{q^2+i\epsilon} + \left(-\frac{i\eta^{\mu\rho}}{q^2+i\epsilon} \times i\Pi^{\lambda\rho}(q) \times -\frac{i\eta^{\lambda\nu}}{q^2+i\epsilon}\right) + \cdots$$
(1.88)

where

$$i\Pi_{\lambda\rho}(1) = -(-ie)^2 \int \frac{d^4k}{(2\pi)^4} \operatorname{Tr} \gamma_\rho \frac{i}{\slashed{k}-\slashed{q}-m+i\epsilon} \gamma_\lambda \frac{i}{\slashed{k}-m+i\epsilon} \quad (1.89)$$

is called the vacuum polarization tensor. By simple Lorentz invariance, the structure of $\Pi_{\lambda\rho}(q)$ must be

$$\Pi_{\lambda\rho}(q) = -\eta_{\lambda\rho}\left(K + q^2\Pi_1(q^2)\right) + q_\lambda q_\rho \Pi_2(q^2) \ . \quad (1.90)$$

However, something is clearly wrong here. If we temporarily keep only the term in K and iterate the vacuum polarization terms we would find an effective photon propagator of the form

$$iD_F^{\mu\nu}(q) \cong -i\frac{\eta^{\mu\nu}}{q^2+i\epsilon} + (-i\frac{\eta^{\mu\rho}}{q^2+i\epsilon}) \times (i\eta_{\rho\lambda}K) \times (-i\frac{\eta^{\lambda\nu}}{q^2+i\epsilon}) + \cdots$$
$$= -i\frac{\eta^{\mu\nu}}{q^2+i\epsilon}\left(1 + K\frac{1}{q^2+i\epsilon} + K^2\left(\frac{1}{q^2+i\epsilon}\right)^2 + \cdots\right) \quad (1.91)$$
$$= -i\frac{\eta^{\mu\nu}}{q^2-K+i\epsilon} \ .$$

$K \neq 0$ then would signify a non-zero photon mass, $m_\gamma^2 = K$, in clear contradiction to experiment. The condition $K = 0$ would seem to be a requirement on the form of the vacuum polarization tensor.

A further stricture may be found by considering Figure VIII.7, which represents the exchange of a virtual photon between the electron current and a current induced in the vacuum due to the presence of the photon field $A^\rho(q)$.

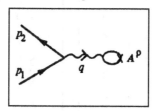

Fig. VIII.7: Vacuum polarization modification of the interaction of an electron with an external electromagnetic field.

Since $\Pi_{\lambda\rho}(q)$ characterizes the proportionality between the induced current and external potential it is called the polarization tensor. However, we know that this amplitude must be invariant under the gauge change

$$A^\rho(q) \to A^\rho(q) + \lambda q^\rho \tag{1.92}$$

so we require

$$q^\rho \Pi_{\rho\lambda}(q) = 0 \ , \quad i.e., \quad \Pi_1(q^2) = \Pi_2(q^2) \ . \tag{1.93}$$

Thus we may write

$$\Pi_{\rho\lambda}(q) = \Pi(q^2)\left(-\eta_{\rho\lambda}q^2 + q_\rho q_\lambda\right) \tag{1.94}$$

as the most general form of the polarization tensor.

This is as far as one can go with arguments based upon physics alone — the explicit form of $\Pi(q^2)$ must be found from calculation. As before, we can employ either dimensional or Pauli–Villars methods, but an interesting difference arises.

First consider the dimensional approach, for which we find

$$\Pi_{\lambda\rho}(q) = ie^2 \left(\mu^2\right)^{2-\frac{d}{2}} \int \frac{d^d k}{(2\pi)^d} \operatorname{Tr} \gamma_\rho(\slashed{k} - \slashed{q} + m)\gamma_\lambda(\slashed{k} + m)$$
$$\times \frac{1}{((k-q)^2 - m^2)(k^2 - m^2)} \tag{1.95}$$
$$= i4e^2 \left(\mu^2\right)^{2-\frac{d}{2}} \int \frac{d^d k}{(2\pi)^d} \frac{1}{((k-q)^2 - m^2)(k^2 - m^2)}$$
$$\times \left[(k-q)_\rho k_\lambda + (k-q)_\lambda k_\rho - \eta_{\rho\lambda}\left(k \cdot (k-q) - m^2\right)\right] \ .$$

The integration may be performed directly, using Eq. 1.25, yielding

$$\Pi_{\lambda\rho}(q) = \left(-q^2 \eta_{\rho\lambda} + q_\rho q_\lambda\right) 4e^2 \left(\mu^2\right)^{2-\frac{d}{2}} \Gamma\left(2 - \frac{d}{2}\right) \frac{1}{(4\pi)^{d/2}}$$
$$\times \int_0^1 dy\, 2y(1-y)\left(m^2 - y(1-y)q^2\right)^{\frac{d}{2}-2} \tag{1.96}$$
$$= \left(-q^2 \eta_{\rho\lambda} + q_\rho q_\lambda\right) \frac{\alpha}{3\pi} \left(\frac{2}{\epsilon} - \gamma - \ln\frac{m^2}{4\pi\mu^2} + \frac{q^2}{5m^2} + \ldots\right)$$

which clearly is of the proper form.

Things are not so simple in the Pauli–Villars case, for a naive calculation performed by setting $q = 0$ yields

$$\Pi_{\lambda\rho}(q=0) = i4e^2 \int \frac{d^4k}{(2\pi)^4} \frac{2k_\lambda k_\rho - \eta_{\rho\lambda}\left(k^2 - m^2\right)}{\left(k^2 - m^2 + i\epsilon\right)^2} \left(-\frac{\Lambda^2}{k^2 - \Lambda^2}\right)^2$$
$$= -\eta_{\rho\lambda}\left(\frac{\alpha}{\pi}\Lambda^2 + \ldots\right) \quad (1.97)$$

which is obviously quadratically divergent and corresponds to $K \neq 0$. Clearly we must be more careful since introduction of a cutoff has destroyed the gauge invariance condition

$$q^\lambda \Pi_{\lambda\rho}(q) = 0 \ . \quad (1.98)$$

The conventional way around this dilemma is to subtract from the naive expression for vacuum polarization a corresponding quantity with the electron mass m^2 replaced by some heavy value Λ^2

$$\Pi_{\rho\lambda}(q) \to \overline{\Pi}_{\rho\lambda}(q) = \Pi_{\rho\lambda}(q)\Big|_{m^2} - \Pi_{\rho\lambda}(q)\Big|_{\Lambda^2} \ . \quad (1.99)$$

It is not useful to spend much time speculating on the meaning of this cutoff Λ^2 since it is certainly unphysical and is introduced only in order to allow finite results at each stage of the calculation. Final (and physical) answers must be independent of Λ.

In any case, *defining* the vacuum polarization via Eq. 1.99 and then carrying out the integration using Eq. 1.25 we find

$$\overline{\Pi}_{\rho\lambda}(q) = \left(q^2 \eta_{\rho\lambda} - q_\rho q_\lambda\right) 4e^2 \frac{1}{(4\pi)^2} \int_0^1 dy\, 2y(1-y) \ln \frac{m^2 - q^2 y(1-y)}{\Lambda^2}$$
$$= \left(q^2 \eta_{\rho\lambda} - q_\rho q_\lambda\right) \frac{\alpha}{3\pi} \left(-\ln \frac{\Lambda^2}{m^2} - \frac{q^2}{5m^2} + \ldots\right) \qquad q^2 \ll m^2 \quad (1.100)$$

which agrees with the result obtained dimensionally provided we make the replacement (*cf.* Eq. 1.38)

$$\ln \frac{\Lambda^2}{m^2} \longleftrightarrow \frac{2}{\epsilon} - \gamma - \ln \frac{m^2}{4\pi\mu^2} \ . \quad (1.101)$$

We now can write the complete form for the radiatively modified electron-photon vertex by combining Eq. 1.63 and Eq. 1.96

$$\text{Amp} = -ieA^\lambda(q)\bar{u}(p_2)\left[\gamma_\lambda\left(1 - \frac{\alpha}{3\pi}\left(\frac{2}{\epsilon} - \gamma - \ln\frac{m^2}{4\pi\mu^2}\right)\right.\right.$$
$$\left.\left. + \frac{\alpha}{3\pi}\frac{q^2}{m^2}\left(\ln\frac{m}{2\Delta E} + \frac{11}{24} - \frac{1}{5}\right)\right) - \frac{\alpha}{8\pi m}[\slashed{q},\gamma_\lambda]\right]u(p_1) \quad (1.102)$$

where we have eliminated the vacuum polarization term involving $q_\lambda q_\rho$ since

$$\bar{u}(p_2) \not{q} \, u(p_1) = \bar{u}(p_2) (\not{p}_1 - \not{p}_2) u(p_1) = 0 \ . \tag{1.103}$$

Although Eq. 1.102 appears unsatisfactory since a divergence has crept back into the scattering amplitude, this is only illusory. Thus consider scattering of our electron probe by a Coulombic "potential"

$$A^\lambda(q) = -\frac{Ze}{\vec{q}^{\,2}} 2\pi\delta(q^0)\delta^{\lambda 0} \tag{1.104}$$

for which the radiatively corrected scattering amplitude becomes, in the $q \to 0$ limit

$$\text{Amp} \underset{q\to 0}{\sim} i\frac{Ze^2}{\vec{q}^{\,2}} 2\pi\delta(q^0)\bar{u}(p_2)\gamma_0 u(p_1) \left(1 - \frac{\alpha}{3\pi}\left(\frac{2}{\epsilon} - \gamma - \ln\frac{m^2}{4\pi\mu^2}\right)\right) \ . \tag{1.105}$$

It should now be clear what the solution to our problem is. Just as in the case of mass, where the quantity m which appears in the Dirac equation is *not* the physical or renormalized mass m_{phys}

$$m_{\text{phys}} = m + \delta m \tag{1.106}$$

the same is true of charge! The quantity e which appears in the Dirac equation is *not* the physical or renormalized quantity e_R measured in Coulomb scattering, defined by

$$\text{Amp} \underset{q\to 0}{\sim} i\frac{Ze_R^2}{\vec{q}^{\,2}} 2\pi\delta(q^0)\bar{u}(p_2)\gamma^0 u(p_1) \ . \tag{1.107}$$

Instead we have

$$e_R^2 \equiv e^2 \left(1 - \frac{\alpha}{3\pi}\left(\frac{2}{\epsilon} - \gamma - \ln\frac{m^2}{4\pi\mu^2}\right)\right) \tag{1.108}$$

and it is this quantity which is given experimentally by the relation

$$\alpha_R = \frac{1}{4\pi} e_R^2 \approx \frac{1}{137} \ . \tag{1.109}$$

When the scattering amplitude in Eq. 1.102 is written in terms of the *renormalized* charge α_R everything is finite—the proper form of the radiatively corrected vertex becomes

$$\text{Amp} = -ie_R A_R^\lambda(q)\bar{u}(p_2)\left[\gamma_\lambda\left(1 + \frac{\alpha_R}{3\pi}\frac{q^2}{m^2}\left(\ln\frac{m}{2\Delta E} + \frac{11}{24} - \frac{1}{5}\right)\right)\right.$$
$$\left. - \frac{\alpha_R}{8\pi m}[\not{q},\gamma_\lambda]\right]u(p_1) \ . \tag{1.110}$$

Before seeing how this vertex can be confronted with an experimental test, it is important to realize that this form has been calculated in the limit $q^2 \ll m^2$. It is also of interest to examine the case $-q^2 \gg m^2$ in which case (*cf.* Eq. 1.96)

$$\text{Amp} \underset{q^2 \gg m^2}{\sim} -ieA^\lambda(q)\bar{u}(p_2)\gamma_\lambda\left(1 - \frac{\alpha}{3\pi}\left(\frac{2}{\epsilon} - \gamma - \ln\frac{-q^2}{4\pi\mu^2}\right) + \ldots\right)u(p_1) \ . \tag{1.111}$$

—the effective vertex retains the same form, but the coupling constant becomes a function of q^2

$$\alpha(q^2) = \alpha\left(1 - \frac{\alpha}{3\pi}\left(\frac{2}{\epsilon} - \gamma - \ln\frac{-q^2}{4\pi\mu^2}\right)\right) . \tag{1.112}$$

In fact this relation can be used in order to *define* the renormalized coupling constant. If at some momentum transfer q_1^2 one *measures* a value $\alpha(q_1^2)$

$$\alpha(q_1^2) \equiv \alpha\left(1 - \frac{\alpha}{3\pi}\left(\frac{2}{\epsilon} - \gamma - \ln\frac{-q_1^2}{4\pi\mu^2}\right)\right) \tag{1.113}$$

then the value of the effective coupling at a different momentum transfer q_2^2 can be written as

$$\begin{aligned}
\alpha(q_2^2) &= \alpha\left(1 - \frac{\alpha}{3\pi}\left(\frac{2}{\epsilon} - \gamma - \ln\frac{-q_2^2}{4\pi\mu^2}\right)\right) \\
&= \alpha(q_1^2)\left(1 + \frac{\alpha}{3\pi}\ln\frac{-q_2^2}{4\pi\mu^2} - \frac{\alpha}{3\pi}\ln\frac{-q_1^2}{4\pi\mu^2}\right) \\
&= \alpha(q_1^2)\left(1 + \frac{\alpha}{3\pi}\ln\frac{q_2^2}{q_1^2}\right) .
\end{aligned} \tag{1.114}$$

Thus the effective coupling strength increases as the momentum transfer rises and $\alpha(q^2)$ is often called the "running" coupling constant. This result is consistent with the picture that at high q^2 the scattering electron is able to penetrate the cloud of virtual e^+e^- pairs which surround a given charge and thus get closer to its "bare" value. This is clear too if the running coupling constant relation is written in a full "leading logarithm" expansion [BjD 64]

$$\alpha(q_2^2) = \frac{\alpha(q_1^2)}{1 - \frac{\alpha(q_1^2)}{3\pi}\ln\frac{q_2^2}{q_1^2}} \tag{1.115}$$

which has a singularity at

$$q_2^2 = q_1^2 \exp\left(\frac{3\pi}{\alpha(q_1^2)}\right) . \tag{1.116}$$

However, such speculation is taking the expression for the effective charge far beyond the simple perturbative regime wherein its validity can be trusted.

The Lamb Shift

Having gone to considerable effort to evaluate the precise form of the radiatively corrected electron-photon vertex, it is important to see how this result can be experimentally verified. In principle, one could perform a high precision scattering experiment, although this would not be easy. Far more precise is to test these radiative corrections via their effect on the Lamb shift.

For an electron in the Coulomb field of a nucleus of charge Z we have

$$A^\lambda = 2\pi\delta(q_0)\left(-\frac{Ze}{\vec{q}^2}, \vec{0}\right) \tag{1.117}$$

and an effective interaction

$$T_{fi} \sim -\frac{Ze^2}{|\vec{q}|^2}\left(1 - \frac{\alpha_R}{3\pi}\frac{|\vec{q}|^2}{m^2}\left(\ln\frac{m}{2\Delta E} + \frac{11}{24} - \frac{1}{5}\right) + \ldots\right) u^\dagger(p_2)u(p_1) \quad (1.118)$$

where we have temporarily omitted the anomalous moment term. Taking the Fourier transform (cf. Eq. VII.10.72), the effective electromagnetic interaction becomes

$$V_{\text{eff}}(r) = -\frac{Z\alpha_R}{r} + \frac{4}{3}\frac{Z\alpha_R^2}{m^2}\delta^3(r)\left(\ln\frac{m}{2\Delta E} + \frac{11}{24} - \frac{1}{5}\right) + \ldots \quad . \quad (1.119)$$

The radiative corrections lead to an effective delta function interaction between electron and nucleus in addition to the usual Coulombic component. This interaction, which includes radiative effects associated with exchange of virtual photons with energy $k > \Delta E$, leads via simple first order perturbation theory to a splitting between the $2P_{1/2}$ and $2S_{1/2}$ levels of a hydrogen-like atom

$$\begin{aligned} E_{2S_{1/2}} - E_{2P_{1/2}}\bigg|_{k>\Delta E} &= \frac{4}{3}\frac{Z\alpha_R^2}{m^2}\left(\ln\frac{m}{2\Delta E} + \frac{11}{24} - \frac{1}{5}\right)|\psi_{2S}(0)|^2 \\ &= \frac{Z^4\alpha_R^5}{6\pi}m\left(\ln\frac{m}{2\Delta E} + \frac{11}{24} - \frac{1}{5}\right) \quad . \end{aligned} \quad (1.120)$$

To this must be appended the contribution from soft photons with $k < \Delta E$ in order to obtain the full contribution to the Lamb shift. However, if we choose $\Delta E \ll Z\alpha m$ this component of the calculation is completely non-relativistic and we can use the result already calculated in Ch. IV

$$E_{2S_{1/2}} - E_{2P_{1/2}}\bigg|_{k<\Delta E} = \frac{Z^4\alpha_R^5}{6\pi}m\ln\frac{\Delta E}{|E - E_0|_{AV}} \quad . \quad (1.121)$$

Adding Eqs. 1.120 and 1.121, we find the full result

$$E_{2S_{1/2}} - E_{2P_{1/2}} = \frac{Z^4\alpha_R^5}{6\pi}m\left(\ln\frac{m}{2|E - E_0|_{AV}} + \frac{11}{24} - \frac{1}{5}\right) \quad (1.122)$$
$$+ \mu_{\text{anom}} \text{ contribution } .$$

Using the Bethe estimate $|E - E_0|_{AV} = 8.3\alpha_R^2 m$ we can easily evaluate this expression, yielding

$$\begin{aligned} E_{2S_{1/2}} - E_{2P_{1/2}} &= (945.5 + 62.3 - 27.2)\,\text{MHz} + \mu_{\text{anom}} \text{ contribution} \\ &= 980.6\,\text{MHz} + \mu_{\text{anom}} \text{ contribution } . \end{aligned} \quad (1.123)$$

In order to complete the calculation, we must include the energy shift due to the anomalous moment. Taking the non-relativistic limit, we note that

$$\bar{u}(p_2)\sigma_{0k}u(p_1) \cong i\chi_2^\dagger\left(\frac{\vec{p}_1 - \vec{p}_2}{2m} - i\vec{\sigma}\times\frac{\vec{p}_1 + \vec{p}_2}{2m}\right)_k \chi_1 \quad (1.124)$$

so that the anomalous moment interaction

$$\frac{\alpha_R}{4\pi} \frac{e}{2m} F^{\mu\nu} \bar{u}(p_2) \sigma_{\mu\nu} u(p_1) \tag{1.125}$$

is equivalent to the effective potential

$$\begin{aligned} V_{\text{eff}} &= \frac{\alpha_R}{2\pi} \frac{e}{2m} \left(\frac{1}{2m} \vec{\nabla} \cdot \vec{E} + 2\frac{1}{2m} \vec{\sigma} \cdot \vec{L} \frac{1}{r} E(r) \right) \\ &= Z \frac{\alpha_R}{2\pi} \frac{e}{(2m)^2} \left(-e\delta^3(\vec{r}) + 2\vec{\sigma} \cdot \vec{L} \frac{e}{4\pi r^3} \right) \ . \end{aligned} \tag{1.126}$$

The effect of this potential on the $2S_{1/2}$, $2P_{1/2}$ states can then be calculated using simple perturbation theory

$\delta^3(r)$ term :
$$\Delta E_{2P_{1/2}} = 0$$
$$\Delta E_{2S_{1/2}} = \frac{Z\alpha_R^2}{2m^2} |\psi_{2S_{1/2}}(0)|^2 = \frac{Z^4 \alpha_R^5}{16\pi} m \tag{1.127}$$

$\vec{\sigma} \cdot \vec{L}$ term:
$$\Delta E_{2P_{1/2}} = \frac{Z\alpha_R^2}{4\pi m^2} \cdot -2 \left\langle \frac{1}{r^3} \right\rangle_{2P_{1/2}} = -\frac{Z^4 \alpha_R^5}{48\pi} m$$
$$\Delta E_{2S_{1/2}} = 0 \ . \tag{1.128}$$

The piece of the Lamb shift due to the anomalous moment then is

$$E_{2S_{1/2}} - E_{2P_{1/2}} = \frac{Z^4 \alpha_R^5}{12\pi} m = 67.9 \,\text{MHz} \tag{1.129}$$

which when added to the previously calculated components yields

$$E_{2S_{1/2}} - E_{2P_{1/2}} = 1048.5 \,\text{MHz} \ . \tag{1.130}$$

Finally, if small corrections for the finite proton size and higher order effects are included, one finds a theoretical prediction

$$\left. E_{2S_{1/2}} - E_{2P_{1/2}} \right|^{\text{th}} = (1057.845 \pm 0.003) \,\text{MHz} \tag{1.131}$$

in spectacular agreement with the experimental number

$$E_{2S_{1/2}} - E_{2P_{1/2}} = (1057.833 \pm 0.004) \,\text{MHz} \ . \tag{1.132}$$

Obviously, each of the components of this calculation — the vacuum polarization, the anomalous electron moment, the charge structure, $-27\,\text{MHz}$, $+68\,\text{MHz}$, $+62\,\text{MHz}$, respectively — gives a contribution to the final result which is much larger than the experimental error $\sim 4\,\text{KHz}$ and we may conclude that the presence of each of these effects is strongly confirmed!

Appendix VIII.1

Our goal in this appendix is to evaluate integrals of the form

$$\int \frac{d^d k}{(2\pi)^d} \frac{[1; k_\mu; k_\mu k_\nu]}{(k^2 - m_2^2 + i\epsilon)^\ell \left((k-q)^2 - m_1^2 + i\epsilon\right)^n} \equiv J_{;\mu;\mu\nu}^{n,\ell}(q) \ . \tag{1.133}$$

First consider the case $\ell = n = 1$, since higher order results can be obtained via successive differentiation with respect to m_1^2 and/or m_2^2. Also we Euclideanize the problem by rotating the k_0 contour onto the imaginary axis. Then using the identity

$$\int_0^\infty d\alpha \, e^{-\alpha\sigma} = -\frac{1}{\sigma} e^{-\alpha\sigma} \Big|_0^\infty = \frac{1}{\sigma} \tag{1.134}$$

we find

$$^E J_{;\mu;\mu\nu}^{1,1}(q) = \int_0^\infty d\alpha \int_0^\infty d\beta \int \frac{d^d k}{(2\pi)^d} \\ \times \exp - \left[\alpha \left((k-q)^2 + m_1^2\right) - \beta \left(k^2 + m_2^2\right)\right] [1; k_\mu; k_\mu k_\nu] \tag{1.135}$$

where the superscript indicates that we are employing the Euclideanized form. Using the results

$$\int_{-\infty}^\infty dk \, e^{-ak^2} = \left(\frac{\pi}{a}\right)^{1/2} , \quad \int_{-\infty}^\infty dk \, k^2 e^{-ak^2} = \frac{1}{2a} \left(\frac{\pi}{a}\right)^{1/2} \tag{1.136}$$

we can write in d-dimensions

$$\int d^d k \, e^{-ak^2} [1; k_\mu; k_\mu k_\nu] = \left(\frac{\pi}{a}\right)^{d/2} \left[1; 0; \frac{1}{2a} \delta_{\mu\nu}\right] . \tag{1.137}$$

Since

$$\exp - [\alpha \left((k-q)^2 + m_1^2\right) - \beta \left(k^2 + m_2^2\right)] \\ = \exp - \left[(\alpha + \beta)\left(k - \frac{\alpha}{\alpha+\beta} q\right)^2\right] \exp - \left(\frac{\alpha\beta}{\alpha+\beta} q^2 + \alpha m_1^2 + \beta m_2^2\right) \tag{1.138}$$

we find, changing variables to $s = k - \frac{\alpha}{\alpha+\beta} q$

$$^E J_{;\mu;\mu\nu}^{1,1}(q) \\ = \int_0^\infty d\alpha \int_0^\infty d\beta \int \frac{d^d s}{(2\pi)^d} \exp - \left[(\alpha+\beta)s^2 + \left(\frac{\alpha\beta}{\alpha+\beta} q^2 + \alpha m_1^2 + \beta m_2^2\right)\right] \\ \times \left[1; \left(s + \frac{\alpha}{\alpha+\beta} q\right)_\mu ; \left(s + \frac{\alpha}{\alpha+\beta} q\right)_\mu \left(s + \frac{\alpha}{\alpha+\beta} q\right)_\nu\right] \\ = \int_0^\infty d\alpha \int_0^\infty d\beta \exp - \left(\frac{\alpha\beta}{\alpha+\beta} q^2 + \alpha m_1^2 + \beta m_2^2\right) \left(\frac{\pi}{\alpha+\beta}\right)^{\frac{d}{2}} \frac{1}{(2\pi)^d} \\ \times \left[1; \frac{\alpha}{\alpha+\beta} q_\mu; \left(\frac{\alpha}{\alpha+\beta}\right)^2 q_\mu q_\nu + \frac{1}{2(\alpha+\beta)} \delta_{\mu\nu}\right] . \tag{1.139}$$

The α, β integration may be most easily performed by changing variables to ρ, x where
$$\alpha = \rho x \qquad \beta = \rho(1-x) \ . \tag{1.140}$$
The Jacobian is found to be
$$J = \begin{vmatrix} \dfrac{\partial \alpha}{\partial x} & \dfrac{\partial \alpha}{\partial \rho} \\ \dfrac{\partial \beta}{\partial x} & \dfrac{\partial \beta}{\partial \rho} \end{vmatrix} = \begin{vmatrix} \rho & x \\ -\rho & 1-x \end{vmatrix} = \rho \tag{1.141}$$
so that
$$\int_0^\infty d\alpha \int_0^\infty d\beta \ldots = \int_0^\infty d\rho\, \rho \int_0^1 dx \ldots \tag{1.142}$$
The ρ integration can now be done via [GrR 65]
$$\int_0^\infty d\rho\, \rho^{\nu-1} e^{-\rho a} = \Gamma(\nu) a^{-\nu} \tag{1.143}$$
yielding
$$\begin{aligned}
{}^E J^{1,1}_{;\mu;\mu\nu}(q) &= \int_0^\infty d\rho\, \rho \int_0^1 dx \exp -\rho\left(q^2 x(1-x) + xm_1^2 + (1-x)m_2^2\right) \\
&\quad \times \left(\frac{\pi}{\rho}\right)^{d/2} \frac{1}{(2\pi)^d} \left[1; xq_\mu; x^2 q_\mu q_\nu + \frac{1}{2\rho}\delta_{\mu\nu}\right] \\
&= \int_0^1 dx\, \frac{1}{(4\pi)^{d/2}} \Gamma\left(2-\frac{d}{2}\right) \left(q^2 x(1-x) + xm_1^2 + (1-x)m_2^2\right)^{\frac{d}{2}-2} \\
&\quad \times \left[1; xq_\mu; x^2 q_\mu q_\nu + \frac{1}{2}\delta_{\mu\nu} \frac{q^2 x(1-x) + xm_1^2 + (1-x)m_2^2}{1-\frac{d}{2}}\right] .
\end{aligned} \tag{1.144}$$
Since
$$\begin{aligned}
\frac{\Gamma(\ell)}{(k^2+m_2^2)^\ell} &= (-1)^{\ell-1} \frac{d^{\ell-1}}{d(m_2^2)^{\ell-1}} \frac{1}{k^2+m_2^2} \\
\frac{\Gamma(n)}{((k-q)^2+m_1^2)^n} &= (-1)^{n-1} \frac{d^{n-1}}{d(m_1^2)^{n-1}} \frac{1}{(k-q)^2+m_1^2}
\end{aligned} \tag{1.145}$$
we can generalize this result to
$$\begin{aligned}
{}^E J^{n,\ell}_{;\mu;\mu\nu}(q) &= (-1)^{\ell+n} \frac{\Gamma\left(n+\ell-\frac{d}{2}\right)}{\Gamma(n)\Gamma(\ell)} \frac{1}{(4\pi)^{d/2}} \\
&\quad \times \int_0^1 dx \left(q^2 x(1-x) + xm_1^2 + (1-x)m_2^2\right)^{\frac{d}{2}-n-\ell} x^{n-1}(1-x)^{\ell-1} \\
&\quad \times \left[1; xq_\mu; x^2 q_\mu q_\nu + \frac{1}{2}\delta_{\mu\nu} \frac{q^2 x(x-1) + xm_1^2 + (1-x)m_2^2}{n+\ell-1-\frac{d}{2}}\right] ,
\end{aligned} \tag{1.146}$$

the Minkowski space version of which is quoted in Eq. 1.25.

PROBLEM VIII.1.1

Vacuum Polarization and Dispersion Relations

We have found by explicit calculation the form of the vacuum polarization tensor
$$\Pi_{\mu\nu}(q) = \left(-q^2\eta_{\mu\nu} + q_\mu q_\nu\right)\Pi(q^2)$$
with
$$\Pi(q^2) = \frac{\alpha}{3\pi}\ln\frac{\Lambda^2}{m^2} - \frac{2\alpha}{\pi}\int_0^1 dy\, y(1-y)\ln\left(1 - \frac{q^2}{m^2}y(1-y)\right)$$
$$\equiv \frac{\alpha}{3\pi}\ln\frac{\Lambda^2}{m^2} + \Pi_f(q^2) \ .$$

i) Show that $\Pi(q^2)$ is an analytic function of q^2 with a branch cut extending from $4m^2 < q^2 < \infty$ and with a discontinuity given by
$$\text{disc } \Pi(q^2) = 2i\,\text{Im } \Pi(q^2) = 2\alpha R(q^2)$$
where
$$R(q^2) = \frac{1}{3}\sqrt{\frac{q^2 - 4m^2}{q^2}}\,\frac{2m^2 + q^2}{q^2}$$
is related to the rate for radiative pair creation via
$$\sum_f (2\pi)^4 \delta^4(q - p_f)\langle f|J_\mu^{\text{em}}|0\rangle^*\langle f|J_\nu^{\text{em}}|0\rangle = \left(-q^2\eta_{\mu\nu} + q_\mu q_\nu\right)2\alpha R(q^2)$$

ii) That such a relation must exist is clear from unitarity of the S-matrix. That is, since $S^\dagger S = 1$ we must have, using
$$S = 1 + iT$$
$$i\langle 0|T - T^\dagger|0\rangle = -\sum_n |\langle n|T|0\rangle|^2 \ .$$
Evaluation of this relation to second order in e^2 is just the expression relating $R(q^2)$ and disc $\Pi(q^2)$ derived in part i). Show this.

iii) Use Cauchy's theorem and the result of part i) to evaluate $\Pi_f(q^2)$. Demonstrate that
$$\Pi_f(q^2) = \frac{\alpha q^2}{\pi}\int_{4m^2}^\infty ds\, R(s)\frac{1}{s(s - q^2 - i\epsilon)}$$

iv) Show that our original form for $\Pi_f(q^2)$ can be written in this way by first changing variables to $x = 1 - 2y$ and integrating by parts
$$\Pi_f(q^2) = -\frac{\alpha}{2\pi}\int_0^1 dx\,\frac{d}{dx}\left(x - \frac{1}{3}x^3\right)\ln\left(1 - \frac{q^2(1 - x^2)}{4m^2 - i\epsilon}\right)$$
$$= \frac{\alpha}{2\pi}\int_0^1 dx\, 2x\left(x - \frac{1}{3}x^3\right)\frac{q^2}{4m^2 - q^2(1-x) - i\epsilon} \ .$$
Finally, change variables again to $s = 4m^2/1 - x^2$ and show that the dispersive result of part iii) obtains.

VIII.2 SPINLESS PARTICLES: ELECTROMAGNETIC INTERACTIONS

Armed with experience from the Dirac formalism, we return to the Klein-Gordon equation in order to show how covariant perturbation theory may be formulated for spinless particles.

Free Propagator

In the previous chapter (*cf.* Sect. VII.8) we identified the free Klein-Gordon propagator as

$$i\Delta_F^{(0)}(x_2 - x_1) = \int \frac{d^4q}{(2\pi)^4} e^{-iq\cdot(x_2-x_1)} \frac{i}{q^2 - m^2 + i\epsilon} \quad . \tag{2.1}$$

Writing

$$q^2 - m^2 + i\epsilon = (q_0 - \omega_q + i\epsilon)(q_0 + \omega_q - i\epsilon) \tag{2.2}$$

where

$$\omega_q = \sqrt{\vec{q}^{\,2} + m^2} \tag{2.3}$$

we can perform the q_0 integration by contour methods yielding

$$\begin{aligned} i\Delta_F^{(0)}(x_2 - x_1) = \theta(t_2 - t_1) &\int \frac{d^3q}{(2\pi)^3} \frac{1}{2\omega_q} f_q^{(+)}(x_2) f_q^{(+)*}(x_1) \\ + \theta(t_1 - t_2) &\int \frac{d^3q}{(2\pi)^3} \frac{1}{2\omega_q} f_q^{(-)}(x_2) f_q^{(-)*}(x_1) \end{aligned} \tag{2.4}$$

where $f_q^{\pm}(x) = \exp(\mp i\omega_q t \pm i\vec{q}\cdot\vec{x})$ are positive/negative energy solutions to the Klein-Gordon equation, respectively. We see explicitly that $\Delta_F^{(0)}$ involves a sum over positive energy solutions if $t_2 > t_1$ and negative energy solutions if $t_2 < t_1$, just as found for the Dirac case. The orthogonality conditions for these solutions, (*cf.* Sect. VI.1) read

$$\begin{aligned} \int f_p^{(\pm)*}(x) i\overleftrightarrow{\partial_0} f_q^{(\pm)}(x) d^3x &= \pm 2\omega_q (2\pi)^3 \delta^3(\vec{p} - \vec{q}) \\ \int f_p^{(\pm)*}(x) i\overleftrightarrow{\partial_0} f_q^{(\mp)}(x) d^3x &= 0 \end{aligned} \tag{2.5}$$

where the symbol $\overleftrightarrow{\partial_0}$ denotes

$$a\overleftrightarrow{\partial_0} b \equiv a\partial_0 b - (\partial_0 a)b \tag{2.6}$$

and guarantees that the propagator performs the task expected — taking positive energy solutions forward in time and negative energy solutions backward

$$\begin{aligned} \int i\Delta_F^{(0)}(x_2 - x_1) i\overleftrightarrow{\partial_0} f_p^{(+)}(x_1) d^3x_1 &= f_p^{(+)}(x_2)\theta(t_2 - t_1) \\ \int i\Delta_F^{(0)}(x_2 - x_1) i\overleftrightarrow{\partial_0} f_p^{(-)}(x_1) d^3x_1 &= -f_p^{(-)}(x_2)\theta(t_1 - t_2) \quad . \end{aligned} \tag{2.7}$$

Electromagnetic Interactions

Study of the free propagator is not our purpose here — rather we wish to examine the effect of electromagnetic interactions on the form of the Klein–Gordon wavefunction. In the presence of a vector potential $A_\mu(x)$ the modified version of the Klein–Gordon equation for a particle of charge e can be found via the "minimal" substitution

$$i\partial_\mu \to i\partial_\mu - eA_\mu(x) \tag{2.8}$$

so that

$$\left((\partial_\mu + ieA_\mu(x))(\partial^\mu + ieA^\mu(x)) + m^2\right)\phi(x) = 0 \ . \tag{2.9}$$

Writing Eq. 2.9 as

$$(\Box + m^2)\phi(x) = \left(-ie\{\partial^x_\mu, A^\mu(x)\} + e^2 A^2(x)\right)\phi(x), \tag{2.10}$$

we may find a formal solution to Eq. 2.9 by use of the Green's function, *i.e.*, the propagator

$$\phi(x) = \phi^{(0)}(x) - \int d^4y\, \Delta_F^{(0)}(x-y) \left(-ie\{\partial^y_\mu, A^\mu(y)\} + e^2 A^2(y)\right)\phi(y) \tag{2.11}$$

where $\phi^{(0)}(x)$ is a solution of the homogeneous (free) Klein–Gordon equation

$$(\Box + m^2)\phi^{(0)}(x) = 0 \ . \tag{2.12}$$

Eq. 2.11 may be solved iteratively to produce a perturbative solution for $\phi(x)$. The first iteration yields

$$\phi^{(1)}(x) = \phi^{(0)}(x) + i\int d^4y\, i\Delta_F^{(0)}(x-y)\left(-ie\{\partial^y_\mu, A^\mu(y)\} + e^2 A^2(y)\right)\phi^{(0)}(y) \ . \tag{2.13}$$

Note that we find interaction terms of first *and* second order in the electromagnetic coupling e. In order to find the full corrected wavefunction to second order in e, we must iterate once more then to find

$$\phi(x) = \phi^{(0)}(x) - ie\int d^4y\, i\Delta_F^{(0)}(x-y)\{i\partial^y_\mu, A^\mu(y)\}\phi^{(0)}(y)$$
$$+ ie^2\int d^4y\, i\Delta_F^{(0)}(x-y)A^2(y)\phi^{(0)}(y) + (-ie)^2\int d^4y\int d^4z\, i\Delta_F^{(0)}(x-y)$$
$$\times \{i\partial^y_\mu, A^\mu(y)\} i\Delta_F^{(0)}(y-z)\{i\partial^z_\nu, A^\nu(z)\}\phi^{(0)}(z) + \mathcal{O}(e^3) \ . \tag{2.14}$$

Suppose we begin at time $t_1 \to -\infty$ with a positive energy solution $f_{p_1}^{(+)}(x)$ and ask for the overlap of this wavefunction at time $t_2 \to +\infty$ with positive energy solution $f_{p_2}^{(+)}(x)$. The result to first order in e is

$$\text{Amp} = \int d^3x\, f_{p_2}^{(+)*}(x) i\overleftrightarrow{\partial_0^x} f_{p_1}^{(+)}(x)$$
$$- ie\int d^3x\int d^4y\, f_{p_2}^{(+)*}(x) i\overleftrightarrow{\partial_0^x} i\Delta_F^{(0)}(x-y)\{i\partial^y_\mu, A^\mu(y)\} f_{p_1}^{(+)}(y) \ . \tag{2.15}$$

Recalling that the propagator acting on the free particle wavefunction at x gives simply the corresponding solution at point y, i.e.,

$$\int d^3 x\, f_{p_2}^{(+)*}(x) i \overleftrightarrow{\partial_0^x} i\Delta_F^{(0)}(x-y) = f_{p_2}^{(+)*}(y) \tag{2.16}$$

we have then

$$\begin{aligned}
\text{Amp} &= (2\pi)^3 2\omega_{p_1} \delta^3(\vec{p}_2 - \vec{p}_1) - ie \int d^4 y\, f_{p_2}^{(+)*}(y) \{i\partial_\mu^y, A^\mu(y)\} f_{p_1}^{(+)}(y) + \cdots \\
&= (2\pi)^3 2\omega_{p_1} \delta^3(\vec{p}_2 - \vec{p}_1) - ie\, (p_2 + p_1)_\mu \int d^4 y\, e^{i(p_2 - p_1)\cdot y} A^\mu(y) \\
&= (2\pi)^3 2\omega_{p_1} \delta^3(\vec{p}_2 - \vec{p}_1) - ie\, (p_2 + p_1)_\mu\, \tilde{A}^\mu(p_2 - p_1)
\end{aligned} \tag{2.17}$$

where

$$\tilde{A}_\mu(q) = \int d^4 y\, e^{iq\cdot y} A_\mu(y) \tag{2.18}$$

is the Fourier transform of the vector potential.

Coulomb Scattering

As an elementary example, consider the case of scattering from a Coulomb potential associated with a heavy particle of charge $-Ze$ located at the origin

$$\tilde{A}^\mu(q) = \int d^4 y\, e^{iq\cdot y} \delta^{\mu 0} \frac{-Ze}{4\pi |\vec{y}|} = 2\pi \delta(q_0) \frac{-Ze}{\vec{q}^{\,2}} \delta^{\mu 0}\,. \tag{2.19}$$

From Eq. 2.17 the transition amplitude is

$$\text{Amp} = 2\pi i \delta(E_2 - E_1) \frac{Ze^2}{(\vec{p}_1 - \vec{p}_2)^2} (E_2 + E_1) \equiv -2\pi i \delta(E_2 - E_1) T_{fi} \tag{2.20}$$

which leads to a scattering cross section[†]

$$\begin{aligned}
d\sigma &= \frac{1}{2E_1 v_1} \int \frac{d^3 p_2}{(2\pi)^3} \frac{1}{2E_2} 2\pi \delta(E_2 - E_1) |T_{fi}|^2 \\
&= \frac{1}{2p_1} \frac{p_2^2}{2E_2} \frac{1}{(2\pi)^2} \frac{1}{dE_2/dp_2} d\Omega\, |T_{fi}|^2 \\
&= d\Omega \frac{Z^2 \alpha^2}{4 p_1^2 \beta^2 \sin^4 \frac{1}{2}\theta}\,.
\end{aligned} \tag{2.21}$$

This result agrees with the corresponding spin-1/2 result, Eq. VII.10.14, except for the factor $(1 - \beta^2 \sin^2(\theta/2))$, the physics of which was explored in Problem VII.10.1. Of course, in the non-relativistic limit — $\beta \approx (p/m) \ll 1$ — the Rutherford cross-section is reproduced.

[†] Here the factors of $\frac{1}{2E_1}, \frac{1}{2E_2}$ arise because of the normalization choice given in Eq. 2.5 for the spin zero states.

Compton Scattering

As a second example consider Compton scattering from a spin zero target of charge e and four momentum p_μ of a photon of four-momentum k_μ, polarization ϵ_μ into a final photon with k'_μ, ϵ'_μ and final spinless particle with four-momentum p'_μ. Calculation of this amplitude requires the correction to the wavefunction to second order in $eA_\mu(x)$. From the e^2A^2 term in Eq. 2.14 we find

$$\begin{aligned}
\text{Amp} &= ie^2 \int d^4y\, f_{p'}^{(+)*}(y)\, 2\epsilon^{*\prime} \cdot \epsilon\, e^{i(k'-k)\cdot y} f_p^{(+)}(y) \frac{1}{\sqrt{2\omega_k 2\omega_{k'}}} \\
&= i(2\pi)^4 \delta^4(p' + k' - p - k)\, 2e^2 \epsilon^{*\prime} \cdot \epsilon \frac{1}{\sqrt{2\omega_k 2\omega_{k'}}}
\end{aligned} \qquad (2.22)$$

which corresponds to the seagull diagram shown in Figure VIII.8. (The factor of two arises because *either* vector potential can generate the final state or the initial state photon wavefunction.)

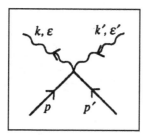

Fig. VIII.8: *The seagull diagram contributing to Compton scattering from a spinless particle.*

There exist two additional ways by which the scattering can occur to $\mathcal{O}(e^2)$ via the last component of Eq. 2.14. The two processes differ depending upon whether the outgoing photon interaction occurs first or last, as shown in Figure VIII.9.

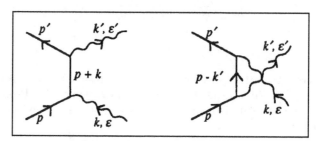

Fig. VIII.9: *Pole contributions to Compton scattering from a spinless particle.*

VIII.2 SPINLESS PARTICLES: ELECTROMAGNETIC INTERACTIONS

The corresponding amplitudes are given by

$$
\begin{aligned}
\text{Amp} &= \frac{(-ie)^2}{\sqrt{2\omega_k 2\omega_{k'}}} \int d^4y \int d^4z \int \frac{d^4q}{(2\pi)^4} \bigg(f_{p'}^{(+)*}(y) \\
&\quad \times (p'+q)\cdot \epsilon^{*} e^{ik'\cdot y} e^{-iq\cdot(y-z)} \frac{i}{q^2-m^2} (p+q)\cdot \epsilon\, e^{-ik\cdot z} f_p^{(+)}(z) \\
&\quad + f_{p'}^{(+)*}(y)\,(p'+q)\cdot \epsilon\, e^{-ik\cdot y} e^{-iq\cdot(y-z)} \frac{i}{q^2-m^2}(p+q)\cdot \epsilon^{*} e^{ik'\cdot z} f_p^{(+)}(z) \bigg) \\
&= -i(2\pi)^4 \delta^4(p+k-p'-k')\, e^2 \bigg(\frac{\epsilon^{*}\cdot(2p'+k')\,\epsilon\cdot(2p+k)}{(p+k)^2-m^2} \\
&\quad + \frac{\epsilon\cdot(2p'-k)\,\epsilon^{*}\cdot(2p-k')}{(p-k')^2-m^2} \bigg) \frac{1}{\sqrt{2\omega_k 2\omega_{k'}}} \,.
\end{aligned}
\tag{2.23}
$$

The full Compton scattering amplitude is found by adding Eqs. 2.22 and 2.23

$$
\text{Amp} = -i(2\pi)^4 \delta(p+k-p'-k')\, \epsilon^{\mu}\epsilon^{\nu*} T_{\mu\nu}
\tag{2.24}
$$

with

$$
T_{\mu\nu} = \frac{e^2}{\sqrt{2\omega_k 2\omega_{k'}}} \left(-2\eta_{\mu\nu} + \frac{(2p'+k')_\nu (2p+k)_\mu}{2p\cdot k} - \frac{(2p'-k)_\mu (2p-k')_\nu}{2p\cdot k'} \right)
\tag{2.25}
$$

We can verify gauge invariance via

$$
k^\mu T_{\mu\nu} = -\frac{e^2}{\sqrt{2\omega_k 2\omega_{k'}}} \left[2k_\nu - (2p'+k')_\nu + (2p-k')_\nu \frac{2p'\cdot k}{2p\cdot k'} \right] \,.
\tag{2.26}
$$

Since $2p'\cdot k = -(p'-k)^2 + m^2 = -(p-k')^2 + m^2 = 2p\cdot k'$ we have

$$
k^\mu T_{\mu\nu} = \frac{e^2}{\sqrt{2\omega_k 2\omega_{k'}}} 2(k-p'-k'+p)_\nu = 0 \quad \text{q.e.d.}
\tag{2.27}
$$

The differential scattering cross-section is calculated in the usual fashion. In the laboratory frame (*i.e.*, rest frame of the target) we find for unpolarized photons

$$
\begin{aligned}
d\sigma_{\text{lab}} &= \frac{1}{2m} \int \frac{d^3p'}{(2\pi)^3} \frac{1}{2E'} \frac{d^3k'}{(2\pi)^3} (2\pi)^4 \delta^4(p+k-p'-k') \\
&\quad \times \frac{1}{2} \sum_\lambda \sum_{\lambda'} \left| \epsilon_\lambda^\mu \epsilon_{\lambda'}^{\nu*} T_{\mu\nu} \right|^2 \\
&= \int \frac{1}{4mE'} \frac{1}{(2\pi)^2} k'^2 dk'\, d\Omega_{\hat{k}'}\, \delta\!\left(m+k^0-k^{0'} - \sqrt{m^2+(\vec{k}-\vec{k}')^2} \right) \\
&\quad \times \frac{1}{2} \sum_\lambda \sum_{\lambda'} \left| \epsilon_\lambda^\mu \epsilon_{\lambda'}^{\nu*} T_{\mu\nu} \right|^2 \,.
\end{aligned}
\tag{2.28}
$$

If we employ a gauge choice such that $\epsilon_\mu, \epsilon'_\mu$ have vanishing time components in the laboratory frame then
$$\epsilon \cdot p = \epsilon' \cdot p = 0 \tag{2.29}$$
and the matrix element simplifies to
$$\epsilon^\mu \epsilon^{\nu*'} T_{\mu\nu} = \frac{-e^2}{\sqrt{2\omega_k 2\omega_{k'}}} \left(2\epsilon \cdot \epsilon^{*'}\right) . \tag{2.30}$$
The cross-section becomes
$$\frac{d\sigma}{d\Omega_{\hat{k}'}} = \frac{\alpha^2}{4m^2} \frac{k'}{k} \frac{1}{1 + \frac{k}{m}(1 - \cos\theta)} \frac{1}{2} \sum_{\lambda,\lambda'} \left|2\epsilon \cdot \epsilon^{*'}\right|^2 . \tag{2.31}$$
The polarization sum was performed previously in Sect. IV.5 where we found
$$\frac{1}{2} \sum_{\lambda,\lambda'} \left|\epsilon \cdot \epsilon^{*'}\right|^2 = \frac{1}{2}(1 + \cos^2\theta) . \tag{2.32}$$
The spin zero Compton scattering cross-section is finally
$$\begin{aligned}\frac{d\sigma}{d\Omega_{\hat{k}'}} &= \frac{\alpha^2}{m^2} \frac{k'}{k} \frac{1}{1 + \frac{k}{m}(1 - \cos\theta)} \frac{1}{2}(1 + \cos^2\theta) \\ &= \frac{\alpha^2}{2m^2} \frac{k'^2}{k^2}(1 + \cos^2\theta)\end{aligned} \tag{2.33}$$
where we have used the relation (*cf.* Eq. VII.10.117)
$$\frac{k'}{k} = \frac{1}{1 + \frac{k}{m}(1 - \cos\theta)} . \tag{2.34}$$
The cross section is similar but not identical to the corresponding result for spin-1/2 Compton scattering. However, in the non-relativistic limit — $k/m \ll 1$ — we have $k \approx k'$ and the differential cross-section reduces to the Thomson form—Eq. IV.5.30—as expected.

PROBLEM VIII.2.1

Gravitational Scattering of Spinless Particles

As we saw in the previous chapter, the equation of motion of the photon field $A_\mu(x)$ is
$$\Box A_\mu(x) = j_\mu(x)$$
where $j_\mu(x)$ is the current density, and the corresponding term in the interaction Hamiltonian which couples $A_\mu(x)$ and $j_\mu(x)$ is given by
$$H_{\text{int}} = \int d^3x\, j_\mu(x) A^\mu(x) .$$

VIII.2 SPINLESS PARTICLES: ELECTROMAGNETIC INTERACTIONS

A similar formalism can be used to describe gravity. In this case the gravtion field, which carries spin two, is described by a symmetric second rank tensor $h_{\mu\nu}(x)$ and obeys the equation of motion

$$\Box h_{\mu\nu}(x) = -16\pi G \left(T_{\mu\nu}(x) - \frac{1}{2}\eta_{\mu\nu}\mathrm{Tr}T \right)$$

where G is the gravitational constant and $T_{\mu\nu}(x)$ is the energy-momentum tensor. The corresponding term in the interaction Hamiltonian is

$$H_{\mathrm{int}} = \frac{1}{2}\int d^3x\, T_{\mu\nu}(x) h^{\mu\nu}(x)$$

where for a scalar field $\phi(x)$ we have

$$\langle p_2 | T_{\mu\nu}(x) | p_1 \rangle = e^{i(p_2-p_1)\cdot x}\left[p_{2\mu}p_{1\nu} + p_{1\mu}p_{2\nu} - \eta_{\mu\nu}(p_1\cdot p_2 - m^2) \right] \frac{1}{\sqrt{4E_1 E_2}}\ .$$

Here the relation between the graviton field $h_{\mu\nu}$ and the metric $g_{\mu\nu}$ is

$$g_{\mu\nu} = \eta_{\mu\nu} + h_{\mu\nu}\ .$$

i) Consider a very massive particle M (which could, for example, be the sun) so that

$$\langle p_2' | T_{\mu\nu}(x) | p_1' \rangle \simeq e^{i(p_2'-p_1')\cdot x} M \delta_{\mu 0}\delta_{\nu 0}\ .$$

Using Fourier transform methods, solve the equation of motion for $h_{\mu\nu}(x)$ and show that

$$h_{00}(x) = -\frac{2GM}{|\vec{x}|} \qquad h_{ij}(x) = -\frac{2GM}{|\vec{x}|}\delta_{ij} \qquad h_{\mu 0} = 0$$

when the mass M is placed at the origin of coordinates.

ii) Consider the scattering of a much lighter particle of mass m by M. Show that the scattering amplitude is given by

$$\mathrm{Amp} = \frac{-1}{\sqrt{4E_1 E_2}}\int d^4x \int d^4y \int \frac{d^4q}{(2\pi)^4} e^{i(p_2'-p_1')\cdot x} M(2\delta^{\mu 0}\delta^{\nu 0} - \eta^{\mu\nu}) e^{iq\cdot(x-y)}$$

$$\times \frac{4\pi G}{q^2}e^{i(p_2-p_1)\cdot y}\left[p_{2\mu}p_{1\nu} + p_{1\mu}p_{2\nu} - \eta_{\mu\nu}(p_1\cdot p_2 - m^2)\right]$$

$$= -(2\pi)^4\delta^4(p_1 + p_1' - p_2 - p_2')\frac{8\pi GM^2}{(p_1-p_2)^2}\frac{(4E_1 E_2 - 2m^2)}{2M\sqrt{4E_1 E_2}}$$

$$\equiv (2\pi)^4\delta^4(p_1 + p_1' - p_2 - p_2')T$$

iii) Evaluate the Fourier transform of the static limit of the transition amplitude and show that the mutual interaction of m and M is described by the potential energy

$$V(r) = \int \frac{d^3q}{(2\pi)^3} e^{-i\vec{q}\cdot\vec{r}} T(\vec{q}) = -\frac{GMm}{r}$$

which is the conventional Newtonian result.

iv) Calculate the differential scattering cross section using the transition amplitude T and show that

$$d\sigma = \frac{E_1}{p_1} \int \frac{d^3 p_2}{(2\pi)^3} 2\pi \delta(E_2 - E_1) |T|^2 = (2GM)^2 \frac{(2E_1^2 - m^2)^2}{\vec{q}^4} d\Omega$$

$$= \frac{(2GM)^2}{16 p_1^4 \sin^4 \frac{\theta}{2}} (2E_1^2 - m^2)^2 d\Omega$$

v) We can connect with the corresponding classical scattering result by using the relation between impact parameter and scattering angle

$$2\pi b \, db = d\sigma \ .$$

Show that in the massless—$m = 0$—scattering limit this relation can be integrated to become

$$\theta = \frac{4GM}{b}$$

for small angle scattering. This is the relationship derived by Einstein and has been tested via starlight deflection measurements during eclipses. (Of course, these experiments involved photons of unit spin rather than spinless particles as considered above. However, general relativity requires that all massless particles, regardless of spin, must follow identical trajectories.)

VIII.3 PATH INTEGRALS AND QUANTUM FIELD THEORY*

In the case of non-relativistic quantum mechanics we found that the theory could be formulated equally well via traditional wavefunction methods and path integral techniques. In the relativistic analog we have seen how a covariant perturbation series can be generated within the context of a wavefunction approach. It is also possible to derive such a series from a path integral framework, and we shall end our formal presentation in this book by outlining how this is accomplished. The field theoretic methods which we describe are now widespread and underlie much of contemporary work in particle, nuclear and condensed matter physics. Consequently there are a number of texts which survey this material in depth [see, e.g. Ra 81] and it is not our purpose to attempt a detailed development of field theory via path integral methods. Nevertheless, it is interesting and important to see how the simple quantum mechanical techniques can be generalized and we thus present here a brief summary.

An advantage of the functional approach to quantum mechanics discussed in Chapter I is that it can be taken over rather directly to quantum field theory. An important difference is that instead of trajectories $x(t)$ which pick out a particular point in space at a given time, one must deal with fields $\phi(\vec{x}, t)$ which are defined at *all* points in space at a given time t. Instead of a sum $\int \mathcal{D}[x(t)]$ over trajectories one has a sum $\int [d\phi(x)]$ over all possible field configurations. Nevertheless, the analogy is rather direct. The formal transition from quantum mechanics to field theory can be accomplished by partitioning spacetime — both space *and* time — into a set of tiny four-dimensional cubes of volume $\delta t\, \delta x\, \delta y\, \delta z$. Within each cube one takes the field

$$\phi(x_i, y_j, z_k, t_\ell) \tag{3.1}$$

as a constant. Derivatives are defined in terms of differences between fields in neighboring blocks, *e.g.*

$$\left. \partial_t \phi \right|_{x_i, y_j, z_k, t_\ell} \simeq \frac{1}{\delta t} [\phi(x_i, y_j, z_k, t_\ell + \delta t) - \phi(x_i, y_j, z_k, t_\ell)] \ . \tag{3.2}$$

The Lagrange density is easily found

$$\left. \mathcal{L}(\phi, \partial_\mu \phi) \right|_{x_i, y_j, z_k, t_\ell} \simeq \mathcal{L}[\phi(x_i, y_j, z_k, t_\ell), \partial_\mu \phi(x_i, y_j, z_k, t_\ell)] \tag{3.3}$$

and the corresponding action can be written as

$$S \simeq \sum_{ijk\ell} \delta x\, \delta y\, \delta z\, \delta t\, \mathcal{L}[\phi(x_i, x_j, z_k, t_\ell)\, \partial_\mu \phi(x_i, y_j, z_k, t_\ell)] \ . \tag{3.4}$$

The field theoretic analog of the path integral can then be constructed by summing over all possible field values in each cell

$$F \sim \prod_{ikj\ell} \int_{-\infty}^{\infty} d\phi(x_i, y_j, z_k, t_\ell) \exp iS[\phi(x_i, y_j, z_k, t_\ell), \partial_\mu \phi(x_i, y_j, z_k, t_\ell)] \ . \tag{3.5}$$

Formally, in the limit in which the cell size is taken to zero Eq. 3.5 is written as

$$F \sim \int [d\phi(x)] \exp iS[\phi, \partial_\mu \phi] \ . \tag{3.6}$$

By analogy with the quantum mechanical situation, since the time integration in the action, Eq. 3.4, is from $-\infty$ to $+\infty$, it is suggestive that this amplitude is to be identified with the vacuum-to-vacuum amplitude

$$\langle 0|0\rangle \sim N \int [d\phi(x)] \exp iS[\phi, \partial_\mu \phi] \ . \tag{3.7}$$

Generally quantum field theory is formulated in terms of vacuum expectation values of time ordered products of fields — the Green's functions of the theory — and it is conventional to fix the normalization constant N by dividing out the vacuum-to-vacuum amplitude. Thus we have

$$G^{(n)}(x_1, x_2 \ldots x_n) = \frac{\langle 0|T(\phi(x_1)\ldots\phi(x_n))|0\rangle}{\langle 0|0\rangle} \tag{3.8}$$

and by analogy to the quantum mechanical case one is led to the path integral definition

$$G^{(n)}(x_1, x_2, \ldots x_n) = \frac{\int [d\phi(x)] \exp iS[\phi, \partial_\mu \phi] \, \phi(x_1) \ldots \phi(x_n)}{\int [d\phi(x)] \exp iS[\phi, \partial_\mu \phi]} . \tag{3.9}$$

These Green's functions can most straightforwardly be evaluated by use of a generating functional (*cf.* Sect. I.6)

$$W[j] = N \int [d\phi(x)] \exp\left(iS[\phi, \partial_\mu \phi] + i \int d^4x j(x)\phi(x)\right) \tag{3.10}$$

in terms of which[†]

$$G^{(n)}(x_1, x_2, \ldots x_n) = (-i)^n \frac{1}{W[0]} \frac{\delta^n}{\delta j(x_1) \ldots \delta j(x_n)} W[j]\bigg|_{j=0} . \tag{3.11}$$

For our purpose we shall deal only with the propagator or two-point function

$$G^{(2)}(x_1, x_2) = (-i)^2 \frac{1}{W[0]} \frac{\delta^2}{\delta j(x_1)\delta j(x_2)} W[j]\bigg|_{j=0} . \tag{3.12}$$

Consider the application of this formalism to free scalar field theory. For simplicity we consider a neutral particle so that the field ϕ may be taken to be hermitian. The Lagrangian density then is given by

$$\mathcal{L}^{(0)}(x) = \frac{1}{2}\partial_\mu \phi(x)\partial^\mu \phi(x) - \frac{1}{2}m^2\phi^2(x) . \tag{3.13}$$

That this is the appropriate form can be verified by use of the Euler–Lagrange relation

$$\partial_\mu \frac{\delta \mathcal{L}^{(0)}}{\delta \partial_\mu \phi(x)} - \frac{\delta \mathcal{L}^{(0)}}{\delta \phi(x)} = 0 \tag{3.14}$$

which yields the Klein-Gordon equation

$$(\Box + m^2)\phi(x) = 0. \tag{3.15}$$

The generating functional $W^{(0)}[j]$ is given by

$$W^{(0)}[j] = N \int [d\phi(x)] \exp i \int d^4x \left(\frac{1}{2}\partial_\mu \phi(x)\partial^\mu \phi(x) - \frac{1}{2}m^2\phi^2(x) + j(x)\phi(x)\right) , \tag{3.16}$$

[†] Here functional differentiation is defined via

$$\frac{\delta j(y)}{\delta j(x)} = \frac{\delta}{\delta j(x)} \int d^4x \delta^4(y-x) j(x) = \delta^4(y-x) .$$

where, in order to make the integral convergent for large ϕ^2 it is necessary to give the mass a negative imaginary part

$$m^2 \to m^2 - i\epsilon \ . \tag{3.17}$$

Integrating by parts, Eq. 3.16 becomes

$$W^{(0)}[j] = \int [d\phi(x)] \exp\left(-\frac{i}{2} \int d^4x \int d^4y\, \phi(x)\mathcal{O}(x-y)\phi(y) + \int d^4x\, j(x)\phi(x)\right) \tag{3.18}$$

where

$$\mathcal{O}(x-y) = (\Box_x + m^2 - i\epsilon)\, \delta^4(x-y) \equiv \langle x\,|\Box + m^2 - i\epsilon|\,y\rangle \ . \tag{3.19}$$

Finally, defining a shifted field†

$$\phi'(x) = \phi(x) - \int d^4y\, \mathcal{O}^{-1}(x-y) j(y) \tag{3.20}$$

we obtain

$$\begin{aligned} W^{(0)}[j] &= N \int [d\phi(x)] \exp\left(-\frac{i}{2}\int d^4x \int d^4y \left(\phi'(x)\mathcal{O}(x-y)\phi'(y) - j(x)\mathcal{O}^{-1}(x-y)j(y)\right)\right) \\ &= \left[N \int [d\phi'(x)]\exp\left(-\frac{i}{2}\int d^4x \int d^4y\, \phi'(x)\mathcal{O}(x-y)\phi'(y)\right)\right] \\ &\quad \times \exp\frac{i}{2}\int d^4x \int d^4y\, j(x)\mathcal{O}^{-1}(x-y) j(y) \end{aligned} \tag{3.21}$$

where we have used

$$[d\phi(x)] = [d\phi'(x)] \ . \tag{3.22}$$

---

† Here the inverse operator is defined via the relation

$$\int d^4x\, \mathcal{O}^{-1}(z-x)\mathcal{O}(x-y) = \delta^4(z-y)$$

and is given by

$$\mathcal{O}^{-1}(z-x) = \frac{1}{\Box_z + m^2 - i\epsilon}\delta^4(z-x) \equiv \left\langle z\,\left|\frac{1}{\Box + m^2 - i\epsilon}\right|\,x\right\rangle \ .$$

In Eq. 3.21 we recognize the factor in brackets as $W^{(0)}[0]$. Thus

$$W^{(0)}[j] = W^{(0)}[0] \exp \frac{i}{2} \int d^4x \int d^4y \, j(x) \mathcal{O}^{-1}(x-y) j(y) \;, \qquad (3.23)$$

and we determine the scalar propagator as[†]

$$\begin{aligned} G^{(2)}(x_1, x_2) &= (-i)^2 \frac{\delta^2}{\delta j(x_1) \delta j(x_2)} \frac{1}{W^{(0)}[0]} W^{(0)}[j]\bigg|_{j=0} \\ &= -i\mathcal{O}^{-1}(x_1 - x_2) \qquad (3.24) \\ &= i\int \frac{d^4k}{(2\pi)^4} \frac{e^{-ik\cdot(x_1-x_2)}}{k^2 - m^2 + i\epsilon} \equiv i\Delta_F^{(0)}(x_1 - x_2) \end{aligned}$$

in agreement with the result found in the previous section.

Propagator and Electromagnetic Interactions

Now consider the effect of an external electromagnetic field described by a vector potential $A_\mu(x)$. We can no longer employ a hermitian field. Rather we must introduce a complex field ϕ, since particle and antiparticle fields differ. It is convenient to represent the complex field $\phi(x)$ in terms of a linear superposition of hermitian fields $\phi_1(x), \phi_2(x)$ via

$$\phi(x) = \sqrt{\frac{1}{2}}(\phi_1(x) + i\phi_2(x)) \qquad \phi^*(x) = \sqrt{\frac{1}{2}}(\phi_1(x) - i\phi_2(x)) \qquad (3.25)$$

Using the minimal substitution

$$\partial_\mu^x \to \partial_\mu^x + ieA_\mu(x) \equiv D_\mu \qquad (3.26)$$

the Lagrangian becomes

$$\mathcal{L}(x) = (D_\mu^x \phi(x))^* D_x^\mu \phi(x) - m^2 \phi^*(x)\phi(x) \;. \qquad (3.27)$$

and the calculation of the generating functional proceeds as before yielding

$$W[j, j^*] = N \int [d\phi(x)][d\phi^*(x)] \exp i \int d^4x \, \big((D_\mu^x \phi(x))^* D_x^\mu \phi(x) - m^2 \phi^*(x)\phi(x) \\ + j^*(x)\phi(x) + j(x)\phi^*(x) \big) \;. \qquad (3.28)$$

[†] The explicit form for the operator $\mathcal{O}^{-1}(x_1 - x_2)$ is obtained by insertion of a complete set of momentum states

$$\frac{1}{\Box_{x_1} + m^2 - i\epsilon} \delta^4(x_1 - x_2) = -\int \frac{d^4k}{(2\pi)^4} \frac{e^{-ik\cdot(x_1-x_2)}}{k^2 - m^2 + i\epsilon} \;.$$

We can at this point emulate the free particle calculation but with the modified operator

$$\tilde{\mathcal{O}}(x-y) = \left(D_\mu^x D_x^\mu + m^2 - i\epsilon\right)\delta^4(x-y) \ . \tag{3.29}$$

The Green's function becomes

$$\begin{aligned} G^{(2)}(x_1, x_2) &= (-i)^2 \frac{\delta^2}{\delta j(x_2)\delta j^*(x_1)} \frac{1}{W[0,0]} W[j, j^*]\bigg|_{j=j^*=0} \\ &= -i\tilde{\mathcal{O}}^{-1}(x_1 - x_2) \\ &= \left\langle x_1 \left| \frac{i}{[i\partial - eA(\hat{x})]_\mu [i\partial - eA(\hat{x})]^\mu - m^2 + i\epsilon} \right| x_2 \right\rangle \ . \end{aligned} \tag{3.30}$$

Here

$$\begin{aligned} \langle x_1 | [i\partial - eA(\hat{x})]_\mu [i\partial - eA(\hat{x})]^\mu | x_2 \rangle \\ = \left(-\Box_{x_1} - ie\{A_\mu(x_1), \partial^\mu_{x_1}\} + e^2 A^2(x_1)\right) \delta^4(x_1 - x_2) \end{aligned} \tag{3.31}$$

and it is no longer possible to calculate the propagator exactly. Nevertheless, a perturbative expansion in powers of e can be generated using the identity

$$\frac{1}{\hat{A} - \hat{B}} = \frac{1}{\hat{A}} + \frac{1}{\hat{A}} \hat{B} \frac{1}{\hat{A}} + \frac{1}{\hat{A}} \hat{B} \frac{1}{\hat{A}} \hat{B} \frac{1}{\hat{A}} + \ldots \tag{3.32}$$

We find then

$$\begin{aligned} -iG^{(2)}(x_1 - x_2) = \bigg\langle x_2 \bigg| & \frac{-1}{\Box + m^2 - i\epsilon} + \frac{-1}{\Box + m^2 - i\epsilon} e\{A_\mu(\hat{x}), i\partial^\mu\} \frac{-1}{\Box + m^2 - i\epsilon} \\ & - \frac{-1}{\Box + m^2 - i\epsilon} e^2 A^2(\hat{x}) \frac{-1}{\Box + m^2 - i\epsilon} + \frac{-1}{\Box + m^2 - i\epsilon} e\{A_\mu(\hat{x}), i\partial^\mu\} \frac{-1}{\Box + m^2 - i\epsilon} \\ & \times e\{A_\nu(\hat{x}), i\partial^\nu\} \frac{-1}{\Box + m^2 - i\epsilon} + \ldots \bigg| x_2 \bigg\rangle, \end{aligned} \tag{3.33}$$

and using completeness we can write

$$\begin{aligned} G^{(2)}(x_1, x_2) = i\Delta_F^{(0)}(x_1, x_2) - i\int d^4z \, i\Delta_F^{(0)}(x_1, z) \left(ie\{A_\mu(z), \partial_z^\mu\}\right. \\ \left. - e^2 A^2(z)\right) i\Delta_F^{(0)}(z, x_2) + (-i)^2 \int d^4z \int d^4y \, i\Delta_F^{(0)}(x_1, z) \, ie\{A_\mu(z), \partial_z^\mu\} \\ \times i\Delta_F^{(0)}(z-y) ie\{A_\nu(y), \partial_y^\nu\} i\Delta_F^{(0)}(y - x_2) + \mathcal{O}(e^3) \ . \end{aligned} \tag{3.34}$$

This propagator may now be used to study the interaction of a charged scalar particle with an electromagnetic field. As an example, consider Compton scattering from a spinless particle of charge e and mass m. Using for $A_\mu(x)$ the photon field

$$A_\mu(x) = \epsilon_\mu \frac{1}{\sqrt{2\omega_k}} e^{-ik\cdot x} + \epsilon_\mu^* \frac{1}{\sqrt{2\omega_k}} e^{ik\cdot x} \tag{3.35}$$

and recalling that $\Delta_F^{(0)}$ propagates a (positive energy) free field solution via

$$\int d^3x_1\, i\Delta_F^{(0)}(x_2,x_1)\, i\overleftrightarrow{\partial}_0 \phi_p^{(0)(+)}(x_1) = \phi_p^{(0)(+)}(x_2) \tag{3.36}$$

we have

$$\begin{aligned}
\mathrm{Amp}\left(\pi_{p_1}^+ \gamma_1 \to \pi_{p_2}^+ \gamma_2\right) &= \int d^3x\, \phi_{p_2}^{(0)(+)}(x)\phi^{(+)}(x) \\
&= \frac{e^2}{\sqrt{2\omega_1 2\omega_2}} \Bigg(\int d^4x\, e^{i(p_2+k_2-p_1-k_1)\cdot x} 2i\epsilon_1 \cdot \epsilon_2^* \\
&\quad - \int d^4x \int d^4y \int \frac{d^4s}{(2\pi)^4} e^{i(p_2+k_2-s)\cdot x} \epsilon_2^* \cdot (p_2+s) \\
&\quad \times \frac{i}{s^2 - m^2 + i\epsilon} e^{i(s-p_1-k_1)\cdot y} \epsilon_1 \cdot (p_1+s) \\
&\quad - \int d^4x \int d^4y \int \frac{d^4s}{(2\pi)^4} e^{i(p_2-k_1-s)\cdot x} \epsilon_1 \cdot (p_2+s) \\
&\quad \times \frac{i}{s^2 - m^2 + i\epsilon} e^{i(s-p_1+k_2)\cdot y} \epsilon_2^* \cdot (p_1+s) \Bigg) \\
&\equiv -i(2\pi)^4 \delta^4(p_1+k_1-p_2-k_2)\, \epsilon_1^\mu \epsilon_2^{\nu*} T_{\mu\nu}
\end{aligned} \tag{3.37}$$

where

$$\begin{aligned}
T_{\mu\nu} &= \frac{e^2}{\sqrt{2\omega_1 2\omega_2}} \Bigg(-2\eta_{\mu\nu} + (2p_1+k_1)_\mu (2p_2+k_2)_\nu \frac{1}{(p_1+k_1)^2 - m^2} \\
&\quad + (2p_1-k_2)_\mu (2p_2-k_1)_\nu \frac{1}{(p_1-k_2)^2 - m^2} \Bigg)
\end{aligned} \tag{3.38}$$

as before. The connection with relativistic Feynman diagrams shown in Figs. VIII.8,9 is clear.

Fermions

Thus far, our development of quantum field theory has been based upon the simple example of scalar fields. For completeness it is important to treat also the case of fermion fields wherein the requirements of antisymmetry impose important modifications to the functional integration techniques thus far developed. The key to the treatment of anticommuting fields is the use of so-called Grassmann variables. Thus, while ordinary c-number quantities (hereafter denoted by roman letters) $a, b, \ldots$ commute with one another

$$[a,a] = [a,b] = [a,c] = \ldots = 0 \ . \tag{3.39}$$

Grassmann numbers (hereafter denoted by Greek letters) $\alpha, \beta, \ldots$ anticommute even though they are c-number quantities

$$\{\alpha,\alpha\} = \{\alpha,\beta\}, = \{\alpha,\gamma\} = \ldots 0 \ . \tag{3.40}$$

VIII.3 PATH INTEGRALS AND QUANTUM FIELD THEORY*

This means that the square of a Grassmann quantity must vanish

$$\alpha^2 = \beta^2 = \gamma^2 = \ldots = 0 \tag{3.41}$$

and that any function of Grassmann variables must have a very simple expansion

$$\begin{aligned} f(\alpha) &= f_0 + f_1 \alpha \\ g(\alpha, \beta) &= g_0 + g_1 \alpha + g_2 \beta + g_3 \alpha \beta \end{aligned} \tag{3.42}$$

Differentiation is defined correspondingly via

$$\begin{aligned} \left\{ \frac{d}{d\alpha}, \alpha \right\} &= \left\{ \frac{d}{d\beta}, \beta \right\} = \ldots = 1 \\ \left\{ \frac{d}{d\alpha}, \beta \right\} &= \left\{ \frac{d}{d\alpha}, \gamma \right\} = \ldots = 0 \end{aligned} \tag{3.43}$$

Thus

$$\begin{aligned} \frac{d}{d\alpha} f(\alpha) &= f_1 \\ \frac{d}{d\beta} g(\alpha, \beta) &= g_2 - g_3 \alpha \end{aligned}, \tag{3.44}$$

and second derivatives have the property

$$\frac{d^2}{d\alpha d\alpha} = 0 \ . \tag{3.45}$$

We must also define the concept of Grassmann integration. Since we demand that integration be translation invariant

$$\int d\alpha \, f(\alpha) = \int d\alpha \, f(\alpha + \beta) \tag{3.46}$$

we require

$$\int d\alpha \, f_1 \beta = 0 \quad i.e., \quad \int d\alpha = 0 \ . \tag{3.47}$$

We normalize the diagonal integral via

$$\int d\alpha \, \alpha = 1 \ , \tag{3.48}$$

so that

$$\int d\alpha \, f(\alpha) = f_1 \ . \tag{3.49}$$

The formalism for treating Fermi fields can now be developed in parallel to that for the scalar field case. Using the free field Lagrangian density

$$\mathcal{L}_0 \left(\bar{\psi}(x), \psi(x) \right) = \bar{\psi}(x) \left(i \not{\nabla}_x - m \right) \psi(x) \tag{3.50}$$

the generating functional for the free spin 1/2 field becomes

$$W[\eta,\bar\eta] = \int [d\psi(x)][d\bar\psi(x)] \exp\left(i\int d^4x \int d^4y\, \bar\psi(x)\mathcal{O}(x-y)\psi(y)\right.$$
$$\left. + i\int d^4x\, \bar\eta(x)\psi(x) + i\int d^4x\, \bar\psi(x)\eta(x)\right) \qquad (3.51)$$

where

$$\mathcal{O}(x-y) = (i\,\slashed\nabla_x - m + i\epsilon)\,\delta^4(x-y) \qquad (3.52)$$

and $\bar\eta(x), \eta(x)$ are Grassmann fields. Changing variables to

$$\psi'(x) = \psi(x) + \int d^4y\, \mathcal{O}^{-1}(x-y)\eta(y)$$
$$\bar\psi'(x) = \bar\psi(x) + \int d^4y\, \bar\eta(y)\mathcal{O}^{-1}(y-x) \qquad (3.53)$$

we find that an alternative form for the generating functional is

$$W[\eta,\bar\eta] = \int [d\psi'(x)][d\bar\psi'(x)] \exp\left(i\int d^4x \int d^4y\, [\bar\psi'(x)\mathcal{O}(x-y)\psi'(y)\right.$$
$$\left. - \bar\eta(x)\mathcal{O}^{-1}(x-y)\eta(y)]\right) \qquad (3.54)$$
$$= W[0,0]\exp\left(-i\int d^4x \int d^4y\, \bar\eta(x)\mathcal{O}^{-1}(x-y)\eta(y)\right).$$

The two particle Green's function is given by

$$G^{(2)}(x_1,x_2) = (-i)^2 \frac{1}{W[0,0]} \frac{\delta^2 W[\eta,\bar\eta]}{\delta\eta(x_2)\delta\bar\eta(x_1)}\bigg|_{\eta=\bar\eta=0}$$
$$= i\mathcal{O}^{-1}(x_1-x_2) = \int \frac{d^4k}{(2\pi)^4} e^{-ik\cdot(x_1-x_2)} \frac{i}{\slashed k - m + i\epsilon} \qquad (3.55)$$

which is the usual Feynman propagator (*cf.* Sect. VII.8).

Electromagnetic effects may be included via the minimal replacement

$$\nabla_\mu \to \nabla_\mu + ieA_\mu \qquad (3.56)$$

whereby the propagator becomes

$$iS_F^A(x_1,x_2) = \left\langle x_1 \left| \frac{i}{i\slashed\nabla - e\slashed A(\hat x) - m + i\epsilon} \right| x_1 \right\rangle \qquad (3.57)$$

which is exact but no longer soluble. As in the bosonic case using completeness we can develop a perturbation series

$$iS_F^A(x_1,x_2) = iS_F^{(0)}(x_1,x_2) - ie\int d^4y\, iS_F^{(0)}(x_1,y)\,\slashed A(y)\,iS_F^{(0)}(y,x_2)$$
$$+ (-ie)^2 \int d^4y \int d^4z\, iS_F^{(0)}(x_1,y)\,\slashed A(y)\,iS_F^{(0)}(y,z)\,\slashed A(z)\,iS_F^{(0)}(z,x_2) + \mathcal{O}(e^3) \qquad (3.58)$$

which is identical to the form found by wavefunction methods in Sect. VII.9.

VIII.4 PION EXCHANGE AND STRONG INTERACTIONS*

As our final topic we examine a semi-realistic problem wherein the techniques which have been developed above may be applied—the pion-nucleon interaction. First, however, it is useful to present a brief introduction.

Pions and the Nuclear Interaction

We have seen that the idea of the electromagnetic interaction being due to photon exchange is in spectacular agreement with experiment. It is suggestive that other interactions may be viewed similarly. This thought occurred to Yukawa during the early 1930's when he was considering the form of the "strong interaction" responsible for nuclear binding. That a powerful nuclear glue must exist is clear from the fact that, considering the protons alone, there exists a repulsive interaction

$$U \sim \frac{Z(Z-1)\alpha}{2R_{av}}$$

where $\frac{1}{2}Z(Z-1)$ is the number of proton pairs and R_{av} is some average proton separation. As Z increases, this repulsive energy rapidly becomes larger and eventually overwhelms the attractive strong nucleon-nucleon interaction, leading to the observed instability of nuclei with $Z > 83$. In fact, a very good fit to all measured nuclear binding energies is obtained from the expression [Fe 62]

$$\text{B.E.} = a_1 A - a_2 A^{2/3} - a_3 \frac{Z(Z-1)}{A^{\frac{1}{3}}} - a_4 \frac{(A/2 - Z)^2}{A} \tag{4.1}$$

where A is the atomic mass. The empirical values of these constants are found to be

$$a_1 \approx 15 \text{MeV} \quad a_2 \approx 14 \text{MeV} \quad a_3 \approx 0.63 \text{MeV} \quad a_4 \approx 83 \text{MeV}. \tag{4.2}$$

Here a_3 represents the aforementioned Coulomb repulsion — recall that the empirical nuclear radius is proportional to $A^{1/3}$ — while a_1 and a_2 denote volume and surface binding effects. The final term a_4 accounts for the experimental feature that even-even nuclei tend to be especially stable, because all spins are paired. In Figure VIII.10 is shown the well-known curve of binding energy per nucleon vs. atomic number which displays the qualitative features of this formula.

An important difference exists between the strong interaction, which gives rise to a_1 and the Coulomb interaction which generates the a_3 term—the Coulomb energy increases roughly as the number of proton *pairs* while that due to the strong interaction has a *linear* dependence on the number of nucleons. This is called "saturation" of the nuclear binding energy and is explained by the feature that the strong nucleon-nucleon force is short-ranged — if nucleons are separated by more than a nucleon diameter (*i.e.*, $\sim$ one fermi) or so the potential rapidly vanishes. Thus once a given nucleon is surrounded by a full complement of others, as shown in Figure VIII.11, there is no further increase in the binding energy per nucleon — this is the origin of the saturation. Of course, there must be a correction for those

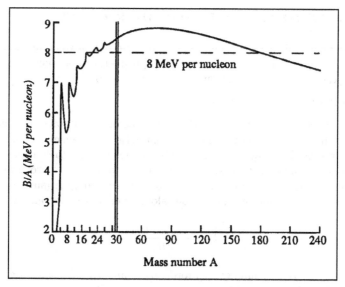

Fig. VIII.10: The familiar curve of binding energy versus atomic mass number.

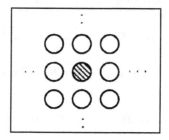

Fig. VIII.11: Because of the short range of the nucleon-nucleon interaction, once a nucleon is completely surrounded by others the nuclear force is "saturated."

nucleons which are on the nuclear surface, since they are not completely surrounded, and this is the origin of the term of opposite sign involving a_2.

Yukawa's explanation of the saturation phenomenon was that the nuclear force involves the exchange of particles with *mass*, as opposed to the electromagnetic interaction which involves the exchange of *massless* photons. The reason that massive exchange leads to a finite-ranged interaction can be understood from the Heisenberg uncertainty principle. Emission of a particle of mass m by a nucleon involves violation of energy conservation by an amount $\sim m$. Such a process, while not allowed classically, is permitted to occur in quantum mechanics but only over a time period δt where $\delta t \sim \frac{1}{\Delta E} \sim \frac{1}{m}$. During this interval the exchanged particle travels a distance $(c)\delta t \sim \frac{1}{m}$ and this is the origin of the finite range. From the observed value $\delta x \sim 1\,\mathrm{fm}$, Yukawa predicted

$$m \sim \frac{1}{\delta x} \sim \frac{\hbar c}{1\,\mathrm{fm}} \sim 200\,\mathrm{MeV} \tag{4.3}$$

for the particle mass. In 1938 a particle having about half this mass was discovered. However, this entity was observed to travel through matter for long distances before interacting, so it cannot be associated with the strong nuclear force. (The particle, having a mass of 105 MeV, is now called the muon and like the electron interacts only via the weak, electromagnetic and gravitational interactions.) In 1947, however, a particle having mass 140 MeV and which interacts strongly with ordinary matter was found. This is the so-called pi meson or pion and is identified as the particle predicted by Yukawa a decade earlier.

Now let's be more quantitative. The pion is a particle carrying zero spin but *negative* parity. (The latter can be understood in the quark model wherein the pion is composed of an S-wave spin-1/2 particle–antiparticle pair which, as shown in the previous chapter, must have negative parity.) The corresponding interaction amplitude between pion and nucleon a is then given by

$$\mathrm{Amp}^{(0)} = ig_a \int d^4x\, \phi_b(x)\bar{g}_a(x)\gamma_5 f_a(x) \tag{4.4}$$

where here g_a is the strong coupling constant and the γ_5 interaction is required in order that the full amplitude be *even* under a parity transformation. This amplitude is the pseudoscalar particle analog of the electromagnetic $j_\mu A^\mu$ interaction given in Eq. VII.10.30. As in that case, in order to deal with the strong interaction *between* a pair of nucleons a and b we must determine the value of the pion field $\phi_b(x)$ generated by particle b at the location of particle a. The differential equation which determines this field is given in analogy with Eq. VII.10.35 by

$$\left(\Box_x + m_\pi^2\right)\phi_b(x) = g_b j_b(x) \tag{4.5}$$

where

$$j_b(x) = \bar{g}_b(x)\gamma_5 f_b(x) \tag{4.6}$$

is the pseudoscalar transition density. We can solve this differential equation by use of Green's function methods. If $\Delta_F^{(0)}(x-y)$ is the solution to the differential equation

$$\left(\Box_x + m_\pi^2\right)\Delta_F^{(0)}(x-y) = -\delta^4(x-y) \tag{4.7}$$

and is written as a Fourier transform

$$\Delta_F^{(0)}(x-y) = \int \frac{d^4q}{(2\pi)^4} e^{-iq\cdot(x-y)} \Delta_F^{(0)}(q) \tag{4.8}$$

we require

$$(\Box_x + m_\pi^2) \Delta_F^{(0)}(x-y) = \int \frac{d^4q}{(2\pi)^4} e^{-iq\cdot(x-y)} \Delta_F^{(0)}(q) (m_\pi^2 - q^2)$$
$$= -\int \frac{d^4q}{(2\pi)^4} e^{-iq\cdot(x-y)} , \quad (4.9)$$

i.e.,

$$\Delta_F^{(0)}(q) = \frac{1}{q^2 - m_\pi^2 + i\epsilon} \quad (4.10)$$

where we have substituted $m^2 \to m^2 - i\epsilon$ as before in order to define the integration contour. This Green's function is the propagator for the pion field.

The pion field $\phi_b(x)$ is then

$$\phi_b(x) = ig_b \int d^4y\, i\Delta_F^{(0)}(x-y) j_b(y) \quad (4.11)$$

and the nucleon a- nucleon b interaction amplitude becomes

$$\text{Amp} = (ig_a)(ig_b) \int d^4x\, d^4y\, j_a(x) i\Delta_F^{(0)}(x-y) j_b(y)$$
$$= i^2 g_a g_b \int d^4x\, d^4y\, \bar{g}_a(x)\gamma_5 f_a(x) i\Delta_F^{(0)}(x-y) \bar{g}_b(y)\gamma_5 f_b(y) \quad (4.12)$$

as shown diagrammatically in Figure VIII.12

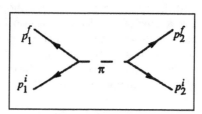

Fig. VIII.12: *Feynman diagram representation of the nucleon-nucleon interaction arising from pion exchange.*

Normalizing in a box of volume $V = 1$ the transition amplitude is

$$\text{Amp} = (ig_a ig_b) \frac{M}{\sqrt{E_1^i E_1^f}} \frac{M}{\sqrt{E_2^i E_2^f}} \int d^4x \int d^4y\, e^{i(p_1^f - p_1^i)\cdot x}$$
$$\times \bar{u}(p_1^f)\gamma_5 u(p_1^i) \int \frac{d^4q}{(2\pi)^4} e^{-iq\cdot(x-y)} \frac{i}{q^2 - m_\pi^2 + i\epsilon} \quad (4.13)$$
$$\times e^{i(p_2^f - p_2^i)\cdot y} \bar{u}(p_2^f)\gamma_5 u(p_2^i) .$$

Performing the x, y and q integrations as before we find the nucleon-nucleon scattering amplitude to be

$$\text{Amp} = -i(2\pi)^4 \delta^4 \left(p_1^i + p_2^i - p_1^f - p_2^f\right) T_{fi} \tag{4.14}$$

with

$$T_{fi} = g_a g_b \frac{M^2}{\sqrt{E_1^i E_2^i E_1^f E_2^f}} \bar{u}(p_1^f)\gamma_5 u(p_1^i) \frac{1}{\left(p_1^f - p_1^i\right)^2 - m_\pi^2 + i\epsilon} \bar{u}(p_2^f)\gamma_5 u(p_2^i) \ . \tag{4.15}$$

Yukawa Interaction

The physics of pion exchange can be gleaned by going to the non-relativistic limit and taking the Fourier transform in order to determine the effective nucleon-nucleon potential. In this limit $q_0 \sim \mathcal{O}(v/c) \times |\vec{q}|$ can be neglected and we need to evaluate

$$V(\vec{x}) = -g_a g_b \int \frac{d^3 q}{(2\pi)^3} e^{i\vec{q}\cdot\vec{x}} \frac{1}{\vec{q}^2 + m_\pi^2} \ . \tag{4.16}$$

Integrating over solid angle we find

$$V(x) = -\frac{g_a g_b}{4\pi^2 ix} \int_0^\infty dq\, q \frac{1}{\vec{q}^2 + m_\pi^2} \left(e^{iqx} - e^{-iqx}\right) \ . \tag{4.17}$$

The q integration may be extended from $-\infty$ to $+\infty$ provided we multiply by one-half since the integrand is an even function of q. The first component of Eq. 4.17 can then be evaluated by contour methods, closing the integration path by a large semicircle in the upper half plane. Noting

$$\vec{q}^2 + m_\pi^2 = (|\vec{q}| + im_\pi)(|\vec{q}| - im_\pi) \tag{4.18}$$

we find

$$\frac{1}{2}\int_{-\infty}^{\infty} dq\, q \frac{1}{\vec{q}^2 + m_\pi^2} e^{iqx} = 2\pi i \times \frac{1}{2} im_\pi \times \frac{1}{2im_\pi} e^{-m_\pi x} = \frac{i\pi}{2} e^{-m_\pi x} \tag{4.19}$$

and an identical result obtains for the e^{-iqx} integral. The full result is

$$V(\vec{x}) = -\frac{g_a g_b}{4\pi x} \exp(-m_\pi x) \tag{4.20}$$

and this from is called the "Yukawa potential." As $m_\pi \to 0$ Eq. 4.20 reproduces the Coulomb interaction. However, for non-zero mass the exponential factor leads to a strong damping of the potential for $x \gg \frac{1}{m_\pi}$ as required by uncertainty principle arguments.

The full pion exchange potential is more complex, since the pseudoscalar density has a non-trivial non-relativistic limit

$$\bar{u}(p_1^f)\gamma_5 u(p_1^i) \sim \chi_f^{(1)\dagger} \frac{\vec{\sigma}\cdot\left(\vec{p}_1^f - \vec{p}_1^i\right)}{2M} \chi_i^{(1)} = \chi_f^{(1)\dagger} \frac{\vec{\sigma}\cdot\vec{q}}{2M} \chi_i^{(1)} \ . \tag{4.21}$$

Thus the pion interaction becomes

$$V_\pi(\vec{x}) = -\frac{g_a g_b}{4M^2}\chi_f^{(1)\dagger}\vec{\sigma}\cdot\vec{\nabla}_x\chi_i^{(1)}\chi_f^{(2)\dagger}\vec{\sigma}\cdot\vec{\nabla}_x\chi_i^{(2)}\frac{1}{4\pi x}\exp(-m_\pi x), \qquad (4.22)$$

and possesses a somewhat complex spin-dependent form. Nevertheless, by studies of the structure of the deuteron and of nucleon-nucleon scattering the presence of the so-called OPEP or one-pion exchange potential has been strongly confirmed [Er 84], and the strong interaction coupling constant g has been determined to be

$$g_{\pi^0 pp} \sim 13.5 \ . \qquad (4.23)$$

This is indeed then a strong interaction, with a coupling strength g which is nearly a factor of 50 larger than the electric charge e! This means also that, unlike the electromagnetic case, a diagram wherein a *pair* of pions are exchanged as in Figure VIII.13 is as large or larger than the single pion exchange diagram! For this reason a perturbative approach to nucleon-nucleon scattering is unwarranted and physicists are still unable to perform reliable calculations of strong interaction processes.

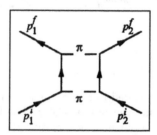

Fig. VIII.13: *A two-pion exchange diagram.*

Isospin

Before ending this discussion we note one additional feature of the pion-nucleon interaction—our previous discussion has been simplified in that in reality there exist three *different* pi mesons with almost identical masses and with charges $+$, $-$, and 0. By charge conservation there exist then four different kinds of nucleon-pion interactions

$$p \to p\pi^0 \ , \quad n \to n\pi^0 \ , \quad p \to n\pi^+ \ , \quad n \to p\pi^- \ , \qquad (4.24)$$

and while in principle these could be four different and distinct numbers, it is found experimentally that the couplings are related in the fashion

$$g_{\pi^+ pn} = g_{\pi^- np} = \sqrt{2}\, g_{\pi^0 pp} = -\sqrt{2}\, g_{\pi^0 nn} \ . \qquad (4.25)$$

This relationship is most conveniently expressed by representing the nucleon as a two-component spinor N, with

$$p = \begin{pmatrix} 1 \\ 0 \end{pmatrix} \qquad n = \begin{pmatrix} 0 \\ 1 \end{pmatrix} \qquad (4.26)$$

VIII.4 PION EXCHANGE AND STRONG INTERACTIONS*

and the pion as a vector quantity with

$$\pi^\pm = \frac{1}{\sqrt{2}}(\pi_x \pm i\pi_y) \qquad \pi^0 = \pi_z \;. \tag{4.27}$$

It is then ascertained that the pion-nucleon couplings can be succinctly given by the form

$$\text{Amp} = g_{\pi^0 pp}\bar{N}\gamma_5\vec{\tau}\cdot\vec{\pi}N \tag{4.28}$$

where

$$\tau_x = \begin{pmatrix} 0 & 1 \\ 1 & 0 \end{pmatrix} \qquad \tau_y = \begin{pmatrix} 0 & -i \\ i & 0 \end{pmatrix} \qquad \tau_z = \begin{pmatrix} 1 & 0 \\ 0 & -1 \end{pmatrix} \tag{4.29}$$

are the familiar Pauli matrices. Then

$$\begin{aligned}
\text{Amp}(p \to \pi^0 p) &= g_{\pi^0 pp}\bar{p}\gamma_5 p(1\;0)\tau_z\begin{pmatrix}1\\0\end{pmatrix} \\
&= g_{\pi^0 pp}\bar{p}\gamma_5 p \\
\text{Amp}(n \to \pi^0 n) &= g_{\pi^0 pp}\bar{n}\gamma_5 n(1\;0)\tau_z\begin{pmatrix}0\\1\end{pmatrix} \\
&= -g_{\pi^0 nn}\bar{n}\gamma_5 n \\
\text{Amp}(p \to n\pi^+) &= g_{\pi^0 pp}\bar{n}\gamma_5 p(0\;1)\frac{1}{\sqrt{2}}(\tau_x - i\tau_y)\begin{pmatrix}1\\0\end{pmatrix} \\
&= \sqrt{2}\,g_{\pi^0 pp}\bar{n}\gamma_5 p \\
\text{Amp}(n \to p\pi^-) &= g_{\pi^0 pp}\bar{p}\gamma_5(1\;0)\frac{1}{\sqrt{2}}(\tau_x + i\tau_y)\begin{pmatrix}0\\1\end{pmatrix} \\
&= \sqrt{2}\,g_{\pi^0 pp}\bar{p}\gamma_5 n \;.
\end{aligned} \tag{4.30}$$

The similarity of this formalism to manipulations with spin has given rise to the term "isospin" for this property and the $\vec{\tau}$ are called the Pauli isospin matrices. A concise form for the one pion exchange potential is then

$$V_\pi(x) = -\frac{g_{\pi^0 pp}^2}{4M^2}\chi_f^{(1)\dagger}\vec{\tau}_1\vec{\sigma}\cdot\vec{\nabla}_x\chi_i^{(1)}\cdot\chi_f^{(2)\dagger}\vec{\tau}_2\vec{\sigma}\cdot\vec{\nabla}_x\chi_i^{(2)}\frac{1}{4\pi x}\exp(-m_\pi x) \tag{4.31}$$

Effective Lagrangians and Field Theory

Finally, we can combine the idea of effective Lagrangians and functional integration in order to look at the nucleon-nucleon interaction in an alternative fashion. Imagine that we are evaluating the full functional integral which describes a set of nucleon fields described by the iso-spinors $\psi(x), \bar{\psi}(x)$ and pion fields described by the iso-vector $\vec{\phi}(x)$

$$F \sim N\int [d\psi_{p,n}]\,[d\bar{\psi}_{p,n}]\,[d\vec{\phi}]\exp iS \;. \tag{4.32}$$

Here the action is given by

$$S = \int d^4x \left(\mathcal{L}_0\left(\bar{\psi},\psi\right) + \frac{1}{2}\partial_\mu\vec{\phi}\cdot\partial^\mu\vec{\phi} - \frac{1}{2}m_\pi^2 \vec{\phi}\cdot\vec{\phi} + ig\bar{\psi}(x)\gamma_5\vec{\tau}\psi(x)\cdot\vec{\phi}(x) \right) \quad (4.33)$$

where

$$\mathcal{L}_0\left(\bar{\psi},\psi\right) = \sum_{p,n} \bar{\psi}_i(x)\left(i\not{\nabla} - M_N\right)\psi_i(x) \quad (4.34)$$

is the free Lagrangian for the Dirac field.

The conventional picture of nuclei is in terms of nucleon degrees of freedom only, without explicit reference to mesons. This is equivalent to integrating out the mesonic degrees of freedom in Eq. 4.32 leaving a functional integral in terms of nucleons only, involving an effective Lagrangian which includes the influence of the mesons within the nucleon system. Writing

$$\vec{j}(x) \equiv ig\bar{\psi}(x)\gamma_5\vec{\tau}\psi(x) \quad \text{and} \quad \mathcal{O}_B(x-y) = (\Box_x + m_\pi^2)\delta^4(x-y) \quad (4.35)$$

we can easily perform the functional integration over $[d\vec{\phi}]$ using the result of Eq. 3.21, yielding

$$F \sim N \int [d\psi_{p,n}]\,[d\bar{\psi}_{p,n}]\, \exp iS_{\text{eff}} \quad (4.36)$$

where

$$S_{\text{eff}} = \int d^4x\,\bar{\psi}\,(i\not{\nabla} - M_N)\,\psi + \int d^4x \int d^4y\,\vec{j}(x)\mathcal{O}_B^{-1}(x-y)\cdot\vec{j}(y)$$
$$= \int d^4x\,\bar{\psi}(i\not{\nabla} - M_N)\psi - ig^2 \int d^4x \int d^4y\,\bar{\psi}(x)\gamma_5\vec{\tau}\psi(x)i\Delta_F^{(0)}(x-y)\cdot\bar{\psi}(y)\gamma_5\vec{\tau}\psi(y)$$
$$(4.37)$$

is the effective Lagrangian. We see that the free nucleon action is modified by a term representing a nucleon-nucleon interaction (potential) generated by the exchange of the pseudoscalar mesons between nucleons. Actually, the form of the potential is much more complex than the simple one-pion exchange component given in Eq. 4.37 and the exchange of many heavier mesons — ρ, ω, ϕ, etc. — must also be included. When this is done one obtains a rather successful picture of the nucleon-nucleon interaction as an effective Lagrangian generated by single meson exchange between nucleons. The pion component, being a pseudoscalar, is the simplest and may be isolated because of its long-range (recall $m_\pi \ll m_\rho, m_\omega, \ldots$). However, each piece is important in constructing the full effective interaction.

Equivalently we can choose to integrate out the nucleon fields, leaving an effective action expressed in terms of pion fields only. In order to perform this functional integration it is necessary to develop an additional piece of formalism. Imagine that we must perform the functional integral

$$F[\mathcal{O}] = N \int [d\phi(x)] \exp \frac{i}{2} \int d^4x\,\phi(x)\mathcal{O}(x)\phi(x) \quad (4.38)$$

where ϕ represents a hermitian scalar field. Representing the discretized $\phi(x)$ as a column vector

$$\phi = (\phi(x_1), \phi(x_2) \ldots \phi(x_N)) \tag{4.39}$$

we can write

$$F[\mathcal{O}] = N \int d\phi_1 \ldots d\phi_N \exp \frac{i}{2} \tilde{\phi} \mathcal{O} \phi \tag{4.40}$$

where the tilde indicates the transpose. Now diagonalize the matrix $\mathcal{O}$, yielding

$$F[\mathcal{O}] = N \int d\phi'_1 \ldots d\phi'_N \exp \frac{i}{2} \sum_{j=1}^{N} \phi_j'^2 \lambda_j = N \prod_{j=1}^{N} \int_{-\infty}^{\infty} d\phi'_j \exp \frac{i}{2} \phi_j'^2 \lambda_j \tag{4.41}$$

where λ_j are the eigenvalues of $\mathcal{O}$ and the prime denotes that we are in the diagonal basis. Performing the integrations, we have

$$F[\mathcal{O}] = N \prod_{j=1}^{N} \left(\frac{2\pi i}{\lambda_j}\right)^{\frac{1}{2}} = \frac{\text{const.}}{(\det \mathcal{O})^{\frac{1}{2}}} \tag{4.42}$$

where $\det \mathcal{O}$ denotes the product of eigenvalues. In the case of a charged scalar field we have

$$\begin{aligned} F[\mathcal{O}] &= N \int [d\phi(x)][d\phi^*(x)] \exp i \int d^4x \phi^*(x) \mathcal{O}(x) \phi(x) \\ &= N \int [d\phi_1(x)][d\phi_2(x)] \exp \frac{i}{2} \sum_{j=1}^{2} \phi_j(x) \mathcal{O}(x) \phi_j(x) = \frac{\text{const.}}{\det \mathcal{O}} \end{aligned} \tag{4.43}$$

using the hermitian decomposition given in Eq. 4.27.

Similarly we can develop this formalism within the context of Fermi fields by using matrix notation and using the discrete sets $\{\alpha_1, \ldots, \alpha_n\}$ and $\{\bar{\alpha}_1, \ldots, \bar{\alpha}_n\}$ of Grassmann variables. A class of integrals which commonly arises in a functional framework is then

$$W[M] = \int d\bar{\alpha}_n \ldots d\bar{\alpha}_1 d\alpha_n \ldots d\alpha_1 \exp i \bar{\alpha} M \alpha \ . \tag{4.44}$$

In the simple 2×2 case we have

$$\begin{aligned} W[M] = \int d\bar{\alpha}_2 \, d\bar{\alpha}_1 d\alpha_2 d\alpha_1 \, [&1 + i\bar{\alpha}_i M_{ij} \alpha_j \\ &+ \bar{\alpha}_2 \bar{\alpha}_1 \alpha_2 \alpha_1 (M_{11} M_{22} - M_{12} M_{21})] \ . \end{aligned} \tag{4.45}$$

Only the final term survives the integration, and we obtain

$$W[M] = \det M \ . \tag{4.46}$$

In fact this result generalizes to an arbitrary $n \times n$ system [Le 82], yielding just the inverse of the result which obtains for Bose fields

Fermi $\quad W[M] = \int d\bar{\alpha}_n d\bar{\alpha}_n \ldots d\alpha_1 d\bar{\alpha}_1 \exp \bar{\alpha} M \alpha = \det M$

Bose $\quad W[M] = \int da_n da_n^* \ldots da_1 da_1^* \exp -a^* M a = (\det M)^{-1} \times \text{const.}$

(4.47)

Taking this formalism over to the case of Fermi *fields* $\psi(x), \bar{\psi}(x)$, since such fields always enter the Lagrangian quadratically, the functional integral can be performed exactly

$$W[\mathcal{O}_D] = \int [d\psi(x)] [d\bar{\psi}(x)] \exp \int d^4x \, \bar{\psi}(x) \mathcal{O}_D \psi(x)$$
$$= N \det \mathcal{O}_D \, .$$
(4.48)

We can now attack the problem at hand—finding an effective Lagrangian for pions in the presence of nucleons. Carrying out the functional integration over nucleons in Eq. 4.32 we find

$$F \sim N \int [d\vec{\phi}] \det \mathcal{O}_D \exp i \int d^4x \left(\frac{1}{2} \partial_\mu \vec{\phi}(x) \cdot \partial^\mu \vec{\phi}(x) - \frac{1}{2} m_\pi^2 \vec{\phi} \cdot \vec{\phi} \right)$$
(4.49)

where

$$\mathcal{O}_D(x-y) = \left(i \not{\nabla}_x - M_N + i g \gamma_5 \vec{\tau} \cdot \vec{\phi}(x) \right) \delta^4(x-y)$$
(4.50)

We may choose the constant N by demanding that there be no effect in the absence of a pion nucleon coupling—i.e., $N = \det^{-1} \mathcal{O}_D^{(0)}$ where

$$\mathcal{O}^{(0)}(x-y) = (i \not{\nabla}_x - M_N) \delta^4(x-y)$$
(4.51)

is the free nucleon operator. We may put this result in the form of an effective Lagrangian via

$$F \sim \int [d\vec{\phi}] \exp i S_{\text{eff}}[\vec{\phi}]$$
(4.52)

with

$$S_{\text{eff}}[\vec{\phi}] = \int d^4x \left(\frac{1}{2} \partial_\mu \vec{\phi} \cdot \partial^\mu \vec{\phi} - \frac{1}{2} m_\pi^2 \vec{\phi} \cdot \vec{\phi} \right) - i \ln \left(\det \mathcal{O}_D / \det \mathcal{O}_D^{(0)} \right).$$
(4.53)

Since

$$\ln \det \mathcal{O}_D = \ln \prod_j \lambda_j = \sum_j \ln \lambda_j = \text{Tr} \ln \mathcal{O}_D$$
(4.54)

we can write

$$S_{\text{eff}}[\vec{\phi}] = \int d^4x \left(\frac{1}{2} \partial_\mu \vec{\phi} \cdot \partial^\mu \vec{\phi} - \frac{1}{2} m_\pi^2 \vec{\phi} \cdot \vec{\phi} \right) - i \text{Tr} \ln \mathcal{O}_D / \mathcal{O}_D^{(0)}.$$
(4.55)

Eq. 4.55 is exact but not of great utility since we cannot explicitly evaluate $\text{Tr}\ln\mathcal{O}_D/\mathcal{O}_D^{(0)}$. However, noting that

$$\mathcal{O}_D/\mathcal{O}_D^{(0)}(x-y) = \left(1 - \frac{i}{i\slashed{\nabla}_x - M_N}g\gamma_5\vec{\tau}\cdot\vec{\phi}(x)\right)\delta^4(x-y) \quad (4.56)$$

we can develop a perturbative evaluation

$$\begin{aligned}\text{Tr}\ln\mathcal{O}_D/\mathcal{O}_D^{(0)} &= \text{Tr}\ln\left(1 + \frac{i}{i\slashed{\nabla} - M_N}g\gamma_5\vec{\tau}\cdot\vec{\phi}(\hat{x})\right) = \text{Tr}\left(\frac{i}{i\slashed{\nabla} - M_N}g\gamma_5\vec{\tau}\cdot\vec{\phi}(\hat{x})\right.\\
&\left.-\frac{1}{2}\frac{i}{i\slashed{\nabla} - M_N}g\gamma_5\vec{\tau}\cdot\vec{\phi}(\hat{x})\frac{i}{i\slashed{\nabla} - M_N}g\gamma_5\vec{\tau}\cdot\vec{\phi}(\hat{x}) + \ldots\right)\\
&= \int d^4x\, \text{tr}\, iS_F^{(0)}(x,x)g\gamma_5\vec{\tau}\cdot\vec{\phi}(x)\\
&\quad - \int d^4x\frac{1}{2}\text{tr}\,g\gamma_5\vec{\tau}\cdot\vec{\phi}(x)\int d^4y\, iS_F^{(0)}(x,y)\gamma_5\vec{\tau}\cdot\vec{\phi}(y)iS_F^{(0)}(y,x) + \ldots\end{aligned}$$
(4.57)

where we have separated the Tr operation into a product over Lorentz – $\int d^4x$ – and Dirac – tr – indices. The first order term vanishes since $\text{tr}\,S_F\gamma_5 = 0$. However, the next piece can be written as

$$-\frac{1}{2}\int d^4x\int d^4y\,\phi_i(x)\phi_j(y)e^{-iq\cdot(x-y)}\Pi_{ij}(q) \quad (4.58)$$

with

$$\Pi_{ij}(q) = g^2\int\frac{d^4s}{(2\pi)^4}\text{tr}\,\gamma_5\tau_i\frac{i}{\slashed{s}-M_N}\gamma_5\tau_j\frac{i}{\slashed{s}-\slashed{q}-M_N} \quad (4.59)$$

being the pionic analog of the vacuum polarization. We see here the effect of the presence of nucleon coupling on the pion effective Lagrangian is to modify the pion propagator within the medium. In a realistic nuclear situation we would not utilize the free nucleon propagator, of course, but would have to have a sum over particle-hole states, making the problem much more complex than we wish to deal with here. There also are generated higher order effects in ϕ^3, ϕ^4, etc. that we shall not explore further. Notice that the sign of the $N\bar{N}$ contribution to Π_{ij} is *opposite* to that which would obtain if the nucleons were bosons, due to the difference between $\det\mathcal{O}$ and $\det^{-1}\mathcal{O}$. This "extra" minus sign for a fermion closed loop is associated with the Grassman variables need to deal with anticommuting fields and is the origin of the Feynman rule #5 given in Sect. VII.9 on an ad hoc basis.

This example indicates clearly the power of the effective Lagrangian method within the framework of functional techniques. However, we do not have the space to pursue this subject further and we end our discussion here.

PROBLEM VIII.4.1

Pion Nucleon Scattering

i) Using the pion-nucleon interaction given in Eq. 4.28 evaluate using perturbation theory the amplitude for π^+p scattering to order g^2. Draw the Feynman diagram which corresponds to this amplitude.

ii) Use the result of part i) to calculate the differential cross section for the scattering of a π^+ meson by an unpolarized proton in the center of mass system. Show that at threshold

$$\frac{d\sigma}{d\Omega} \approx \frac{g^4}{4\pi^2}\left(\frac{M_p}{M_p + m_\pi}\right)^2 \frac{1}{(2M_p - m_\pi)^2} \;.$$

iii) The coupling constant g is known to have the size—$g^2/4\pi \sim 13.5$. How does the above cross section calculated using this value compare with the experimental result at very low kinetic energy (≤ 20 MeV)

$$\frac{d\sigma}{d\Omega}^{\text{exp}} \simeq 1.5 \times 10^{-27} \text{cm}^2 \;?$$

Explain the reason for any discrepancy.

PROBLEM VIII.4.2

Anomalous Moments and Meson Cloud Effects

We have seen in Sect. VIII.1 that the magnetic moment of the electron as measured in Bohr magnetons is altered from its natural Dirac value of two by electromagnetic corrections. In the same way the magnetic moments of proton and neutron are altered by strong interaction corrections, giving rise to anomalous magnetic moments for these particles also. However, because the strong interaction is involved the size of such moments can be substantial—of the order of the Dirac value itself!

i) By considering the scattering of a proton and neutron by an external electromagnetic field, compute the static magnetic moment of the neutron and proton taking into account lowest order mesonic effects—i.e., the diagrams shown below.

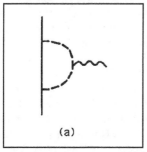 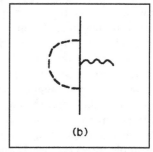

(a)   (b)

Show that in the static limit—$M_N \gg m_\pi$—one has

$$\kappa_p = \frac{g^2_{\pi^0 pp}}{16\pi^2} \qquad \kappa_n = -\frac{g^2_{\pi^0 pp}}{4\pi^2}$$

where κ is the anomalous moment measured in Bohr magnetons.

ii) Compare the meson cloud values of the anomalous moment with the experimental values

$$\kappa_p = 1.79 \qquad \kappa_n = -1.91 \;.$$

Do you expect these values to agree? Why or why not?

PROBLEM VIII.4.3

Effective Lagrangian for the Constant Magnetic Field

Consider a charged scalar field interacting with an external magnetic field $\vec{B} = B\hat{e}_z$ for which the corresponding Klein-Gordon equation reads

$$(D^2 + m^2)\phi(x) = 0$$

where $D_\mu = \partial_\mu + ieA_\mu$ is the covariant derivative and $A_\mu(x)$ is the vector potential associated with the magnetic field. The effective action is then given by

$$e^{iS_{\text{eff}}(B)} = \frac{\int [d\phi(x)][d\phi^*(x)] e^{i \int d^4x \phi^*(x)(D^2+m^2)\phi(x)}}{\int [d\phi(x)][d\phi^*(x)] e^{i \int d^4x \phi^*(x)(\Box+m^2)\phi(x)}}$$

or

$$S_{\text{eff}}(B) = i\text{Trln}\frac{D^2 + m^2}{\Box + m^2} .$$

Of course, the operation "Trln" applied to a differential operator is not a trivial one and the purpose of this problem is to evaluate this quantity for the case at hand.

i) Demonstrate that

$$\ln\frac{a}{b} = \int_0^\infty \frac{ds}{s}(e^{-bs} - e^{-as})$$

so that

$$S_{\text{eff}}(B) = i\text{Tr}\int_0^\infty \frac{ds}{s} e^{-m^2 s}(e^{-\Box s} - e^{-D^2 s}) .$$

In order to evaluate the trace we require a complete set of solutions to the equations

$$D^2 \phi_n(x, y, z, t) = \lambda_n \phi_n(x, y, z, t)$$
$$\Box \phi_n(x, y, z, t) = \kappa_n \phi_n(x, y, z, t) .$$

Then we may write

$$S_{\text{eff}}(B) = i\sum_n \int_0^\infty \frac{ds}{s} e^{-m^2 s}(e^{-\kappa_n s} - e^{-\lambda_n s}) .$$

ii) Show that when $\vec{B}=0$ the eigenstates are given by

$$\phi(x, y, z, t) = \exp i(k_x x + k_y y + k_z z - k_t t)$$

with eigenvalues

$$\kappa_n = -k_t^2 + k_x^2 + k_y^2 + k_z^2 .$$

iii) For the gauge choice

$$A_\mu = (0, Bx\hat{e}_y)$$

show that the eigenstates in the presence of the magnetic field become

$$\phi(x,y,z,t) = \exp i(k_z z + k_y y - k_t t)\psi_n(x - \frac{k_y}{eB})$$

where $\psi_n(x)$ is an eigenstate of the harmonic oscillator Hamiltonian with frequency

$$\omega = eB$$

and that the corresponding eigenvalues are given by

$$\lambda_n = -k_t^2 + k_z^2 + eB(2n+1) \ .$$

Now rotate to Euclidean space

$$k_t \to ik_0$$

so that

$$\kappa_n^2 = k_0^2 + k_x^2 + k_y^2 + k_z^2$$
$$\lambda_n^2 = k_0^2 + k_z^2 + eB(2n+1)$$

and evaluate the trace using box quantization. Taking a box with sides L_1, L_2, L_3 and a time interval T, we have

$$\kappa: \quad \sum_n \to L_1 L_2 L_3 T \int_{-\infty}^{\infty} \frac{d^4k}{(2\pi)^4}$$

$$\lambda: \quad \sum_n \to L_2 L_3 T \int_0^{eBL_1} dk_y \int_{-\infty}^{\infty} \frac{dk_0 dk_z}{(2\pi)^2} \sum_{n=0}^{\infty}$$

where the integration on k_y is over all values with $x' = x - k_y/eB$ kept positive.

iv) Now evaluate the effective action

$$S_{\text{eff}}(B) = L_1 L_2 L_3 T \int_0^{\infty} \frac{ds}{s} \int_{-\infty}^{\infty} \frac{dk_0 dk_z}{(2\pi)^2} e^{-(m^2+k_0^2+k_z^2)s}$$

$$\times \left[\frac{eB}{2\pi} \sum_{n=0}^{\infty} e^{-eB(2n+1)s} - \int_{-\infty}^{\infty} \frac{dk_x dk_y}{(2\pi)^2} e^{-(k_x^2+k_y^2)s} \right]$$

and show that

$$S_{\text{eff}}(B) = L_1 L_2 L_3 T \frac{1}{16\pi^2} \int_0^{\infty} \frac{ds}{s^3} e^{-m^2 s} \left(\frac{eBs}{\sinh eBs} - 1 \right) \ .$$

The "physics" of this result can be seen via an alternative derivation—the effective action is simply the shift in the vacuum energy due to the presence of the magnetic field times the interaction time. Since the zero point energy is given by

$\sum_n \omega_n$ (i.e., $\frac{1}{2}\omega_n$ associated with the positively charged states and $\frac{1}{2}\omega_n$ with the negatively charged states) we have

$$S_{\text{eff}}(B) = T \sum_n (\omega_n(B) - \omega_n(0)) \ .$$

v) Show that
$$\omega_n(B) = \sqrt{k_z^2 + eB(2n+1) + m^2}$$
$$\omega_n(0) = \sqrt{k_x^2 + k_y^2 + k_z^2 + m^2}$$

so that
$$S_{\text{eff}}(B) = L_2 L_3 T \int_{-\infty}^{\infty} \frac{dk_z}{2\pi}$$
$$\times \left[\int_0^{eBL_1} \frac{dk_x}{2\pi} \sum_n \sqrt{k_z^2 + eB(2n+1) + m^2} \right.$$
$$\left. -L_1 \int_{-\infty}^{\infty} \frac{dk_y dk_x}{(2\pi)^2} \sqrt{k_x^2 + k_y^2 + k_z^2 + m^2} \right] \ .$$

vi) Use the representation
$$\sqrt{a} = \frac{1}{2\sqrt{\pi}} \int_0^{\infty} \frac{ds}{s^{\frac{3}{2}}} e^{-as}$$
to prove the identity of the two expressions for $S_{\text{eff}}(B)$.

Epilogue

We have covered a great deal of territory, from basic quantum mechanics to an introduction to quantum field theory. A wide spectrum of individual topics has been examined, in many ways a seeming eclectic potpourri—scattering theory, perturbation and adiabatic techniques, functional and wavefunction approaches in both the relativistic and nonrelativistic domains, etc. Despite this range of subject matter, it is my hope that a sense of unity has also emerged, with the notion of even the most advanced (quantum field theory) topics being only a generalization of the simple path integral techniques which underlaid our beginning presentation on propagator methods, with the Dirac equation being generated by a suggestive generalization of the Schrödinger equation, with the photon exchange picture being strongly founded in the solution of the basic Maxwell equations, etc. A further indication of this unity is our examination of some problems, such as the electric dipole transition rate, by several differing procudres or the use of both relativistic and nonrelativistic techniques to study the Lamb shift. I find this ability to examine the same result from new and differing points of view a real sense of understanding and satisfaction, and it is my hope in writing these notes that other students of the subject may experience the same.

Notation

Schrödinger Equation

The Schrödinger equation for a spinless, spin one-half particle of charge e and mass m is

$$S = 0: \quad \left[\frac{1}{2m}(\vec{p} - e\vec{A})^2 + e\phi\right]\psi(\vec{x}, t) = i\frac{\partial}{\partial t}\psi(\vec{x}, t)$$

$$S = \frac{1}{2}: \quad \left[\frac{\vec{\sigma} \cdot (\vec{p} - e\vec{A})\vec{\sigma} \cdot (\vec{p} - e\vec{A})}{2m} + e\phi\right]\psi(\vec{x}, t) = i\frac{\partial}{\partial t}\psi(\vec{x}, t)$$

where $\vec{p} = -i\vec{\Delta}$ is the momentum operator in co-ordinate space. The projection operator for a spinor $\chi(\hat{s})$ aligned along the direction $\hat{s}$ is

$$\chi_i(\hat{s})\chi_j^\dagger(\hat{s}) = \frac{1}{2}[1 + \vec{\sigma} \cdot \hat{s}]_{ij}$$

so that absolute squares of spin one-half matrix elements can be evaluated via conventional trace techniques [Me 70]

$$|\chi^\dagger(\hat{s}')\mathcal{O}\chi(\hat{s})|^2 = \frac{1}{4}\text{Tr}\,(1 + \vec{\sigma} \cdot \hat{s}')\mathcal{O}(1 + \vec{\sigma} \cdot \hat{s}')\mathcal{O}^\dagger \;.$$

Useful identities in this regard are

$$\text{Tr}\, 1 = 2$$

$$\text{Tr}\, \vec{\sigma} \cdot \vec{a} = 0$$

$$\text{Tr}\, \vec{\sigma} \cdot \vec{a}\vec{\sigma} \cdot \vec{b} = 2\vec{a} \cdot \vec{b}$$

plus the Pauli relation

$$\vec{\sigma} \cdot \vec{a}\vec{\sigma} \cdot \vec{b} = \vec{a} \cdot \vec{b} + i\vec{\sigma} \cdot \vec{a} \times \vec{b} \;.$$

Transition rates: nonrelativistic

Fermi's golden rule for the decay rate from a particle in a state $|a\rangle$ to an n-particle final state $|f\rangle$ is

$$d\Gamma_a = \int \frac{d^3p_1}{(2\pi)^3} \cdots \frac{d^3p_n}{(2\pi)^3}(2\pi)^4\delta^4(p_a - p_1 \ldots - p_n)|T_{fi}|^2$$

where the T-matrix element T_{fi} is

$$T_{fi} = V_{fi} + \sum_n \frac{V_{fn}V_{ni}}{E_a - E_n} + \sum_{\ell n} \frac{V_{f\ell}V_{\ell n}V_{ni}}{(E_a - E_\ell)(E_a - E_n)} + \ldots \;.$$

NOTATION

The scattering cross section for two particles a and b is

$$d\sigma = \frac{1}{|\vec{v}_a - \vec{v}_b|} \int |T_{fi}|^2 \frac{d^3 p_1}{(2\pi)^3} \cdots \frac{d^3 p_n}{(2\pi)^3} (2\pi)^4 \delta^4(p_a + p_b - p_1 \ldots - p_n) .$$

Relativistic notation

Four-vectors such as spacetime (t, x, y, z) or energy-momentum (E, p_x, p_y, p_z) are denoted by a contravariant symbol

$$x^\mu = (t, \vec{x}) \qquad p^\mu = (E, \vec{p})$$

while the corresponding covariant quantities are obtained via use of the metric tensor

$$x_\mu = \eta_{\mu\nu} x^\nu = (t, -\vec{x}) \qquad p_\mu = \eta_{\mu\nu} p^\nu = (E, -\vec{p})$$

where

$$\eta_{\mu\nu} = \begin{pmatrix} 1 & 0 & 0 & 0 \\ 0 & -1 & 0 & 0 \\ 0 & 0 & -1 & 0 \\ 0 & 0 & 0 & -1 \end{pmatrix} .$$

Contraction of two four-vectors is denoted by the use of a dot symbol or by the use of repeated contravariant and covariant idices

$$A \cdot B = A^\mu B_\mu = A^\mu \eta_{\mu\nu} B^\nu = A^0 B^0 - \vec{A} \cdot \vec{B} .$$

The momentum operator in co-ordinate notation is given by

$$p^u = i \frac{\partial}{\partial x_\mu} = i \partial^\mu = i \nabla^\mu = i(\frac{\partial}{\partial t}, -\vec{\nabla})$$

and transforms as a four-vector so that, e.g.

$$\partial^\mu A_\mu = \frac{\partial A^0}{\partial t} + \vec{\nabla} \cdot \vec{A}$$

is a Lorentz invariant.

Dirac Notation

The Dirac matrices γ^μ are defined via

$$\gamma^0 = \begin{pmatrix} 1 & 0 \\ 0 & -1 \end{pmatrix} \qquad \gamma^i = \begin{pmatrix} 1 & \sigma_i \\ -\sigma_i & 0 \end{pmatrix}$$

where σ_i is a Pauli matrix. The γ^μ satisfy the anticommutation relations

$$\gamma^\mu \gamma^\nu + \gamma^\nu \gamma^\mu = 2\eta^{\mu\nu} .$$

Also appearing are the combinations

$$\sigma^{\mu\nu} = \frac{i}{2}[\gamma^\mu, \gamma^\nu] \quad \text{and} \quad \gamma_5 = -i\gamma^0\gamma^1\gamma^2\gamma^3 \ .$$

The inner product of the gamma matrices with an ordinary four-vector is often denoted by the use of a slash

$$\gamma^\mu A_\mu = \slashed{A} = \gamma^0 A^0 - \vec{\gamma}\cdot\vec{A} \ .$$

The Dirac equation for a particle of charge e and mass m reads then

$$(i\slashed{\nabla} - e\slashed{A} - m)\psi(x) = 0$$

and has the plane-wave solutions (for $A_\mu = 0$)

$$\text{positive energy:} \quad \psi(x) = u(p,s)e^{-ip\cdot x}$$
$$\text{negative energy:} \quad \psi(x) = v(p,s)e^{ip\cdot x}$$

where the four-component spinors $u(p,s)$ and $v(p,s)$ satisfy the algebraic equations

$$(\slashed{p}-m)u(p,s) = 0 \qquad (\slashed{p}+m)v(p,s) = 0 \ .$$

If s^μ is a four-vector which has the form $(0,\hat{s})$ in the rest frame, where $\hat{s}$ is a unit-vector along the spin-direction, we find

$$u_a(p,s)\bar{u}_b(p,s) = \left[\frac{\slashed{p}+m}{2m}\cdot\frac{1+\slashed{s}\gamma_5}{2}\right]_{ab}$$

$$v_a(p,s)\bar{v}_b(p,s) = \left[\frac{\slashed{p}-m}{2m}\cdot\frac{1+\slashed{s}\gamma_5}{2}\right]_{ab} \ .$$

Here the adjoint spinors $\bar{u}(p,s)$ and $\bar{v}(p,s)$ are defined via

$$\bar{u}(p,s) = u^\dagger(p,s)\gamma^0 \qquad \bar{v}(p,s) = v^\dagger(p,s)\gamma^0$$

and satisfy the normalization conditions

$$\bar{u}(p,s)u(p,s) = 1 \qquad \bar{v}(p,s)v(p,s) = -1 \ .$$

Trace methods

The absolute square of a Dirac matrix element can be evaluated by use of trace methods

$$|\bar{u}(p',s')\Gamma u(p,s)|^2 = \frac{1}{16m^2}\text{Tr}\,(\slashed{p}'+m)(1+\slashed{s}'\gamma_5)\Gamma(\slashed{p}+m)(1+\slashed{s}\gamma_5)\bar{\Gamma}$$

where

$$\bar{\Gamma} \equiv \gamma^0\Gamma^\dagger\gamma^0$$

and the traces arising therein can be evaluated using the identities

$$\text{Tr}\, 1 = 4$$

$$\text{Tr}\, \gamma_5 = 0$$

$$\text{Tr}\, \Gamma = 0 \quad \text{if } \Gamma \text{ contains an odd number of } \gamma_u\text{'s}$$

$$\text{Tr}\, \slashed{A}\slashed{B} = 4A \cdot B$$

$$\text{Tr}\, \slashed{A}\slashed{B}\slashed{C}\slashed{D} = 4(A \cdot B\, C \cdot D + A \cdot D\, B \cdot C - A \cdot C\, B \cdot D)$$

$$\text{Tr}\, \gamma_5 \slashed{A}\slashed{B} = 0$$

$$\text{Tr}\, \gamma_5 \slashed{A}\slashed{B}\slashed{C}\slashed{D} = 4i\epsilon_{\alpha\beta\gamma\delta}A^\alpha B^\beta C^\gamma D^\delta \ .$$

Other useful identities include

$$\gamma_\mu \slashed{A} \gamma^\mu = -2\slashed{A}$$

$$\gamma_\mu \slashed{A}\slashed{B} \gamma^\mu = 4A \cdot B$$

$$\gamma_\mu \slashed{A}\slashed{B}\slashed{C} \gamma^\mu = -2\slashed{C}\slashed{B}\slashed{A}$$

Klein-Gordon Equation

For spinless particles of charge e and mass m the relativistic wave equation is

$$[(i\partial - eA)_\mu (i\partial - eA)^\mu - m^2]\phi(x) = 0$$

which (for $A_\mu = 0$) has the plane wave solutions

$$\text{positive energy:} \quad \phi_p^{(+)}(x) = e^{-ip\cdot x}$$

$$\text{negative energy:} \quad \phi_p^{(-)}(x) = e^{ip\cdot x}$$

which satisfy the normalization conditions

$$\int d^3x \, \phi_p^{(\pm)*}(x) i \overleftrightarrow{\partial_0} \phi_q^{(\pm)}(x) = \pm 2E_p (2\pi)^3 \delta(\vec{p} - \vec{q}) \ .$$

Transition rates: relativistic

The decay rate for a particle a into a combination of Dirac and Klein-Gordon particles is given by

$$d\Gamma = N_a \int |M_{fi}|^2 \frac{d^3p_1}{(2\pi)^3} N_1 \frac{d^3p_2}{(2\pi)^3} N_2 \ldots \frac{d^3p_n}{(2\pi)^3} N_n (2\pi)^4 \delta^4(p_a - p_1 - p_2 \ldots - p_n)$$

where $N_i = \frac{1}{2E_i}, \frac{m_i}{E_i}$ for Dirac, Klein-Gordon particles respectively and $M_{fi} = \langle f|T|i\rangle$ is the invariant transition amplitude for the process. The corresponding expression for the scattering cross-section of two particles a and b is

$$d\sigma = \frac{N_a N_b}{|\vec{v}_a - \vec{v}_b|} \int |M_{fi}|^2 \frac{d^3p_1}{(2\pi)^3} N_1 \frac{d^3p_2}{(2\pi)^3} N_2 \ldots \frac{d^3p_n}{(2\pi)^3} N_n (2\pi)^4 \delta^4(p_a + p_b - p_1 \ldots - p_n)$$

where the normalization factors N_i are as above.

References

In the text, papers are cited by using the first two initials of the first author, plus the first initial of remaining authors, plus the year of publication. For example, the first reference is cited as [AhB 59].

Aharonov, Y. and Bohm, D., "Significance of Electromagnetic Potentials in the Quantum Theory," Phys. Rev. **115**, 485 (1959).

Aharonov, Y. and Casher, A., "Topological Quantum Effects for Neutral Particles," Phys. Rev. Lett. **53**, 319 (1984).

Baym, G., *Lectures in Quantum Mechanics*, W.A. Benjamin, Reading, MA (1969).

Berry, M.V., "Quantum Phase Factors Accompanying Adiabatic Change," Proc. Roy. Soc. **A392**, 45 (1984).

Bethe, H., "Electromagnetic Shift of Energy Levels," Phy. Rev. **72**, 339(L) (1947).

Bethe, H., Brown, L.M., and Stehn, J.R., "Numerical Value of the Lamb Shift," Phys. Rev. **77**, 370 (1950).

Blinder, S.M., "Evolution of a Gaussian Wavepacket," Am. J. Phys. **36**, 525 (1968).

Bloch, F. and Nordsieck, A., "Note on the Radiation Field of the Electron," Phys. Rev. **52**, 54 (1937).

Bohm, D., *Quantum Theory*, Prentice-Hall, Englewood Cliffs, N.J. (1951).

Born, M. and Oppenheimer, R., "Zur Quantenthorie der Molekeln," Ann. der Phys. **84**, 457 (1927).

Brillouin, L., "Notes on Undulatory Mechanics," J. Phys. **7**, 353 (1926).

Cimmins, A. et al., "Observation of the Topological Aharonov-Casher Phase Shift by Neutron Interferometry," Phys. Rev. Lett. **63**, 380 (1989).

Coleman, S., *Aspects of Symmetry*, Cambridge, New York (1985).

Das, A. and Melissinos, A.C., *Quantum Mechanics: A Modern Introduction*, Gordon and Breach, New York (1986).

Dirac, P.A.M, "Quantized Singularities in the Electromagnetic Field," Proc. Roy. Soc. **A133**, 60 (1931).

Dirac, P.A.M., *The Principles of Quantum Mechanics*, Clarendon, New York (1958).

Elizade, E. and Romeo, A., "Essentials of the Casimir Effect and its Computation," Am. J. Phys. **59**, 711 (1991).

Ericson, T.E.O., "Nuclear Low Energy Tests of OPEP," Comm. Nuc. Part. Phys. **13**, 157 (1984).

Fermi, E., *Nuclear Physics*, Univ. of Chicago Press, Chicago (1962).

Feynman, R.P. and Wheeler, J.A., "Classical Electrodynamics in Terms of Direct Interparticle Action," Rev. Mod. Phys **21**, 425 (1949).

Feynman, R.P., *Quantum Electrodynamics*, W.A. Benjamin, Reading, MA (1961).

Feynman, R.P., Leighton, R.B. and Sands, M., *The Feynman Lectures on Physics*, Addison-Wesley, Reading, MA (1964).

Feynman, R.P. and Hibbs, A.R., *Quantum Mechanics and Path Integrals*, McGraw-Hill, New York (1965).

Fierz, M., "Zum Theorie magnetisch Gelander Teilchen," Helv. Phys. Acat., **17**, 27 (1944).

Goldberger, M.L. and Watson, K.M., *Collision Theory*, Wiley, New York (1964).

Gradshtein, I.S. and Ryzhik, I.M., *Tables of Integrals, Series, and Products*, Academic Press, New York (1980).

Holstein, B.R. and Swift, A.R., "Spreading Wave Packets—a Cautionary Note," Am. J. Phys. **40**, 829 (1972).

Holstein, B.R., "Semiclassical Treatment of Above Barrier Scattering," Am. J. Phys **52**, 321 (1984).

Holstein, B.R., "Semiclassical Treatment of the Double Well," Am. J. Phys. **56**, 338 (1988).

Itzykson, C. and Zuber, J., *Quantum Field Theory*, McGraw-Hill, New York (1980).

Jackson, J.D., *Classical Electrodynamics*, Wiley, New York (1962).

Kaufmann, W.B., "Strong Interaction Effects in Hadronic Atoms," Am. J. Phys. **45**, 735 (1977).

Kramers, H., "Wellenmechanik und halbzahlige Quantisierung," Zeit. Phys. **39**, 828 (1926).

Lamb, W.E. and Retherford, R.C., "Fine Structure of the Hydrogen Atom by a Microwave Method," Phys. Rev. **72**, 241 (1947).

Lamb, W.E. and Retherford, R.C., "Fine Structure of the Hydrogen Atom. Part I," Phys. Rev. **79**, 549 (1950).

Landau, L.D. and Lifshitz, E., *Quantum Mechanics: Nonrelativistic Theory*, Pergamon, New York (1977).

Marinov, M.S., "Path Integrals in Quantum Theory: an Outlook of Basic Concepts," Phys. Rept. **60**, 1 (1980).

Matthews, J. and Walker, R.L., *Mathematical Methods of Physics*, W.A. Benjamin, Reading, MA (1964).

McGervey, J.D., *Introduction to Modern Physics*, Academic Press, New York (1971).

McLaughlin, D.W., "Complex Time, Contour Independent Path Integrals and Barrier Penetration," J. Math. Phys. **13**, 1099 (1972).

Merzbacher, E., *Quantum Mechanics*, Wiley, New York (1970).

Morse, P. and Feshbach, H., *Methods of Theoretical Physics*, McGraw-Hill, New York (1953).

Preston, M.A., *Physics of the Nucleus*, Addison-Wesley, Reading, MA (1962).

Ramond, P., *Field Theory: a Modern Primer*, Benjamin-Cummings, Reading, MA (1981).

Sakurai, J.J., *Advanced Quantum Mechanics*, Addison-Wesley, Reading, MA (1967).

Sakurai, J.J., *Modern Quantum Mechanics*, Benjamin-Cummings, Reading, MA (1985).

Schulman, L.S., *Techniques and Applications of Path Integration*, Wiley, New York (1981).

Semon, M.D., "Experimental Verification of an Aharonov-Bohm Effect in Rotating Reference Frames," Found. Phys. **12**, 49 (1982).

Shalit, A. de, *Preludes in Theoretical Physics*, No. Holland, Amsterdam (1966).

Spaarnay, M.J., "Measurements of Attractive Forces Between Flat Plates," Physica, **24**, 751 (1958).

Titchmarsh, E.C., *Introduction to the Thoery of Fourier Integrals*, Clarendon, New York (1948).

Tonomura, A. et al, "Observations of Aharonov-Bohm Effect by Electron Holography," Phys. Rev. Lett. **48**, 1443 (1982).

Welton, T.A., Phys. Rev. **74**, 1157 (1948).

Wentzel, G., "Eine Verallgemeinerung der Quantenbedingungen fur die Zwecke der Wellenmechanik," Zeit. Phys. **39**, 828 (1926).

Wertheim, G.K., *Mössbauer Effect: Principles and Applications*, Academic Press, New York (1964).

Zimmerman, J.E. and Mercerau, R., "Compton Wavelength of Superconducting Electrons," Phys. Rev. Lett **14**, 887 (1965).

Index

Adiabatic approximation, 232
Aharonov-Bohm effect (*see* Bohm-Aharonov effect)
Aharonov-Casher effect, 303
Airy functions, 110, 212
Alignment, 143
Anharmonic oscillator, 49
Annihilation operator, 15, 41, 45, 126, 135
Anomalous dispersion, 188
Anomalous magnetic moment, 303, 367, 422
Anti-Stokes line, 162
Antiparticle, 263, 321
Axial vector, 90, 293

Barrier penetration, 220, 231
Bernoulli numbers, 129
Berry's phase, 247
Bilinear forms, 291
Bohm-Aharonov effect, 111
Born approximation, 59, 61, 68, 70
Born-Oppenheimer approximation, 250
Bose symmetry, 81
Breit-Fermi interaction, 360

Casimir effect, 127
Classical radius of the electron, 161, 193
Cauchy's theorem, 191, 394
Causality, 56, 189, 261, 372
Center of mass coordinates, 73, 82
Centrifugal potential, 94
Charge conjugation, 322
Charge quantization, 120, 255
Chirality, 320
Clausius-Mossotti relation, 197
Clebsch-Gordon coefficient, 140, 147
Complex energy, 203
Compton scattering, 158
  relativistic, 363, 398, 407
Convection current, 319
Coulomb gague, 118, 120, 123
Coulomb interaction, 357, 411, 415
Covariance,
  Dirac equation, 281
  Klein-Gordon equation, 259
Creation operator, 15, 41, 45, 126
Cross section,
  differential, 56, 71
  total, 66
  two-body, 76
Current conservation, 105, 210, 290, 357
Current density,
  Dirac equation, 289, 305
  Klein-Gordon equation, 262, 310

Darwin term, 276, 299, 359
Determinant methods, 20, 229, 419
Dielectric constant, 195
Differential scattering cross section (*see* Cross section: differential)
Dirac equation, 281
  Hamiltonian form, 282
  two-component form, 281
Dirac sea, 323, 324
Dispersion relations, 186, 394
Dual tensor, 104

Effective Lagrangians, 195, 417
Electric dipole approximation, 134, 135
Electric dipole moment, 320, 360
Electric dipole radiation, 138, 143, 149
Electric quadrupole radiation, 148
Electromagnetic decay rate, 133, 152
Energy-momentum tensor, 401
Equivalence principle, 109, 201
Euler-Lagrange equation, 117, 404

Fermi's golden rule, 58, 69, 133, 135, 149, 173, 209
Feynman diagram, 29, 31, 59, 158, 166, 340
Feynman rules, 421
Fine structure constant, 103, 130, 166, 277, 305, 361
Forced harmonic oscillator(*see* Harmonic oscillator: forced
Form factor, 62, 362, 380
Frequency space, 12, 31, 51, 59
Functional differentiation, 42, 404
Functional methods, 41, 46, 245
Functional perturbation theory, 46, 49

Galilean invariance, 77, 110
Gauge invariance, 105, 382, 387, 399
Gauge transformation, 105, 249, 256

Generating functional, 41, 48, 245, 404, 410
Grassmann variable, 408, 419
Green's function, 37, 43, 47, 199, 332, 336, 352, 396, 403, 413
Gyromagnetic ratio, 281, 297, 319, 351

Hamilton's principle, 10, 102, 118
Harmonic oscillator,
  Dirac solution, 17
  forced, 25, 36, 48, 231, 238, 245
  matrix elements, 40, 44
  propagator, 14
  time-dependent, 246
  wavefunction, 15, 17, 40
Heisenberg representation, 327
Helicity, 311, 315
Hermite polynomail, 18, 23
Hyperfine interaction, 361

Imaginary time, 40, 221
Index of refraction, 186, 194, 198, 201
Infrared catastrophe, 381
Interaction representation, 27
Isospin, 416

Klein's paradox, 269
Klein-Gordon equation, 261
  two-component form, 265
Klein-Nishina formula, 162, 365
Kramers-Heisenberg dispersion formula, 158, 161, 174

Lagrange's equation of motion, 99
Laguerre polynomial, 36, 231, 239
Lamb shift, 178, 183, 195, 301, 389
Large time limit, 39
Legendre polynomials, 92, 95
Line shape, 149, 154, 174
Lorentz condition, 105, 118, 352
Lorentz force law, 101
Lorentz transformation, 103, 117, 259, 285, 297

Magnetic dipole radiation, 147
Magnetic monopole, 120–122, 252–254, 293
Magnetization current, 319
Maxwell Lagrangian, 116
Maxwell equations, 103–107, 119, 226, 352

Maxwell tensor, 103
Mesonic atom, 274–280
Minimal substitution, 263, 406
Multiple scattering technique, 226
Mössbauer effect, 155

Normal dispersion, 188
Number operator, 15, 125

One-photon exchange interaction, 353, 357
One-pion exchange potential (OPEP), 416
Optical theorem, 64–69, 95, 188, 193
Oscillator,
  (see anharmonic oscillator)
  (see harmonic oscillator)

Pair production, 268, 330, 346
Parity, 87, 140, 152, 293, 323, 413
Partial wave amplitude, 92, 94
Path integral, 6, 18, 20, 402
  Hamiltonian form, 14
  Pauli exclusion principle, 83
  perturbation theory, 49
  scattering, 70
Pauli-Villars regularization, 372
Perturbation theory,
  functional (see Functional perturbative theory)
  time-dependent (see Time-dependent perturbative theory)
Photoelectric effect, 126
Photon, 126, 130, 131
Pion, 411
Pion-exchange interaction, 411
Pion-nucleon scattering, 421
Pionic atom, 276, 279
Planck distribution, 139
Polar vector, 144, 293
Polarizability, 195
Polarization, 141, 162
Pole diagram, 160, 363
Propagator,
  adiabadic, 233
  Dirac, 301
  forced oscillator, 39
  free particle, 7
  harmonic oscillator, 15

Klein-Gordon, 395
linear potential, 23–24
semiclassical, 215
WKB, 218
Pseudovector (see Axial vector)
Pseudoscalar, 292, 323, 413

Q-factor, 155
Quadrupole operator, 148
Quantum field theory, 402

Rabi's formula, 35
Radiation damping, 165
Radiation field quantization, 123
Radiative corrections, 369
Radiative decay, 133, 135
  angular distribution, 140
Rayleigh scattering, 161, 163
Recoil effect, 177
Reflection coefficient, 224, 272
Renormalization, 169, 324, 369
Residue theorem, 12
Resonant scattering, 164, 173, 194
Rosenbluth cross section, 368
Rotation matrix, 149
Rotational invariance, 87, 89, 255
Runge-Lenz vector, 301
Running coupling constant, 389
Rutherford scattering (see Scattering, Coulomb)

S-Matrix, 52, 87, 92, 94
  unitarity, 67, 94
Saturation of nuclear force, 412
Scattering amplitude, 51, 54, 58
  bosons, 79
  direct, 80, 83
  exchange, 80, 83
  fermions, 83
  identical particle, 78
  partial wave expansion, 92
  phase shift, 94
  two-body, 74
Scattering matrix, 85, 91
Scattering,
  Coulomb, 81, 346, 366, 397
  gravitational, 62, 400
Schrödinger equation, 1, 2, 281
Schrödinger representation, 28–29
Seagull diagram, 159, 164, 325, 326, 398

Selection rules, 140, 146–148
Self-energy, 170, 201, 371
Semiclassical technique, 211
Semiempirical mass formula, 413
Spatial inversion, 87, 140, 288, 323
Spherical Bessel functions, 93, 96
Spin-orbit interaction, 359
Spreading (wavepacket), 3–5
Stationary phase approximation, 55, 218, 222
Stokes line, 162
Superconductor, 115

T-Matrix, 52, 76, 86, 428
Theta function, 3
Thomas precession, 297
Thomas-Reiche-Kuhn sum rule, 194
Thomson scattering, 324, 328
Time development operator, 2, 3, 27
Time ordered product, 42, 403
Time reversal, 87–91
Time-dependent perturbation theory, 27
Topological phase (see Berry's phase)
Trace techniques, 349, 355
Transition amplitude, 29, 31, 44, 51, 57, 59
Translation invariance, 88, 409
Transmission coefficient, 228, 273
21 cm radiation, 361, 368
Two-body scattering (see Scattering, two-body)
Two-slit diffraction, 112–113

Vacuum energy, 126, 130, 135, 138, 424
Vacuum polarization, 279, 384, 394, 421
Vector potential, 104–106
  quantized, 121, 134
Vertex correction, 377
Virial theorem, 300

WKB approximation, 211
  connection formulae, 212
Wave function renormalization, 370
Wigner-Eckart theorem, 140, 147, 149
Wigner-Weisskopf technique, 149, 178

Yukawa interaction, 415

Zeeman effect, 280
Zero point energy (see Vacuum energy)
Zitterbewegung, 269, 271, 299, 326, 328